DEMATIACEOUS HYPHOMYCETES

This and other publications of the
Commonwealth Agricultural Bureaux
can be obtained through any major bookseller or
direct from
Commonwealth Agricultural Bureaux,
Central Sales, Farnham Royal, Bucks., England

DEMATIACEOUS
HYPHOMYCETES

BY

M. B. ELLIS
B.SC., PH.D.(LOND.)

Principal Mycologist,
Commonwealth Mycological Institute, Kew

COMMONWEALTH MYCOLOGICAL INSTITUTE
KEW, SURREY, ENGLAND
1971

First published 1 July 1971 by the
Commonwealth Mycological Institute
Kew, Surrey, England

SBN   85198   027   9

Printed in Great Britain by The Eastern Press, Ltd., London and Reading

CONTENTS

001318

CONTENTS

INTRODUCTION

This book is about many common and some less common hyphomycetes with dark conidia or conidiophores. It aims to make their identification easier and in particular to enable students and others who do not have direct access to a large herbarium to name a number of their own collections and isolates to species level.

Interest in this group of fungi is increasing. It includes many important pathogens of plants and animals, many species which cause spoilage of food and deterioration of paper and paper products, plastics, textiles, timber, etc. and others which are used in various industrial processes. Plant pathologists, soil microbiologists, chemists, allergy units, mushroom growers, veterinary surgeons, industrial firms, universities and various government research departments need to identify them or to have them identified. I hope that interest will increase even more with the production of this book.

When I was introduced to the hyphomycetes 25 years ago by that great mycologist and stimulating investigator E. W. Mason relatively little was known about them and we had the same difficulty with their identification that many people still seem to experience. About 50 per cent of the specimens sent to the Commonwealth Mycological Institute from all over the world are hyphomycetes and the naming of these at CMI is now a much simpler matter than it was in 1945. A number of eminent mycologists have helped to make this possible and references to their work appear throughout the text. My own notes and drawings made during the past quarter of a century are incorporated: the descriptions and figures are all based on fresh specimens and slides in our herbarium matched whenever this has been possible against type or authenticated collections. I am most grateful to my former colleague Dr. K. Pirozynski for allowing me to use his beautiful illustrations of *Hiospira, Beltrania* and allied genera, *Circinotrichum, Gyrothrix* and *Chaetochalara* and I could not resist using a delightful drawing of *Acrospeira mirabilis* made by the late Dr. S. P. Wiltshire.

A question often asked the professional mycologist is, " How do you set about naming a specimen? " It is surprisingly difficult to give a satisfactory answer. I suppose the truth of the matter is that after making or having had made slide preparations which demonstrate adequately the structure of the fungus, he or she relies mainly on experience gained by examining many thousands of specimens and isolates, but partly also on information gleaned from colleagues and from what available monographs and other standard works of reference there are. Experience cannot be passed on but having it enables one to draw attention to characters of special taxonomic significance which seem obvious once they have been pointed out. Much time and thought has been given to the preparation of the key to 295 genera presented here. Some hyphomycetes with quite striking peculiarities are nevertheless often difficult to track down. To assist with the identification of these the key is followed by a list of characters which are of particular diagnostic value because they are common to only a few genera or species. At the end of the book there is a substratum index and a list of the very common plurivorous species.

The dematiaceous hyphomycetes are a fascinating group of organisms about which even now really very little is known. They occur everywhere in abundance, are easy to collect, grow and examine, and provide excellent material for research projects. Studies of conidium development by means of time-lapse photography and of fine structure by means of the electron microscope have been carried out with only a few of them but all such studies have been of absorbing interest.

In hyphomycetes the vegetative or assimilative **mycelium**, whether it is immersed in the substrate or grows on the surface, is made up of septate hyphae, cylindrical filaments with walls enclosing usually multinucleate protoplasm which continually lays down new wall material at the growing apex. Branching may take place close behind the apex or further back. The septa are discs each generally with a simple pore in the middle.

Hyphae sometimes become aggregated to form **sclerotia.** Each of these has a rind of rather small, dark, thick-walled, tightly interwoven hyphae. Sometimes pulvinate **stromata** are formed which are often prosenchymatous, i.e. made up of elongated hyphal elements still recognisable as such. They may be pseudo-parenchymatous, however, superficially resembling the parenchyma of flowering plants. One or more of the stroma cells may become very much enlarged and bear numerous conidiophores. This is a peculiarity of the genus *Camptomeris*.

Hyphae in some genera regularly form **hyphopodia,** short lateral, sometimes lobed appressoria which adhere firmly to the surface of the host plant. From each of these a narrow filament penetrates the cuticle. **Stomatopodia** are appressoria produced from lateral branches above or in a stoma. Hyphae are modified in various ways. Sometimes they become bristle-like, often thick-walled and are then called **setae.** Occasionally setae are capitate having a bulb-like swelling at the apex and resembling drumsticks: a good example of this is seen in *Sporoschisma.*

When they bear conidia hyphae are called **conidiophores.** Sometimes all the hyphae are morphologically very similar to one another; those which bear conidia are then referred to as **micronematous** conidiophores. Conidiophores which are morphologically very different from purely vegetative hyphae are said to be **macronematous**; they are usually erect. There are cases where conidiophores differ only slightly from other hyphae; they are often ascending but seldom erect and may be termed **semi-macronematous.**

Where the wall of a hypha is thick it is often seen to have two layers. The outer or primary wall is continuous and frequently has granular electron-dense material which may be concentrated in the outer region. Dark colour is frequently due to the deposition of melanin. The inner or secondary wall surrounds individual cells as these are delimited by septa; in some genera it has been found to be electron-transparent.

Cells which produce conidia are called **conidiogenous cells**; they may be **integrated** (Fig. 1 K) that is incorporated in the main axis or branches of the conidiophore where they are either terminal or intercalary, or they may be

discrete (Fig. 1 L) when they often have a distinctive shape, being ampulliform, lageniform, spherical, etc. Discrete conidiogenous cells are sometimes subtended by short branches which themselves have been given various names.

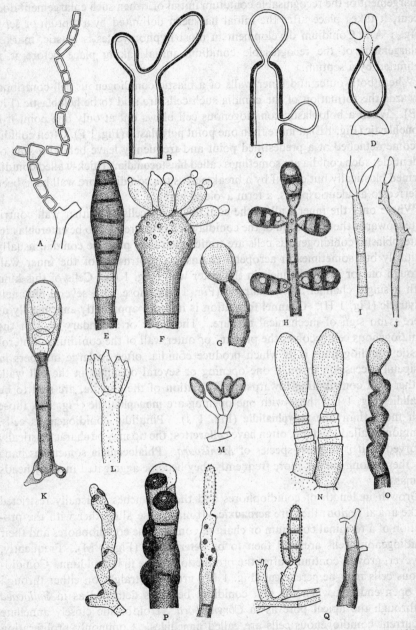

FIG. 1. Conidiogenous cells: A, thallic; B, holoblastic; C, enteroblastic tretic; D, enteroblastic phialidic; E, monoblastic; F, polyblastic; G, monotretic; H, polytretic; I, monophialidic; J, polyphialidic; K, integrated; L, discrete; M, determinate; N, percurrent; O, sympodial; P, cicatrized; Q, denticulate.

In by far the greatest number of hyphomycetes both outer and inner walls of the conidiogenous cells are continuous with the two walls of the conidia. The term **thallic** is used to describe conidium development where there is no enlargement of the recognisable conidium initial or, when such enlargement does occur, it takes place after the initial has been delimited by a septum or septa (Fig. 1 A). Conidium development in most hyphomycetes is blastic, marked enlargement of the recognisable conidium initial taking place before it is delimited by a septum.

Where both outer and inner walls of a blastic conidiogenous cell contribute towards the formation of the conidia such cells are said to be **holoblastic** (Fig. 1 B). When a holoblastic conidiogenous cell blows out at only one point it is **monoblastic** (Fig. 1 E) at more than one point **polyblastic** (Fig. 1 F). Often conidia become detached at a predestined point and frequently leave behind a scar or a denticle; such conidia are sometimes called **blastoconidia.** Thick-walled conidia formed blastically but cut off by a break across the conidiophore wall have been referred to as aleuriospores, a term avoided here.

Where only the inner wall of the conidiogenous cells or neither wall contributes towards the formation of the conidia such cells are said to be **enteroblastic.** Enteroblastic conidiogenous cells are **tretic** when they produce conidia, usually solitarily but sometimes in acropetal chains, by protrusion of the inner wall through one or more channels in the outer wall (Fig. 1 C). Cells of this kind with a single channel are **monotretic** (Fig. 1 G), those with several channels **polytretic** (Fig. 1 H). Channel formation is induced apparently enzymically as there is no sign of mechanical rupture. The inner or secondary wall of the conidiogenous cell becomes the primary or outer wall of the conidium. Enteroblastic conidiogenous cells which produce conidia, often in large numbers in basipetal succession through one opening or several openings in the cell wall, neither wall contributing towards the formation of the conidia, are said to be **phialidic** (Fig. 1 D); those with one opening are **monophialidic** (Fig. 1 I), those with more than one **polyphialidic** (Fig. 1 J). Phialidic conidiogenous cells, commonly called **phialides**, often have **collarettes**; the tip may be characteristically recurved as it is in most species of *Menispora*. **Phialoconidia** sometimes hang together in long chains, more frequently they become aggregated in slimy heads or masses.

Growth in length of conidiophores and their branches is usually restricted to the apical region: they are **acroauxic.** It may cease altogether with the production of a terminal conidium or chain of conidia; the conidiophores and their conidiogenous cells are said then to be **determinate** (Fig. 1 M). Frequently, however, growth continues after the production of the first conidium. Conidiogenous cells may be **percurrent** (Fig. 1 N), growing straight on either through the open end left when the first conidium becomes detached as in *Spilocaea* or through the apical pore as in *Corynespora*. Holoblastic closely annellate percurrent conidiogenous cells are called **annellides.** Commonly proliferation is **sympodial** (Fig. 1 O), the main axis elongating by growth of a succession of apices each of which develops behind and to one side of the previous apex where growth ceased with the production of a conidium or conidia. Conidiophores sometimes grow on beyond a conidium-bearing region: where this occurs the terminal sterile part is often setiform.

In *Coniosporium* and *Trimmatostroma* growth in length of the conidiophore takes place by the laying down of cross walls and elongation of cells just behind the apex. Such conidiophores are said to be **meristematic**. In a small group of fungi, which includes *Arthrinium*, the conidiophores are **basauxic**, each consisting of a mother cell and an extensible filament arising from within it which may be conidiogenous. In known examples one or more conidia or sterile cells are borne on the filament. The apex of the extensible filament becomes incorporated in the first-formed conidium or sterile cell; subsequent lateral conidia and sterile cells, when present, arise blastically in succession towards the base. With reference to the mother cell only the terminal conidium or sterile cell is holoblastic.

Using characters outlined so far dematiaceous hyphomycetes can be divided into the six groups shown in the following table and this is the method of classification adopted in this book.

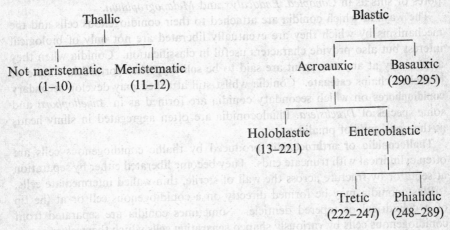

Conidiophores are branched or unbranched. Branching is occasionally dichotomous or trichotomous; sometimes short branches are arranged in verticils; collar-like branches which encircle the main axis are characteristic of the genus *Gonytrichum*; in *Acarocybe* and *Acarocybella* descending branches are formed. Frequently branching is restricted to the apical region; the unbranched lower part is then referred to as the **stipe** and the upper branched part as the **head** which when brush-like is said to be penicillate.

Vesicular swellings occur sometimes at the apex of the conidiophore or at intervals along it; these swellings, often called **ampullae**, are either conidiogenous cells on the surface of which conidia develop more or less simultaneously as in *Gonatobotryum* or they bear a number of short branches or discrete conidiogenous cells as in *Aspergillus*. Thick dark transverse septa are found in *Arthrinium* and allied genera.

Conidiophores are said to be **mononematous** when they are solitary or caespitose, **synnematous** when numerous threads or filaments are tightly adpressed or fused along most of their length and splay out only at the apex. Relatively short conidiophores closely investing the surface of pulvinate stromata form **sporodochia**.

Hyphomycete **conidia** are usually simple, occasionally branched, and are of many different shapes, acerose, allantoid, biconic, cheiroid, clavate, cuneiform, cylindrical, doliiform, ellipsoidal, falcate, fusiform, helicoid, lenticular, limoniform, lunate, navicular, obclavate, obovoid, obpyriform, obturbinate, ovoid, pyriform, reniform, sigmoid, spathulate, spherical, T-, U-, V-, Y-shaped, etc. They are sometimes **appendiculate** with delicate **setulae**, frequently transversely septate or pseudoseptate, less frequently muriform. Differently coloured cells are not uncommon, the end ones in *Curvularia* for example being often much paler than intermediate ones. In *Beltrania* and allied genera each conidium has a transverse median hyaline band. Thick, dark transverse bands or septa occur in some genera, e.g. *Endophragmia*. In *Hermatomyces* pale cells surround darker central ones. The conidium wall may be smooth, rugose, verrucose or echinulate, in *Hiospira* it is reticulate. Sometimes there are well-defined germ pores or slits as in *Conoplea*, *Endocalyx* and *Melanographium*.

The ways in which conidia are attached to their conidiogenous cells and the mechanisms by which they are eventually liberated are not only of biological interest but also provide characters useful in classification. Conidia when they arise singly at any one point are said to be **solitary**, when formed in simple or branched chains **catenate**. Conidia whilst still attached may develop **secondary conidiophores** on which secondary conidia are formed as in *Annellophora* and some species of *Drechslera*. Phialoconidia are often aggregated in slimy heads at the open ends of phialides.

Thalloconidia or **arthroconidia** produced by thallic conidiogenous cells are often cylindrical with truncate ends. They become liberated either by separation at septa or by fracture across the wall of sterile, thin-walled intermediate cells.

Blastoconidia may be formed directly on a conidiogenous cell or at the tip of a cylindrical or tapered denticle. Sometimes conidia are separated from conidiogenous cells by variously shaped **separating cells** which themselves sometimes bear denticles. Conidiogenous cells bearing denticles are said to be **denticulate** (Fig. 1 Q). Flat or protuberant scars, often with a minute pore in the centre, are formed quite frequently on the surface of conidiogenous cells where conidia become detached and are of diagnostic value. Conidiogenous cells with scars are said to be **cicatrized** (Fig. 1 P). Scars may be small and flat as in *Periconiella*, very broad and thin as in *Pleiochaeta* or dark and prominent as in *Fusicladium*.

Phialoconidia are sometimes truly **endogenous** being produced entirely within the phialide, e.g. in *Sporoschisma*; often they are **semi-endogenous** forming partly inside the tip of the phialide, or **acrogenous**, where the contents pass through the opening, the conidia taking shape immediately outside. Rarely they are **exogenous**, the contents of the conidiogenous cell protruding beyond the opening and the conidia developing, sometimes in several columns a little way above the opening as in *Chloridium*.

KEY TO GENERA

26. Conidia neither cruciately septate, nor deeply constricted at septa . . . 27
    Conidia cruciately septate or deeply constricted at septa 28
27. Conidia flattened, with rows of cells radiating fanwise from the hilum

    Conidia not flattened . . . . *Mycoenterolobium* (19)
28. Mycelium with hyphopodia . . . . *Monodictys* (34)
    Mycelium without hyphopodia . . . . . . *Sarcinella* (18)
29. Conidia cruciately septate 29
    Conidia not cruciately septate . . . . . *Tetracoccosporium* (43)
30. Setae present, conidia spherical, colourless or pale . . . . *Chuppia* (17)
    Setae absent . . . . *Botryotrichum* (31)
31. Conidiogenous cells tapered to a point 31
    Conidiogenous cells denticulate, denticles narrow, cylindrical . . *Acremoniella* (42)
    Conidiogenous cells neither tapered nor denticulate . . . *Asteromyces* (41)
32. Hyphae often with thick, dark septa 32
    Hyphae without thick, dark septa 33
33. Conidia ellipsoidal or navicular, often with a germ slit 34
    Conidia spherical, with a minute pore . . . . . *Mammaria* (38)
34. Colonies pulvinate, rust coloured . . . . . *Gilmaniella* (29)
    Colonies not pulvinate . . . . . . *Allescheriella* (28)
35. With an accompanying *Chalara* state 35
    Without a *Chalara* state . . . . . . *Chalaropsis* (30)
36. Conidia obpyriform or obturbinate, mucronate, in tight clusters . *Echinobotryum* (40)
    Conidia otherwise 36
37. Parasitic on *Clypeolella* and *Schiffnerula* 37
    Thermophilic, without phialides . . . . . *Acremoniula* (25)
    Neither parasitic nor thermophilic, with phialides . . . *Thermomyces* (26)
38. Conidia forked, V- or U-shaped . . . . . . *Humicola* (27)
    Conidia with 4 divergent arms . . . . . . *Hirudinaria* (21)
    Conidia several times branched,branches hooked at ends . *Tetraposporium* (23)
    Conidia with pyriform central cell and 2–3 divergent pluriseptate branches *Casaresia* (46)

    Conidia with stalk cell and 4 divergent arms . . . . *Ceratosporium* (22)
    Conidia with 4 columns of cells terminating in divergent, tapering, septate *Tripospermum* (49)
    appendages

    Conidia cheiroid . . . . *Tetraploa* (20)
39. Chains of conidia arising directly from cells of a stroma . . *Dictyosporium* (24)
    Chains of conidia not arising directly from cells of a stroma . . *Spilodochium* (54)
40. Conidia helicoid, coiled in 3 planes 40
    Conidia not helicoid . . . . *Helicodendron* (48)
41. Conidia 0-septate 41
    Conidia mostly 1-septate . . . . *Alysidium* (51)
    Conidia mostly pluriseptate . . . . *Bispora* (52)
42. Conidiogenous cells mostly determinate or percurrent . . . *Taeniolella* (53)
    Conidiogenous cells sympodial 43
43. Conidiogenous cells monoblastic, terminal 88
    Conidiogenous cells polyblastic or where monoblastic intercalary . . 44
44. Conidiogenous cells not cicatrized 72
    Conidiogenous cells cicatrized with a single scar at the apex . . . 45
45. Conidia catenate 71
    Conidia solitary 46
46. Chains acropetal 51
    Chains basipetal 47
47. Conidia without septa 50
    Conidia with septa 48
48. Mycelium superficial, with hyphopodia 49
    Mycelium immersed, without hyphopodia . . . . . *Ampullifera* (56)
49. Not forming sporodochia, conidia dry . . . . . *Xylohypha* (55)
    Forming sporodochia, conidia slimy or encrusted . . . *Heteroconium* (57)
50. Sporodochia without setae, conidia 2-septate . . . . *Septotrullula* (59)
    Sporodochia with setae, conidia 1-septate . . . . *Bactrodesmiella* (62)
    . . . . *Trichodochium* (91)

51. Conidiophores mostly unbranched 52
    Conidiophores branched 69
52. Forming pulvinate sporodochia 53
    Not forming sporodochia 57
53. Conidiophores very slender with a swollen vesicle at the apex, conidia clathrate or muri-
    form, sometimes corniculate . . . . . *Oncopodium* (68)
    Conidiophores very slender without a vesicle 54
    Conidiophores stouter 55
54. Conidia transversely septate, often with cells unequally coloured and thick, dark bands
    at the septa . . . . *Bactrodesmium* (61)
    Conidia muriform . . . . *Berkleasmium* (63)
    Conidia cheiroid, pendent . . . . . . *Cryptocoryneum* (65)
55. Conidia 1-septate, sporodochia on leaf veins, elongated . . *Oedothea* (88)
    Conidia 1–2-septate, with radiating hyphal plates . . . *Pollaccia* (86)
    Conidia muriform, corniculate . . . . . . *Petrakia* (93)
    Conidia with transverse and often also longitudinal or oblique septa . . 56
56. No stroma. Conidia rostrate, middle cells with longitudinal or oblique septa
    . . . . *Dictyodesmium* (92)
    . . . . *Stigmina* (89)
    Well-defined stroma always present 58
57. Conidiophores not thin-walled and collapsing or calyciform at the apex . . 67
    Conidiophores thin-walled and collapsing or calyciform at the apex . . 59
58. Setae present 60
    Setae absent . . . . *Septosporium* (64)
59. Setae subulate, conidia muriform . . . . .
    Setae dumb-bell-shaped, flat-topped, conidia with pale peripheral cells surrounding dark
    central ones . . . . *Hermatomyces* (67)
    . . . . *Kostermansinda* (69)
60. Conidiophores synnematous, conidia muriform 61
    Conidiophores mononematous 62
61. Conidia not branched or appendaged 66
    Conidia branched or appendaged 66
62. Conidia 0-septate, spherical with a protuberant cylindrical peg at the base
    . . . . *Domingoella* (70)
    Conidia 0-septate, spherical or ellipsoidal without a protuberant peg
    . . . . *Acrogenospora* (71)
    Conidia muriform, not helicoid, no secondary conidia . . *Acrodictys* (77)
    Conidia muriform, helicoid, with secondary conidia . . *Xenosporium* (78)
    Conidia 0–1-septate, colonies with radiating hyphal plates . . *Spilocaea* (87)
    Conidia with 1 or more septa or pseudosepta, colonies without radiating hyphal plates
    63
63. Mycelium superficial, hyphae thick, with simple or lobed stomatopodia *Septoidium* (72)
    Mycelium superficial or immersed, without stomatopodia . . 64
64. Mycelium superficial, with hyphopodia . . . *Clasterosporium* (74)
    Mycelium superficial with hemispherical lateral cells resembling hyphopodia but without
    haustoria . . . . *Hansfordiellopsis* (76)
    Mycelium without hyphopodia or lateral cells . . . 65
65. Secondary conidia borne on short annellate secondary conidiophores at the apex of
    primary conidia . . . . . . *Annellophora* (75)
    No secondary conidia . . . . . . *Sporidesmium* (73)
66. Conidia Y-shaped . . . . . . *Iyengarina* (79)
    Conidia bifurcate or with several upwardly directed branches . *Ceratosporella* (81)
    Conidia with thick, subulate, septate, lateral appendages arising from the basal cells
    . . . . *Teratosperma* (80)
    Conidia with a stalk cell and usually 3 conical, septate arms . *Triposporium* (82)
    Conidia each consisting of a 1–4-celled stalk surmounted by a group of cells bearing
    3 divergent arms . . . . . . *Actinocladium* (83)
    Conidia complex with lateral protuberances bearing incurved, claw-like processes
    . . . . *Arachnophora* (84)
    Conidia branched with stalked attachment organs . . . *Grallomyces* (85)

67. Conidiophores often with a cup at the apex, conidia frequently with cells unequally
coloured and dark bands at the septa . . . . . *Endophragmia* (94)
Tip of conidiophore swollen, rounded, thin-walled, sometimes collapsing and becoming
cupulate 68
68. Conidia spathulate, transversely septate . . . *Phragmospathula* (100)
Conidia not spathulate; transversely septate or pseudoseptate . *Deightoniella* (101)
Conidia muriform . . . . *Stemphylium* (102)
69. Branching verticillate, conidia complex with dark brown central and pale lateral cells

Branching irregular, conidia 4-septate, caudate . . . *Physalidium* (99)
Conidiophores with a few short branches at the apex . . *Exosporiella* (90)
70. Conidia 0-septate, spherical 70
Conidia circinate, rostrate, septate . . . *Staphylotrichum* (32)
Conidia pendulous, caudate, muriform . . . *Circinoconis* (97)
Conidia transversely septate, verrucose . . . *Pendulispora* (98)
Conidia transversely septate, smooth, cells often unequally coloured, with thick, dark
bands at the septa . . . . . . *Brachysporiella* (96)
71. Scars thick, cupulate, conidiophores straight . . . *Hansfordiella* (103)
Scars flat, broad, conidiophores becoming strongly curved . . *Fusicladiella* (104)
72. Conidia not helicoid or strongly curved 73
Conidia helicoid or strongly curved 84
73. Conidiophores without swollen intercalary or terminal ampullae, although in *Pachnocybe*
individual threads in the synnemata are often clavate . . . 74
Conidiophores with terminal or terminal and intercalary ampullae . . 81
74. Conidiophores synnematous . . . . *Pachnocybe* (111)
Conidiophores mononematous 75
75. Conidia in pairs at the apex of the conidiophore . . *Microclava* (66)
Conidia not in pairs 76
76. Conidiogenous cells terminal and intercalary 77
Conidiogenous cells terminal 79
77. Setae present 78
Setae absent, conidia septate, catenate . . . *Septonema* (58)
78. Conidia catenate . . . . . . *Lacellina* (106)
Conidia solitary . . . . . . *Herposira* (105)
79. Conidiophores dichotomously or trichotomously branched . . . 80
Conidiophores irregularly branched . . . . *Polyscytalum* (115)
80. Conidia 0-septate, spherical or subspherical . . . *Polypaecilum* (114)
Conidia 1-septate . . . . . . *Balanium* (113)
81. Ampullae terminal on stipe or branches 82
Ampullae terminal and intercalary 83
82. Conidiophores simple, ampullae very large, conidia septate . *Cephaliophora* (109)
Conidiophores branched towards the apex, conidia mostly 0-septate . *Botrytis* (110)
83. Conidia 0-septate . . . . . *Gonatobotryum* (107)
Conidia 1-3-septate . . . . . . *Oedemium* (108)
84. Conidiophores synnematous . . . . *Trochophora* (112)
Conidiophores mononematous 85
85. Walls of conidiophores and conidia coarsely reticulate . . *Hiospira* (119)
Walls not reticulate 86
86. Conidia coiled in 3 planes to form an ellipsoidal or cylindrical spore body

Conidia coiled in 1 plane . . . . . *Helicoon* (116)
87. Conidia colourless or brightly coloured, filaments hygroscopic . *Helicosporium* (118)
Conidia brown or dull-coloured, filaments not hygroscopic . . *Helicoma* (117)
88. Conidiophores with terminal and intercalary cups . . *Endophragmiopsis* (95)
Conidiophores without cups 89
89. Conidiogenous cells neither clearly denticulate nor cicatrized . . 90
Conidiogenous cells denticulate 90
Conidiogenous cells cicatrized 92
90. Conidia 1-septate 117
Conidia 0-septate . . . . . *Hadronema* (122)
. 91

91. No stroma. Conidiophores mononematous, scattered, conidia without a germ slit
Virgariella (120)
Stroma present. Conidiophores caespitose or synnematous, conidia often with germ slit
Melanographium (121)
92. Conidiophores not arising from radially lobed basal cells 93
Conidiophores and setae when present arising from radially lobed basal cells . 112
93. Conidia aggregated in slimy heads . . . . . . *Cacumisporium* (127)
Conidia solitary, dry 94
94. Mycelium with hyphopodia 95
Mycelium without hyphopodia 96
95. Conidia curved . . . . *Curvulariopsis* (126)
Conidia straight . . . . *Mitteriella* (144)
96. Conidia corniculate, often trigonous, muriform . . . . *Oncopodiella* (132)
Conidia reniform, small, 0-septate . . . . . . *Virgaria* (133)
Conidia commonly sigmoid, 3-septate . . . . . . *Nakataea* (140)
Conidia falcate or lunate, 0-septate . . . . . . *Idriella* (123)
Conidia circinate or curved, septate . . . . . *Helicomina* (131)
Conidia helicoid, very tightly coiled and simulating dictyospores . *Helicorhoidion* (138)
Conidia otherwise 97
97. Conidiophores nodose with terminal and intercalary swellings . . . 98
Conidiophores not nodose 99
98. Conidiophores not branched . . . . *Cordana* (124)
Conidiophores branched, nodose swellings often proliferating as short lateral branches
Gonatophragmium (137)
99. Denticles mostly broad, conical 100
Denticles narrow, tapered to a point 101
Denticles cylindrical 102
100. Stroma present, young conidiophores often percurrent . *Pseudocercospora* (130)
Mycelial cells in hypodermis and epidermis swollen . . . *Pyriculariopsis* (128)
101. Conidiophores unbranched, dark . . . . . *Pleurophragmium* (125)
Conidiophores branched, pale . . . . . . *Phaeodactylium* (136)
102. Denticles neither thread-like nor cut off by septa to form separating cells . 103
Denticles thread-like or cut off by septa to form separating cells . . 104
103. Conidiophores synnematous . . . . *Phaeoisaria* (135)
Conidiophores mononematous often branched di- or trichotomously . *Dicyma* (134)
104. Denticles long, narrow, thread-like 105
Denticles short or very short 107
105. Conidia short, not appendiculate 106
Conidia long, cylindrical, often appendiculate . . . . *Camposporium* (143)
106. Conidia usually pendulous, smooth, often with 1 or more cells paler than the others
Brachysporium (141)
Conidia not pendulous, often verruculose or echinulate . *Scolecobasidium* (142)
107. Conidia septate . . . . *Pyricularia* (139)
Conidia not septate 108
108. Conidiophores often torsive, conidia frequently with a pore or germ slit *Conoplea* (149)
Conidiophores not torsive, conidia without pores or germ slit . . . 109
109. Conidiogenous cells repeatedly geniculate 110
Conidiogenous cells not repeatedly geniculate 111
110. Conidiophores synnematous . . . . *Dematophora* (146)
Conidiophores mononematous . . . . . . *Geniculosporium* (145)
111. Conidiophores smooth, sometimes with sterile setiform branches . *Hansfordia* (148)
Conidiophores often verrucose, without setiform branches . *Nodulisporium* (147)
112. Conidia not biconic or turbinate, without a hyaline transverse median band
Hemibeltrania (155)
Conidia biconic or turbinate with a hyaline transverse median band . 113
113. Setae absent . . . . *Pseudobeltrania* (154)
Setae present 114
114. Conidiophores branched, the upper part setiform 115
Conidiophores simple or when branched with the upper part not setiform . 116
115. Conidia biconic, short beaked . . . . . . *Beltraniopsis* (151)
Conidia turbinate or biconic, often caudate . . . . *Beltraniella* (152)

116. Conidia biconic, free end spicate or apiculate . . . . . *Beltrania* (150)
     Conidia turbinate, base drawn out to a fine point . . . . *Ellisiopsis* (153)
117. Conidia solitary, rarely in very short chains 118
     Conidia catenate, usually in long, often branched chains . . . 143
118. Conidia normally with longitudinal and oblique as well as transverse septa . 119
     Conidia normally without septa or with transverse septa only . . 123
119. Conidiophores synnematous . . . . . . *Sclerographium* (160)
     Conidiophores mononematous 120
120. Conidiophores unbranched 121
     Conidiophores often branched 122
121. Conidiophores long, dark, closely septate, stroma none or rudimentary
                                              Dactylosporium (163)
     Conidiophores short, stroma erumpent . . . . . *Thyrostromella* (180)
122. Conidiophores loosely branched . . . . . . *Sirosporium* (181)
     Conidiophores each with a stipe and a complex head of branches
                                              Mystrosporiella (187)
123. Stroma none or rudimentary, if present prosenchymatous . . . 124
     Stroma present, often well developed, usually pseudoparenchymatous . 134
124. Conidia mostly appendiculate 125
     Conidia not appendiculate 126
125. Conidia each with a single lateral appendage arising from the basal cell
                                              Centrospora (165)
     Conidia each with several, sometimes branched appendages arising from the apical cell
                                              Pleiochaeta (164)
126. Mostly parasitic on superficial leaf ascomycetes 127
     Not parasitic on leaf ascomycetes 128
127. Scars usually numerous, conspicuous . . . . . *Spiropes* (159)
     Scars few, flattened against the side of the conidiophore, not conspicuous
                                              Eriocercospora (158)
128. Conidiophores unbranched or only occasionally branched 129
     Conidiophores branched 133
129. Conidiophores caespitose 130
     Conidiophores scattered 131
130. Conidia mostly 1-septate, smooth . . . . . . *Passalora* (166)
     Conidia with numerous septa, verrucose . . . . *Stenellopsis* (167)
131. Scars large, often dark and prominent, conidiophores stout, mostly thick-walled
                                              Pseudospiropes (161)
     Scars small, conidiophores mostly rather slender 132
132. Ramo-conidia frequently present . . . . . *Rhinocladiella* (157)
     Ramo-conidia absent . . . . *Veronaea* (156)
133. Conidiophores loosely and irregularly branched . . . *Haplariopsis* (184)
     Conidiophores each composed of a stipe and a more or less complex head of branches
                                              Periconiella (186)
134. Conidia 0-septate, spherical or subspherical 135
     Conidia 0- or sometimes 1-septate, broadly fusiform . . . *Fusicladium* (171)
     Conidia mostly cuneiform, 1-septate, scars on conidiophore unilateral
                                              Polythrincium (178)
     Conidia mostly with 2 or more septa, or where 1-septate not cuneiform . 136
135. Stroma superficial, pulvinate, conidia very small, on palms . *Pseudoepicoccum* (169)
     Stroma erumpent, conidia larger . . . . . . *Hadrotrichum* (170)
136. Stroma with a large, spongy, columnar superficial part . . *Podosporiella* (182)
     Stroma with 1 or a few cells greatly enlarged and bearing numerous conidiophores
                                              Camptomeris (179)
     Stroma otherwise 137
137. Conidiophores always or commonly synnematous 138
     Conidiophores mononematous, mostly caespitose 139
138. Scars well-developed, mycelium partly superficial . . *Annellophragmia* (162)
     Scars thin, mycelium all immersed . . . . . *Phaeoisariopsis* (168)
139. Setae present . . . . *Schizotrichum* (175)
     Setae absent 140

140. Conidiophores frequently with a thickened band along one side, becoming incurved
  Cercosporidium (177)
  Conidiophores without a thickened band, not incurved 141
141. Stroma subcuticular, erumpent, scars few, thin, flat, conidia often with thick, dark septa
  Prathigada (176)
  Stroma not subcuticular, scars conspicuous 142
142. Conidia colourless or pale, smooth, often subulate or narrowly obclavate
  Cercospora (174)
  Conidia colourless or pale, often verruculose, short, commonly 1-septate
  Asperisporium (172)
  Conidia olivaceous or reddish brown, multiseptate, verrucose . *Verrucispora* (173)
143. Conidiophores much branched . . . . . . *Mycovellosiella* (189)
  Conidiophores unbranched or not much branched 144
144. Conidiophores with unilateral nodose swellings which may proliferate as short lateral
  branches . . . . *Fulvia* (191)
  Conidiophores without unilateral swellings 145
145. Parasitic on *Cercospora*, no stroma . . . . . *Cladosporiella* (188)
  Not parasitic on *Cercospora*, stroma often present 146
146. Mycelium mostly superficial, verruculose . . . . . *Stenella* (190)
  Mycelium usually immersed 147
147. Conidia usually with a small, distinctly protuberant scar at each end
  Cladosporium (193)
  Conidia without a small protuberant scar at each end . . *Phaeoramularia* (192)
148. Conidia dry 149
  Conidia aggregated in slimy heads or masses 168
149. Conidiogenous cells monoblastic 150
  Conidiogenous cells polyblastic 155
150. Conidiogenous cells determinate, conidia solitary 151
  Conidiogenous cells percurrent, closely annellate, conidia catenate . . 153
151. Conidiophores synnematous . . . . . . *Paathramaya* (195)
  Conidiophores mononematous 152
152. Conidiogenous cells spherical or subspherical, conidiophores without a sterile setiform
  apex . . . . *Nigrospora* (194)
  Conidiogenous cells curved, tapering to a point, conidiophores often with a sterile setiform
  apex . . . . *Zygosporium* (196)
153. Conidiophores mononematous . . . . . . *Scopulariopsis* (198)
  Conidiophores synnematous 154
154. Conidiophores without setiform branches . . . . *Doratomyces* (199)
  Conidiophores with setiform branches . . . . . *Trichurus* (200)
155. Conidia solitary 156
  Conidia catenate 164
156. Conidiogenous cells with dark, prominent scars . . . *Zygophiala* (183)
  Conidiogenous cells denticulate 157
  Conidiogenous cells without prominent scars or denticles 159
157. Conidiogenous cells umbellately arranged over the swollen apex of the conidiophore
  Pseudobotrytis (215)
  Conidiogenous cells not umbellately arranged 158
158. Conidiogenous cells usually in verticils, recurved, resembling cocks' combs
  Costantinella (217)
  Conidiogenous cells clustered . . . . . *Kumanasamuha* (207)
159. Conidiogenous cells catenate . . . . . . *Sadasivania* (208)
  Conidiogenous cells not catenate 160
160. Conidia multiseptate, constricted at septa . . . . *Dwayabeeja* (205)
  Conidia 2-septate, conidiogenous cells often in pairs or verticils
  Spondylocladiopsis (185)
  Conidia 0-(rarely 1-) septate 161
161. Without setae, conidiogenous cells determinate . . . *Wardomyces* (39)
  Without setae, conidiogenous cells sympodial, branches in verticils
  Verticicladium (220)
  With setae, conidiogenous cells percurrent, conidia often acerose, arranged in a ring below
  the apex of the conidiogenous cell 162

162. Conidiophores encasing lower part of setae . . . . *Ceratocladium* (214)
    Conidiophores not encasing lower part of setae 163
163. Setae unbranched . . . . *Circinotrichum* (212)
    Setae repeatedly branched at the apex . . . . . *Gyrothrix* (213)
164. Conidia mostly septate, constricted at septa . . . . *Torula* (204)
    Conidia 0-septate 165
165. Conidiophores with short unciform lateral branches, walls verruculose or spinulose
                                               Trichobotrys (206)
    Conidiophores without short unciform lateral branches . . . 166
166. Conidiophores without terminal and intercalary ampullae . . *Periconia* (209)
    Conidiophores with terminal and intercalary ampullae . . . 167
167. Setae absent . . . . *Haplobasidion* (210)
    Setae present . . . . *Lacellinopsis* (211)
168. Sporodochial 169
    Not sporodochial, conidiophores scattered, mononematous or synnematous, each with a stipe and head 170
169. Setae absent, conidia mostly branched, cheiroid . . . *Cheiromycella* (197)
    Setae present, conidia 0-septate, held together for a long time in chains and connected by narrow isthmi . . . . *Wiesneriomyces* (216)
170. Conidiophores synnematous 171
    Conidiophores mononematous 172
171. Conidia 0-septate . . . . *Graphium* (202)
    Conidia septate . . . . *Arthrobotryum* (203)
172. Conidia 2-septate . . . . *Sterigmatobotrys* (221)
    Conidia 0-septate 173
173. Conidiogenous cells determinate . . . . . . *Haplographium* (218)
    Conidiogenous cells percurrent, closely annellate . . . *Leptographium* (201)
    Conidiogenous cells sympodial . . . . *Verticicladiella* (219)
174. Conidiogenous cells tretic 175
    Conidiogenous cells phialidic 196
175. Conidiogenous cells monotretic 176
    Conidiogenous cells polytretic 184
176. Conidiophores semi-macronematous, forming arched loops . . *Piricauda* (222)
    Conidiophores macronematous, not forming arched loops . . . 177
177. Conidiophores made up of ascending and descending threads . . . 178
    Conidiophores without descending threads 179
178. Conidiogenous cells integrated . . . . . . *Acarocybella* (223)
    Conidiogenous cells discrete . . . . *Acarocybe* (229)
179. Conidiophores mononematous 180
    Conidiophores synnematous 183
180. Conidiophores unbranched . . . . *Corynespora* (224)
    Conidiophores branched 181
181. Conidiogenous cells arranged in verticils . . . . *Spondylocladiella* (230)
    Conidiogenous cells not in verticils 182
182. Conidia long, pseudoseptate . . . . . . *Corynesporella* (226)
    Conidia shorter, septate . . . . *Dendryphiopsis* (227)
183. Conidia solitary . . . . *Podosporium* (225)
    Conidia catenate . . . . *Didymobotryum* (228)
184. Conidiogenous cells determinate 185
    Conidiogenous cells sympodial 188
185. Conidiogenous cells clavate . . . . . . *Blastophorella* (235)
    Conidiogenous cells lobed, conidiophores dichotomously or trichotomously branched
                                            Dichotomophthora (231)
    Conidiogenous cells not clavate or lobed 186
186. Conidiophores branched, conidia catenate . . . . *Diplococcium* (234)
    Conidiophores unbranched, conidia mostly solitary 187
187. Conidia pseudoseptate, mostly obclavate, developing laterally, often in verticils through minute channels beneath septa . . . . *Helminthosporium* (232)
    Conidia short, 0–3-septate . . . . . . *Spadicoides* (233)
188. Conidiophores synnematous . . . . . *Dendrographium* (236)
    Conidiophores mononematous 189

189. Conidiophores with terminal and intercalary nodose swellings . *Dendryphiella* (246)
     Conidiophores not nodose 190
190. Conidiophores branched at the apex to form a stipe and head . *Dendryphion* (247)
     Conidiophores not branched at the apex 191
191. Conidia frequently with longitudinal and oblique as well as transverse septa . . 192
     Conidia with transverse septa or pseudosepta only 193
192. Conidia frequently catenate, mostly obclavate and rostrate . . *Alternaria* (244)
     Conidia mostly solitary, not obclavate or rostrate, commonly ellipsoidal or obovoid
     Ulocladium (245)
     . 194
193. Conidia pseudoseptate 194
     Conidia septate 195
194. Stroma usually well-developed, conidia mostly obclavate with a prominent dark scar at
     the base . . . . *Exosporium* (237)
     Stroma present in a few species, conidia usually cylindrical or fusiform, rarely obclavate,
     common on Gramineae . . . . *Drechslera* (238)
195. Conidia short, with 3 or more septa, often curved, with end cells frequently paler than
     intermediate cells . . . . *Curvularia* (239)
     Conidia limoniform, 2-septate, central cell large and very dark, end cells pale, very short
     Brachydesmiella (240)
     Conidia oblong, rounded at the ends, 3-septate, on Cyperus . *Duosporium* (241)
     Conidia ellipsoidal, constricted at the septa, thin-walled, echinulate
     Acroconidiella (242)
     Conidia obclavate, clearly rostrate, with a large, dark scar at the base
     Phaeotrichoconis (243)
196. Conidiogenous cells almost always monophialidic 197
     Conidiogenous cells regularly polyphialidic 228
197. Conidia all endogenous becoming extruded 198
     Conidia semi-endogenous or acrogenous (occasionally also endogenous in *Phialocephala*)
     203
     Conidia exogenous 227
198. Not sporodochial, conidiophores scattered or caespitose, conidia mostly catenate . 199
     Sporodochial, conidia catenate or in slimy heads 202
199. Setae absent, conidia cylindrical or oblong, truncate at each end . . *Chalara* (249)
     Setae present in most species 200
200. Setae subulate . . . . *Chaetochalara* (250)
     Setae when present capitate 201
201. Conidia cylindrical or oblong, septate . . . . . *Sporoschisma* (248)
     Conidia cuneiform, 0-septate . . . . *Catenularia* (253)
202. Conidiophores rarely branched, conidia catenate . . . . *Bloxamia* (251)
     Conidiophores always much branched, conidia aggregated in slimy heads
     Cystodendron (252)
203. Sporodochial 204
     Not sporodochial 205
204. Sporodochia pulvinate, hard, gelatinous, black, shining; phialides with long, narrow,
     curved necks . . . . *Agyriella* (262)
     Sporodochia sessile or stalked, a viscid green to black conidial mass is surrounded by a
     white zone of hyphae from which setae project in some species; phialides without necks
     Myrothecium (280)
205. Stroma extensive, consisting of a floor and roof separated by a cavity and connected by
     stromatic columns . . . . *Cryptostroma* (260)
     Little or no stroma formed 206
206. Conidiophores micronematous or semi-macronematous 207
     Conidiophores macronematous 211
207. Conidia 1–3-septate, catenate, usually with slipped chains . . *Fusariella* (264)
     Conidia 0-septate, aggregated in slimy heads or masses 208
208. Setae present 209
     Setae absent 210
209. Conidiogenous cells discrete, close together forming a palisade, conidia with setulae
     Mahabalella (257)
     Conidiogenous cells integrated, terminal and intercalary, with projecting lateral collarettes,
     conidia without setulae . . . . *Bahupaathra* (256)

210. Conidiogenous cells integrated and terminal or discrete, with well-defined collarettes
    Phialophora (261)
    Conidiogenous cells integrated, intercalary, without distinct collarettes
    Aureobasidium (255)
211. Conidiophores synnematous 212
    Conidiophores mononematous 214
212. Conidiophore threads encasing the lower part of a large erect seta
    Menisporopsis (259)
    Conidiophore threads not encasing a seta 213
213. Synnemata fringed with setae, conidia 0-septate . . . *Saccardaea* (281)
    Synnemata not fringed with setae, conidia 3-septate with pronounced longitudinal ridges
    or striations . . . . *Virgatospora* (282)
214. Conidiogenous cells integrated, terminal . . . . . *Gliomastix* (258)
    Conidiogenous cells discrete, variously arranged 215
215. Phialides not in verticils, upper part of conidiophore usually sterile and often setiform
    216
    Phialides in verticils 217
    Phialides forming a more or less complex head at the apex of a stipe . . 219
216. Phialides in rows obscured by a shield of sterile cells, conidia falcate, 1-septate
    Cryptophiale (272)
    Phialides borne on short, encircling collar branches . . . *Gonytrichum* (266)
    Phialides mostly recurved at the tip, conidia often curved with a setula at each end
    Menispora (265)
    Phialides ampulliform or lageniform on short branches arising below the middle of the
    conidiophore, conidia cylindrical . . . . . . *Chaetopsina* (263)
    Phialides pale with dark collarettes, conidia spherical or subspherical
    Angulimaya (267)
217. Phialides cylindrical, rounded at the apex, each with a minute opening and no collarette
    Stachylidium (271)
    Phialides ampulliform or lageniform, often with distinct collarettes . . 218
218. Conidia curved, often falcate, upper part of conidiophore usually sterile and setiform
    Zanclospora (270)
    Conidia allantoid, ellipsoidal or cylindrical, upper part of conidiophore not sterile and
    setiform . . . . *Verticillium* (269)
219. Conidiophores long, finely echinulate, bearing only a few branches near the apex, conidia
    in long chains, limoniform, echinulate or verruculose with spiral bands
    Acrophialophora (268)
    Conidiophores with more complex branching at the apex 220
220. Conidiophores each with long, sterile setiform branches in a whorl below the head
    Gliocephalotrichum (279)
    Conidiophores without a whorl of setiform branches below the head . . 221
221. Conidiophores each swollen at the apex with a spherical or clavate vesicle, the surface of
    which is covered by numerous phialides or short branches bearing phialides . 222
    Conidiophores without vesicles 223
222. Conidia catenate, dry . . . . *Aspergillus* (276)
    Conidia aggregated in slimy heads . . . . . . *Custingophora* (277)
223. Conidiophore stipes unbranched or occasionally branched, each stipe or branch bearing
    at its apex a crown of phialides 224
    Heads often made up of several series of branches, phialides arranged penicillately 226
224. Phialides lageniform, conidia catenate, rather large, ellipsoidal or limoniform
    Phialomyces (275)
    Phialides mostly clavate, pyriform or ellipsoidal, each rounded at the apex and with a
    very small opening 225
225. Conidia catenate, mostly spherical . . . . . . *Memnoniella* (274)
    Conidia aggregated in large, slimy heads, seldom spherical . . *Stachybotrys* (273)
226. Stipe with a terminal head and often a number of similar heads borne laterally and
    alternately, conidia catenate, dry . . . . . . *Thysanophora* (278)
    Conidia often catenate but aggregated in slimy heads . . . *Phialocephala* (254)

227. Conidiophores macronematous, unbranched, conidiogenous cells integrated, terminal, more or less cylindrical, frequently percurrent, conidia 0-septate, aggregated in slimy masses and often formed in long columns arising from cytoplasm which projects a little way above the opening of the phialide . . . . *Chloridium* (283)

Conidiophores semi-macronematous, conidiogenous cells often ampulliform or lageniform, conidia 0–3-septate . . . . . . *Exophiala* (284)

228. Conidiophores synnematous, conidia falcate . . . *Harpographium* (287)

Conidiophores mononematous 229

229. Conidiophores unbranched or rarely branched 230

Conidiophores branched 231

230. Setae absent, conidia cylindrical, straight . . . . *Cylindrotrichum* (285)

Setae often present, conidia frequently curved, falcate, with a fine setula at each end *Codinaea* (286)

231. Branches in verticils . . . . *Selenosporella* (289)

Conidiophores with a number of lateral branches just above the base, upper part sterile, setiform . . . . *Chaetopsis* (288)

232. Conidia terminal . . . . *Spegazzinia* (290)

Conidia terminal and lateral 233

233. Conidia muriform or cruciately septate . . . . *Dictyoarthrinium* (295)

Conidia without septa 234

234. Setae present . . . . *Cordella* (292)

Setae absent 235

235. Stroma absent . . . . *Arthrinium* (291)

Stroma present 236

236. Stroma arising from an annulus, expanding into a fringed funnel packed with a black mass of conidia . . . . *Endocalyx* (294)

Stroma erumpent . . . . *Pteroconium* (293)

CHARACTERS OF PARTICULAR DIAGNOSTIC VALUE BECAUSE THEY ARE COMMON TO ONLY A FEW GENERA OR SPECIES

(Numbers in parenthesis are the serial numbers of genera)

VEGETATIVE

Hyphae arched with stalked attachment organs: *Grallomyces* (85).

—— bearing hyphopodia: *Ampullifera* (56), *Ampulliferina* (3), *Ceratophorum* (16), *Clasterosporium* (74), *Curvulariopsis* (126), *Endophragmiopsis* (95), *Mitteriella* (144), *Sarcinella* (18).

—— with dark transverse septa or bands: *Arthrinium* (291), *Gilmaniella* (29), *Mammaria* (38).

Setae present, separate (but see also under conidiophores with setiform apices or branches):

(A) Simple, usually subulate: *Bahupaathra* (256), *Beltrania* (150), *Beltraniella* (152), *Botryotrichum* (31), *Ceratocladium* (214), *Chaetochalara* (250), *Codinaea* (286), *Cordella* (292), *Ellisiopsis* (153), *Herposira* (105), *Lacellina* (106), *Lacellinopsis* (211), *Mahabalella* (257), *Menispora* (265), *Menisporopsis* (259), *Myrothecium* (280), *Oedemium* (108), *Saccardaea* (281), *Schizotrichum* (175), *Septosporium* (64), *Trichodochium* (91), *Wiesneriomyces* (216).

(B) Branched: *Gyrothrix* (213), *Oedemium* (108).

(C) Capitate: *Catenularia* (253), *Hermatomyces* (67), *Sporoschisma* (248).

Stromata cerebriform—convoluted like a brain: *Cerebella* (37).

—— with one or more cells greatly enlarged and bearing numerous conidiophores: *Camptomeris* (179).

—— with floor and roof connected by columns: *Cryptostroma* (260).

REPRODUCTIVE

Conidiophores

Apex setiform: *Angulimaya* (267), *Beltraniella* (152), *Beltraniopsis* (151), *Chaetopsina* (263), *Chaetopsis* (288), *Costantinella terrestris* (217), *Cryptophiale* (272), *Gonytrichum macrocladum* (266), *Hansfordia* (148), *Helicoma* (117), *Helicosporium* (118), *Hiospira* (119), *Kumanasamuha* (207), *Menispora* (265), *Periconia atropurpurea*, *P. lateralis*, *P. hispidula* (209), *Spondylocladiopsis* (185), *Trichobotrys* (206), *Zanclospora* (270), *Zygosporium* (196).

—— swollen, vesicular: *Aspergillus* (276), *Asteromyces* (41), *Botrytis* (110), *Cephaliophora* (109), *Custingophora* (277), *Haplobasidion* (210), *Haplographium*

24

(218), *Kostermansinda* (69), *Lacellinopsis* (211), *Oncopodium* (68), *Pseudobotrytis* (215).

Base radially lobed: *Beltrania* (150), *Beltraniella* (152), *Beltraniopsis* (151), *Ellisiopsis* (153), *Hemibeltrania* (155), *Pseudobeltrania* (154).

Branching dichotomous or trichotomous: *Balanium* (113), *Botrytis* (110), *Circinoconis* (97), *Dichotomophthora* (231), *Dicyma* (134), *Oedemium* (108), *Phaeodactylium* (136), *Polypaecilum* (114).

—— verticillate or with conidiogenous cells in verticils: *Physalidium* (99), *Selenosporella* (289), *Spondylocladiella* (230), *Spondylocladiopsis* (185), *Stachylidium* (271), *Verticillium* (269), *Zanclospora* (270).

——, with collar branches encircling the stipe: *Gonytrichum* (266).

——, with descending branches: *Acarocybe* (229), *Acarocybella* (223).

——, with setiform branches: *Chaetopsis* (288), *Gliocephalotrichum* (279), *Gonytrichum macrocladum* (266), *Hansfordia* (148), *Trichurus* (200).

Curved strongly, often thickened along one side: *Camptomeris* (179), *Cercosporidium* (177), *Fuscicladiella* (104).

Nodose—with terminal and intercalary swellings: *Cladosporium oxysporum* group (193), *Cordana* (124), *Dendryphiella* (246), *Gonatobotryum* (107), *Haplobasidion* (210), *Lacellinopsis* (211), *Oedemium* (108), *Stemphylium* (102).

——, with unilateral swellings: *Fulvia* (191), *Gonatophragmium* (137).

Torsive—spirally twisted: *Conoplea* (149), *Deightoniella* (101), *Lacellinopsis spiralis* (211), *Periconia funerea* (209), *Polythrincium* (178), *Zygophiala* (183).

Septa thick and very dark or with dark bands: *Arthrinium* (291), *Cordella* (292), *Dictyoarthrinium* (295), *Mammaria* (38), *Pteroconium* (293).

Walls coarsely reticulate: *Hiospira* (119).

Conidiogenous cells

Acuminate—tapered to a point: *Acremoniella* (42), *Tetracoccosporium* (43), *Zygosporium* (196).

Lobed: *Dichotomophthora* (231).

Recurved at the tip; *Menispora* (265).

Umbellately arranged: *Pseudobotrytis* (215).

Conidia

Appendiculate—with one or more setulae: *Beltrania* (150), *Camposporium* (143), *Centrospora* (165), *Codinaea* (286), *Mahabalella* (257), *Menispora* (265), *Menisporopsis* (259), *Pleiochaeta* (164).

Biconic: *Beltrania* (150), *Beltraniella* (152), *Beltraniopsis* (151), *Pseudobeltrania* (154).

Branched—either forked or with thick, septate arms or processes: *Actinocladium* (83), *Arachnophora* (84), *Casaresia* (46), *Ceratosporella* (81), *Ceratosporium* (22), *Cheiromycella* (197), *Cryptocoryneum* (65), *Dictyosporium* (24), *Grallomyces* (85), *Hirudinaria* (21), *Iyengarina* (79), *Teratosperma* (80), *Tetraploa* (20), *Tetraposporium* (23), *Tripospermum* (49), *Triposporium* (82).

Caudate—with a tail at the base: *Beltraniella* (152), *Exosporiella* (90), *Pendulispora* (98).

Cheiroid—shaped rather like a hand: *Cheiromycella* (197), *Cryptocoryneum* (65), *Dictyosporium* (24).

Complex isthmospores: *Isthmospora* (10).

Corniculate—with short horn-like projections: *Acrodictys brevicornuta* (71), *Oncopodiella* (132), *Oncopodium* (48), *Petrakia* (93), *Teratosperma* (80).

Cuneiform—wedge-shaped: *Catenularia* (253), *Graphium* (202), *Leptographium* (201), *Polythrincium* (178).

Endogenous: *Bloxamia* (251), *Catenularia* (253), *Chaetochalara* (250), *Chalara* (249), *Cystodendron* (252), *Phialocephala*, occasionally (254), *Sporoschisma* (248).

Falcate: *Codinaea* (286), *Cryptophiale* (272), *Harpographium* (287), *Idriella* (123), *Zanclospora* (270).

Helicoid or circinate—in the form of a coil or spiral: *Circinoconis* (97), *Helicoma* (117), *Helicomina* (131), *Helicosporium* (118), *Hiospira* (119), *Trochophora* (112), *Troposporella* (47), *Xenosporium* (78).

——, coiled in several planes to form an ellipsoidal spore body: *Helicodendron* (48), *Helicoon* (116), *Helicorhoidion* (138).

Lenticular—shaped like a biconvex lens: *Arthrinium* (291), *Cordella* (292), *Endocalyx* (294), *Hermatomyces* (67), *Pteroconium* (293), *Stephanosporium* (7).

Lobed: *Isthmospora* (10), *Physalidium* (99), *Pteroconium* (293), *Spegazzinia* (290), *Thermomyces stellatus* (26).

Lunate: *Idriella* (123).

Oblong or cylindrical, truncate at the ends: *Ampulliferina* (3), *Bahusakala* (2), *Bloxamia* (251), *Chaetochalara* (250), *Chalara* (249), *Coremiella* (5), *Scytalidium* (1), *Septotrullula* (59), *Sporoschisma* (248), *Thielaviopsis* (4).

Reniform—kidney-shaped: *Melanographium* (121), *Stachybotrys nephrospora* (273), *Virgaria* (133).

Sigmoid: *Curvularia deightonii* (239), *Nakataea* (140), *Spiropes scopiformis* (159), *Sporidesmium inflatum* (73).

Spathulate: *Phragmospathula* (100).

T-shaped: *Scolecobasidium terreum* (142).

U- or V-shaped: *Hirudinaria* (21).

Y-shaped: *Iyengarina* (79).

With a hyaline median transverse band: *Beltrania* (150), *Beltraniella* (152), *Beltraniopsis* (151), *Ellisiopsis* (153), *Pseudobeltrania* (154).

Septa always or almost always 2: *Acrospeira* (35), *Brachydesmiella* (240), *Pollaccia* (86), *Pyricularia* (139), *Spadicoides obovata* (233), *Spondylocladiella* (230), *Spondylocladiopsis* (185), *Sterigmatobotrys* (221).

—— often cruciately arranged: *Dictyoarthrinium sacchari* (295), *Sarcinella* (18), *Spegazzinia* (290), *Tetracoccosporium* (43), *Ulocladium* (245).

—— very dark or with dark bands: *Bactrodesmium* (61), *Bispora* (52), *Brachysporiella* (96), *Didymobotryum* (228), *Diplococcium* (234), *Endophragmia* (94), *Prathigada* (176), *Spadicoides* (233).

Walls coarsely reticulate: *Hiospira* (119).

—— with spirally arranged markings: *Acrophialophora* (268), *Kumanasamuha* (207).

—— striately marked or with longitudinal ridges: *Bahusakala* (2), *Hadronema* (122), *Myrothecium brachysporum* and *M. striatisporum* (280), *Stachybotrys cylindrospora* (273), *Trichodochium* (91), *Virgatospora* (282).

—— with a longitudinal germ slit: *Conoplea* (149), *Endocalyx* (294), *Mammaria* (38), *Melanographium* (121), *Wardomyces* (39).

THE GENERA

1. SCYTALIDIUM

Scytalidium Pesante, 1957, *Annali Sper. agr.*, N.S., **11**, Suppl.: CCLXI–CCLXV.

Colonies effuse, dark blackish brown. *Mycelium* immersed and superficial. *Hyphae* smooth, some narrow, cylindrical, colourless, others thicker, pale to mid brown with occasional darker swollen cells and often thick, very dark brown septa. The hyphae often lie parallel to one another and may be closely adpressed forming bundles. *Stroma* none. *Setae* and *hyphopodia* absent. *Conidiophores* micronematous, mononematous or sometimes synnematous, branched or unbranched, straight or flexuous, colourless or brown, smooth. *Conidiogenous cells* fragmenting and forming arthroconidia, integrated, intercalary, determinate, cylindrical, doliiform or ellipsoidal. *Conidia* catenate, separating, dry, schizogenous, simple, smooth, 0–septate, with septa sometimes thick and very dark, of two kinds: (1) colourless, thin-walled, cylindrical or oblong, truncate at each end, (2) broader, mid or dark brown, thick-walled, oblong, doliiform or broadly ellipsoidal.

Type species: Scytalidium lignicola Pesante.

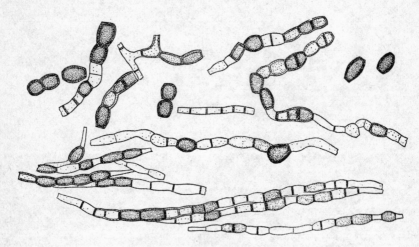

FIG. 2. *Scytalidium lignicola* (×650).

Scytalidium lignicola Pesante, 1957, *Annali Sper. agr.*, N.S., **11**, Suppl.: CCLXI–CCLXV.

(Fig. 2)

Hyphae 1·5—6µ thick except for swollen cells which may be up to 10µ thick. *Conidia* (1) hyaline, 6—10 × 1—3µ (2) brown, 6—15 × 5—10µ

Isolated from wood of *Pinus* and *Platanus*, roots of *Vitis* and from soil; Cyprus, Great Britain, India, Italy, Rhodesia.

2. BAHUSAKALA

Bahusakala Subramanian, 1958, *J. Indian bot. Soc.*, **37**: 63.

Colonies effuse, fuscous. *Mycelium* mostly superficial. *Stroma* none. *Setae* and *hyphopodia* absent. *Conidiophores* macronematous or semi-macronematous, mononematous, flexuous, irregularly branched, brown or dark brown, smooth. *Conidiogenous cells* integrated, intercalary and terminal, determinate, fragmenting to form arthroconidia. *Conidia* catenate, simple, brown or dark brown, 0-, 1- or pluri-septate, often rough-walled with striate markings or rugose, cylindrical or oblong, truncate or rounded at the ends.

Type species: Bahusakala olivaceonigra (Berk. & Br.) Subram.

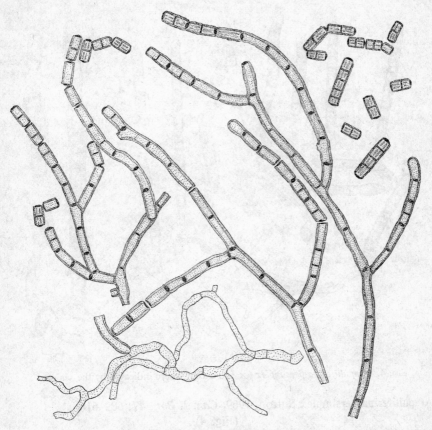

FIG. 3. *Bahusakala olivaceonigra* (× 650).

Bahusakala olivaceonigra (Berk. & Br.) Subram., 1958, *J. Indian bot. Soc.*, **37**: 61–63.

Septonema olivaceo-nigrum Berk. & Br., 1873, *J. Linn. Soc.*, **14**: 90.

(Fig. 3)

Conidiophores branched and variable in length, 3–5μ thick. *Conidia* mostly 1-septate, occasionally 0-septate or with 2–5 septa, 5–24μ long, 4–6μ thick.

On *Agave*, Ceylon; on *Yucca gloriosa* and unknown Liliaceae, India.

3. AMPULLIFERINA

Ampulliferina Sutton, 1969, *Can. J. Bot.*, **47**: 609.

Colonies effuse, brown to brownish black. *Mycelium* superficial. *Stroma* none. *Setae* absent. *Hyphopodia* present. *Conidiophores* macronematous, mononematous, short, straight, brown, smooth. *Conidiogenous cells* integrated, intercalary and terminal, determinate, fragmenting. *Conidia* catenate, formed by fragmentation through the middle of thick, dark septa, cylindrical with truncate ends except the terminal conidium which is rounded at its apex, 1-septate, brown, smooth.

Type species: Ampulliferina persimplex Sutton.

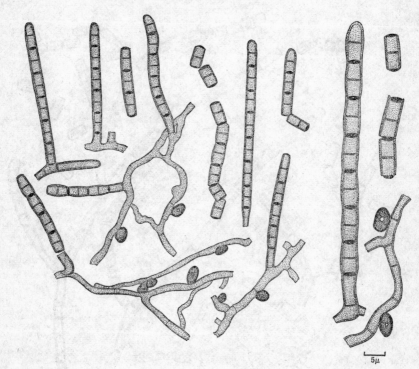

FIG. 4. *Ampulliferina persimplex* (× 650 except where indicated by the scale).

Ampulliferina persimplex Sutton, 1969, *Can. J. Bot.*, **47**: 609–616.

(Fig. 4)

Hyphae pale brown to brown, smooth, 2–5µ thick. *Hyphopodia* spherical, ellipsoidal or clavate, rarely lobed, greyish brown, 3–7µ diam. *Conidia* 10–12 × 4–5µ.

On fallen dead leaves of *Ledum groenlandicum*; Canada.

4. THIELAVIOPSIS

Thielaviopsis Went, 1893, *Meded. Proefstn SuikRiet W. Java* ' Kagok ', **5**: 4.

Stilbochalara Ferdinandsen & Winge, 1910, *Bot. Tidsskr.*, **30**: 220–222.

Hughesiella Batista & Vital, 1956, *Anais Soc. Biol. Pernamb.*, **14**: 141–144.

Colonies effuse, grey, olivaceous, dark blackish brown or black, velvety or pulverulent. *Mycelium* partly superficial, partly immersed. *Stroma* none. *Setae* and *hyphopodia* absent. *Conidiophores* macronematous, mononematous, unbranched or irregularly branched, straight or flexuous, hyaline or pale brown, smooth. *Conidiogenous cells* fragmenting to form arthroconidia, terminal and intercalary, determinate, cylindrical. *Conidia* catenate, dry, schizogenous, often seceding with difficulty, simple, doliiform, ellipsoidal, obovoid or oblong, mid to very dark brown, smooth in most species, 0-septate. In *T. basicola* the short chains of conidia remain intact for a long time and resemble multiseptate conidia. There is also a phialidic *Chalara* state and often a *Ceratocystis* state.

Type species: Thielaviopsis state of *Ceratocystis paradoxa* (Dade) C. Moreau = *T. ethaceticus* Went.

KEY

Conidia doliiform, obovoid or ellipsoidal, longer than wide   .   *Ceratocystis paradoxa*
Conidia short oblong, wider than long   .   .   .   .   *basicola*

Thielaviopsis state of **Ceratocystis paradoxa** (Dade) C. Moreau, 1952, *Revue Mycol.*, **17**: 22.

(Fig. 5 A)

Colonies dark blackish brown to black. *Conidiophores* colourless to pale brown, up to 50 × 4–6µ. *Arthroconidia* catenate, doliiform, ellipsoidal or obovoid, mid to very dark brown, smooth, rather thick-walled, sometimes with a hyaline longitudinal slit, mostly 10–25 × 8–16µ. *Phialides* up to 200µ long, 8–10µ thick in the broadest part, tapering to 3–4µ. *Phialoconidia* in long chains, at first cylindrical, colourless, becoming ellipsoidal and pale to mid golden brown, mostly 7–14 × 3–6µ.

Common and widespread in the tropics on many different plants, especially important hosts being *Ananas, Cocos, Elaeis, Musa, Saccharum* and *Theobroma* in all of which it causes disease [CMI Distribution Map, 142].

Thielaviopsis basicola (Berk. & Br.) Ferraris, 1912, Flora Italica, Hyphales: 233–234.

(Fig. 5 B)

Colonies grey or olivaceous, often velvety. *Conidiophores* up to 50 × 6–9µ. *Arthroconidia* usually in chains of 4–8 remaining together for a long time and resembling large multiseptate conidia, finally seceding, but with difficulty, oblong or short cylindrical, only the apical one in each chain rounded at the end, others truncate at both ends, mid to dark golden brown, smooth, 7–12µ long, 10–17µ wide. *Phialides* up to 100µ long, 5–8µ thick in the broadest part, tapering to 3–4µ. *Phialoconidia* cylindrical, truncate at the ends, hyaline, 7–17 × 2·5–4·5µ.

Causes black root rot of tobacco (*Nicotiana*) and occurs on many other plants including *Cichorium, Citrus, Crotalaria, Gloxinia, Lupinus, Lycopersicon,*

Pisum, Primula and *Viola,* sometimes causing disease; cosmopolitan [CMI Distribution Map, 218].

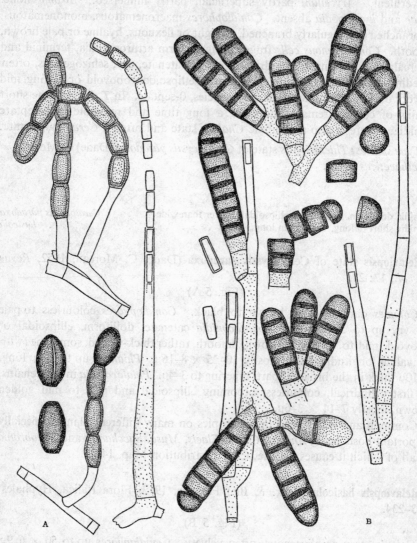

Fig. 5. A, *Thielaviopsis* state of *Ceratocystis paradoxa*; B, *T. basicola* (× 650).

5. COREMIELLA

Coremiella Bubák & Krieger, 1912, *Annls mycol.,* **10**: 52–53.

Colonies effuse, olivaceous to dark blackish brown. *Mycelium* immersed. *Stroma* pseudoparenchymatous. *Conidiophores* macronematous, usually synnematous forming rather loose coremia with compound stipe and head but sometimes mononematous and caespitose, flexuous, irregularly or subdichotomously branched, pale to mid olivaceous brown or brown, smooth. *Conidio-*

genous cells fragmenting, integrated, terminal, determinate, cylindrical; transverse septa are laid down in basipetal succession and neighbouring or alternate cells develop thick inner walls, intermediate cells lose their contents and their lateral walls which remain thin collapse and eventually break liberating conidia. *Conidia* catenate, schizogenous, simple, pale to mid olivaceous brown or brown, 0-septate, smooth, oblong or cubical, often with a minute papilla in the middle of each end and a small frill at the periphery.

Type species: Coremiella cubispora (Berk. & Curt.) M. B. Ellis = *C. cystopoides* Bubák & Krieger.

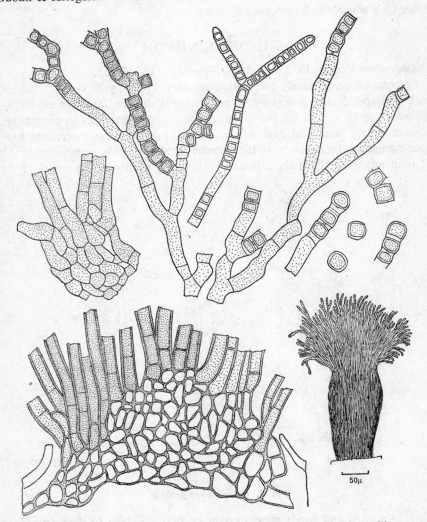

Fig. 6. *Coremiella cubispora* (× 650 except where indicated by the scale).

Coremiella cubispora (Berk. & Curt.) M. B. Ellis comb. nov.
  Cladosporium cubisporum Berk. & Curt., 1875, apud Berk. in *Grevillea*, **3**: 107.

D.H.—2

Coremiella ulmariae (Mac Weeney) Mason, 1953, apud Hughes in *Can. J. Bot.*, **31**: 640.

C. cystopoides Bubák & Krieger, 1912, *Annls mycol.*, **10**: 52–53.

(Fig. 6)

Coremia up to 800μ high, 200μ broad. *Conidiophores* 4–9μ thick. *Conidia* 5–9 × 4–9μ.

On dying and dead leaves, stems and branches of herbaceous plants, shrubs and trees; Europe, E. & S. Africa, U.S.A. A number of collections have been made in Great Britain on marsh and fen plants such as *Eleocharis palustris*, *Filipendula ulmaria* and *Lythrum salicaria*.

6. OIDIODENDRON

Oidiodendron Robak, 1932, *Nyt Mag. Naturvid.*, **71**: 243–255.

Colonies effuse, variously coloured, often grey, brown, yellow or olivaceous; with one species *O. rhodogenum* a red pigment diffuses into culture media. *Mycelium* partly superficial, partly immersed; hyphae sometimes forming ropes. *Stroma* none. *Setae* and *hyphopodia* absent. *Conidiophores* macronematous, mononematous, branched, with the branches hyaline or pale, sometimes almost at right angles to the main axis and each other, mostly in the upper region

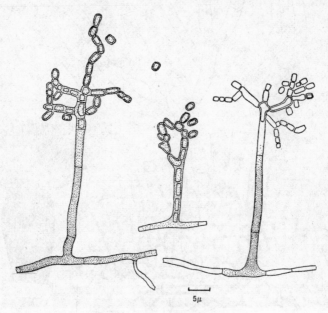

5μ

FIG. 7. *Oidiodendron griseum.*

forming a stipe and head; stipe straight or flexuous, pale to dark brown or olivaceous brown, smooth or verrucose. *Conidiogenous cells* integrated, terminal on branches, determinate, cylindrical, fragmenting to form arthroconidia, producing a number of short segments by basipetal septation, the conidia then maturing from the tip back towards the main axis. *Conidia* catenate, dry, schizogenous, separating readily, simple, cylindrical, doliiform, ellipsoidal, spherical or subspherical, hyaline, greyish green, brown or olivaceous brown, smooth or verruculose, 0-septate.

Lectotype species: *Oidiodendron tenuissimum* (Peck) Hughes = *O. fuscum* Robak.

Barron, G. L., *Can. J. Bot.*, **40**: 589–607, 1962.

The two species of *Oidiodendron* most frequently received at the CMI are *O. tenuissimum* and *O. griseum*.

Oidiodendron griseum Robak, 1934, apud Melin & Nannf. in *Svenska SkogsFör. Tidskr.*, 3/4: 440.

(Fig. 7)

Colonies grey to olivaceous brown, often raised in the middle, wrinkled and sometimes clearly zonate. *Conidiophores* olivaceous to blackish brown, smooth, up to 100μ long, 1·5–2μ thick. *Conidia* oblong or ellipsoidal, pale greyish green, smooth or sometimes minutely verruculose 2–3·5 × 1·5–2μ.

Isolated from litter, soil and wood pulp; Europe, N. America, Trinidad.

Oidiodendron tenuissimum (Peck) Hughes, 1958, *Can. J. Bot.*, **36**: 790.

Colonies pale grey to blackish brown, sometimes slightly raised in the centre. *Conidiophores* brown or olivaceous brown, smooth or verrucose, up to 300μ long, 1·5–2·5μ thick. *Conidia* globose, subglobose or ellipsoidal, the outer wall dark and distinctly verruculose, 2–4 × 1·5–2·5μ, linked in chains by narrow connectives and then resembling beads on a string.

Isolated from litter, soil, wood and bark, probably the commonest species; Europe, Nigeria, N. America.

7. STEPHANOSPORIUM

Stephanosporium Dal Vesco, 1961, *Allionia*, **7**: 181–193.

Colonies rather slow-growing, effuse, purplish grey to black. *Mycelium* partly superficial, partly immersed, sometimes forming ropes. *Stroma* none. *Setae*

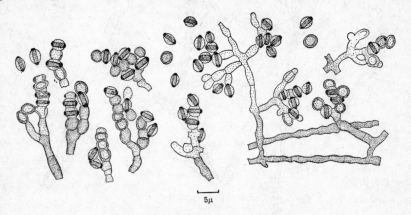

5μ

FIG. 8. *Stephanosporium cereale.*

and *hyphopodia* absent.   *Conidiophores* macronematous, mononematous, branched like a tree; stipe straight or flexuous, mid to dark brown, smooth. *Conidiogenous cells* integrated, terminal on branches, determinate, torulose, fragmenting to form arthroconidia, producing a number of short segments by basipetal septation the conidia then maturing from the tip back towards the main axis, the outer wall remaining continuous for some time surrounding the chain as a thin membrane. *Conidia* catenate, dry, when mature lying at different angles and often at right angles to one another, schizogenous, simple, lenticular though not excessively flat, brown or greyish brown with a dark, very distinctive equatorial band or girdle which protrudes slightly at each end, smooth, 0-septate.

Type species: Stephanosporium cereale (Thüm.) Swart = *S. atrum* Dal Vesco.

Stephanosporium cereale (Thüm.) Swart, 1965, *Trans. Br. mycol. Soc.*, **48**: 459–461.

(Fig. 8)

Conidiophores sometimes up to 50μ long, usually 10–20μ long, 1·5–2·5μ thick. *Conidia* 3–5 × 2–3μ.

On dead leaves and culms of *Bambusa* and *Secale* and isolated from air, paper, soil and textiles; Europe, Pakistan, Philippines, N. America.

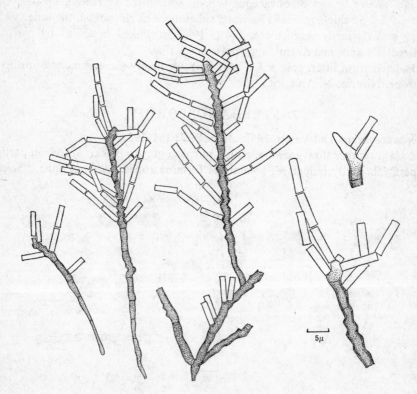

Fɪɢ. 9. *Sympodiella acicola* (× 650 except where indicated by the scale).

8. SYMPODIELLA

Sympodiella Kendrick, 1958, *Trans. Br. mycol. Soc.*, **41**: 519–521.

Colonies effuse, dark blackish brown, hairy. *Mycelium* superficial, hyphae anastomosing to form a network. *Stroma* none. *Setae* and *hyphopodia* absent. *Conidiophores* macronematous, mononematous, scattered, solitary, unbranched, straight or flexuous, geniculate, dark brown, paler near the apex, smooth. *Conidiogenous cells* discrete, solitary, determinate, cylindrical. The first conidiogenous cell is produced by modification of the conidiophore apex, a growing point develops behind and to one side of this and almost immediately forms another conidiogenous cell. The conidiophore thus increases in length sympodially but forms conidiogenous cells instead of conidia; these increase in length and eventually fragment to form chains of arthroconidia. *Conidia* catenate, dry, schizogenous, simple, cylindrical with truncate ends, hyaline, smooth, 0-septate.

Type species: Sympodiella acicola Kendrick.

Sympodiella acicola Kendrick, 1958, *Trans. Br. mycol. Soc.*, **41**: 519–521.
(Fig. 9)

Conidiophores up to 300 × 2–4μ. *Conidia* cylindrical, very slightly expanded and truncate at each end, 7–14 × 2–2·5μ.

On rotten leaves of *Pinus*; Great Britain.

9. WALLEMIA

Wallemia Johan-Olsen, 1887, *Forh. Vidensk. Selsk. Krist.*, 1887, No. 12: 1–20.

Colonies usually slow-growing, often irregular in shape, sometimes fan-like

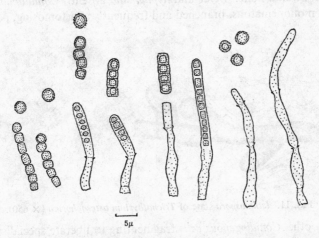

FIG. 10. *Wallemia sebi.*

or stellate, sometimes humped, frequently brown or reddish brown but may be orange, cream or dark blackish brown. *Mycelium* partly immersed, partly superficial. *Stroma* none. *Setae* and *hyphopodia* absent. *Conidiophores* macronematous, slender, mononematous, closely packed together but not forming synnemata or regular, pulvinate sporodochia, usually unbranched, straight or flexuous, subhyaline, smooth, each consisting of a phialide-like lower part with a dark collarette and beyond this a protuberant upper part which becomes septate basipetally and finally fragments to form conidia. Conidiophores frequently proliferate percurrently through the collarettes and a number of these can often be seen at different levels. *Conidiogenous cells* fragmenting to form modified arthroconidia, integrated, terminal, cylindrical. *Conidia* catenate, dry, schizogenous, simple, at first hyaline, cubical, smooth, becoming spherical or subspherical, minutely verruculose, straw-coloured individually, brown in mass, 0-septate.

Type species: Wallemia sebi (Fr.) v. Arx = *W. ichthyophaga* Johan-Olsen.

Wallemia sebi (Fr.) v. Arx, 1970, The Genera of Fungi sporulating in pure Culture: 166.
Sporendonema sebi Fries, 1832, Syst. mycol., **3**: 435.

(Fig. 10)

Conidiophores varying in length according to the amount of proliferation, 1–3µ thick. *Conidia* 2–3·5µ diam.

Most frequently isolated from foodstuffs including jam, bread, cakes, salted fish, milk and fats but also occurs in air, hay, textiles and man; cosmopolitan.

10. ISTHMOSPORA

Isthmospora F. L. Stevens, 1918, *Bot. Gaz.*, **65**: 244.
Colonies black, hyperparasitic on *Meliola* and similar fungi. *Mycelium* superficial. *Stroma* none. *Setae* and *hyphopodia* absent. *Conidiophores* micronematous, mononematous, branched and frequently anastomosing, pale to dark

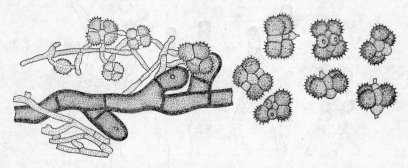

FIG. 11. *Isthmospora* state of *Trichothyrium asterophorum* (× 650).

brown, smooth. *Conidiogenous cells* fragmenting to liberate specialised arthroconidia sometimes called isthmospores, integrated, intercalary, determinate; initially narrow connecting hyphae they later become lageniform or subulate and finally parallel and paired, one on each side of the conidium. *Conidia* solitary, schizogenous, dry, brown, smooth or echinulate, complex, sarciniform, lobed.

Type species: Isthmospora state of *Trichothyrium asterophorum* (Berk. & Br.) Höhnel = *I. spinosa* F. L. Stevens.

Hughes, S. J., *Mycol. Pap.*, **50**: 77–94, 1953.

Isthmospora state of **Trichothyrium asterophorum** (Berk. & Br.) Höhnel, 1909, *Sber. Akad. Wiss. Wien*, **118**: 1482; for full synonymy see Hughes in *Mycol. Pap.* **50**.

(Fig. 11)

Hyphae often compacted to form a plate covering the Melioline colony and extending some way beyond it. *Conidia* complex, lobed, sarciniform, dark brown, echinulate, 15–21 × 12–17μ.

Very common on *Meliola* and similar fungi in the tropics.

11. CONIOSPORIUM

Coniosporium Link ex Fries; Link, 1809, *Magazin Ges. naturf. Freunde Berl.*, **3** (1): 8 [as ' *Conisporium* ']; Fries, 1821, Syst. mycol., **1**: XL.

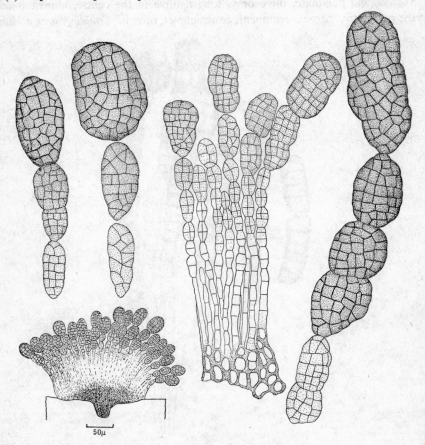

FIG. 12. *Coniosporium olivaceum* (× 650 except where indicated by the scale).

Sirodesmium de Notaris, 1849, *Memorie R. Accad. Sci. Torino*, 2 Ser., **10**: 348–350.

Bonordeniella Penzig & Saccardo, 1902, *Malpighia*, **15**: 259–260.

Sporodochia pulvinate, punctiform, olive to black. *Mycelium* immersed. *Stroma* present. *Setae* and *hyphopodia* absent. *Conidiophores* macronematous, meristematic, growth in length taking place by the laying down of cross walls and elongation of cells just behind the apex, straight or flexuous, closely packed together, unbranched, smooth, rugulose or verrucose. *Conidiogenous cells* integrated, terminal, meristematic, more or less cylindrical, fragmenting. *Conidia* catenate, schizogenous, dry, simple, ellipsoidal, oblong, obovoid, pyriform or subspherical, pale to dark brown, with transverse and usually also longitudinal septa, often muriform, smooth or verrucose.

Type species: Coniosporium olivaceum Link ex Fr.

Coniosporium olivaceum Link ex Fr.; Link, 1809, Fries, 1832 (refs as for genus). For synonymy see Hughes, *Can. J. Bot.*, **36**: 754, 1958.
(Fig. 12)

Sporodochia pulvinate, olive or yellowish olive in the centre, almost black at the periphery. *Stroma* erumpent, conspicuous, brown. *Conidiophores* arising

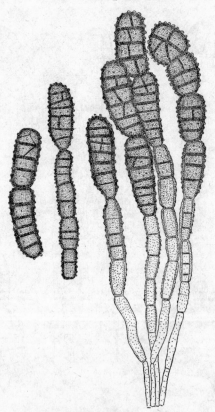

FIG. 13. *Coniosporium* state of *Hysterium insidens* (× 650).

from surface of stroma, up to 50μ long, 2–4μ thick, subhyaline to pale brown, smooth or rugulose. *Conidia* in long chains, ellipsoidal, pyriform or subspherical, muriform, pale to mid dark brown, smooth, 15–63 × 10–34 (mostly 20–45 × 15–25)μ.

On dead wood; specimens seen from Czechoslovakia and U.S.A.

Coniosporium state of **Hysterium insidens** Schw., 1832, *Trans. Am. phil. Soc.* N.S., **4**: 244. For synonymy see Hughes, *Can. J. Bot.*, **36**: 754, 1958.

(Fig. 13)

Sporodochia pulvinate, black or almost black, up to 1·5 cm. diam. *Stroma* erumpent, conspicuous, outer part brown, inner part subhyaline. *Conidiophores* arising from surface of stroma, up to 50μ long, 2–5μ thick, subhyaline near the base, pale brown above, smooth to verrucose. *Conidia* in long chains, oblong rounded at the ends, or ellipsoidal, at first with transverse septa only but later with longitudinal septa also, brown to dark brown, verrucose, 13–38 × 7–17 (commonly 20–30 × 7–11)μ.

Common on decorticated wood, e.g. fence posts; Canada, Europe, S. Africa, U.S.A.

12. TRIMMATOSTROMA

Trimmatostroma Corda, 1837, Icon. Fung., **1**: 9.

Sporodochia pulvinate, sometimes confluent, dark blackish brown to black. *Mycelium* immersed. *Stroma* present, erumpent, often large, brown, pseudoparenchymatous. *Setae* and *hyphopodia* absent. *Conidiophores* macronematous or semi-macronematous, meristematic, growth in length taking place by a laying down of cross walls and elongation of cells just behind the apex, unbranched or occasionally loosely branched, usually short, straight or flexuous, closely packed together forming sporodochia, pale to mid brown, smooth rugulose or verruculose. *Conidiogenous cells* integrated, terminal, roughly cylindrical, fragmenting. *Conidia* formed in branched, basipetal chains, schizogenous, dry, simple, branched, forked or lobed, straight or flexuous, very variable in shape, cylindrical rounded at the apex, ellipsoidal, clavate, pyriform, subspherical, etc., pale to dark brown or olivaceous brown, smooth or verruculose, with transverse and often longitudinal or oblique septa.

Type species: Trimmatostroma salicis Corda.

Hughes, S. J., *Can. J. Bot.*, **31**: 627–629, 1953.

Trimmatostroma salicis Corda, 1837, Icon. Fung., **1**: 9–10.

(Fig. 14 A)

Conidia curved or bent, cylindrical rounded at the ends or clavate, often forked or variously branched, with up to 13 transverse and sometimes 1 or a few longitudinal or oblique septa, pale to mid brown or olivaceous brown. walls and septa darker, smooth or verruculose, 12–38 × 4–10μ.

Common on twigs and branches of *Salix*; Europe.

Trimmatostroma betulinum (Corda) Hughes, 1953, *Can. J. Bot.*, **31**: 628.
Coniothecium betulinum Corda, 1837, Icon. Fung., **1**: 32.
(Fig. 14 B)

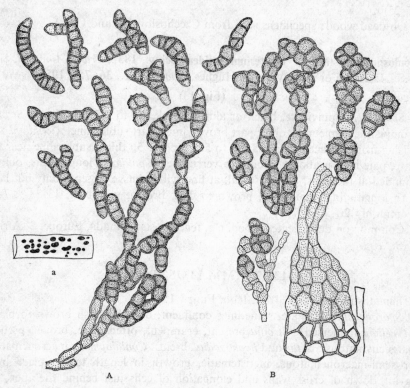

FIG. 14. A, *Trimmatostroma salicis*; B, *T. betulinum* (a, habit sketch; other figures × 650).

Stroma usually very large. *Conidia* extremely variable in shape, subspherical, clavate, etc., sometimes lobed, with 1 or several transverse and longitudinal or oblique septa, pale to mid brown, smooth or verruculose, 5–20 × 5–14µ.

On twigs, branches and occasionally leaves of *Betula*, also isolated from pine litter; Europe.

13. USTILAGINOIDEA

Ustilaginoidea Brefeld, 1895, Untersuchungen, **12**: 194–205.

Fructifications in the ovaries of individual grains of grasses, transforming them into large, very dark olive green or sometimes orange velvety masses: when cut open the inner part is seen to be bright orange towards the surface, almost white in the centre. *Mycelium* partly superficial, partly immersed. *Stroma* always present, prosenchymatous. *Setae* and *hyphopodia* absent. *Conidiophores* micronematous, mononematous, sparingly branched, flexuous, intertwined, hyaline, smooth. *Conidiogenous cells* polyblastic, integrated, intercalary, determinate, cylindrical, denticulate; denticles short, cylindrical. *Conidia* solitary, dry, pleurogenous, simple, spherical or subspherical, pale to dark olive green, verruculose, 0-septate.

Type species: Ustilaginoidea virens (Cooke) Takahashi = *U. oryzae* (Pat.) Bref.

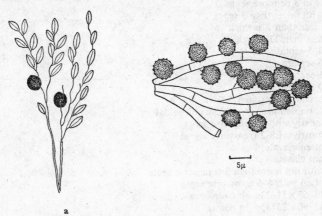

FIG. 15. *Ustilaginoidea virens* (a, habit sketch).

Ustilaginoidea virens (Cooke) Takahashi, 1896, *Bot. Mag., Tokyo,* **10** (2): 16–20.
(Fig. 15)

Fructifications the same length as the normal grain but usually at least twice as fat: the outer dark olive green mass is composed mostly of conidia. *Conidiophores* 2–2·5μ thick. *Conidia* spherical or somewhat compressed, dark olive, verruculose, 4–5μ diam.

On *Oryza* and occasionally other grasses; widely distributed [CMI Distribution Map, 347]. Another species **U. ochracea** P. Henn. which occurs on many grasses, especially Paniceae in the tropics has rather larger conidia (6–8μ diam.).

14. PITHOMYCES

Pithomyces Berkeley & Broome, 1873, *J. Linn. Soc.,* **14**: 100.
Scheleobrachea Hughes, 1958, *Can. J. Bot.,* **36**: 802.

Colonies punctiform or effuse, yellow, olive green, brown or black. *Mycelium* all or mostly superficial. *Stroma* none. *Setae* and *hyphopodia* absent. *Conidiophores* micronematous or semi-macronematous, mononematous, branched, straight or flexuous, subhyaline or pale olive to brown, smooth or verruculose. *Conidiogenous cells* monoblastic or polyblastic, integrated, intercalary and sometimes also terminal, determinate, cylindrical, denticulate; denticles cylindrical, sometimes quite long. *Conidia* solitary, pleurogenous or acropleurogenous, dry, simple, detached through fracture of the denticle, a part of which often remains attached to the base of the conidium, ellipsoidal, clavate, limoniform, obovoid, oblong rounded at the ends, pyriform or obpyriform, straw-coloured to dark blackish brown, smooth, echinulate or verruculose, with 0–13 transverse and often 1 or more oblique or longitudinal septa.

Type species: Pithomyces flavus Berk. & Br.

Ellis, M. B., *Mycol. Pap.,* **76**: 7–19, 1960; **103**: 38–42, 1965.

KEY

Mature conidia mostly without longitudinal septa 1
Mature conidia mostly with longitudinal septa 5
1. Conidia with 3 or more septa 2
   Conidia with fewer than 3 septa 4
2. Conidia more than 11µ thick 3
   Conidia less than 11µ thick . . . . *atro-olivaceus*
3. Conidia 3–5-septate . . . . *flavus*
   Conidia 6-septate . . . . *elaeidicola*
4. Conidia 1-septate . . . . *cupaniae*
   Conidia 0-septate . . . . *africanus*
5. Conidia broadly ellipsoidal 6
   Conidia variable in shape, seldom ellipsoidal 7
6. Conidia mostly with 3 transverse septa . . . . *chartarum*
   Conidia mostly with 2 transverse septa . . . . *maydicus*
7. Conidia often clavate 8
   Conidia not clavate 9
8. Conidia with not more than 3 transverse septa . . . . . *sacchari*
   Conidia often with 4–6 transverse septa . . . . . . *cynodontis*
9. Conidia 15–53×11–33µ on *Crataegus* . . . . *quadratus*
   Conidia 45–90×25–45µ on palms . . . . *pulvinatus*

Pithomyces chartarum (Berk. & Curt.) M. B. Ellis, 1960, *Mycol. Pap.*, **76**: 13–15.
(Fig. 16 A)

Colonies at first punctiform, black, up to 0·5 mm. diameter, later sometimes becoming confluent. *Conidiophores* micronematous, branched and anastomosing, pale olive, smooth or occasionally verruculose, 2–5µ thick; denticles 2–10 × 2–3·5µ. *Conidia* broadly ellipsoidal, with 3–4 (mostly 3) transverse septa, the middle cells usually divided by longitudinal septa, often constricted at the septa, mid to dark brown when mature, echinulate or verruculose, 18–29 × 10–17µ; a small piece of the denticle invariably remains attached to the base of the conidium.

On paper and common on dead leaves of many (more than 50) different plants, especially frequent on fodder grasses, isolated from air and soil; cosmopolitan. It is apparently the cause of facial eczema in sheep in New Zealand.

Conidial measurements, substrata and geographical distribution of 10 other species of *Pithomyces* are given below.

P. atro-olivaceus (Cooke & Harkness) M. B. Ellis (Fig. 16 B): 14–35 × 7–10µ, on bark of *Acacia*; U.S.A.

P. flavus Berk. & Br. (Fig. 16 C): 28–45 × 15–26µ, on decaying leaf sheaths of monocotyledons, including *Oncosperma*; Ceylon.

P. elaeidicola M. B. Ellis (Fig. 16 D): 33–55 × 15–24µ, on dead spathe and severed leaf rachis of *Elaeis*; Sierra Leone.

P. cupaniae (Syd.) M. B. Ellis (Fig. 16 E): 10–20 × 3–5µ, on *Albizia*, *Anisophyllea*, *Carpodinus*, *Clitandra*, *Cupania*, *Funtumia*, *Hymenocardia*, *Jasminum*, *Milletia*, *Rapanea* and *Sorindeia*; Costa Rica, Sierra Leone, Uganda, Zambia.

P. africanus M. B. Ellis (Fig. 16 F): 10–19 × 5–8µ, on *Borassus* and *Hyphaene*; Ghana, Sierra Leone.

P. maydicus (Sacc.) M. B. Ellis (Fig. 16 G): 12–20 × 6–12µ, on *Andropogon*, *Bridelia*, *Calopogonium*, *Centrosema*, *Chasmopodium*, *Cinnamomum*, *Coix*,

Colocasia, Elaeis, Gladiolus, Oryza, Pueraria, Saccharum, Sorghum, Vanda, Vitex and *Zea*; Ghana, Guinea, Guyana, Hong Kong, India, Jamaica, Malaya, Netherlands New Guinea, New Caledonia, Philippines, Sabah, Sierra Leone.

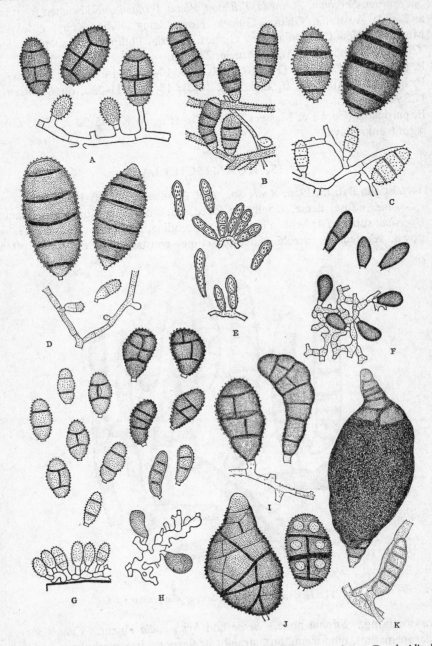

FIG. 16.   *Pithomyces* species; A, *chartarum*; B, *altro-olivaceus*; C, *flavus*; D, *elaeidicola*; E, *cupaniae*; F, *africanus*; G, *maydicus*; H, *sacchari*; I, *cynodontis*; J, *quadratus*; K, *pulvinatus* (× 650).

P. sacchari (Speg.) M. B. Ellis (Fig. 16 H): 12–25 × 5–15μ, on *Anacardium, Ananas, Andropogon, Areca, Aristida, Arundinella, Borassus, Cajanus, Camellia, Coix, Cynodon, Elaeis, Gladiolus, Imperata, Musa, Nicotiana, Ophiurus, Saccharum, Sorghum, Themeda, Triticum, Vigna, Withania,* and isolated from air and soil; Australia, Ghana, Guinea, Hong Kong, India, Kenya, Malaya, Mauritius, New Caledonia, New Guinea, Nigeria, Philippines, Sabah, Sudan, Trinidad, Uganda, U.S.A., Venezuela, Zambia.

P. cynodontis M. B. Ellis (Fig. 16 I): 20–55 × 10–25μ, on *Cynodon*; Zambia.

P. quadratus (Atk.) M. B. Ellis (Fig. 16 J): 15–53 × 11–23μ, on *Crataegus*; U.S.A.

P. pulvinatus (Cooke & Massee) M. B. Ellis (Fig. 16 K): 45–90 × 25–45μ, on palm trunks; Java.

15. HORMISCIELLA

Hormisciella Batista, 1956, *Anais Soc. Biol. Pernamb.*, **14**: 100–101.

Colonies effuse, dense, woolly or felted, dark olivaceous brown to black. *Mycelium* superficial; hyphae repent and ascending, thick, brown or olivaceous brown, verruculose to echinulate, sometimes constricted at the septa, often

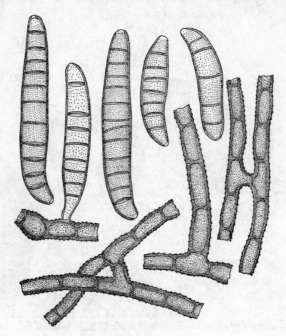

FIG. 17. *Hormisciella* state of *Strigopodia resinae* (× 650).

anastomosing. *Stroma* none. *Setae* and *hyphopodia* absent. *Conidiophores* micronematous, mononematous, straight or flexuous, irregularly branched and anastomosing, brown or olivaceous brown, verruculose to echinulate. *Conidiogenous cells* monoblastic, integrated, intercalary, determinate, cylindrical or doliiform. *Conidia* solitary, dry, pleurogenous, simple, straight, curved or sigmoid, cylindrical rounded at the apex and truncate at the base, or fusiform, mid pale to dark brown or olivaceous brown, often slightly paler at the apex, smooth, multiseptate.

Type species: Hormisciella atra Batista.

Hormisciella state of **Strigopodia resinae** (Sacc. & Bres.) Hughes, 1965, *Can. J. Bot.*, **46**: 1100–1104; with full synonymy.

(Fig. 17)

Hyphae 5–11μ thick. *Conidia* 5–14-septate, 50–120μ long, 11–14μ thick in the broadest part, 3–4μ wide at the truncate base.

On conifers, associated with resinous exudates; Europe, N. America.

16. CERATOPHORUM

Ceratophorum Saccardo, 1880, *Michelia*, **2**: 22.

Colonies hypophyllous, effuse, dark brown to black, velvety. *Mycelium* partly superficial partly immersed, joined by hyphae passing through stomata. *Stroma* none. *Setae* absent. *Hyphopodia* present on superficial hyphae, sometimes continuous but more often stalked; head cell lobed, brown, stalk cell cylindrical, pale. *Conidiophores* micronematous or semi-macronematous, mononematous, straight or flexuous, branched, pale, smooth. *Conidiogenous cells* integrated, intercalary, monoblastic, determinate, sometimes with short broad denticles. *Conidia* solitary, pleurogenous, simple, obclavate, rostrate, often coiled at the apex, truncate at the base, dark reddish brown, basal cell and beak pale, smooth, pseudoseptate.

Type species: Ceratophorum helicosporum (Sacc.) Sacc.

Ellis, M. B., *Mycol. Pap.*, **70**: 13–16, 1958.

Hughes, S. J., *Mycol. Pap.*, **36**: 16–39, 1951.

KEY

Conidia 100-200μ long, with 11-23 pseudosepta . . . . . *helicosporum*
Conidia mostly less than 100μ long, with 7-11 pseudosepta . . . *uncinatum*

Ceratophorum helicosporum (Sacc.) Sacc., 1880, *Michelia*, **2**: 22.

(Fig. 18 A)

Superficial *hyphae* pale brown, 1–3μ thick. *Hyphopodia* few, lateral and terminal, head cell 15–19 × 13–18μ; stalk cell 3–12 × 3–6μ. *Conidia* 100–200μ long, 12–15μ thick in the broadest part, 2–3μ at apex, 5–7μ at base.

On leaves of oak (*Quercus robur*); Italy.

Ceratophorum uncinatum (Clinton) Sacc., 1886, *Syll. Fung.*, **4**: 396.

(Fig. 18 B)

Superficial *hyphae* pale brown, 1–3μ thick. *Hyphopodia* few; head cell 13–21 × 8–18μ, stalk cell 5–10 × 4–5μ. *Conidiophores* 5–7 × 4–6μ. *Conidia* 75–110μ long, 10–13μ thick in broadest part, 2–3μ at apex, 4–6μ at base.

On leaves of oak (*Quercus macrocarpa*); U.S.A.

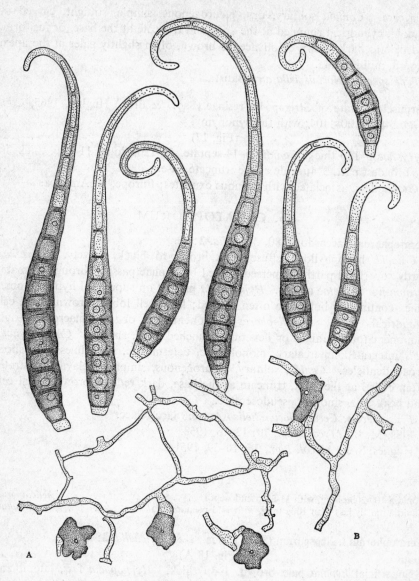

FIG. 18. A, *Ceratophorum helicosporum*; B, *C. uncinatum* (× 650).

17. CHUPPIA

Chuppia Deighton, 1965, *Mycol. Pap.*, **101**: 32.

Colonies orbicular or effuse, dark blackish brown. *Mycelium* superficial, densely aggregated and closely coiled around the upper part of leaf hairs. *Stroma* none. *Setae* and *hyphopodia* absent. *Conidiophores* micronematous, flexuous, irregularly branched, golden brown to dark olive brown, smooth. *Conidiogenous cells* monoblastic, integrated, intercalary, determinate, cylindrical or doliiform, denticulate, denticles very broad conical. *Conidia* solitary, pleuro-genous, simple, ellipsoidal, subspherical or irregular in shape, dark golden brown, muriform and strongly constricted at the septa (sarciniform), smooth to verruculose, with protuberant hilum.

Type species: Chuppia sarcinifera Deighton.

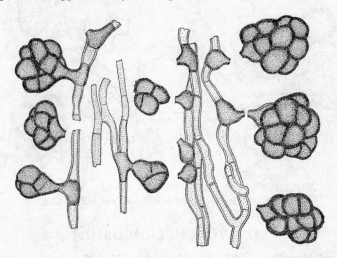

Fig. 19. *Chuppia sarcinifera* (× 650).

Chuppia sarcinifera Deighton, 1965, *Mycol. Pap.*, **101**: 32–34.

(Fig. 19)

Conidia up to 48 × 30μ. On leaves of *Solanum*, Venezuela.

18. SARCINELLA

Sarcinella Saccardo, 1880, *Michelia*, **2**: 31.

Colonies effuse, pale to dark blackish brown. *Mycelium* superficial, composed of a network of thick branched and anastomosing hyphae. *Stroma* none. *Setae* absent. *Hyphopodia* present, often hemispherical. *Conidiophores* micro-nematous or semi-macronematous, unbranched or branched, straight or flexuous, brown, smooth. *Conidiogenous cells* monoblastic, integrated, terminal or inter-calary, determinate, cylindrical. *Conidia* solitary, dry, acrogenous or pleuro-genous, simple, subspherical or irregularly sarciniform, dark brown or reddish brown, smooth, muriform, septa sometimes cruciately arranged, deeply con-stricted at the septa.

Type species: Sarcinella state of *Schiffnerula pulchra* (Sacc.) Petrak = *S. heterospora* Sacc.

Sarcinella state of **Schiffnerula pulchra** (Sacc.) Petrak, 1928, *Annls mycol.*, **26**: 397.

(Fig. 20)

Hyphae and *conidiophores* brown, 4–8µ thick. *Hyphopodia* 10–12 × 8–10µ. *Conidia* 20–28 × 16–24µ.

On leaves of *Fraxinus* and *Ligustrum*; Europe, N. America.

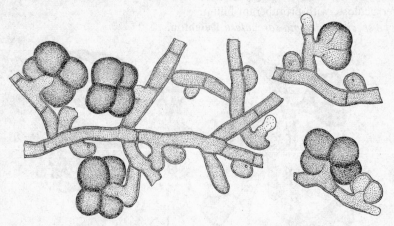

FIG. 20. *Sarcinella* state of *Schiffnerula pulchra* (× 650).

19. MYCOENTEROLOBIUM

Mycoenterolobium Goos, 1970, *Mycologia*, **62**: 171–175.

Colonies effuse, black. *Mycelium* partly immersed, partly superficial. *Stroma* none. *Setae* and *hyphopodia* absent. *Conidiophores* micronematous, mononematous, branched or unbranched, hyaline or subhyaline, smooth. *Conidiogenous cells* monoblastic, integrated, terminal or intercalary, determinate, cylindrical, ampulliform or subspherical, rarely denticulate; denticles when formed cylindrical. *Conidia* solitary, dry, acrogenous or pleurogenous, flattened in one plane, variable in shape, sometimes lobed, dark brown, smooth, muriform with rows of cells radiating fanwise from the hilum.

Type species: Mycoenterolobium platysporum Goos.

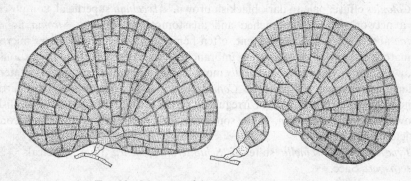

FIG. 21. *Mycoenterolobium platysporum* (× 650).

Mycoenterolobium platysporum Goos, 1970, *Mycologia*, **62**: 171–175.
(Fig. 21)

Conidiophores 1·5–5μ thick. *Conidia* black and shining by reflected light, dark brown and looking rather like scallop shells when viewed under a microscope by transmitted light, 50–80 × 70–130μ (20–25μ thick).

On decaying wood of *Araucaria*; Hawaii.

20. TETRAPLOA

Tetraploa Berkeley & Broome, 1850, *Ann. Mag. nat. Hist.*, 2, **5**: 459.

Colonies effuse, brown or dark greyish brown. *Mycelium* superficial. *Conidiophores* micronematous, branched and anastomosing to form a network, flexuous,

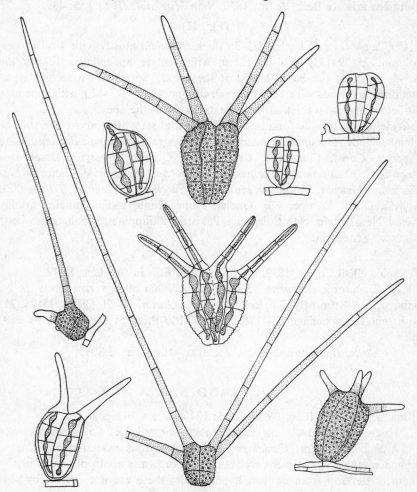

FIG. 22. *Tetraploa aristata* (× 650).

hyaline to pale yellowish brown, often verruculose. *Conidiogenous cells* monoblastic or occasionally polyblastic, integrated, intercalary, determinate, cylindrical. *Conidia* solitary, dry, pleurogenous, appendaged, brown, verruculose or verrucose, muriform; in mature conidia there are shallow furrows between

4 (or rarely 3) columns of cells which develop independently, tend to diverge from one another apically and terminate each in a septate setiform appendage.

Type species: Tetraploa aristata Berk. & Br.

Ellis, M. B., *Trans. Br. mycol. Soc.*, **33**: 246–251, 1949.

KEY

Conidia mostly with 2 or 4 cells in each column . . . . . . *aristata*
Conidia with 4–8, mostly 5–6 cells in each column . . . . . . *ellisii*

Tetraploa aristata Berk. & Br., 1850, *Ann. Mag. nat. Hist.*, 2, **5**: 459.
(Fig. 22)

Conidiophores and *hyphae* 1·5–3µ thick. *Conidia* mostly with 4 cells to each column, 25–39 (32) × 14–29 (21)µ, with septate appendages 12–80µ long, 4.5–8µ thick at the base, 2–3·5µ at the apex. Sometimes a second type of conidium is formed with 2 cells to each column, 8–18 × 7–12µ, with appendages 80–330µ long, 3–6µ thick at the base and 1–2µ at the apex.

Widespread, usually found on leaf bases and stems just above the soil, host plants include *Ammophila, Anadelphia, Andropogon, Axonopus, Bambusa, Carex, Cladium, Cocos, Cortaderia, Cymbopogon, Cyperus, Dactylis, Deschampsia, Erianthus, Euchlaena, Gynerium, Heteropogon, Ischaemum, Juncus, Musa, Phaseolus, Phoenix, Phormium, Phragmites, Saccharum, Sorghum, Triticum, Zea*; Bolivia, Cuba, Europe, Fiji, Ghana, Hong Kong, India, Jamaica, Malaya, Nepal, New Britain, New Caledonia, Pakistan, Philippines, Sabah, Sierra Leone, Uganda, Venezuela.

Tetraploa ellisii Cooke, 1879, apud Cooke & Ellis in *Grevillea*, **8**: 12.

Conidiophores and *hyphae* 2·5–4µ thick. *Conidia* often verrucose at the base only, with 4–8 (mostly 5–6) cells to each column, 30–51 (39) × 15–26 (21)µ, with septate appendages 24–178µ long, 7–10µ thick at the base, 2–4µ at the apex.

On *Chloris, Dactylis* and *Zea*; Argentina, Rhodesia, U.S.A.

21. HIRUDINARIA

Hirudinaria Cesati, 1856, *Hedwigia*, **1**: 104 and tab. 14 figs 1–3.

Colonies effuse, black. *Mycelium* mostly superficial. *Stroma* none. *Setae* and *hyphopodia* absent. *Conidiophores* micronematous or semi-macronematous, mononematous; sometimes very short, pale brown, smooth processes or thick denticles develop from swollen hyphal cells; these are not cut off by septa. *Conidiogenous cells* monoblastic or polyblastic, integrated, intercalary, determinate, cylindrical or doliiform, denticulate. *Conidia* solitary, dry, pleurogenous, forked, often V-shaped or U-shaped, mid to dark brown, verrucose; branches subulate, multiseptate.

Type species: Hirudinaria macrospora Ces.

Hughes, S. J., *Mycol. Pap.*, **39**: 11–24, 1951.

Hirudinaria macrospora Ces., 1856, *Hedwigia*, **1**: 104 and tab. 14 figs 1–3.
Klotzsch Herb. myc. Ed.nov.: 269.
(Fig. 23)

Colonies hypophyllous. *Mycelium* composed of a network of branched and anastomosing, 1·5–2μ thick hyphae with here and there slightly swollen pale

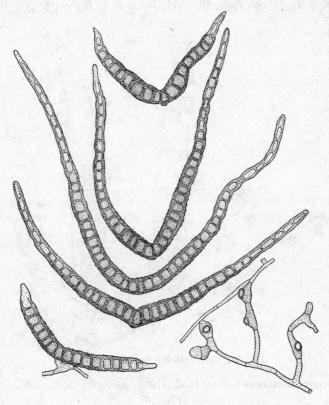

FIG. 23. *Hirudinaria macrospora* (× 650).

cells which bear conidia on very short (2–3μ long) processes (denticles) not cut off by a septa. *Conidia* with arms straight or curved, often divergent, up to 150μ long, 6–8μ thick, dark brown and verrucose near the base, tapering gradually to the paler, smoother, 2–3μ wide apex.

On leaves of *Crataegus*; Europe, N. America.

22. CERATOSPORIUM

Ceratosporium Schweinitz, 1832, *Trans. Am. phil. Soc.*, N.S., **4**: 300.

Colonies effuse, brown to black. *Mycelium* partly superficial, partly immersed. *Stroma* none. *Setae* and *hyphopodia* absent. *Conidiophores* micronematous, flexuous, irregularly branched with the branches anastomosing and often at right-angles, brown, smooth. *Conidiogenous cells* integrated, intercalary, mono-blastic, determinate, cylindrical, denticulate; denticles cylindrical or broad conical. *Conidia* solitary, pleurogenous, branched, mid to dark brown, smooth, with a swollen, pyriform central cell and 2–3 divergent, pluriseptate branches. Narrow curved, hyaline, 0-septate *secondary conidia* sometimes formed (Hughes, 1964).

Type species: Ceratosporium fuscescens Schw.

Hughes, S. J., *Mycol. Pap.*, **39**: 1–11, 1951; *N.Z. Jl Bot.*, **2**: 305–309, 1964.

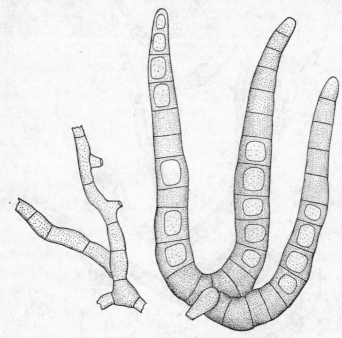

FIG. 24. *Ceratosporium fuscescens* (× 650).

Ceratosporium fuscescens Schw., 1832, *Trans. Am. phil. Soc.*, N.S., **4**: 300.
(Fig. 24)

Hyphae brown, 5–7µ, thick forming a network. *Conidia* composed of a central cell 9–14µ thick in its broadest part, 3–5µ at the base and 2–3 divergent, 10–19-septate branches up to 230µ long, 14–22µ thick below, tapering to 5–10µ. In *C. rilstonii* Hughes the conidia have similar measurements but the branches are intertwined and closely adpressed.

On dead wood and bark; Canada, Great Britain, New Zealand, U.S.A.

Ceratosporium productum Petch, 1906, *Ann. R. bot. Gdns Peradeniya*, **3**: 9.

Hyphae pale brown, 4–5µ thick. *Conidia* mid pale brown, branches 10–14-septate, up to 180µ long, 11–15µ thick at the base, tapering to 3·5–5·5µ.

On dead branches of *Hevea brasiliensis*; Ceylon.

23. TETRAPOSPORIUM

Tetraposporium Hughes, 1951, *Mycol. Pap.*, **46**: 25–28.

Colonies effuse, thin, hyperparasitic. *Mycelium* superficial. *Stroma* none.

Setae and *hyphopodia* absent.   *Conidiophores* micronematous or semi-macronematous, mononematous; usually very short, pale brown, smooth processes or thick denticles develop from swollen hyphal cells but are not cut off by septa. *Conidiogenous cells* monoblastic, integrated, intercalary, determinate, cylindrical or doliiform, denticulate; denticles stout, cylindrical. *Conidia* solitary, dry, pleurogenous, branched, with usually 4 arms, pale olivaceous or brown, smooth; arms subulate, multiseptate.

Type species: Tetraposporium asterinearum Hughes.

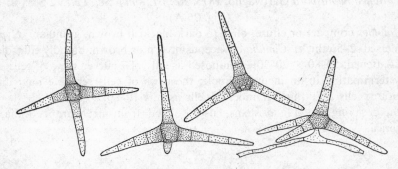

FIG. 25. *Tetraposporium asterinearum* (× 650).

Tetraposporium asterinearum Hughes, 1951, *Mycol. Pap.*, **46**: 25–26.

(Fig. 25)

Hyphae 1–2μ thick, with fertile cells thickened to 3–4μ. *Conidia* with 1 terminal and 3 lateral arms each up to 40μ long, 5–6μ thick at the base, tapering to 1·5–2μ.

On *Asterina* on *Funtumia* and *Homalium*; Ghana, Sierra Leone.

24. DICTYOSPORIUM

Dictyosporium Corda, 1836, Weitenweber's Beiträge zur gesammten Natur-und Heilwissenschaften: 87.

Speira Corda, 1837, Icon. Fung., **1**: 9.

Cattanea Garovaglio, 1875, *Rc. Ist. lomb. Sci. Lett.*, 2 Ser., **8**: 125.

Colonies effuse or compact, olive, brown or black, often granular. *Mycelium* mostly immersed. *Stroma* none. *Setae* and *hyphopodia* absent. *Conidiophores* micronematous, mononematous, flexuous, irregularly branched, hyaline to brown, smooth. *Conidiogenous cells* monoblastic, integrated, terminal and sometimes also intercalary, determinate, cylindrical, doliiform, spherical or subspherical. *Conidia* solitary, dry, acrogenous and sometimes also pleurogenous, branched, cheiroid, in most species flattened in one plane, olive to brown, smooth, multiseptate.

Type species: Dictyosporium elegans Corda.

Damon, S. C., *Lloydia*, **15**: 110–124, 1952.

K<small>EY</small>

Conidia not flattened in one plane, each row of cells terminating in a hook .   *heptasporum*
Conidia flattened in one plane.   1
1. Conidia 50–80×24–31μ .   .   .   .   *elegans*
   Conidia usually less than 50μ long   2
2. Conidia clearly cheiroid, rows always of different lengths   .   .   .   *toruloides*
   Conidia oblong or irregular, rows often of the same length   .   .   .   *oblongum*

Dictyosporium heptasporum (Garov.) Damon, 1952, *Lloydia*, **15**: 118.
   Cattanea heptaspora Garovaglio, 1875, *Rc. Ist. lomb. Sci. Lett.*, 2 Ser., **8**: 125.
(Fig. 26 A)

Colonies compact or effuse, olive to dark blackish brown, granular. *Hyphae* immersed, 2–5μ thick. *Conidia* olivaceous brown or brown, broadly ellipsoidal, not flattened, 50–80 × 20–30μ, composed usually of 7 curved rows of cells each row terminating in an incurved hook; the rows of cells separate only under pressure; cells 5–7μ thick, number of cells usually 70–80.

On decaying wood and stems, also isolated from air; Europe, India, N. America.

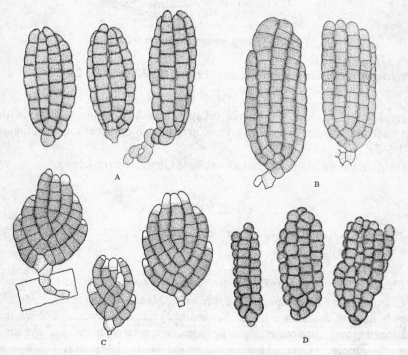

F<small>IG</small>. 26. *Dictyosporium* species: A, *heptasporum*; B, *elegans*; C, *toruloides*; D, *oblongum*;
(× 650).

Dictyosporium elegans Corda, 1836, Weitenweber's Beiträge . . .: 87.
(Fig. 26 B)

Colonies effuse, blackish brown, granular. *Hyphae* 2–4μ thick. *Conidia* golden or reddish brown, 50–80 × 24–31μ, flattened in one plane, composed usually of 5–6 rows of cells all approximately the same length; cells 4–9µ thick, total number of cells 51–96.

On dead wood and herbaceous plants, found several times on barley and oat stubble; Europe, E. & W. Africa, N. America.

Dictyosporium toruloides (Corda) Guéguen, 1905, *Bull. trimest. Soc. mycol. Fr.*, **21**: 101.

Speira toruloides Corda, 1837, Icon Fung., **1**: 9.

(Fig. 26 C)

Colonies effuse, black, granular, sometimes less than 1 mm. diam. but may encircle a stem and extend along it for up to 2 cm. *Hyphae* hyaline to pale brown, 2–5µ thick. *Conidia* borne singly at the ends of hyphae where these break through to the surface; they are flattened in one plane and composed of 6–8 rows of cells which are of unequal length, the middle ones being longer than those at each side. The rows of cells develop as a branched system and each conidium is shaped rather like a hand with the fingers held close together. The cells are smooth and mostly olive to brown; the terminal one in each row is often hyaline. Width of conidium in its broadest part 25–34µ, length of longest row of cells 38–45µ, thickness of cells 5–7µ, total number of cells 36–51.

Common on wood and dead herbaceous stems; Europe, N. America, Pakistan.

Dictyosporium oblongum (Fuckel) Hughes, 1958, *Can. J. Bot.*, **36**: 762.

Speira oblonga Fuckel, 1870, Symb. mycol.: 349.

(Fig. 26 D)

Colonies effuse, black, granular. *Conidia* oblong or irregular, 30–50 × 12–30µ, brown, composed usually of 3–6 rows of cells of either the same length or different lengths, often strongly constricted at the septa; cells 5–7µ thick, number of cells 22–46.

On dead wood; Europe, N. America.

25. ACREMONIULA

Acremoniula Ciferri, 1962, *Atti Ist. bot. Univ. Lab. crittogam. Pavia*, Ser. 5, **19**: 85.

Colonies effuse, black, overgrowing and parasitic on *Clypeolella* and on *Schiffnerula* and its *Sarcinella* state. *Mycelium* superficial. *Stroma* none. *Setae* and *hyphopodia* absent. *Conidiophores* micronematous or semi-macronematous, mononematous, branched or unbranched, straight or flexuous, colourless, smooth. *Conidiogenous cells* monoblastic, integrated and terminal or discrete, determinate, cylindrical or doliiform. *Conidia* solitary, dry, acrogenous, simple, obovoid, pyriform, ellipsoidal or subspherical, truncate at the base, mid to very dark brown, thick-walled, smooth, 0-septate.

Type species: Acremoniula suprameliola Cif.

Acremoniula sarcinellae (Pat. & Har.) Arnaud ex Deighton, 1969, *Mycol. Pap.*, **118**: 3–5.

(Fig. 27)

Conidiophores 3·5–4·5μ thick. *Conidia* mostly obovoid or pyriform, 9–16 × 9–13μ, truncate base about 4μ wide.

Overgrowing and parasitic on *Clypeolella* and *Schiffnerula* colonies; Africa (East, West and Central), Cuba, India, Panama.

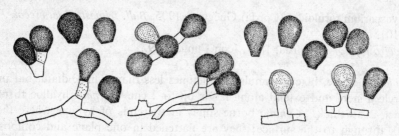

FIG. 27. *Acremoniula sarcinellae* (× 650).

26. THERMOMYCES

Thermomyces Tsiklinsky, 1899, *Annls Inst. Pasteur, Paris*, **13**: 500–505.

Colonies effuse, cottony or velvety, grey, greenish grey, buff, dark blackish brown or black. *Mycelium* partly superficial, partly immersed. *Stroma* none. *Setae* and *hyphopodia* absent. *Conidiophores* micronematous or semi-macronematous, mononematous, unbranched or irregularly branched, straight or flexuous, colourless or brown, smooth. *Conidiogenous cells* monoblastic, integrated and terminal or discrete, determinate, cylindrical or lageniform. *Conidia* solitary, dry, acrogenous, simple, spherical to subspherical or angular and lobed, pale to dark blackish brown, smooth or verrucose, 0-septate. No phialidic state known.

Type species: Thermomyces lanuginosus Tsiklinsky.

Apinis, A. E., *Nova Hedwigia*, **5**: 57–78, 1963.

Pugh, G. J. F., Blakeman, J. P. & Morgan-Jones, G., *Trans. Br. mycol. Soc.*, **47**: 115–121, 1964.

KEY

Conidia spherical or subspherical, verrucose . . . . *lanuginosus*
Conidia angular, lobed . . . . *stellatus*

Thermomyces lanuginosus Tsiklinsky, 1899, *Annls Inst. Pasteur, Paris*, **13**: 500–505.

(Fig. 28 A)

Conidiophores 1·5–2·5μ thick. *Conidia* spherical or subspherical, coarsely verrucose and blackish brown when mature, 7–12μ diam.

Isolated from *Triticum*, *Zea*, mouldy hay, leaf litter, soil and man, thermophilic; Europe, India, Nigeria, N. America.

Thermomyces stellatus (Bunce) Apinis, 1963, *Nova Hedwigia*, **5**: 75.

(Fig. 28 B)

Conidiophores 1·5–2·5μ thick. *Conidia* angular, lobed, smooth, pale to mid brown on greyish brown, 5–10 × 5–9μ.

Isolated from mouldy hay, thermophilic; Great Britain.

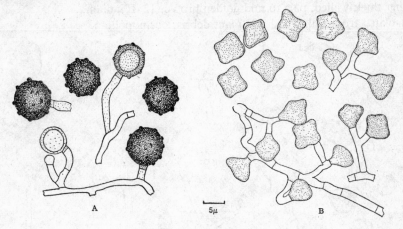

FIG. 28. A, *Thermomyces lanuginosus*; B, *T. stellatus*.

27. HUMICOLA

Humicola Traaen, 1914, *Nyt. Mag. Naturvid.*, **52**: 31–34.

Colonies effuse, cottony, sometimes funiculose, at first white, later pale grey, greyish brown or blackish brown. *Mycelium* superficial and immersed. *Stroma* none. Intercalary *chlamydospores* sometimes formed. *Setae* and *hyphopodia* absent. *Conidiophores* micronematous or semi-macronematous, unbranched or irregularly branched, straight or flexuous, colourless to pale golden brown, smooth. *Conidiogenous cells* monoblastic, integrated, terminal, determinate, cylindrical, doliiform, pyriform or infundibuliform. *Conidia* solitary, dry, acrogenous, simple, typically spherical, occasionally obovoid or pyriform, pale to mid golden brown, usually smooth, 0-septate. *Humicola* also has a phialidic state, the *phialides* being discrete, subulate, colourless, smooth. *Phialoconidia* catenate or in slimy heads, very small, colourless, smooth, 0-septate.

Lectotype species: Humicola fuscoatra Traaen.

Fassatiová, O., *Česká Mykol.*, **18**: 102–108, 1964; **21**: 78–89, 1967.

White, W. L. & Downing, M. H., *Mycologia*, **45**: 951–963, 1953.

KEY

Conidia 6–11μ diam. . . . . *fuscoatra*
Conidia 12–17μ diam. . . . . *grisea*

Humicola fuscoatra Traaen, 1914, *Nyt. Mag. Naturvid.*, **52**: 33–34.

(Fig. 29 A)

Colonies finally greyish brown to blackish brown. *Conidia* rather thin-walled, pale golden brown, 6–11μ diam.

Isolated from soil and plant debris; cosmopolitan.

Humicola grisea Traaen, 1914, *Nyt. Mag. Naturvid.*, **52**: 34.

(Fig. 29 B)

Colonies cottony, white to pale grey, then dark grey, black in reverse. *Con dia* rather thick-walled, pale to mid golden brown, 12–17μ diam.

Isolated from soil, wood and plant debris; cosmopolitan.

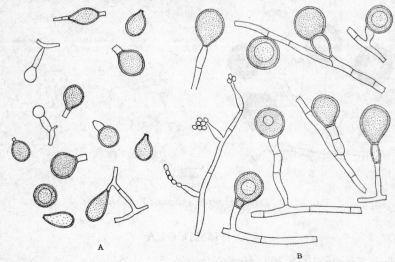

FIG. 29. A, *Humicola fuscoatra*; B, *H. grisea* (× 650).

28. ALLESCHERIELLA

Allescheriella P. Hennings, 1897, *Hedwigia*, **36**: 244.

Colonies pulvinate, round, sometimes confluent, at first white, then yellow,

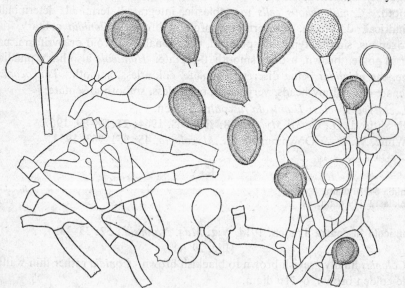

FIG. 30. *Allescheriella crocea* (× 650).

finally rust-coloured. *Mycelium* partly superficial, partly immersed; hyphae interwoven and anastomosing, forming cushions. *Stroma* none. *Setae* and *hyphopodia* absent. *Conidiophores* semi-macronematous, mononematous, mostly branched, straight or flexuous, hyaline to very pale reddish brown, smooth. *Conidiogenous cells* monoblastic, integrated and terminal on stipe and branches or discrete, determinate, cylindrical. *Conidia* solitary, dry, acrogenous, simple, spherical, subspherical, obovoid or ellipsoidal, reddish brown, smooth, thick-walled, 0-septate.

Type species: Allescheriella crocea (Mont.) Hughes = *A. uredinoides* P. Henn. Hughes, S. J., *Mycol. Pap.*, **41**: 1–8, 1951.

Allescheriella crocea (Mont.) Hughes, 1951, apud Baker & Dale in *Mycol. Pap.* **33**: 97.

(Fig. 30)

Hyphae up to 7μ thick. *Conidiophores* 3–5μ thick. *Conidia* 14–28 × 11–19μ. On rotten wood of various trees including *Albizia, Gmelina, Hevea* and *Mangifera*; Brazil, Ceylon, Cuba, Ghana, India, New Caledonia, Nigeria, Sierra Leone, Trinidad, Uganda, Uruguay, U.S.A., Venezuela, Zambia.

29. GILMANIELLA

Gilmaniella Barron, 1964, *Mycologia*, **56**: 514–518.
Adhogamina Subram. & Lodha, 1964, *Antonie van Leeuwenhoek*, **30**: 319–320.
Colonies effuse, at first pale grey, finally dark blackish brown. *Mycelium* superficial and immersed; hyphae at first colourless, smooth, becoming brown and verruculose or finely echinulate, transverse septa often thick and very dark. *Stroma* none. *Setae* and *hyphopodia* absent. *Conidiophores* semi-macronematous, mononematous, straight or flexuous, frequently branched but sometimes unbranched, colourless, smooth. *Conidiogenous cells* monoblastic and polyblastic, integrated, terminal and intercalary or discrete, determinate, cylindrical. *Conidia* solitary, dry, acropleurogenous, simple, spherical, dark brown with a small but very distinct apical pore, smooth, 0-septate.

Type species: Gilmaniella humicola Barron.

Gilmaniella humicola Barron, 1964, *Mycologia*, **56**: 514–518.
Adhogamina ruchera Subram. & Lodha, 1964, *Antonie van Leeuwenhoek*, **30**: 319–320.

(Fig. 31)

Conidiophores up to 40 × 2–3μ. *Conidia* 7–10 (mostly 7–8)μ diam. Isolated from soil and dung; cosmopolitan.

30. CHALAROPSIS

Chalaropsis Peyronel, 1916, *Staz. sper. agr. ital.*, **49**: 585–586.
Colonies effuse, greenish black or greyish black. *Mycelium* immersed and superficial. *Stroma* none. *Setae* and *hyphopodia* absent. *Conidiophores* semimacronematous, flexuous, sympodially or irregularly branched, colourless,

smooth. *Conidiogenous cells* monoblastic or polyblastic, integrated and terminal or discrete, determinate, cylindrical. *Conidia* solitary, acrogenous on short branches, simple, ellipsoidal, obovoid, spherical or subspherical, olive to dark blackish brown, 0-septate, smooth or minutely spinulose.

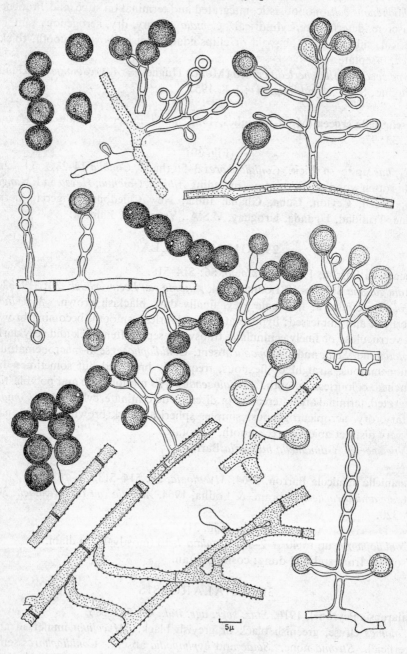

Fig. 31. *Gilmaniella humicola.*

All species so far known also have a *Chalara* state.

Type species: Chalaropsis thielavioides Peyronel.

Hennebert, G. L., *Antonie van Leeuwenhoek*, **33**: 333–340, 1967.

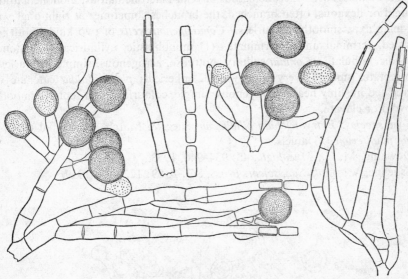

FIG. 32. *Chalaropsis thielavioides* (× 650).

Chalaropsis thielavioides Peyronel, 1916, *Staz. sper. agr. ital.*, **49**: 585–586.
(Fig. 32)

Conidiophores erect or ascending, very variable in length, 4–9μ thick. *Conidia* spherical or subspherical, olivaceous brown, mostly 14–19μ diam. *Chalara* conidiophores with stipes up to 70 × 5–7μ, *phialides* cylindrical or lageniform, up to 60μ long, base 5–6μ wide, neck 3–4μ thick; *conidia* catenate, cylindrical with truncate ends, colourless, smooth, 0-septate, 8–15 × 2·5–4·5μ.

Originally described from stems and roots of lupins; collections have since been made on carrot in Denmark, Great Britain and the Netherlands, elm, poplar and walnut in Great Britain, *Crotalaria juncea* in Sabah and roots of seedling peach trees in Palestine.

Hennebert refers to three other species: *Chalaropsis punctulata* Hennebert isolated from roots of *Lawsonia inermis* in Africa with obovoid or ellipsoidal conidia 12–20 × 8–13 (average 16 × 10)μ; the *Chalaropsis* state of *Ceratocystis radicicola* (Bliss) C. Moreau isolated from date palm (*Phoenix dactylifera*) in U.S.A., also with obovoid or ellipsoidal conidia 11–29 × 9–22 (average 20 × 15)μ and the *Chalaropsis* state of *Ceratocystis variospora* (Davidson) C. Moreau from chestnut oak (*Quercus montana*) in U.S.A., with subspherical conidia mostly 10–14μ diam.

31. BOTRYOTRICHUM

Botryotrichum Saccardo & Marchal, in Marchal, 1885, *Bull. Soc. r. Bot. Belg.*, **24**: 66.

Colonies effuse, at first white, later buff to grey. *Mycelium* mostly superficial. *Stroma* none.   *Setae* unbranched, flexuous, pale to mid brown or greyish brown, often verrucose or encrusted, especially near the base. *Hyphopodia* absent. *Conidiophores* micronematous or semi-macronematous, mononematous, straight or flexuous, often branched, the branches sometimes at right-angles to the main axis, smooth, colourless. *Conidiogenous cells* of two kinds, both integrated, terminal and determinate—(1) monoblastic, cylindrical; (2) monophialidic, subulate. *Conidia* solitary, botryose, acrogenous, simple, colourless, non-septate, smooth, very thick-walled, spherical; *phialoconidia* catenate or aggregated in slimy heads, acrogenous, simple, colourless, non-septate, smooth, obovoid or clavate.

Type species: Botryotrichum piluliferum Sacc. & March. = conidial *Chaetomium piluliferum* J. Daniels.

Downing, M. H., *Mycologia*, **45**: 934–940, 1953.

Estienne, V., *Annls Soc. scient. Brux.*, Sér. 2, **59**: 122–134, 1939.

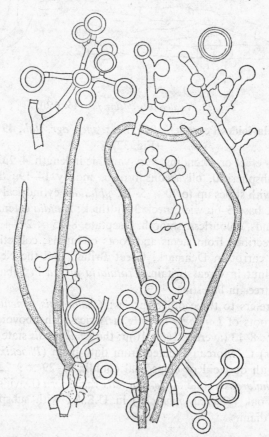

FIG. 33. *Botryotrichum* state of *Chaetomium piluliferum* (× 650).

Botryotrichum state of **Chaetomium piluliferum** J. Daniels, 1961, *Trans. Br. mycol. Soc.*, **44**: 79–86.

B. piluliferum Sacc. & March., 1885, *Bull. Soc. r. Bot. Belg.*, **24**: 66–67.
(Fig. 33)

Hyphae 2–4μ thick. *Setae* up to 250μ long, 3–5μ thick near the base, tapering to 2–3μ. *Conidia* mostly 9–16μ diam.; *phialoconidia* 3–4 × 1·5–2·5μ.

Occasionally found on dead herbaceous plants and frequently isolated from air, canvas, cellophane, dung, paper, soil, etc.; specimens in Herb. IMI from Belgium, Canada, Egypt, Great Britain, Malaya and U.S.A.

32. STAPHYLOTRICHUM

Staphylotrichum Meyer & Nicot, 1957, *Bull. trimest. Soc. mycol. Fr.*, **72**: 318–323.
Botrydiella Badura, 1963, *Allionia*, **9**: 182–184.

Colonies effuse, velvety or cottony, often orange-yellow, rarely olivaceous. *Mycelium* partly superficial, partly immersed. *Stroma* sometimes formed,

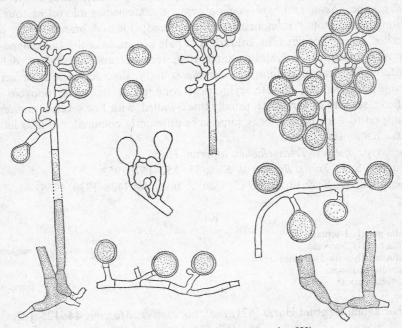

FIG. 34. *Staphylotrichum coccosporum* (× 650).

pseudoparenchymatous. *Setae* and *hyphopodia* absent. *Conidiophores* macronematous and sometimes also micronematous, mononematous. Macronematous conidiophores solitary or caespitose, branched towards the apex forming a stipe and head; stipe straight or flexuous, smooth, lower part dark brown, greyish or olivaceous brown, upper part hyaline or very pale; branches sometimes at right-angles to the main axis and to each other. *Conidiogenous cells* monoblastic, integrated, terminal on stipe and branches or discrete, determinate, cylindrical. *Conidia* solitary, dry, acrogenous, simple, spherical, subspherical, pyriform or ellipsoidal, pale brown, greyish or olivaceous brown, smooth, thick-walled, 0-septate.

D.H.—3

Type species: Staphylotrichum coccosporum Meyer & Nicot.
Maciejowska, Z. & Williams, E. B., *Mycologia*, **55**: 221–225, 1963.

Staphylotrichum coccosporum Meyer & Nicot, 1957, *Bull. trimest. Soc. mycol. Fr.*, **72**: 318–323.

(Fig. 34)

Conidiophores up to 1200μ long, 5–8μ thick near the base, 1–4μ near the apex. *Conidia* 7–14 (mostly 10–12)μ diam.

Isolated from *Saccharum* and from soil; Ceylon, Congo, Costa Rica, Hong Kong, India, Jamaica, N. America.

33. TRICHOCLADIUM

Trichocladium Harz, 1871, *Bull. Soc. impér. Moscow*, **44**: 125–127.

Colonies effuse, grey to black. *Mycelium* partly superficial, partly immersed. *Stroma* none. *Setae* and *hyphopodia* absent. *Conidiophores* micronematous or semi-macronematous, mononematous, scattered, loosely branched or un-branched, straight or flexuous, colourless or pale brown, smooth. *Conidiogenous cells* monoblastic or polyblastic, integrated, terminal and intercalary, deter-minate, cylindrical or doliiform. *Conidia* solitary, dry, acrogenous or acro-pleurogenous, simple, clavate, cylindrical rounded at the apex, obovoid or pyriform, smooth or verrucose, usually thick-walled, with 1 or several transverse septa, mid to dark brown; cells sometimes differently coloured. *Idriella* has a *Trichocladium* state.

Lectotype species: Trichocladium asperum Harz.
Hughes, S. J., *Trans. Br. mycol. Soc.*, **35**: 152–157, 1952.
Kendrick, W. B. & Bhatt, G. C., *Can. J. Bot.*, **44**: 1728–1730, 1968.

KEY

Conidia mostly 1-septate	1
Conidia mostly 2-septate	*pyriforme*
Conidia mostly with 3 or more septa	*opacum*
1. Conidia verrucose	*asperum*
Conidia smooth	*canadense*

Trichocladium asperum Harz, 1871, *Bull. Soc. impér. Moscow*, **44**: 125–127.
(Fig. 35 A)

Colonies effuse, cottony, grey. *Conidiophores* 2–3·5μ thick. *Conidia* mostly acrogenous, predominantly 1-septate, clavate, obovoid or ellipsoidal, narrowed to a truncate base, dark brown, coarsely verrucose, 15–30μ long, 10–15μ thick in the broadest part.

Isolated from *Daucus*, *Picea*, *Pisum*, *Solanum*, beerwort, moss and soil; Australia, Europe, Mauritius, New Zealand, N. America.

Trichocladium canadense Hughes, 1959, *Can. J. Bot.*, **37**: 857–859.
(Fig. 35 B)

Colonies effuse, black. *Conidiophores* 3–5μ thick. *Conidia* acropleurogenous, predominantly 1-septate, occasionally 2-septate, clavate or ellipsoidal, truncate at the base, mid to dark brown, smooth, 15–30μ long, 8–12μ thick in the broadest part.

Isolated from wood of *Acer*, etc.; Canada.

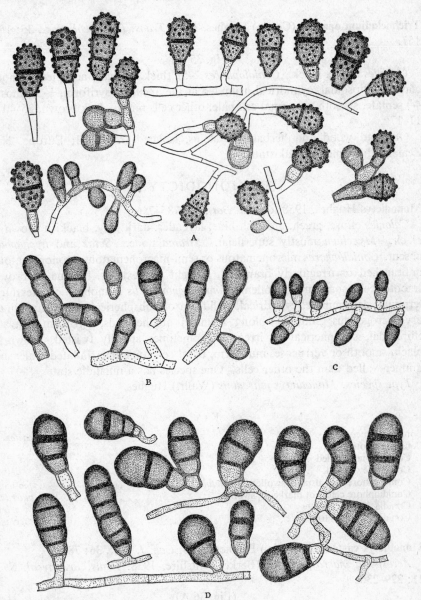

FIG. 35. *Trichocladium* species: A, *asperum*; B, *canadense*; C, *pyriforme*; D, *opacum* (× 650).

Trichocladium pyriforme Dixon, 1968, *Trans. Br. mycol. Soc.*, **51**: 160–164.

(Fig. 35 C)

Colonies effuse, black. *Conidiophores* 1·5–3μ thick. *Conidia* mostly pleurogenous, predominantly 2-septate, clavate or pyriform, proximal cell hyaline or pale, middle cell mid brown, distal cell very dark brown, smooth, 12–20µ long, 6–8µ thick in the broadest part.

Isolated from soil; England and Ireland.

Trichocladium opacum (Corda) Hughes, 1952, *Trans. Br. mycol. Soc.*, **35**: 154–157.

(Fig. 35 D)

Colonies effuse, black. *Conidiophores* 2–5µ thick. *Conidia* acropleurogenous, clavate, ellipsoidal, cylindrical, rounded at the apex, or pyriform, 1–5- (mostly 4-) septate, smooth, proximal cell pale, other cells mid to dark brown, 25–40 × 11–17µ.

On dead wood and herbaceous stems, isolated from soil; Europe, New Zealand, N. America, Pakistan.

34. MONODICTYS

Monodictys Hughes, 1958, *Can. J. Bot.*, **36**: 785–786.

Colonies effuse, green, greenish blue, lavender, dark grey, blackish brown or black. *Mycelium* mostly superficial. *Stroma* none. *Setae* and *hyphopodia* absent. *Conidiophores* micronematous or semi-macronematous, mononematous, unbranched or irregularly branched, straight or flexuous, hyaline to brown, smooth, cells sometimes swollen. *Conidiogenous cells* monoblastic, integrated, terminal, determinate, cylindrical, doliiform or subspherical. *Conidia* solitary dry, acrogenous, simple, oblong rounded at the ends, pyriform, clavate, ellipsoidal, subspherical or irregular, sometimes spirally twisted, brown to black, smooth or verrucose, muriform, basal cell sometimes inflated, paler and thinner-walled than the other cells. One species has a phialidic state.

Type species: Monodictys putredinis (Wallr.) Hughes.

KEY

Conidia verrucose	*castaneae*
Conidia smooth	1
1. Conidia many-celled	2
Conidia few-celled	3
2. Conidiophore cells usually swollen; on *Betula* . . .	*paradoxa*
Conidiophore cells not markedly swollen	*putredinis*
3. Conidia 17–30×15–19µ	*levis*
Conidia 7–14×5–10µ	*glauca*

Monodictys castaneae (Wallr.) Hughes, 1958, *Can. J. Bot.*, **36**: 785.

Acrospeira macrosporoidea (Berk.) Wiltshire, 1938, *Trans. Br. mycol. Soc.*, **21**: 229–236.

(Fig. 36 A)

Colonies lavender to dark grey or black. *Conidia* oblong rounded at the ends, pyriform, clavate, subspherical or irregular, mid to dark reddish brown, usually verrucose, basal cell sometimes paler than the others, 14–40 × 10–25µ.

On dead stems, rotten wood, damp linoleum, paper, sacking, etc.; Europe, N. America.

Monodictys paradoxa (Corda) Hughes, 1958, *Can. J. Bot.*, **36**: 786.

Coniosporium paradoxum (Corda) Mason & Hughes, 1951, *Mycol. Pap.*, **37**: 10–17.

(Fig. 36 B)

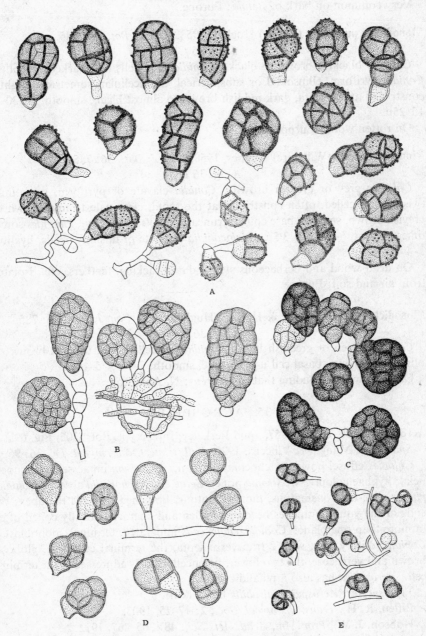

FIG. 36. *Monodictys* species: A, *castaneae*; B, *paradoxa*; C, *putredinis*; D, *levis*; E, *glauca* (× 650).

Colonies effuse, with black, shining clusters of conidia. *Conidiophore* cells nearly always swollen. *Conidia* oblong rounded at the ends, ellipsoidal, pyriform or subspherical, dark olivaceous or blackish olive, smooth, often with one or more of the basal cells paler than the others, 20–43 × 17–30μ.

Very common on bark of *Betula*; Europe.

Monodictys putredinis (Wallr.) Hughes, 1958, *Can. J. Bot.*, **36**: 785.
(Fig. 36 C)

Colonies blackish brown to black. *Conidiophore* cells not markedly swollen. *Conidia* pyriform, ellipsoidal or subspherical, multicellular, sometimes slightly constricted at the septa, dark reddish brown to almost black, smooth, 20–30 × 15–25μ.

On rotten wood; Europe.

Monodictys levis (Wiltshire) Hughes, 1958, *Can. J. Bot.*, **36**: 785.
(Fig. 36 D)

Colonies grey or greyish brown. *Conidia* clavate or pyriform, sometimes twisted, few-celled, often constricted at the septa, clear pale to mid brown or greyish brown, smooth, base conico-truncate, 17–30 × 15–19μ. *Phialides* sometimes present, 6–11 × 2–3·5μ, *phialoconidia* catenate or in slimy heads, hyaline, 2–4 × 1·5–2μ.

On dead wood and herbaceous stems, damp sacking, feathers, etc., isolated from air and soil; Europe.

Monodictys glauca (Cooke & Harkn.) Hughes, 1958, *Can. J. Bot.*, **36**: 785.
(Fig. 36 E)

Colonies green or greenish blue. *Conidia* few-celled, mid to dark brown or olivaceous brown, basal cell usually pale, smooth, 7–14 × 5–10μ.

On dead wood including that of *Quercus*; N. America.

35. ACROSPEIRA

Acrospeira Berk. & Br., 1857, apud Berkeley, Intr. crypt. Bot.: 305, Fig. 69a.
Spirospora Mangin & Vincens, 1920, *Bull. trimest. Soc. mycol. Fr.*, **36**: 96.

Colonies effuse, powdery, chocolate brown. *Mycelium* immersed and superficial; hyphae colourless. *Stroma* none. *Setae* and *hyphopodia* absent. *Conidiophores* semi-macronematous, mononematous, branched towards the apex; the ends of the young branches become swollen and characteristically coiled often in more than one plane. *Conidiogenous cells* integrated, terminal, monoblastic. *Conidia* solitary, dry, with 2 transverse septa, the terminal cell large, globose, brown and verrucose, the two lower, paler cells closely adpressed to the terminal cell. *Acrospeira* has also a phialidic state.

Type species: Acrospeira mirabilis Berk. & Br.

Biffen, R. H., *Trans. Br. mycol. Soc.*, **2**: 17–25, 1903.
Hobson, J. W., *Proc. Am. Acad. Arts Sci.*, **48**: 265–267, 1912.
Peyronel, B., *Bull. trimest. Soc. mycol. Fr.*, **37**: 56–61, 1921.
Wiltshire, S. P., *Trans. Br. mycol. Soc.*, **21**: 233–235, 1938.

Acrospeira mirabilis Berk. & Br., 1857, apud Berkeley, Intr. crypt. Bot.: 305, Fig. 69a; 1861, *Ann. Mag. nat. Hist.*, Ser. 3, **7**: 449.

Spirospora castaneae Mang. & Vincens, 1920, *Bull. trimest. Soc. mycol. Fr.*, **36**: 89–97.

(Full synonymy given by B. Peyronel in *Bull. trimest Soc. mycol. Fr.*, **37**: 56–61, 1921).

(Fig. 37)

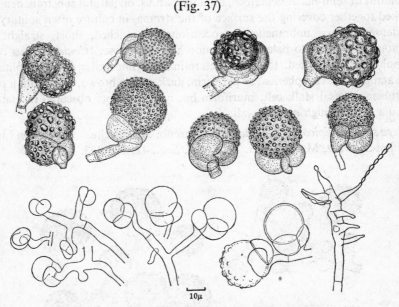

FIG. 37. *Acrospeira mirabilis* (del. S. P. Wiltshire).

Colonies effuse; a loose mass of conidia forms a chocolate-brown powder filling the space between the fleshy cotyledons. Diseased tissue pale ashy-brown. Where only part of a chestnut is diseased a dark brown line marks the limit of the diseased portion. *Mycelium* immersed and superficial, hyphae branched, septate, anastomosing, colourless, smooth, 2–7µ thick. *Conidiophores* arising terminally and laterally on the hyphae and very similar to them, erect or ascending, usually at first with 3 branches towards the apex, later more. The ends of the branches swell slightly and become very distinctly coiled, often in more than one plane. A *conidium* develops in the following manner: transverse septa are laid down dividing the coiled portion into 3 cells, the terminal one increasing greatly in size to 20–35µ and becoming globose, brown and roughly warted. The penultimate cell also becomes verrucose but remains rather paler; the basal cell is usually pale, smooth and clavate or obconical. The spiral twisting is tight and the two lower cells are closely adpressed to the large terminal cell. *Phialides* are readily formed in culture on p.d.a., arising directly on the hyphae and on the swollen ends of conidiophores. *Phialoconidia* produced basipetally in long chains at the tips of phialides, obpyriform, colourless, 2–3µ long.

On fruits of Spanish chestnut sometimes causing considerable losses; England (imported into), Spain, U.S.A.

36. EPICOCCUM

Epicoccum Link ex Schlecht.; Link, 1815, *Magazin Ges. naturf. Freunde Berl.*
7: 32; Schlechtendal, 1824, Synop. Pl. crypt.: 136.

Sporodochia pulvinate, black, not convoluted. *Mycelium* mostly immersed.
Stroma present, pulvinate. *Setae* and *hyphopodia* absent. *Conidiophores* macronematous or semi-macronematous, mononematous, on natural substrata densely
packed together covering the surface of the stroma, in culture often solitary or
in dense clusters, unbranched or occasionally branched, short, straight or
flexuous, colourless to pale brown, smooth or verrucose. *Conidiogenous cells*
monoblastic, integrated, terminal, determinate, cylindrical. *Conidia* solitary,
dry, acrogenous, subspherical or pyriform, dark golden brown, often with a pale
protuberant basal stalk cell, muriform but with the septa obscured in mature
conidia by the rough opaque wall.

Type species: Epicoccum purpurascens Ehrenb. ex Schlecht. = *E. nigrum* Link.
Schol-Schwarz, M. B., *Trans. Br. mycol. Soc.*, **42**: 149–173, 1959.

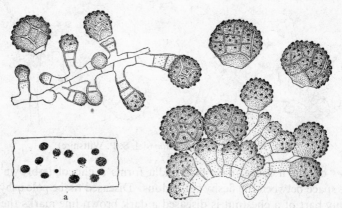

Fig. 38. *Epicoccum purpurascens* (a, habit sketch; other figures × 650).

Epicoccum purpurascens Ehrenb. ex Schlecht., 1824, Synop. Pl. crypt.: 136.
 E. nigrum Link, 1815, *Magazin Ges. naturf. Freunde Berl.*, **7**: 32.
 For full synonymy, see Schol-Schwarz, 1959.

(Fig. 38)

Sporodochia up to 2 mm. diam. *Conidiophores* 5–15 × 3–6µ. Mature *conidia*
most commonly 15–25µ diam. but smaller and much larger (up to 50µ diam.)
conidia are formed sometimes.

An extremely common early secondary invader on all sorts of plants, frequently found on leaf spots with other fungi such as *Periconia byssoides* and
Corynespora cassiicola; cosmopolitan. It has been isolated from air, animals,
foodstuffs, textiles, etc. In culture it produces usually bright orange but sometimes yellow or red pigments and sporulates readily under near u.v. light.

37. CEREBELLA

Cerebella Cesati, 1851, apud Rabenhorst in *Bot. Ztg*, **9**: 669.

Sporodochia much convoluted, dark blackish brown. *Mycelium* partly superficial, partly immersed. *Stroma* present, the surface thrown into deep folds. *Setae* and *hyphopodia* absent. *Conidiophores* semi-macronematous, mononematous, densely packed together covering the surface of the stroma, simple or branched, pale yellow or brown. *Conidiogenous cells* integrated, terminal, monoblastic, determinate. *Conidia* solitary, dry, variable in shape, transversely, obliquely and longitudinally septate, constricted at the septa, smooth or verruculose, brown or dark brown, the basal stalk cell usually pale and protuberant.

Type species: C. *andropogonis* Ces.

Langdon, R. F. N., *Mycol. Pap.*, **61**: 1955.

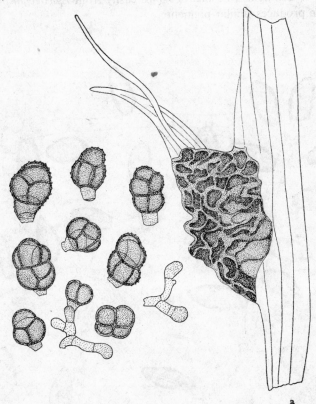

FIG. 39. *Cerebella andropogonis* (a, habit sketch; other figures × 650).

Cerebella andropogonis Ces., 1851, apud Rabenhorst in *Bot. Ztg*, **9**: 669.

Epicoccum andropogonis (Ces.) Schol-Schwarz, 1959, *Trans. Br. mycol. Soc.*, **42**: 171 (32 other synonyms are listed by Langdon).

(Fig. 39)

Sporodochia on spikelets of grasses infected with ergot, cerebriform, dark blackish brown, their size determined by the size of the grass spikelets and the amount of honey-dew available for colonization (Langdon, 1955). *Mycelium* composed of colourless hyphae which form a compact mass filling the space between glumes and enveloping the distal part of the spikelet. *Sporodochia* very strongly convoluted on natural substrata, less strongly in culture. *Conidiophores* short, often branched, pale, smooth, 2–5µ thick. *Conidia* terminal, variable in shape, brown, smooth or verruculose, 15–30 × 12–27µ each composed at maturity of a pale, more or less protuberant stalk cell and a darker spore body made up of 2–10 or sometimes more cells.

Widespread on many different grasses throughout the tropics; often mistaken for one of the Ustilaginales. A red pigment is usually formed in agar cultures of this species. The very variable conidium shape, marked constriction at evident septa and the convoluted surface of sporodochia are characters which enable one to distinguish *C. andropogonis* easily from *Epicoccum purpurascens* which often produces similar pigment.

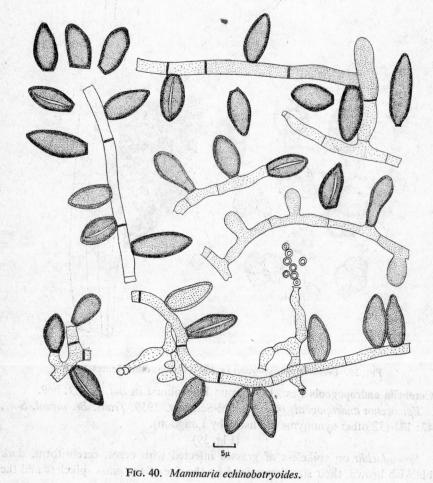

5µ

FIG. 40. *Mammaria echinobotryoides*.

38. MAMMARIA

Mammaria Cesati, 1854, *Bot. Ztg.*, **12**: 190.

Colonies effuse, dark brown to black. *Mycelium* partly superficial, partly immersed; hyphae often with thick, dark transverse septa. *Stroma* none. *Setae* and *hyphopodia* absent. *Conidiophores* micronematous or semi-macronematous, irregularly branched, straight or flexuous, colourless to pale brown, smooth or verrucose often with thick, dark transverse septa. *Conidiogenous cells* monoblastic or polyblastic, integrated, terminal and intercalary, cylindrical. *Conidia* solitary, acropleurogenous, simple, ellipsoidal or navicular, truncate at the base, mid to dark golden brown, smooth, 0-septate, often with an elongated germ slit. There is also a *phialidic* state in which the *conidiogenous cells* are monophialidic, discrete, lageniform, with collarettes; the *conidia* which are sometimes in short chains and become aggregated in slimy heads are semi-endogenous, spherical or subspherical, hyaline, smooth, 0-septate.

Type species: Mammaria echinobotryoides Ces.

Hennebert, G. L., *Trans. Br. mycol. Soc.*, **51**: 756–761, 1968.

Hughes, S. J., *Sydowia*, Beiheft **1**: 359–363, 1957.

Mammaria echinobotryoides Ces., 1854, *Bot. Ztg*, **12**: 190. For synonymy see Hennebert, 1968.

(Fig. 40)

Conidiophores 2–4μ thick. *Conidia* 10–18 × 5–8μ. *Phialides* 6–16 × 3–5·5μ. *Phialoconidia* 1·5–2μ diam.

On wood and commonly isolated from soil; Europe, N. America.

39. WARDOMYCES

Wardomyces Brooks & Hansford, 1923, *Trans. Br. mycol. Soc.*, **8**: 135–137.

Colonies effuse, at first white becoming grey to black when sporulating freely. *Mycelium* mostly superficial. *Stroma* none. *Setae* and *hyphopodia* absent. *Conidiophores* semi-macronematous, mononematous, scattered, straight or flexuous, hyaline or subhyaline, smooth, branched, sometimes penicillately. *Conidiogenous cells* polyblastic, occasionally integrated and terminal but mostly discrete, arranged pencillately, determinate, ampulliform, subspherical, clavate, doliiform or ellipsoidal. *Conidia* solitary, dry, botryose, acropleurogenous, the first one apical and remaining so, the others arising laterally, ellipsoidal, ovoid, oblong rounded at the apex or navicular, often truncate at the base, brown or blackish brown, smooth, each with a longitudinal germ slit, 0-septate.

Type species: Wardomyces anomalus Brooks & Hansf.

Dickinson, C. H., *Trans. Br. mycol. Soc.*, **47**: 321–325, 1964; **49**: 521–522, 1966.

Gams, W., *Trans. Br. mycol. Soc.*, **51**: 798–802, 1968.

Hennebert, G. L., *Can. J. Bot.*, **40**: 1203–1211, 1962; *Trans. Br. mycol. Soc.*, **51**: 753–756, 1968.

KEY

Conidia 0-septate   1
Conidia 1-septate   .   .   .   .   *humicola*
1. Conidia mostly 8–10μ long, navicular, sharply pointed   .   .   .   *pulvinatus*
   Conidia mostly 5–7μ long   2
2. Conidia ovoid to navicular   .   .   .   .   *anomalus*
   Conidia oblong to ellipsoidal   .   .   .   .   *inflatus*

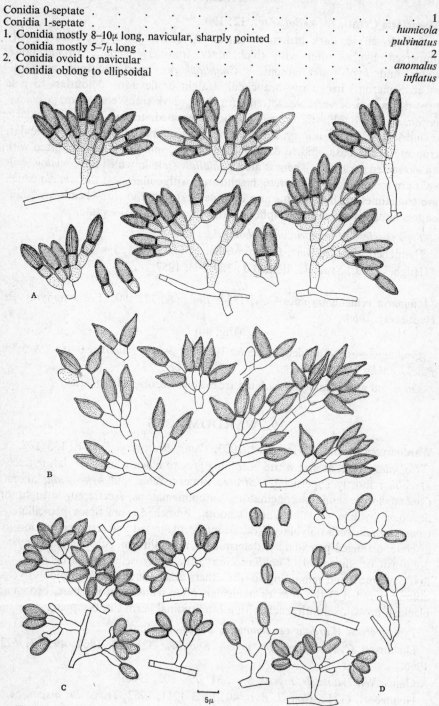

5μ

FIG. 41. *Wardomyces* species: A, *humicola*; B, *pulvinatus*; C, *anomalus*; D, *inflatus*.

Wardomyces humicola Hennebert & Barron, 1962, apud Hennebert in *Can. J. Bot.*, **40**: 1209–1211.

(Fig. 41 A)

Conidiogenous cells doliiform, 3–5 × 2–4μ. *Conidia* navicular, 1-septate, slightly constricted at the septum, 10–13 × 3–4μ.

Isolated from soil; N. America.

Wardomyces pulvinatus (Marchal) Dickinson, 1966, *Trans. Br. mycol. Soc.*, **49**: 521–522.

W. papillatus Dickinson, 1964, *Trans. Br. mycol. Soc.*, **47**: 322.

(Fig. 41 B)

Conidiogenous cells mostly doliiform, 4–6 × 3μ. *Conidia* navicular, sharply pointed at the apex, truncate at the base, 0-septate, mostly 8–10 × 3–4μ.

Isolated from soil; Europe.

Wardomyces anomalus Brooks & Hansf., 1923, *Trans. Br. mycol. Soc.*, **8**: 135–137.

(Fig. 41 C)

Conidiogenous cells doliiform or clavate, 4–7 × 2–4μ. *Conidia* ovoid to navicular, truncate at the base, 0-septate, mostly 5–7 × 3.5–4μ.

Isolated from eggs, meat, soil, etc.; Europe, N. America.

Wardomyces inflatus (Marchal) Hennebert, 1968, *Trans. Br. mycol. Soc.*, **51**: 755–756.

W. hughesii Hennebert, 1962, *Can. J. Bot.*, **40**: 1027.

(Fig. 41 D)

Conidiogenous cells ampulliform, subspherical or clavate, 3–6 × 2–4μ. *Conidia* ellipsoidal or oblong rounded at the apex truncate at the base, 0-septate 5–7 × 3–4μ.

Isolated from *Acer*, *Pteridium* and soil; Europe, N. America.

40. ECHINOBOTRYUM

Echinobotryum Corda, 1831, in Sturm's Deut. Fl., III, 2, **12**: 51.

Colonies effuse, mid to dark olivaceous brown or brown. *Hyphae* partly superficial, partly immersed. *Stroma* none. *Setae* and *hyphopodia* absent. *Conidiophores* semi-macronematous, mononematous, straight or flexuous, unbranched or branched, colourless to pale brown, smooth. *Conidiogenous cells* polyblastic, integrated and terminal or discrete on branches, determinate, clavate, cylindrical or doliiform. *Conidia* solitary, acropleurogenous, in clusters, simple, obpyriform or obturbinate, mucronate, with a broad, flat base, brown or dark brown, smooth or verrucose, 0-septate.

Type species: Echinobotryum state of *Doratomyces stemonitis* (Pers. ex Fr.) Morton & Smith = *E. atrum* Corda.

Hennebert, G. L., *Trans. Br. mycol. Soc.*, **51**: 749–753, 1968.

Echinobotryum state of **Doratomyces stemonitis** (Pers. ex Fr.) Morton & Smith, 1963, *Mycol. Pap.*, **86**: 70–74.

    E. atrum Corda, 1831, in Sturm's Deut. Fl., III, 2, **12**: 51.

(Fig. 42)

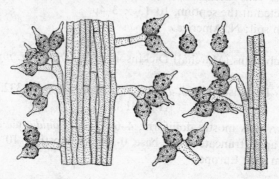

Fig. 42. Echinobotryum state of Doratomyces stemonitis (× 650).

Conidiophores up to 30μ long, 2–5μ thick. *Conidia* 9–14 × 5–8μ.
Common on dead herbaceous stems and wood and isolated from soil, dung etc.; Europe, N. America.

41. ASTEROMYCES

Asteromyces F. & Mme Moreau ex Hennebert, 1962, *Can. J. Bot.*, **40**: 1211–1213.
    Colonies effuse, ochraceous to blackish brown. *Mycelium* partly superficial, partly immersed. *Stroma* none. *Setae* and *hyphopodia* absent. *Conidiophores*

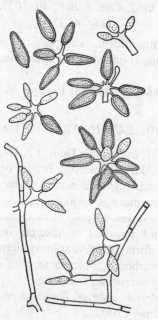

Fig. 43. Asteromyces cruciatus (× 650).

micronematous or semi-macronematous, mononematous, unbranched or branched, straight or flexuous, hyaline to brown or olivaceous brown, smooth. *Conidiogenous cells* polyblastic, integrated, terminal and sometimes also intercalary or discrete, determinate but swelling as lateral conidia develop, cylindrical to clavate, denticulate; denticles numerous, cylindrical, long, narrow. *Conidia* solitary, dry, first formed conidium terminal, subsequent ones lateral, simple, ovoid, obpyriform or obclavate, pale to mid brown or olivaceous brown, smooth, without a germ slit, 0-septate.

Type species: Asteromyces cruciatus F. & Mme Moreau ex Hennebert.

Asteromyces cruciatus F. & Mme Moreau ex Hennebert, 1962, *Can. J. Bot.*, **40**: 1213–1214.

(Fig. 43)

Conidiophores 2–3µ thick. *Conidiogenous cells* swelling to 4–6µ, denticles 3–6 × 0·5–1µ. *Conidia* 10–18µ long, 4–7µ thick in the broadest part.

Isolated from soil of sand-dunes and root surfaces of *Ammophila*; Europe.

42. ACREMONIELLA

Acremoniella Saccardo, 1886, Syll. Fung., **4**: 302.
  Harzia Costantin, 1888, Les Mucedinées Simples: 42.
  Monopodium Delacroix, 1890, *Bull. Soc. mycol. Fr.*, **6**: 99.
  Eidamia Lindau, 1904, Rabenhorst's Krypt-Fl., I, **8**: 123.

Colonies effuse, cottony, at first colourless, later often brown or cinnamon brown. *Mycelium* superficial and immersed. *Stroma* none. *Setae* and *hyphopodia* absent. *Conidiophores* semi-macronematous, mononematous, simple or loosely branched, the branches often being at right angles and usually tapered to a fine point, colourless, smooth, septate, sometimes with 1–3 septa close together just below the apex. *Conidiogenous cells* monoblastic, integrated, terminal. *Conidia* solitary, dry, non-septate, ovoid, obovoid or sub-globose, golden brown or cinnamon brown, smooth or verrucose, double-walled, the inner wall thick, with a pore at the base. *Acremoniella* has also a phialidic state.

Lectotype species: A. atra (Corda) Sacc.
  Mason, E. W., *Mycol. Pap.*, **3**: 29–34, 1933.
  Groves, J. W. & Skolko, A. J., *Can. J. Res.* (C), **24**: 74–77, 1946.

Acremoniella atra (Corda) Sacc., 1886, Syll. Fung., **4**: 302.
  (Full synonymy given by E. W. Mason in *Mycol. Pap.*, **3**: 29–30, 1933.)
  (Fig. 44 A)

Colonies effuse, cottony, at first colourless, later cinnamon brown. *Mycelium* superficial and immersed; hyphae branched, septate, colourless, smooth, 2–7µ thick. *Conidiophores* of two kinds, the ones most frequently found simple or loosely branched, the branches often at right-angles to the stipe or one another and usually tapered to a fine point, colourless, smooth, up to 100µ long, 4–8µ thick at the base, sparingly septate but often with 1–3 septa very close together near the apices of the branches. *Conidia* borne at the tips of the conidiophore and each tapered branch, solitary, dry, 0-septate, ovoid, obovoid or subglobose, golden brown or cinnamon brown, smooth or sometimes with a very slight wrinkling of the exospore, 20–30 × 15–25μ. The *second kind of conidiophore,*

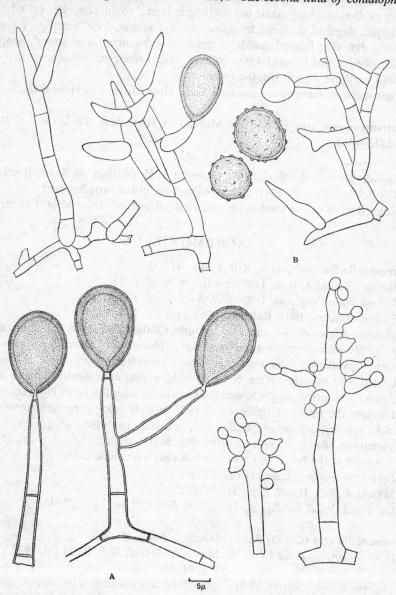

FIG. 44. A, *Acremoniella atra*; B, *A. verrucosa* (× 650 except where indicated by the scale).

more rarely seen, consists typically of a mononematous, colourless, septate stipe 3–6μ thick and up to 90μ long, with a globose terminal vesicle 7–12μ diam. bearing over its surface numerous flask-shaped *phialides* 4–10μ long, 3·5–5μ thick in the broadest part. These conidiophores may grow on beyond the first vesicle, swell again at a higher level and bear more phialides; phialides are also sometimes formed without a vesicle. *Phialoconidia* produced in basipetal succession at the tips of phialides, sometimes catenate, subglobose, 0-septate, colourless, smooth 2–4 × 1·5–3µ.

Recorded on many different plants including bamboo, barley, bean, beet, bracken, carrot, clover, fescue, maize, oats, oil-palm, olive, onion, parsnip, pea, potato, radish, rice, tobacco, also isolated from air, dung, timber and soil; Australia, Canada, Denmark, France, Germany, Great Britain, Ireland, Italy, Japan, Kenya, Rhodesia, Sierra Leone, South Africa, Sweden, U.S.A. Sporulates freely in culture.

Acremoniella verrucosa Tognini, 1896, *Rc. Ist. lomb. Sci. Lett.*, 2 ser. **29**: 864.
Eidamia tuberculata Horne & Jones, 1924, *Ann. Bot.*, **38**: 354.
(Fig. 44 B)
Similar to *A. atra* but with verrucose to tuberculate, much more frequently globose or subglobose conidia 17–26 × 17–23µ.

Recorded on clover, maize, pea, *Pinus patula*, potato, *Sechium edule*, *Taraxacum kok-saghyz*, wheat, also isolated from blotting paper, soil and wood; Australia, Canada, Germany, Great Britain, Italy, Rhodesia, Tanzania and U.S.S.R.

43. TETRACOCCOSPORIUM

Tetracoccosporium Szabó, 1905, *Hedwigia*, **44**: 76–77.
Colonies effuse, grey. *Mycelium* mostly superficial. *Stroma* none. *Setae* and *hyphopodia* absent. *Conidiophores* semi-macronematous, loosely branched, the

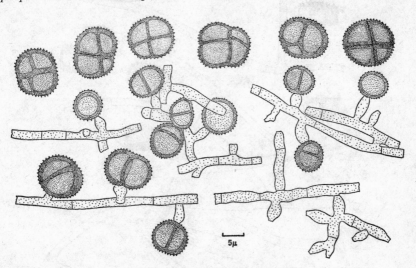

FIG. 45. *Tetracoccosporium paxianum.*

branches often at right-angles to the main axis and to each other, straight or slightly flexuous, hyaline or pale olive, smooth. *Conidiogenous cells* monoblastic, integrated, terminal, determinate, ampulliform or lageniform, sometimes tapered to a point as in *Acremoniella*. *Conidia* solitary, dry, acrogenous, simple, spherical or subspherical, mid to dark brown or olivaceous brown, verruculose or minutely echinulate, divided cruciately by septa at right-angles to one another.

Type species: Tetracoccosporium paxianum Szabó.

Tetracoccosporium paxianum Szabó, 1905, *Hedwigia*, **44**: 76–77.
(Fig. 45)

Conidiophores and *hyphae* 2–3μ thick. *Conidia* 12–18 × 11–13μ.
Isolated from dung and soil; Europe, India.

44. STEMPHYLIOMMA

Stemphyliomma Saccardo & Traverso, 1913, Syll. Fung., **22**: 1394–1395.
Stemphyliopsis Spegazzini, 1910, *Revta Fac. Agron. Vet. Univ. B. Aires*, Ser. 2, **6**: 193–194 (non A. L. Smith, 1901).

Colonies effuse, cottony, olivaceous brown. *Mycelium* superficial, hyphae minutely verruculose or echinulate. *Stroma* none. *Setae* and *hyphopodia* absent.

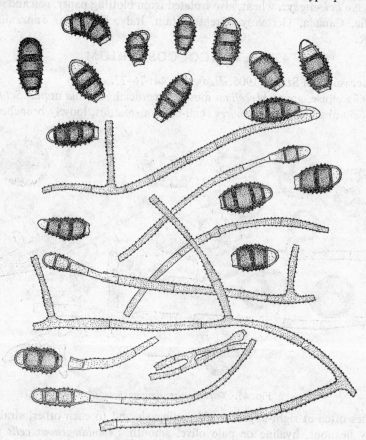

FIG. 46. *Stemphyliomma valparadisiacum* (× 650).

Conidiophores micronematous or semi-macronematous, mononematous, loosely branched, flexuous, golden brown, minutely verruculose or echinulate. *Conidiogenous cells* monoblastic, integrated, terminal on long, repent or ascending branches, determinate or rarely percurrent, cylindrical. *Conidia* solitary, dry, acrogenous, simple, ellipsoidal, verruculose or finely echinulate, transversely septate sometimes constricted at the septa, end cells often subhyaline or very pale, intermediate cells dark brown, becoming detached by a break across the distal part of the conidiogenous cell wall.

Type species: Stemphyliomma valparadisiacum (Speg.) Sacc. & Trav.

Stemphyliomma valparadisiacum (Speg.) Sacc. & Trav., 1913, Syll. Fung., **22**: 1394–1395.

(Fig. 46)

Hyphae, including conidiophores, 2–4μ thick. *Conidia* 19–26 × 7–13μ, 3-septate, septa and walls thick.

On dead leaves of *Puya*; Chile.

45. MUROGENELLA

Murogenella Goos & Morris, 1965, *Mycologia*, **57**: 776–781.

Colonies effuse, olivaceous, rather thin. *Mycelium* partly superficial, partly immersed. *Stroma* none. *Setae* and *hyphopodia* absent. *Conidiophores* micronematous and semi-macronematous, mononematous, unbranched or irregularly

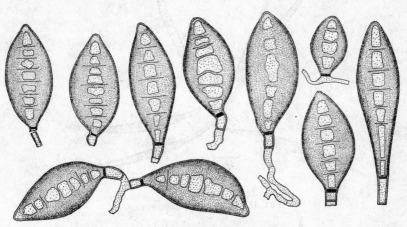

FIG. 47. *Murogenella terrophila* (× 650).

branched, straight or flexuous, colourless to rather pale brown, smooth. *Conidiogenous cells* monoblastic, integrated, terminal or sometimes intercalary, determinate, cylindrical. *Conidia* solitary, dry, acrogenous, occasionally also pleurogenous, simple, ellipsoidal to broadly fusiform, golden brown, smooth, pseudoseptate.

Type species: Murogenella terrophila Goos & Morris.

Murogenella terrophila Goos & Morris, 1965, *Mycologia*, **57**: 776–781.

(Fig. 47)

Conidiophores 2–4µ thick. *Conidia* 25–55 × 15–24µ, 3–7-pseudoseptate. Isolated from soil; N. America.

46. CASARESIA

Casaresia Fragoso, 1920, *Boln R. Soc. esp. Hist. nat.*, **20**: 112–114.
  Ankistrocladium Perrott, 1960, *Trans. Br. mycol. Soc.*, **43**: 556–558.
  Colonies often aquatic, effuse, brown or dark brown. *Mycelium* abundant; hyphae pale to mid brown. *Stroma* none. *Setae* and *hyphopodia* absent. *Conidiophores* semi-macronematous, mononematous, straight or flexuous, unbranched or irregularly branched, brown, smooth. *Conidiogenous cells* integrated, terminal, monoblastic, determinate, cylindrical. *Conidia* solitary, acrogenous, branched, dark reddish brown, smooth; branches usually at right angles to the main axis, multiseptate, each distinctly and sharply hooked at its distal end.
  Type species: Casaresia sphagnorum Frag.
  Petersen, R. H., *Mycologia*, **55**: 18–21, 1963.

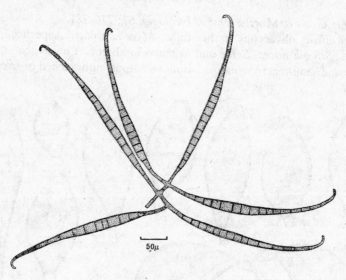

FIG. 48. *Casaresia sphagnorum.*

Casaresia sphagnorum Frag., 1920, *Boln R. Soc. esp. Hist. nat.*, **20**: 113.

(Fig. 48)

Main axis of spore cylindrical, brown, 4–7µ thick. Branches fusiform, dark reddish brown, 300–500µ long, 15–30µ thick in the broadest part, with up to 31 transverse septa.

On *Sphagnum* and on branches and other plant remains submerged in water; Canada, Great Britain, Spain, U.S.A.

47. TROPOSPORELLA

Troposporella Karsten, 1892, *Hedwigia*, **31**: 299.

Colonies scattered, small, pulvinate, fawn or snuff-coloured. *Mycelium* partly superficial, partly immersed, hyphae aggregated and interwoven near the point of origin of the conidiophores. *Stroma* none. *Setae* and *hyphopodia* absent. *Conidiophores* macronematous or semi-macronematous, loosely branched or unbranched, pale brown, smooth. *Conidiogenous cells* monoblastic, integrated, terminal on stipe and branches, determinate, cylindrical. *Conidia* solitary, dry, acrogenous, simple, helicoid, pale to mid golden brown, smooth, with numerous transverse septa, often constricted at the septa.

Type species: Troposporella fumosa Karst.

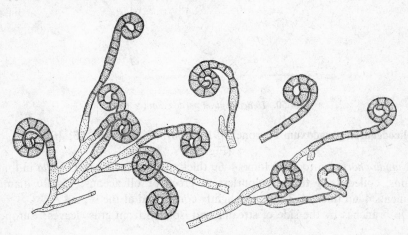

Fig. 49. *Troposporella fumosa* (× 650).

Troposporella fumosa Karst., 1892, *Hedwigia*, **31**: 299.
(Fig. 49)

Conidiophores up to 100μ long, 3–5μ thick. *Conidia* 12–18μ diam., with filament 3–5μ thick, 7–15-septate, coiled 1½ to 2 times.

On bark of *Populus*; Europe, N. America.

48. HELICODENDRON

Helicodendron Peyronel, 1918, *Nuovo G. bot. ital.*, **25**: 462–464.
Helicodesmus Linder, 1925, *Am. J. Bot.*, **12**: 267.

Colonies effuse, at first white, shining, later greenish to olive, grey or brown. *Mycelium* partly immersed, partly superficial. *Stroma* none. *Setae* and *hyphopodia* absent. *Conidiophores* semi-macronematous, mononematous, much branched, flexuous, colourless to brown or olivaceous, smooth. *Conidiogenous cells* polyblastic, integrated, terminal and intercalary, determinate, cylindrical. *Conidia* catenate, dry, acropleurogenous, simple, coiled in 3 planes to form a cylindrical, ovoid or ellipsoidal spore body or with only a few loose coils; filaments colourless to green, olivaceous or brown, smooth, septate.

Type species: Helicodendron paradoxum Peyronel.
Glenn-Bott, Janet I., *Trans. Br. mycol. Soc.*, **38**: 17–30, 1955.
Linder, D. H., *Ann. Mo. bot. Gdn*, **16**: 329–333, 1929.

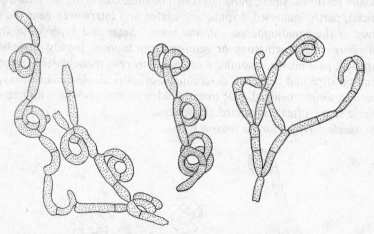

FIG. 50. *Helicodendron paradoxum* (× 650).

Helicodendron paradoxum Peyronel, 1918, *Nuovo G. bot. ital.*, **25**: 462–464.
(Fig. 50)

Conidiophores up to 100µ long, 4–6µ thick, pale olivaceous. *Conidia* in long chains, coiled 1–3½ times, colourless to green or olivaceous, 15–50µ diam.; filaments 4–6µ thick, 1–3-septate, slightly constricted at the septa.

On branches by the side of streams and isolated from grass leaves; Europe.

49. TRIPOSPERMUM

Tripospermum Spegazzini, 1918, *Physis*, **4** (17): 295.

Colonies effuse, brown, blackish brown or black, often crust-like. *Mycelium* superficial, sometimes torulose; hyphae much branched and anastomosing to form a network. *Stroma* none. *Setae* and *hyphopodia* absent. *Conidiophores* semi-macronematous, mononematous, erect or ascending, unbranched or occasionally loosely branched, pale to mid brown or olivaceous brown, smooth. *Conidiogenous cells* monoblastic or polyblastic, integrated, terminal or inter-calary, determinate, cylindrical or doliiform. *Conidia* solitary, dry, acrogenous or pleurogenous, branched, usually made up of a pyriform or ellipsoidal stalk cell and 4 divergent, subulate, multiseptate arms, pale to mid brown or olivaceous brown, smooth. Perfect states often belong to the Capnodiales.

Type species: Tripospermum acerinum (Syd.) Speg.

Hughes, S. J., *Mycol. Pap.*, **46**: 10–22, 1951.

Tripospermum myrti (Lind) Hughes, 1951, *Mycol. Pap.*, **46**: 17–18.
(Fig. 51)

Colonies effuse, black, thin, crust-like. *Hyphae* 4–8µ thick, cells often doliiform. *Conidiophores* up to 90µ long, 4–8µ thick. *Conidia* with stalk cell 6–10 × 4–7µ, arms up to 30µ long, 4–8µ thick at the base, tapering to 1–2µ, 1–4-septate, often constricted at the septa; one of the arms in this species usually lies parallel to the stalk cell.

On twigs of Myrtaceae and *Spartium* and leaves of *Arundinaria* and *Fraxinus;* Europe.

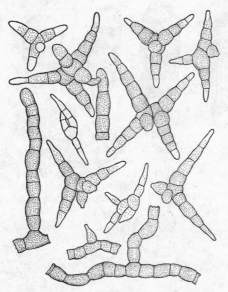

FIG. 51. *Tripospermum myrti* (× 650).

50. ANNELLOPHORELLA

Annellophorella Subram., 1962, *Proc. Indian Acad. Sci.*, Sect. B, **55**: 6.

Mycelium partly superficial, partly immersed, hyphae hyaline to brown. *Stroma* none. *Setae* and *hyphopodia* absent. *Conidiophores* semi-macronematous, mononematous, repent, ascending or erect, simple or branched, sometimes with very short branches, pale to mid brown. *Conidiogenous cells* integrated, terminal and intercalary, monoblastic, percurrent. *Conidia* dry, variable in shape, brown to dark brown, with transverse and often also oblique or longitudinal septa, thick-walled.

Type species: Annellophorella faureae (P. Henn.) M. B. Ellis.

Annellophorella faureae (P. Henn.) M. B. Ellis, 1963, *Mycol. Pap.*, **87**: 13.

Brachysporium faureae P. Henn., 1903, H. Baum's Kunene—Sambesi Expedition, Berlin: 169.

Clasterosporium densum H. & P. Syd., 1912, *Annls mycol.*, **10**: 444.

Annellophorella densa (H. & P. Syd.) Subram., 1962, *Proc. Indian Acad. Sci.*, Sect. B, **55**: 6.
                                         (Fig. 52)

Colonies hypophyllous effuse, dark blackish brown. *Mycelium* partly superficial, partly immersed, linked by hyphae passing through stomata; hyphae hyaline to brown, 1·5–6µ thick. *Conidiophores* differing slightly if at all from the

thicker superficial hyphae, repent, erect or ascending, simple or branched, sometimes with very short branches, straight or flexuous, pale to mid brown, smooth,
stipe and branches percurrent, each with up to 7 successive proliferations. *Conidia*
solitary, straight or curved, variable in shape but often cylindrical or clavate,

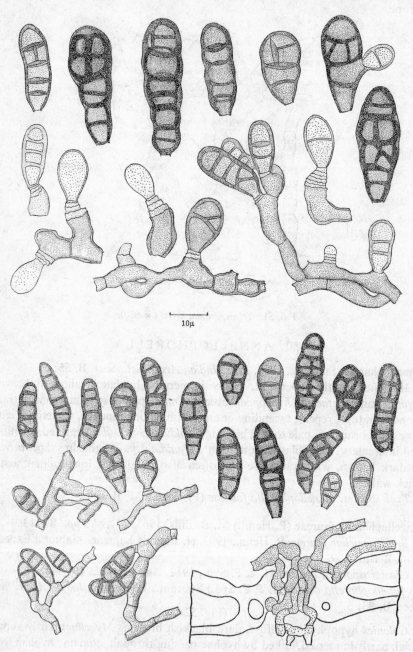

10μ

FIG. 52. *Annellophorella faureae* (× 650 except where indicated by the scale).

rounded at the apex, truncate at the base, brown to dark brown, thick-walled, smooth, with 1–8 transverse and 0–4 longitudinal or oblique septa, 15–40 (30)μ long, 7–14 (10·5)μ thick in the broadest part, 2·5–4μ wide at the base.

On leaves of *Faurea speciosa*; S. Africa.

51. ALYSIDIUM

Alysidium Kunze ex Steudel; Kunze, 1817, in Kunze & Schmidt, Mykol. Hefte, **1**; 11; Steudel, 1824, Nomencl. bot.: 54.

Mycelium partly superficial, partly immersed, hyphae thick. *Stroma* none. *Setae* absent. *Hyphopodia* absent. *Conidiophores* semi-macronematous, often upturned hyphae the same thickness as the rest of the mycelium, branched, pale or dark coloured. *Conidiogenous cells* monoblastic and polyblastic, integrated, terminal. *Conidia* dry, catenate, often in branched chains, spherical, oval, limoniform, ellipsoidal or oblong, pale or dark coloured, smooth, 0-septate.

Type species: Alysidium fulvum Kunze ex Steudel = **Alysidium dubium** (Pers. ex Fr.) M. B. Ellis, comb. nov. basionym *Trichoderma dubium* Pers., Observationes mycol., **1**: 99, 1796; Fries, Syst. mycol., **3**: 216, 1832.

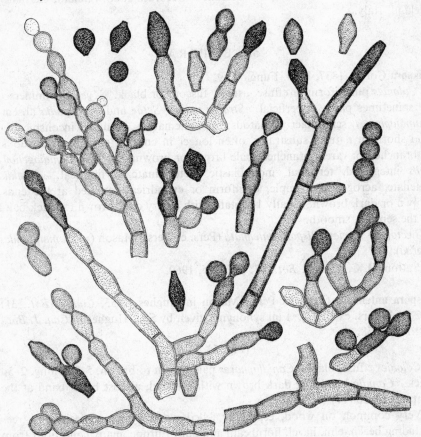

FIG. 53. *Alysidium resinae* (× 650).

Alysidium resinae (Fr.) M. B. Ellis, comb. nov.

Myxotrichum resinae Fr., 1832, Syst. mycol., **3**: 350.

Racodium aterrimum Ehrenb., 1818, Sylv. mycol. berol.: 10, 22.

Acladium aterrimum (Ehrenb.) Hughes, 1958, *Can. J. Bot.*, **36**: 731.

Torula ramosa Fuckel, 1870, Symb. mycol.: 348.

Oidium ramosum (Fuckel) Hughes, 1953, *Can. J. Bot.*, **31**: 591.

Trichosporium nigricans Sacc., 1880, *Michelia*, **2**: 125.

Dematium dimorphum Karst., 1887, *Meddn Soc. Fauna Flora fenn.*, **14**: 91.

(Fig. 53)

Colonies effuse, black. *Mycelium* partly superficial, partly immersed; hyphae branched, septate, brown to dark blackish brown, smooth, 5–11µ thick. *Conidiophores* upturned branches of the mycelium, variable in length and of the same thickness and colour as the hyphae, branched. *Conidia* formed in simple and branched chains terminally and laterally on the conidiophore and its branches, spherical and 7–12 (9·4)µ diam., or limoniform, ellipsoidal or oblong 11–20 (15·6) × 7–14 (9·7)µ, brown to dark blackish brown, smooth, non-septate.

On dead wood of conifers and deciduous trees; England, Finland, Germany, Ireland, Italy.

52. BISPORA

Bispora Corda, 1837, Icon. Fung., **1**: 9.

Colonies punctiform or effuse, usually fuscous or black. *Mycelium* immersed or sometimes partly superficial. *Stroma* none. *Setae* and *hyphopodia* absent. *Conidiophores* semi-macronematous, mononematous, usually inconspicuous and short on natural substrates, often longer in culture, straight or flexuous, unbranched or rarely branched, pale brown or brown, smooth. *Conidiogenous cells* integrated, terminal, monoblastic, determinate, cylindrical. *Conidia* catenate, acrogenous, simple, doliiform or cylindrical rounded at the ends, brown or dark brown, usually 1-septate, with a very dark brown to black band at the septum, smooth.

Lectotype species: Bispora antennata (Pers. ex Pers.) Mason (= *B. monilioides Corda*).

Sutton, B. C., *Can. J. Bot.*, **47**: 612–615, 1969.

Bispora antennata (Pers. ex Pers.) Mason in Hughes, 1953, *Can. J. Bot.*, **31**: 582 [as ' (Pers. ex Fr.) ']. Full synonymy given by S. J. Hughes in *Can. J. Bot.*, **36**: 741, 1958.

(Fig. 54 A)

Colonies effuse, black. *Conidiophores* pale brown to brown, 5–30µ long, 2–5µ thick. *Conidia* brown or dark brown with a broad, almost black band at the septum, 13–20 (17) × 7–8 (7·1)µ.

Very common on wood, especially felled wood of many deciduous trees including beech, elm, hazel, hornbeam and oak; Europe, many collections from Great Britain.

Bispora betulina (Corda) Hughes, 1958, *Can. J. Bot.*, **36**: 740, with full synonymy.
(Fig. 54 B)

Colonies punctiform to effuse, fuscous. *Conidiophores* on natural substrata 5–15 × 3–4μ, much longer in culture. *Conidia* occasionally 2-septate, brown or dark brown with very dark band or bands, 9–14 (11) × 4–5 (4·4)μ.

Common on dead wood; Canada, Europe, U.S.A., many collections from Great Britain.

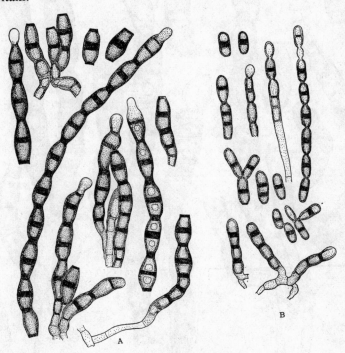

FIG. 54. A, *Bispora antennata*; B, *B. betulina* (× 650).

53. TAENIOLELLA

Taeniolella Hughes, 1958, *Can. J. Bot.*, **36**: 816–818.

Colonies effuse or punctiform and pulvinate, mid to dark brown, olivaceous brown or black. *Mycelium* immersed and also sometimes superficial. *Stroma* none. *Setae* and *hyphopodia* absent. *Conidiophores* semi-macronematous, caespitose or scattered, usually short, unbranched or sparingly branched near the base, flexuous, brown or olivaceous brown, smooth or verruculose. *Conidiogenous cells* monoblastic, integrated, terminal, determinate, cylindrical or doliiform. *Conidia* sometimes solitary but more commonly in long, simple or branched acropetal chains, often seceding with difficulty, acrogenous, simple, straight or flexuous, cylindrical, rounded or truncate at the ends, ellipsoidal, doliiform or obclavate, mid to dark brown or olivaceous brown, smooth or verrucose, occasionally with 1 but usually with 2 or more transverse septa.

Type species: Taeniolella exilis (Karst.) Hughes.

KEY

On dead branches of deciduous trees 1
On senescent leaves of *Plantago media* . . . . *plantaginis*
1. Conidia 12–15µ thick, mostly 1–3-septate . . . . . . *exilis*
    Conidia 7–11µ thick, 3–24-septate . . . . *stilbospora*

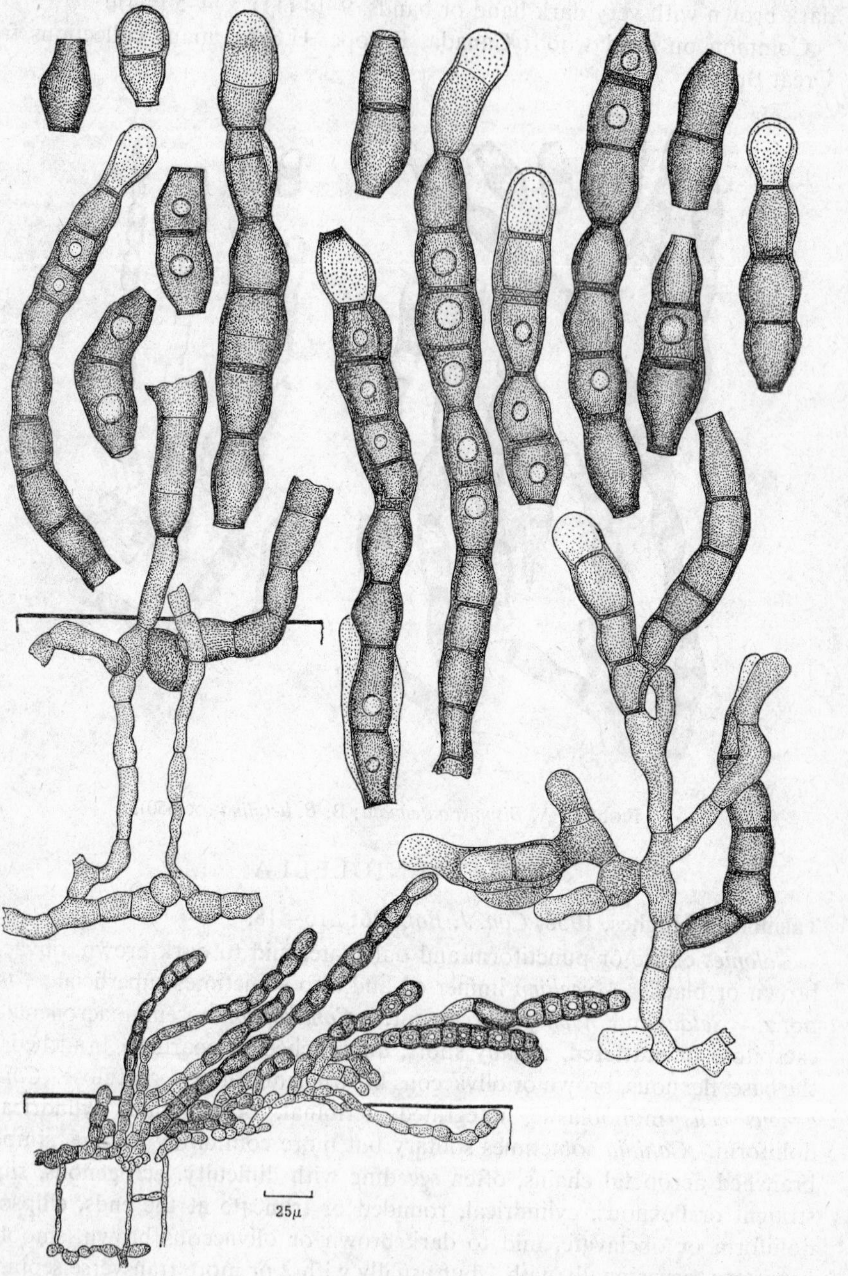

FIG. 55. *Taeniolella exilis* (× 650 except where indicated by the scale).

Taeniolella exilis (Karst.) Hughes, 1958, *Can. J. Bot.*, **36**: 817.
(Fig. 55)

Colonies effuse, dark olivaceous brown to black. *Conidiophores* caespitose, 5–9μ thick. *Conidia* often doliiform, olivaceous brown, smooth, up to 6- but mostly 1–3-septate, 22–75 × 12–15μ.

On wood and bark of *Betula*; Europe, N. America.

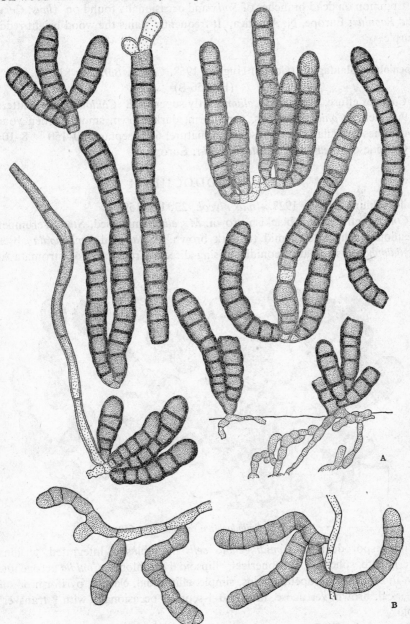

FIG. 56. A, *Taeniolella stilbospora*; B, *T. plantaginis* (× 650).

Taeniolella stilbospora (Corda) Hughes, 1958, *Can. J. Bot.*, **36**: 817.
(Fig. 56 A)

Colonies pulvinate or effuse, dark olivaceous brown to black, velvety. *Conidiophores* caespitose or scattered, 3–5µ thick. *Conidia* straight or flexuous, cylindrical rounded at the apex, often truncate at the base, mid to dark brown, smooth, 3–24-septate, 25–140 × 7–11µ.

Common on dead branches of *Salix* and occasionally found on *Alnus*, *Corylus* and *Populus*; Europe, N. America. It frequently stains the wood bright reddish purple.

Taeniolella plantaginis (Corda) Hughes, 1958, *Can. J. Bot.*, **36**: 817.
(Fig. 56 B)

Colonies effuse, black. *Mycelium* partly superficial. *Conidiophores* scattered, 2–9µ thick. *Conidia* flexuous, vermiform, dark brown, smooth when young, sometimes irregularly verrucose when mature, 6–28-septate, 30–150 × 8–10µ.

On senescent leaves of *Plantago media*; Europe.

54. SPILODOCHIUM

Spilodochium Sydow, 1927, *Annls mycol.*, **25**; 158–159.

Colonies effuse, dark blackish brown. *Mycelium* immersed. *Stroma* erumpent, pseudoparenchymatous, mid to dark brown. *Setae* and *hyphopodia* absent. *Conidiophores* stromatic, conidia arising directly from cells of stromata and

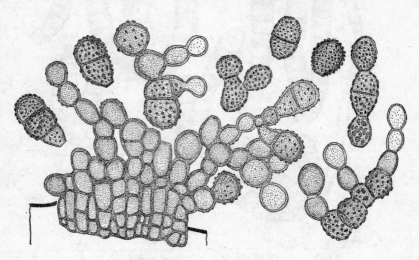

FIG. 57. *Spilodochium vernoniae* (× 650).

forming sporodochia. *Conidiogenous cells* monoblastic, integrated, terminal, determinate, spherical, subspherical, ellipsoidal or oblong. *Conidia* acrogenous, dry, in branched acropetal chains, simple, ellipsoidal, clavate, pyriform or subspherical, brown, verrucose, mostly 0–1-septate, occasionally with 2 transverse septa.

Type species: Spilodochium vernoniae Syd.

Spilodochium vernoniae Syd., 1927, *Annls mycol.*, **25**: 158–159.

(Fig. 57)

Colonies epiphyllous. Characterised by the chains of conidia arising directly from cells of the stroma. *Conidia* 10–28 × 9–17µ.

On leaves of *Vernonia*; Costa Rica, Jamaica.

55. XYLOHYPHA

Xylohypha (Fr.) Mason, 1960, apud Deighton in *Mycol. Pap.*, **78**: 43.

Colonies effuse, powdery, brown, dark blackish brown or black. *Mycelium* immersed. *Stroma* none or rudimentary. *Setae* and *hyphopodia* absent. *Conidio-phores* macronematous, mononematous, scattered or caespitose, unbranched, usually rather short, straight or flexuous, brown, smooth. *Conidiogenous cells* monoblastic, integrated, terminal, determinate, cylindrical or doliiform. *Conidia* dry, in long, unbranched or occasionally branched acropetal chains which break up very readily, acrogenous, simple, ellipsoidal, cylindrical or oblong rounded at the ends, pale to mid brown, smooth, almost always 0-septate, very rarely 1-septate.

Type species: Xylohypha nigrescens (Pers. ex Fr.) Mason.

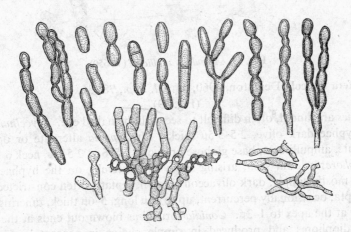

FIG. 58. *Xylohypha nigrescens* (× 650).

Xylohypha nigrescens (Pers. ex Fr.) Mason, 1960, apud Deighton in *Mycol. Pap.*, **78**: 43.

(Fig. 58)

Conidiophores 15–35 × 2·5–4·5µ on natural substrata, sometimes longer in culture. *Conidia* 7–15 (mostly 9–12) × 4–6µ.

Common on dead wood of *Acer, Betula, Cornus, Corylus, Fagus, Fraxinus* (most frequently), *Hedera, Ligustrum, Populus, Salix, Sambucus, Tilia, Viburnum, Vitex*; Europe, New Zealand. When specimens are being collected the conidia often rub off on the hands as a brown powder.

56. AMPULLIFERA

Ampullifera Deighton, 1960, *Mycol. Pap.*, **78**; 36.

Mycelium superficial; hyphae olivaceous. *Stroma* none. *Setae* absent. *Hyphopodia* present, in most species mucronate, ampulliform. *Conidiophores* macronematous, mononematous, usually simple, olivaceous. *Conidiogenous cells* integrated, terminal, monoblastic, occasionally percurrent. *Conidia* catenate or occasionally solitary, ellipsoidal, oblong-ellipsoidal or doliiform, olive, smooth, 0-septate or with 1 or a few transverse septa.

Type species: Ampullifera foliicola Deighton.

Deighton, F. C., *Mycol. Pap.*, **78**: 36–42; **101**: 28–31.

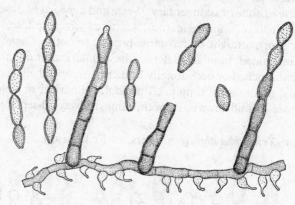

FIG. 59. *Ampullifera foliicola* (× 650).

Ampullifera foliicola Deighton, 1960, *Mycol. Pap.*, **78**: 36.
(Fig. 59)

Colonies arachnoid, often difficult to see with the naked eye. *Mycelium* superficial; hyphae dark olive, 2·5–5·5μ thick. *Hyphopodia* alternate or opposite, mucronate, ampulliform, base globose or ovoid, 2·5–6 × 2·5–4μ, neck 6 × 0·5μ. *Conidiophores* seldom seen, arising singly and laterally on the hyphae, erect, straight, mostly simple, dark olivaceous brown, septate, often constricted at the upper septa, occasionally percurrent, up to 36μ long, 5–6μ thick, tapering rather abruptly at the apex to 1–2μ. *Conidia* formed as blown out ends at the tips of the conidiophores and produced in simple chains in acropetal succession, ellipsoid or oblong-ellipsoid, olivaceous, smooth, 8–15μ long, 4–6μ thick in the broadest part, tapered at each end to 1–2μ.

Associated with lichens and algae on leaves of various plants; Brazil, Ghana, Malaya, Sabah, San Domingo, Sarawak, Sierra Leone.

Three other species of *Ampullifera* have been described; these have been found only once or twice.

57. HETEROCONIUM

Heteroconium Petrak, 1949, *Sydowia*, **3**: 265–266.

Colonies effuse, olivaceous to dark blackish brown. *Mycelium* all or mostly superficial. *Stroma* none. *Setae* and *hyphopodia* absent. *Conidiophores* macronematous, mononematous, unbranched, straight or curved, mid to dark brown or reddish brown, smooth. *Conidiogenous cells* monoblastic, integrated, terminal, percurrent or determinate, cylindrical. *Conidia* catenate, dry, acrogenous, simple, cylindrical with rounded ends or obclavate, mid to dark brown or reddish brown, smooth, multiseptate.

Type species: Heteroconium citharexyli Petrak.

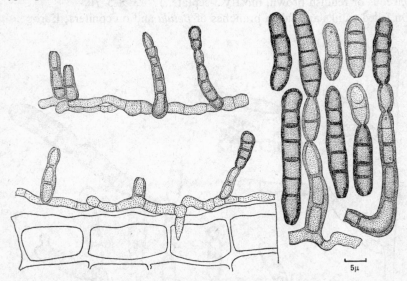

FIG. 60. *Heteroconium citharexyli* (× 650 except where indicated by the scale).

Heteroconium citharexyli Petrak, 1949, *Sydowia*, **3**: 265–266.
(Fig. 60)

Colonies hypophyllous. *Conidiophores* up to 40μ long, 4–7μ thick. *Conidia* 1–10-septate, 10–40 × 4–7μ.

On living leaves of *Citharexylum*; Ecuador.

58. SEPTONEMA

Septonema Corda, 1837, Icon. Fung., **1**: 9.

Colonies effuse, hairy or velvety, pale olivaceous to dark blackish brown. *Mycelium* mostly superficial. *Stroma* none. *Setae* and *hyphopodia* absent. *Conidiophores* macronematous, mononematous, unbranched or with short lateral branches, straight or slightly flexuous, brown, smooth. *Conidiogenous cells* monoblastic or sometimes polyblastic, terminal and intercalary on stipe and branches, integrated, determinate, cylindrical. *Conidia* dry, mostly acrogenous, formed in long, often branched, acropetal chains, seceding readily, simple, cylindrical or oblong, abruptly tapered to the usually truncate ends, ellipsoidal or fusiform, pale to dark olive, brown or reddish brown, smooth, rather thick-walled, with 1 or several transverse septa.

Type species: Septonema secedens Corda.

Hughes, S. J., *Naturalist, Hull*, 1951: 173–176, 1951.

D.H.—4

Septonema secedens Corda, 1837, Icon. Fung., **1**: 9.

(Fig. 61)

Conidiophores cylindrical, pale to mid olivaceous or reddish brown, up to 200μ long, 4–6μ thick. *Conidia* formed in long, usually branched chains, cylindrical or oblong, abruptly tapered to the truncate ends, pale to mid olivaceous or reddish brown, mostly 3-septate, 17–23 × 5–7μ.

On felled trunks and fallen branches of *Betula* and on conifers; Europe.

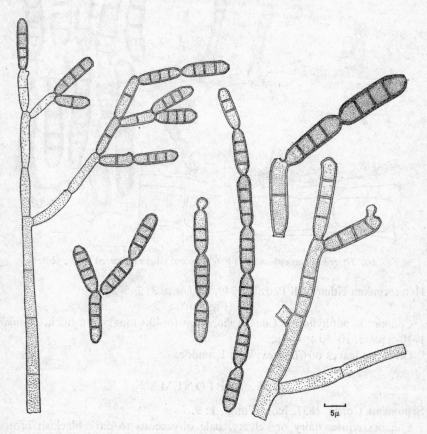

FIG. 61. *Septonema secedens* (× 650 except where indicated by the scale).

59. SEPTOTRULLULA

Septotrullula Höhnel, 1902, *Sber. Akad. Wiss. Wien*, Abt. 1, **111**: 1025–1027.

Sporodochia pulvinate, olivaceous buff or brown. *Mycelium* immersed. *Stroma* present, prosenchymatous. *Setae* and *hyphopodia* absent. *Conidiophores* macronematous, mononematous, slender, crowded, parallel, forming pulvinate sporodochia, unbranched, straight or flexuous, subhyaline or olivaceous, smooth. *Conidiogenous cells* monoblastic, integrated, terminal, determinate, cylindrical. *Conidia* catenate, acrogenous, cylindrical or oblong with truncate

ends, pale olivaceous, smooth with 1 or more transverse septa; the long acropetal chains are slimy, they become encrusted and the conidial mass often
cracks on drying.

Lectotype species: Septotrullula bacilligera Höhnel.

Sutton, B. C. & Pirozynski, K., *Trans. Br. mycol. Soc.*, **48**: 355–357, 1965.

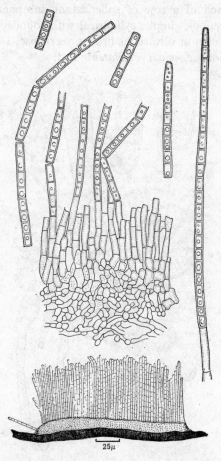

FIG. 62. *Septotrullula bacilligera* (× 650 except where indicated by the scale).

Septotrullula bacilligera Höhnel, 1902, *Sber. Akad. Wiss. Wien*, Abt. 1, **111**:
1026.

(Fig. 62)

Sporodochia 1–2 mm. diam. *Conidiophores* 10–30 × 2–3μ. *Conidia* 1–5-
(mostly 3-) septate, 19–29 × 2–3·5μ.

On bark of *Betula, Fagus* and *Quercus*; Europe.

60. HORMOCEPHALUM

Hormocephalum Sydow, 1939, *Annls mycol.*, **37**: 424–425.

Colonies irregular, effuse, cottony, grey to olivaceous brown. *Mycelium* superficial. *Stroma* none. *Setae* and *hyphopodia* absent. *Conidiophores* macronematous, mononematous, branched, the branches restricted to the apical region forming a stipe and head; stipe straight or flexuous, mid pale brown, lower part smooth, upper part and branches verrucose. *Conidiogenous cells* monoblastic, integrated, determinate, spherical, terminal on branches which are often each composed of a row of spherical or subspherical cells. *Conidia* solitary, dry, acrogenous, simple, cylindrical with rounded ends or ellipsoidal, pale or mid-pale brown or olivaceous brown, verrucose, 1–5-septate.

Type species: Hormocephalum ecuadorense Syd.

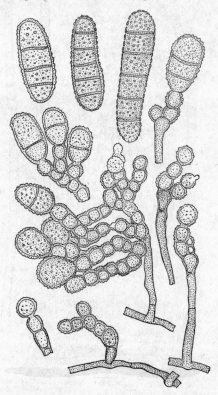

FIG. 63. *Hormocephalum ecuadorense* (× 650).

Hormocephalum ecuadorense Syd., 1939, *Annls mycol.*, **37**: 424–425.
(Fig. 63)

Colonies hypophyllous. *Conidiophores* up to 100μ long; lower part of stipe 2–4μ thick, upper part of stipe and branches 7–9μ thick. *Conidia* 16–60 × 10–16μ.

On leaves of *Aspilia*; Ecuador.

61. BACTRODESMIUM

Bactrodesmium Cooke, 1883, *Grevillea*, **12**: 35.

Sporodochia scattered, punctiform, brown or black. *Mycelium* mostly immersed. *Stroma* none or rudimentary. *Setae* and *hyphopodia* absent. *Conidio-phores* fasciculate, formed at the ends of hyphae where these push through to the surface, narrow, broadening slightly near the apex, simple or branched, colour-less to pale brown, smooth, septate. *Conidiogenous cells* monoblastic, integrated, terminal, determinate, cylindrical. *Conidia* acrogenous, solitary, cells often un-equally coloured with very dark bands at the septa and basal cells much paler than the others, with 3 or more transverse septa and in one species with longi-tudinal or oblique septa in the end cells, smooth, clavate, cylindrical, ellipsoidal or obovoid.

Lectotype species: Bactrodesmium abruptum (Berk. & Br.) Mason & Hughes. Ellis, M. B., *Mycol. Pap.*, **72**; 2–14, 1959; **87**: 42, 1963; **103**: 37–38, 1965. Sutton, B. C., *Can. J. Bot.*, **45**: 1777–1781, 1967.

KEY

Conidia with thick black bands at one or more of the septa 1
Conidia without thick black bands at the septa 5
1. End cells often with longitudinal or oblique septa . . . . . *obliquum*
   End cells without longitudinal or oblique septa 2
2. Conidia more than 30µ broad . . . . *papyricola*
   Conidia less than 30µ broad 3
3. Conidia usually ellipsoidal . . . . *cedricola*
   Conidia clavate 4
4. Conidia 12–17µ broad, with the penultimate cell always much longer than the others
                                                          abruptum
   Conidia 15–25µ broad with the penultimate cell often only slightly longer than the terminal
   cell . . . . *obovatum*
   Conidia 18–23µ broad, with the penultimate cell shorter than the terminal cell
                                                          microleucurum
5. Conidia more than 20µ broad . . . . *atrum*
   Conidia less than 20µ broad 6
6. Conidial cells all approximately the same shade 7
   Conidial cells not all the same shade 8
7. Conidia pale to mid brown, usually 4-septate . . . . *spilomeum*
   Conidia all very pale brown, 5–6 septate . . . . . *pallidum*
8. Basal cells of conidia subhyaline, other cells brown . . . *traversianum*
   Apical and basal cells of conidia subhyaline, other cells brown . . *betulicola*

Bactrodesmium abruptum (Berk. & Br.) Mason & Hughes apud Hughes, 1958, *Can. J. Bot.*, **36**: 738.

Sporidesmium abruptum Berk. & Br., 1965, *Ann. Mag. nat. Hist.*, Ser. 3, **15**: 401.

Clasterosporium abruptum (Berk. & Br.) Sacc., 1886, Syll. Fung., **4**: 389.

(Fig. 64 A)

Sporodochia scattered, punctiform, black, shining, usually 100–300µ but occasionally 500µ diam. *Mycelium* immersed in the substratum; hyphae pale to mid brown, 1–5µ thick. *Conidiophores* fasciculate, formed at the ends of hyphae where these push through to the surface of the substratum, usually unbranched, flexuous, subhyaline to very pale brown, smooth, septate, up to 40µ long, 2–3µ thick at the base, broadening towards the apex to 3–5µ. *Conidia* solitary, terminal, clavate, rounded at the top, truncate at the base, the upper part mid or dark reddish brown, the colour becoming progressively paler towards the basal cell which is usually subhyaline, smooth, 3–7-septate, 32–70µ long, the penultimate cell much longer than any of the others, 12–17µ thick in

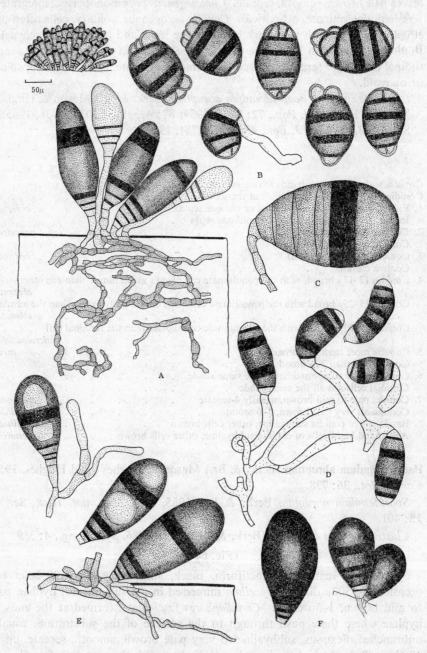

FIG. 64. *Bactrodesmium* species (1): A, *abruptum*; B, *obliquum*; C, *papyricola*; D, *cedricola*; E, *obovatum*; F, *microleucurum* (× 650).

the broadest part, 3–5μ wide at the base, with dark bands at the septa, the upper one thick and black.

On wood and bark of various deciduous trees including ash, oak and sycamore; Great Britain.

Measurements of conidia and details of substrata and distribution of 10 other species of *Bactrodesmium* are given below.

B. obliquum Sutton (Fig. 64 B): 23–25 × 16–22μ; on *Picea glauca*; Canada.

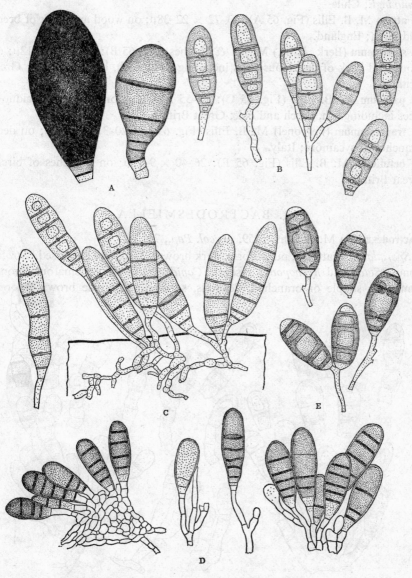

FIG. 65. *Bactrodesmium* species (2): A, *atrum*; B, *spilomeum*; C, *pallidum*; D, *traversianum*; E, *betulicola* (× 650).

B. papyricola C. & M. Moreau ex M. B. Ellis (Fig. 64 C): 50–65 × 32–38μ; on paper; French Guinea.

B. cedricola M. B. Ellis (Fig. 64 D): 20–35 × 9–18μ; on *Cedrus deodara*; W. Pakistan.

B. obovatum (Oudem.) M. B. Ellis (Fig. 64 E): 28–58 × 15–23μ; on wood and bark of deciduous trees including alder, ash, beech, birch, elm and oak; Great Britain, Netherlands.

B. microleucurum (Speg.) M. B. Ellis (Fig. 64 F): 30–48 × 18–25μ; on *Chusquea cummingii*; Chile

B. atrum M. B. Ellis (Fig. 65 A): 43–72 × 22–38μ; on wood and bark of beech and birch; England.

B. spilomeum (Berk. & Br.) Mason & Hughes (Fig. 65 B): 24–43 × 8–12μ; on wood and bark of deciduous trees including ash, beech, birch and elm; Great Britain.

B. pallidum M. B. Ellis (Fig. 65 C): 35–55 × 9–12μ; on wood of deciduous trees including ash, beech and oak; Great Britain.

B. traversiamum (Peyronel) M. B. Ellis (Fig. 65 D): 20–37 × 8–12μ; on dead branches of sycamore; Italy.

B. betulicola M. B. Ellis (Fig. 65 E): 26–40 × 9–15μ; on branches of birch; Great Britain.

62. BACTRODESMIELLA

Bactrodesmiella M. B. Ellis, 1959, *Mycol. Pap.*, **72**: 14.

Sporodochia scattered, punctiform, dark brown. *Mycelium* immersed. *Stroma* none. *Setae* and *hyphopodia* absent. *Conidiophores* macronematous, mononematous, simple or branched, flexuous, subhyaline to pale brown, smooth,

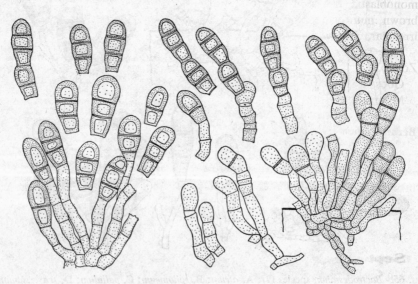

FIG. 66. *Bactrodesmiella masonii* (× 650).

septate. *Conidiogenous cells* integrated, terminal, cylindrical, monoblastic, percurrent. *Conidia* acrogenous, solitary or in chains, simple, pale to mid brown, smooth, usually 2-septate, cylindrical, rounded at the apex and truncate at the base, or clavate.

Type species: Bactrodesmiella masonii (Hughes) M. B. Ellis.

Bactrodesmiella masonii (Hughes) M. B. Ellis, 1959, *Mycol. Pap.*, **72**: 14.
Bactrodesmium masonii Hughes, 1953, *Can. J. Bot.*, **31**: 654.

(Fig. 66)

Sporodochia scattered, punctiform, dark brown, up to 200μ diam. *Mycelium* mostly immersed; hyphae pale, smooth 2–4μ thick. *Conidiophores* fasciculate, simple or branched, flexuous, subhyaline to pale brown, smooth, septate, percurrent, up to 40μ long, 2–4μ thick at the base, broadening gradually to 4–7μ at the apex. *Conidia* formed singly and also basipetally in chains as blown-out ends at the apex of the conidiophore and each successive proliferation, cylindrical to subclavate, rounded at the apex, truncate at the base, pale to mid brown, smooth, 1–2- (usually 2-) septate, 16–26μ long, 8–12μ thick in the broadest part, 4–7μ wide at the base.

On cupules of beech lying on the ground; Great Britain.

63. BERKLEASMIUM

Berkleasmium Zobel in Corda, 1854, *Icon Fung.*, **6**: 4.

Sporodochia punctiform, raised, black, shining. *Mycelium* immersed. *Stroma* prosenchymatous or rudimentary. *Setae* and *hyphopodia* absent. *Conidiophores* macronematous, narrow, closely packed together in the sporodochia, flexuous, unbranched or rarely branched, smooth. *Conidiogenous cells* integrated, terminal, monoblastic, determinate, cylindrical. *Conidia* solitary, acrogenous, simple, brown, muriform, smooth, clavate, ellipsoidal, oblong rounded at the ends or irregular, often with a protruding hilum.

Type species: Berkleasmium concinnum (Berk.) Hughes (= *B. cordaeanum* Zobel).

Goos, R. D., *Can. J. Bot.*, **47**: 503–504, 1969.
Moore, R. T., *Mycologia*, **50**: 686–688, 1959; **51**: 734–739, 1961 (with key).

Berkleasmium concinnum (Berk.) Hughes, 1958, *Can. J. Bot.*, **36**: 740.

(Fig. 67)

Conidiophores straw coloured, up to 30μ long, 2–5μ thick. *Conidia* golden brown, 60–124 (99)μ long, 24–31 (28)μ thick.

On rotten wood; Canada and U.S.A.

64. SEPTOSPORIUM

Septosporium Corda, 1831, in J. Sturm's Deut. Fl., **3** (12): 33–34.

Colonies effuse, dark brown to black, velvety or hairy. *Mycelium* partly superficial, partly immersed; superficial hyphae often thick, brown, smooth or

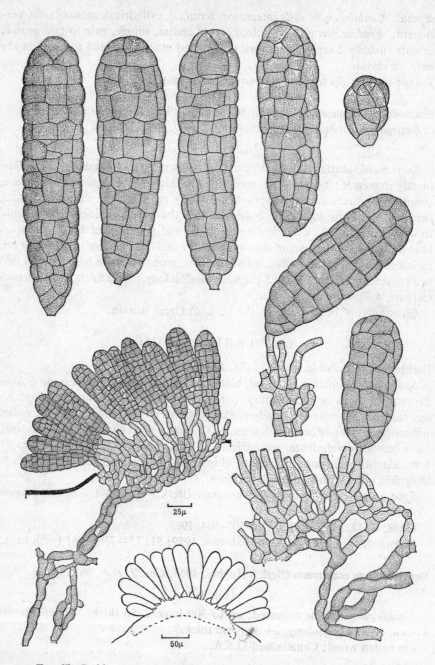

FIG. 67. *Berkleasmium concinnum* (× 650 except where indicated by scales).

verrucose, branched and anastomosing to form a network. *Stroma* none. *Setae* numerous, unbranched, subulate, dark brown to black, smooth, thick-walled. *Hyphopodia* absent. *Conidiophores* macronematous, mononematous, unbranched, usually narrower than the vegetative hyphae, straight or flexuous, subhyaline to pale brown, smooth. *Conidiogenous cells* monoblastic, integrated, terminal, determinate or percurrent, cylindrical. *Conidia* solitary, dry, acrogenous, simple, usually broadly ellipsoidal with a protuberant hilum, rostrate in one species,

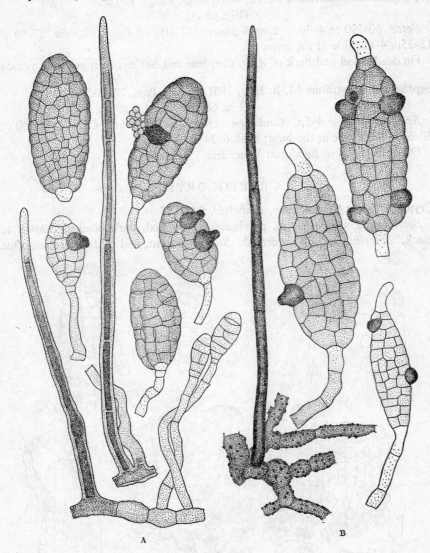

FIG. 68. A, *Septosporium bulbotrichum*; B, *S. rostratum* (× 650).

straw-coloured to dark brown, smooth, muriform; often one or more cells of the conidium swell individually, darken and become transformed into short-beaked pycnidia.

Type species: Septosporium atrum Corda.
Ellis, M. B., *Mycol. Pap.*, **79**: 1–5, 1961.

KEY

Conidia not rostrate . . . . *bulbotrichum*
Conidia rostrate . . . . *rostratum*

Septosporium bulbotrichum Corda, 1837, Icon. Fung., **1**: 12.
(Fig. 68 A)
Setae 60–190 × 4–7μ. *Conidiophores* 12–80 × 2–6μ. *Conidia* 27–80 × 12–35μ, 4–6μ wide at the base.

On dead wood and bark of *Acer, Carpinus* and *Sorbus*; Europe, N. America.

Septosporium rostratum M. B. Ellis, 1961, *Mycol. Pap.*, **79**: 3–5.
(Fig. 68 B)
Setae 150–700 × 4–5μ. *Conidiophores* up to 70 × 4–6μ. *Conidia* 60–150 × 20–37μ, 4–6μ wide at the base; beak 6–24 × 6–7μ.

On dead culms of *Bambusa*; Venezuela.

65. CRYPTOCORYNEUM

Cryptocoryneum Fuckel, 1866, *Hedwigia*, **5** (2): 25.
Sporodochia usually small, pulvinate, flat-topped, dark blackish brown to black. *Mycelium* mostly immersed. *Stroma* present, mid to dark brown, often

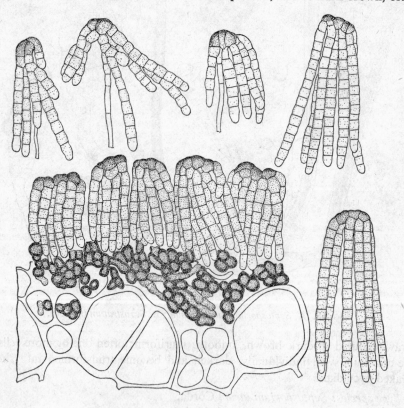

FIG. 69. *Cryptocoryneum condensatum* (× 650).

made up of cells grouped together rather loosely. *Setae* and *hyphopodia* absent. *Conidiophores* macronematous, mononematous, very slender, erect, often obscured by the pendent arms of conidia, straight or flexuous, unbranched, subhyaline or pale brown, smooth. *Conidiogenous cells* monoblastic, integrated, terminal, determinate, cylindrical. *Conidia* solitary, acrogenous, branched, cheiroid, each made up of a small number of swollen, mid to dark brown cap cells from which septate, subhyaline or pale brown arms grow downwards towards the substratum.

Type species: Cryptocoryneum condensatum (Wallr.) Mason & Hughes (= *C. fasciculatum* Fuckel).

Cryptocoryneum condensatum (Wallr.) Mason & Hughes, 1953, apud Rimington in Nat. Hist. Scarborough Distr., **1**: 161.

For full synonymy see Hughes, *Can. J. Bot.*, **36**: 758, 1958.

(Fig. 69)

Conidiophores up to 80μ long, 1–3μ thick. *Conidia* 40–85 × 20–30μ; arms 3–5μ thick with up to 17 transverse septa.

Very common on dead, fallen wood and bark especially of deciduous trees; Europe.

66. MICROCLAVA

Microclava F. L. Stevens, 1917, *Trans. Ill. St. Acad. Sci.*, **10**: 204.

Ommatospora Batista & Cavalcanti, 1964, *Riv. Patol. veg.*, Pavia, Ser. 3, **4**: 565.

Colonies effuse, not visible to the naked eye. *Mycelium* superficial. *Stroma* none. *Setae* and *hyphopodia* absent. *Conidiophores* macronematous, mononematous, unbranched, straight or flexuous, pale, arising laterally on the hyphae,

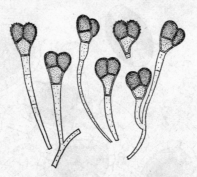

FIG. 70. *Microclava miconiae* (× 650).

erect. *Conidiogenous cells* polyblastic, integrated, terminal, determinate, swollen, ellipsoidal, subspherical or obconical, often subtended by another cell which is also wider than the rest of the conidiophore. *Conidia* ellipsoidal or subspherical with a broad, flat base, borne side by side in pairs or occasionally threes, nonseptate, brown, verrucose, finely echinulate or smooth, only rarely deciduous.

Type species: Microclava miconiae F. L. Stevens.

Deighton, F. C., *Trans. Br. mycol. Soc.*, **52**: 316–321, 1969, describes the type and 2 other species.

Microclava miconiae F. L. Stevens, 1917, *Trans. Ill. St. Acad. Sci.*, **10**: 204.
(Fig. 70)

Hyphae 2–3·5µ thick. *Conidiophores* 30–100µ long, 2·5µ thick at base, expanding to 3·5–4µ. *Conidiogenous cells* 7µ wide, pale olive. *Conidia* broadly ellipsoidal brown, minutely echinulate or sometimes smooth, 6·5–9·5 × 5·5–7µ.

Associated with a Microthyriaceous fungus on leaves of *Miconia laevigata*, Puerto Rico, and overgrowing *Calothyrium jahnii* on the same host, Dominican Republic.

67. HERMATOMYCES

Hermatomyces Spegazzini, 1911, *An. Mus. nac. Hist. nat. B. Aires*, Ser. 3, **13**: 445–446.

Colonies effuse, dark brown or blackish brown, velvety. *Mycelium* partly superficial, partly immersed. *Stroma* none. *Setae* when present broad, cylindrical or dumb-bell-shaped, flat-topped, pale with a brown cap. *Hyphopodia* absent. *Conidiophores* closely packed together, macronematous, mononematous, unbranched or occasionally forked, short, straight or flexuous, pale brown, smooth. *Conidiogenous cells* monoblastic, integrated, terminal, determinate, cylindrical. *Conidia* solitary acrogenous, lenticular, elliptical to almost round in one plane, smooth, muriform, with pale peripheral cells surrounding central dark brown to black cells; black and shining by reflected light.

Type species: Hermatomyces tucumanensis Speg.

Hughes, S. J., *Mycol. Pap.*, **50**: 72–76 and 100, 1953.

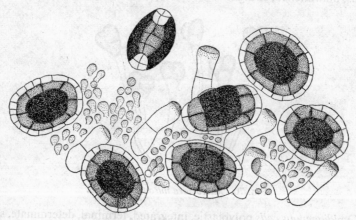

FIG. 71. *Hermatomyces tucumanensis* (× 650).

Hermatomyces tucumanensis Speg., 1911, *An. Mus. nac. Hist. nat. B. Aires*, Ser. 3, **13**: 446.

(Fig. 71)

Fructifications composed of an outer sterile byssus of narrow, erect brown hyphae up to 60μ long surrounding a fertile area. *Conidiophores* up to 12 × 2·5–3μ. *Setae* (paraphyses) up to 45 × 12μ. *Conidia* 29–42 × 19–29μ, 14–16μ thick.

On dead branches of *Alchornea*, *Averrhoa*, *Celtis*, *Coffea*, *Smilax*, etc. and rachides of *Elaeis*; Argentina, Ghana, Sierra Leone.

S. J. Hughes in *Mycol. Pap.*, **50** gives a very full account of this and another species *H. sphaericum* (Sacc.) Hughes.

68. ONCOPODIUM

Oncopodium Saccardo, 1904, *Annls mycol.*, **2**: 19.

Sporodochia gregarious or scattered, punctiform or plate-like, dark brown to black. *Mycelium* all immersed or partly superficial. *Stroma* prosenchymatous, brown erumpent. *Setae* and *hyphopodia* absent. *Conidiophores* macronematous, mononematous, caespitose, forming sporodochia, unbranched, flexuous, hyaline, smooth, narrow except at the apex where there is a vesicular swelling. *Conidiogenous cells* monoblastic, integrated, terminal, determinate, clavate, spherical or subspherical. *Conidia* solitary, dry, acrogenous, simple or corniculate, subglobose or hemispherical, sometimes flattened dorsiventrally, clathrate or muriform, mid to dark brown, sometimes with pale peripheral cells surrounding darker cells, smooth; when detached usually carrying away with them the upper part of the conidiophore.

Type species: Oncopodium antoniae Sacc. & D. Sacc.

Ellis, M. B., *Mycol. Pap.*, **87**: 17–20, 1963.

KEY

Conidia corniculate .   .   .   . *antoniae*
Conidia not corniculate .   .   .   . *panici*

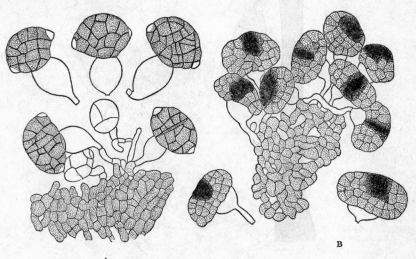

A

B

FIG. 72. A, *Oncopodium antoniae*; B, *O. panici* (× 650).

Oncopodium antoniae Sacc. & D. Sacc., 1904, *Annls mycol.*, **2**; 19.

(Fig. 72 A)

Sporodochia plate-like, black. *Conidiophores* 20–30μ long, 1·5–2μ thick, terminal vesicle 5–15μ diam. *Conidia* subglobose, often somewhat flattened dorsiventrally, at first hyaline, later turning brown except for one marginal cell on each side which usually remains hyaline and often projects as a conical protuberance, smooth, 19–27 (24) × 16–20 (17·5)μ.

On dry branches of *Berberis* ; Italy.

Oncopodium panici H. J. Hudson, 1961, *Trans. Br. mycol. Soc.*, **44**: 406.

(Fig. 72 B)

Sporodochia punctiform, dark brown to black. *Conidiophores* 20–30μ long, 2–4μ thick, terminal vesicle 5–14μ diam. *Conidia* hemispherical to subglobose, flattened dorsiventrally, upper central part somewhat domed, dark brown to black, remainder subhyaline to pale brown, 21–32 (25) × 11–16 (13·6)μ.

On dead culms of *Panicum* and *Saccharum*; Jamaica.

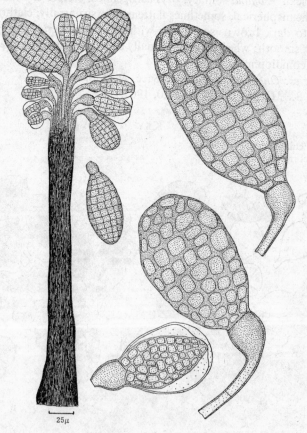

25μ

Fig. 73. *Kostermansinda magna* (× 650 except where indicated by the scale).

69. KOSTERMANSINDA

Kostermansinda Rifai, 1968, *Reinwardtia*, **7**: 376–380.

Colonies effuse, greenish black, with large synnemata clearly visible under a binocular dissecting microscope. *Mycelium* partly superficial, partly immersed. *Stroma* none. *Setae* and *hyphopodia* absent. *Conidiophores* macronematous, synnematous, erect, dark blackish brown; individual threads unbranched, splaying out to form a head, straight or flexuous, smooth, brown, paler and swollen at the apex. *Conidiogenous cells* monoblastic, integrated, terminal, usually determinate, very rarely percurrent, cylindrical or swollen. *Conidia* solitary, dry, acrogenous, simple, broadly ellipsoidal or clavate, smooth, each made up of a large, golden brown, muriform terminal part and a smaller, non-septate, pale basal vesicle. Conidia are liberated by a break across the wall of the conidiogenous cell just below the vesicle.

Type species: Kostermansinda magna (Boedijn) Rifai.

Kostermansinda magna (Boedijn) Rifai, 1968, *Reinwardtia*, **7**: 378–380.
  Sclerographium magnum Boedijn, 1960, *Persoonia*, **1**: 319–320.

(Fig. 73)

Synnemata up to 400 × 60µ, dark blackish brown to almost black; individual threads 3–5µ thick, brown. *Conidia* 55–97µ long, muriform part 25–42µ thick, vesicle 12–16µ diam.

On decaying petioles of palms; Java.

70. DOMINGOELLA

Domingoella Petrak & Ciferri, 1932, *Annls mycol.*, **30**: 339–340.

Colonies effuse, greyish olive, thinly hairy. *Mycelium* superficial, composed of a network of branched and anastomosing, narrow, pale olive hyphae.

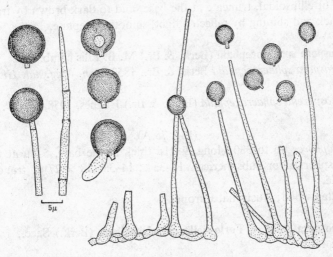

FIG. 74. *Domingoella asterinarum* (× 650 except where indicated by the scale).

Stroma none. *Setae* and *hyphopodia* absent. *Conidiophores* macronematous, mononematous, straight or slightly flexuous, unbranched, cylindrical to subulate, swollen at the base, pale brown, smooth. *Conidiogenous cells* monoblastic, integrated, terminal, percurrent, cylindrical. *Conidia* solitary, dry, acrogenous, simple, spherical with a narrow, cylindrical protuberant peg at the base, 0-septate, pale to mid brown or olivaceous brown, thick-walled, smooth.

Type species: Domingoella asterinarum Petrak & Cif.

Domingoella asterinarum Petrak & Cif., 1932, *Annls mycol.*, **30**: 339–340.
(Fig. 74)

Hyphae 2–3μ thick. *Conidiophores* up to 120μ long, 1–3μ thick, swollen to 4–7μ at the base. *Conidia* 7–10μ diam.

Overgrowing and presumably parasitic on colonies of *Asterina, Asterolibertia, Hysterostomella* and *Linotexis* on many different plants; common throughout the tropics.

71. ACROGENOSPORA

Acrogenospora M. B. Ellis gen. nov.

Deuteromycotina, hyphomycetes. *Coloniae* effusae, fuscae vel atrae, pilosae. *Mycelium* partim superficiale et partim immersum. *Conidiophora* singula ex apice lateribusque hypharum oriunda, erecta, simplicia, recta vel leniter flexuosa, subulata vel cylindrica, septata, basi atrobrunnea, apicem versus pallidiora, laevia, percurrentia. *Conidia* singula, sicca, primo in apice conidiophori et dein proliferationis cujusque successivae oriunda, sphaerica, subsphaerica, obovoidea vel ellipsoidea, basi truncata, brunnea vel atro-brunnea, laevia, 0-septata. Species typica: *Acrogenospora sphaerocephala* (Berk. & Br.) M. B. Ellis.

Colonies effuse, dark blackish brown to black, hairy. *Mycelium* partly superficial, partly immersed. *Stroma* none. *Setae* and *hyphopodia* absent. *Conidiophores* macronematous, mononematous, erect, unbranched, straight or slightly flexuous, subulate or cylindrical, dark brown at the base, paler towards the apex, smooth. *Conidiogenous cells* monoblastic, integrated, terminal, percurrent, cylindrical. *Conidia* solitary, dry, acrogenous, simple, spherical, subspherical, obovoid or ellipsoidal, truncate at the base, mid to dark brown by transmitted light, black and shining by reflected light, smooth, 0-septate.

Acrogenospora sphaerocephala (Berk. & Br.) M. B. Ellis comb. nov.

Monotospora sphaerocephala Berk. & Br., 1859, *Ann. Mag. nat. Hist.*, III, **3**: 361.

Monotosporella sphaerocephala (Berk. & Br.) Hughes, 1958, *Can. J. Bot.*, **36**: 787.

(Fig. 75 A)

Conidiophores up to 380μ long, 9–11μ thick at the base, 5–8μ at the apex. *Conidia* spherical or subspherical, 15–33 × 14–33 (28 × 27)μ; truncate base 5–7μ wide.

On rotten wood; Australia, Europe.

Acrogenospora state of **Farlowiella carmichaeliana** (Berk.) Sacc., 1891, *Syll. Fung.*, **9**: 1101.

(Fig. 75 B)

Conidiophores up to 400μ long, 9–12μ thick at the base, 5–9μ at the apex. *Conidia* broadly ellipsoidal to obovoid, 20–40 × 15–25 (28 × 20)μ; truncate base 5–8μ wide.

On rotten wood; Europe.

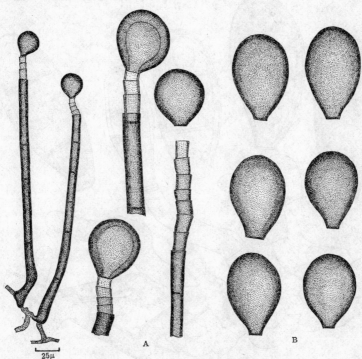

FIG. 75. A, *Acrogenospora sphaerocephala*; B, *A*. state of *Farlowiella carmichaeliana* (× 650 except where indicated by the scale).

72. SEPTOIDIUM

Septoidium Arnaud, 1921, *Annls Épiphyt*, **7**: 106.

Diploidium Arnaud, 1923, *Annls Épiphyt*., **9**: 33.

Colonies usually hypophyllous, effuse, reddish brown, olivaceous brown or black. *Mycelium* superficial; hyphae thick, often golden brown or reddish brown, smooth, branched, intertwined and anastomosing to form a close network. *Stroma* none. *Setae* absent. *Stomatopodia* present, simple or lobed. *Conidiophores* macronematous or semi-macronematous, mononematous, unbranched, straight or flexuous, pale to mid golden brown or reddish brown, smooth. *Conidiogenous cells* monoblastic, integrated, terminal, percurrent, cylindrical. *Conidia* solitary, dry, acrogenous, simple, clavate, cylindrical, rounded at the apex or almost ellipsoidal, always truncate at the base, pale to mid golden brown or reddish brown, smooth, with 1 or several transverse septa. Perfect state where known *Parodiopsis*.

Type species: Septoidium clusiaceae Arn.

Baker, R. E. D., *Mycol. Pap.*, **58**: 1–16, 1955.

Septoidium state of **Parodiopsis hurae** (Arn.) Baker & Dale, 1950, *Mycol. Pap.*, **33**: 9.

(Fig. 76)

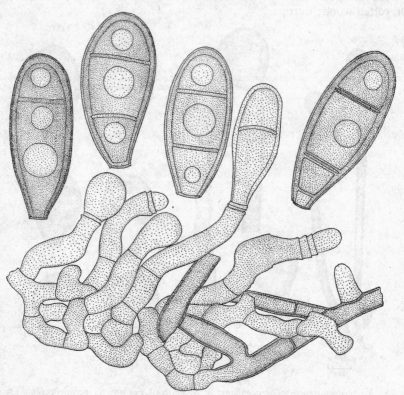

FIG. 76. *Septoidium* state of *Parodiopsis hurae* (× 650).

Conidiophores up to 80μ long, 8–16μ thick. *Conidia* 1–3- (mostly 2-) septate, 55–75 × 22–33μ.

On leaves of *Hura*; Brazil, Dominican Republic, Haiti, San Domingo, Trinidad.

73. SPORIDESMIUM

Sporidesmium Link ex Fries; Link, 1809, *Magazin Ges. naturf. Freunde Berl.*, **3** (1): 41; Fries, 1821, Syst. mycol., **1**: XL.

Podoconis Boedijn, 1933, *Bull. Jard. bot. Buitenz.*, III, **13**: 133–134.

Colonies effuse, olivaceous, brown, grey or black, often hairy or velvety. *Mycelium* superficial or immersed. *Stroma* rarely formed. *Setae* and *hyphopodia* absent. *Conidiophores* macronematous, mononematous, sometimes caespitose, unbranched, straight or flexuous, mid to dark brown. *Conidiogenous cells* monoblastic, integrated, terminal, determinate or percurrent, cylindrical, doliiform or lageniform. *Conidia* solitary, dry, acrogenous, simple, straight, curved

or occasionally sigmoid, cylindrical, fusiform, obclavate, obpyriform or obturbinate, sometimes rostrate, subhyaline, straw-coloured or pale to dark brown,
olivaceous brown or reddish brown, smooth or verruculose, transversely septate
or pseudoseptate.

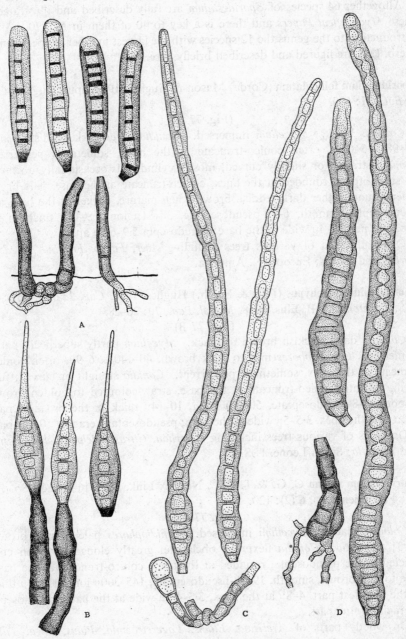

FIG. 77. *Sporidesmium* species (1): A, *folliculatum*; B, *brachypus*; C, *vagum*; D, *adscendens*
(× 650).

Type species: Sporidesmium atrum Link.

Deighton, F. C., *Mycol. Pap.*, **117**: 25–30, 1969.

Ellis, M. B., *Mycol. Pap.*, **70**: 16–84, 1958; **72**: 73–75, 1959; **82**: 45–46, 1961; **87**: 25–31, 1963; **93**: 25–28, 1963; **103**: 43–46, 1965.

Altogether 64 species of *Sporidesmium* are fully described and illustrated in these *Mycological Papers* and there is a key to 40 of them in No. 70. As an introduction to the genus the 12 species with the largest number of specimens in Herb. IMI are figured and described briefly here.

Sporidesmium folliculatum (Corda) Mason & Hughes, 1953, apud Hughes, *Can. J. Bot.*, **31**: 609.

(Fig. 77 A)

Colonies black. *Mycelium* immersed. *Conidiophores* dark reddish brown 40–98 × 5–6·5μ, often conico-truncate at the apex, sometimes percurrent. *Conidia* straight or slightly curved, mostly cylindrical, occasionally obclavate or subfusiform, rounded at the apex, conico-truncate at the base, pale brown when young, rather dark reddish brown when mature, darker at the base and pseudosepta, smooth, 6–12-pseudoseptate, 38–81μ long, 8–11μ thick in the broadest part, 2–4μ wide at the base: pseudosepta 5·9–6·5μ apart.

On dead wood of various trees including *Acer, Fagus, Fraxinus, Hedera, Sorbus* and *Ulmus*; Europe, N. America.

Sporidesmium brachypus (Ellis & Everh.) Hughes, 1958, *Can. J. Bot.*, **36**: 807.
    S. deightonii M. B. Ellis, 1958, *Mycol. Pap.*, **70**: 26–28.

(Fig. 77 B)

Colonies dark blackish brown to black. *Mycelium* partly superficial, partly immersed. *Conidiophores* mid to dark brown, 40–140 × 6–9μ, often conico-truncate at the apex, sometimes percurrent. *Conidia* straight or curved, fusiform, rostrate, conico-truncate at the base, straw-coloured to golden brown, smooth, 5–8-pseudoseptate, 50–90μ long, 10–14μ thick in the broadest part, 1–2μ near the apex, 3·5–5μ wide at the base; pseudosepta averaging 10·4μ apart.

On twigs of various trees including *Averrhoa, Citrus, Dichrostachys, Petrea* and *Thevetia*; Sierra Leone, U.S.A.

Sporidesmium vagum C. G. & T. F. L. Nees ex Link, 1825, in Linné's Sp. Pl., Ed. 4 (Willdenow's), **6** (2): 120.

(Fig. 77 C)

Colonies black. *Mycelium* immersed. *Conidiophores* pale to mid brown, 17–31 × 3·5–6μ. *Conidia* flexuous, obclavate, greatly elongated, often constricted at the pseudosepta, rounded at the apex, conico-truncate at the base, pale to mid brown, smooth, 19–38-pseudoseptate, 145–300μ long, 11–14μ thick in the broadest part, 4–8μ at the apex, 3·5–4·5μ wide at the base; pseudosepta averaging 7·4μ apart.

On woody parts of *Averrhoa, Elaeis, Lagerstroemia, Pinus, Thea, Tilia, Ziziphus*, etc.; Europe, Ghana, Jamaica, Nigeria, Pakistan, Sierra Leone, Tanzania.

Sporidesmium adscendens Berk., 1840, *Ann. Nat. Hist.*, **4**: 291.
(Fig. 77 D)

Colonies black. *Mycelium* mostly immersed. *Conidiophores* mid to dark reddish brown, 20–45 × 7–10µ. *Conidia* flexuous, obclavate, elongated, often constricted at the pseudosepta, rounded at the apex, conico-truncate at the base,

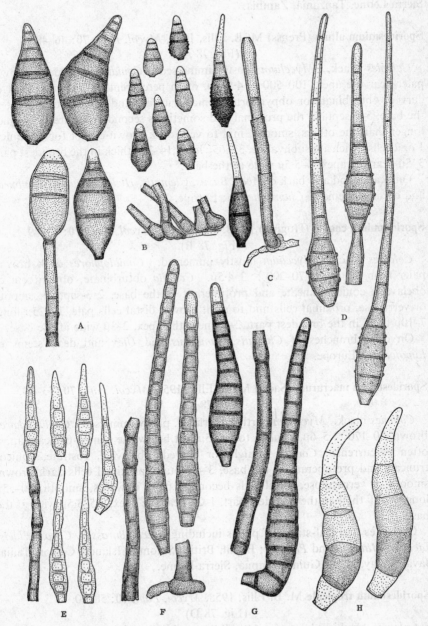

FIG. 78. *Sporidesmium* species (2): A, *altum*; B, *cookei*; C, *macrurum*; D, *tropicale*; E, *lepto-sporum*; F, *eupatoriicola*; G, *pedunculatum*; H, *inflatum* (× 650).

mid to dark reddish brown, basal cell often pale, smooth, 16–62-pseudoseptate, 110–375μ long, 14–20μ thick in the broadest part, 5–10μ at the apex, 5·5–7μ wide at the base; pseudosepta averaging 5·8μ apart.

On woody parts of *Buxus, Carpodinus, Cassia, Elaeis, Heeria, Hippocratea, Nauclea, Nyssa, Quercus, Uapaca, Vangueria,* etc.; Europe, Falkland Islands, Sierra Leone, Tanzania, Zambia.

Sporidesmium altum (Preuss) M. B. Ellis, 1958, *Mycol. Pap.,* **70**: 46–48.
(Fig. 78 A)

Colonies black. *Mycelium* mostly immersed. *Conidiophores* dark brown, paler near the apex, 100–500 × 4·5–6μ, often percurrent. *Conidia* straight or curved, obturbinate or obpyriform, conico-truncate and often protuberant at the base, 5–8-septate, the proximal and sometimes second and third cells much longer than the others, smooth, mid to very dark brown except for the apical 1 or 2 cells which are subhyaline, 35–65μ long, 14–25μ thick in the broadest part, 3–5μ near the apex, 4–5·5μ wide at the base.

On dead wood and bark of *Acer, Buxus, Clematis, Hedera, Prunus, Sambucus* and on dead stems of *Chamaenerion;* Europe.

Sporidesmium cookei (Hughes) M. B. Ellis, 1958, *Mycol. Pap.,* **70**: 48–49.
(Fig. 78 B)

Colonies black. *Mycelium* mostly immersed. *Conidiophores* dark brown, paler near the apex, 70–200 × 3–4·5μ. *Conidia* obturbinate, obpyriform or obclavate, conico-truncate and protuberant at the base, 2–3-septate, smooth or verrucose, proximal cells mid to dark brown, distal cells pale, 17–25μ long, 7–10μ thick in the broadest part, 2–4μ near the apex, 2–3μ wide at the base.

On dead branches of *Clematis, Sambucus* and *Ulex* and dead stems of *Eupatorium;* Europe.

Sporidesmium macrurum (Sacc.) M. B. Ellis, 1958, *Mycol. Pap.,* **70**: 53–54.
(Fig. 78 C)

Colonies black. *Mycelium* partly superficial, partly immersed. *Conidiophores* brown, 30–170 × 5–6μ, swollen to 7–7·5μ just below the conico-truncate apex, often percurrent. *Conidia* straight or curved, obclavate, rostrate, conico-truncate and protuberant at the base, 3–4-septate, proximal cells dark brown, smooth or verruculose, distal cells becoming paler upwards, smooth, 40–55μ long, 9–11μ thick in the broadest part, 1–2μ near the apex, 2·5–3·5μ wide at the base.

On leaves and leaf-stalks of palms including *Areca, Borassus, Cocos, Elaeis, Licuala, Mauritia* and *Phoenix;* Brazil, British Solomon Islands, Ceylon, India, Java, Malaya, New Guinea, Papua, Sierra Leone.

Sporidesmium tropicale M. B. Ellis, 1958, *Mycol. Pap.,* **70**: 58–60.
(Fig. 78 D)

Colonies blackish brown. *Mycelium* partly superficial, partly immersed. *Conidiophores* brown, 40–340 × 5–7μ. *Conidia* straight or curved, obclavate,.

rostrate, conico-truncate at the base, pale to mid brown, proximal part usually verruculose, 7–19-septate, 80–230μ long, 12–15μ thick in the broadest part, 2–4μ near the apex, 3·5–4·5μ wide near the base, septa averaging 10·4μ apart.

On dead branches of many different shrubs and trees; Bolivia, Ceylon, India, Ghana, Jamaica, Malaya, Nigeria, Sierra Leone, U.S.A.

Sporidesmium leptosporum (Sacc. & Roum.) Hughes, 1958, *Can. J. Bot.*, **36**: 808.

(Fig. 78 E)

Colonies dark brown. *Mycelium* mostly immersed. *Conidiophores* mid to dark brown, 35–100 × 3·5–5μ, often percurrent. *Conidia* narrowly obclavate, truncate or conico-truncate at the base, subhyaline to pale straw-coloured, often brown at the base, 5–21-septate or pseudoseptate, 25–90 (rarely—130)μ long, 5–7 (rarely—8)μ thick in the broadest part, 3–4μ wide at the base, septa averaging 5·5–6·5μ apart.

On cupules of *Fagus*, leaves of *Ammophila*, and dead wood and bark of *Averrhoa*, *Buxus*, *Coffea*, *Elaeis*, *Fagus*, *Fraxinus*, *Morus*, *Polygonum*, *Quercus*, *Sambucus*, *Thea*, *Ulmus* and *Uvaria*; Europe, Nepal, Pakistan, Sierra Leone and Venezuela.

Sporidesmium eupatoriicola M. B. Ellis, 1958, *Mycol. Pap.*, **70**; 67.

(Fig. 78 F)

Colonies blackish brown to black. *Mycelium* immersed. *Conidiophores* blackish brown to black, 70–150μ × 5–7μ, often percurrent. *Conidia* obclavate, conico-truncate at the base, dark brown, smooth, 14–31-septate, 60–195μ long, 8–11μ thick in the broadest part, 4–6μ at apex and base; septa averaging 6·2μ apart.

On dead stems of *Eupatorium* and *Filipendula* and branches of *Betula* and *Ochthocosmus;* Europe, Sierra Leone.

Sporidesmium pedunculatum (Peck) M. B. Ellis, 1958, *Mycol. Pap.*, **70**: 67–68.

(Fig. 78 G)

Colonies grey to black. *Mycelium* mostly immersed. *Conidiophores* dark reddish brown to black, 25–130 × 5–8μ, often percurrent. *Conidia* straight or curved, obclavate, conico-truncate at the base, smooth, rather dark reddish brown, distal cell often subhyaline, 9–26-septate, 50–130μ long, 11–13μ thick in the broadest part, 4–7μ near the apex, 4–6μ wide at the base, septa averaging 5·4μ apart.

On dead wood of *Populus*, *Taxus*, etc.; Europe, N. America.

Sporidesmium inflatum (Berk. & Rav.) M. B. Ellis, 1958, *Mycol. Pap.*, **70**: 70–72.

(Fig. 78 H)

Colonies olivaceous grey or olivaceous brown. *Mycelium* superficial. *Conidiophores* pale to mid brown, 80–300 × 5–9μ, often percurrent. *Conidia* usually sigmoid when mature, obclavate or subfusiform, conico-truncate at the base, 3–5-septate, often constricted at the septa, smooth, subhyaline or pale brown except for the second and occasionally the third cell which are mid to dark brown, 45–90μ long, 11–18μ thick in the broadest part, 3–5μ near the apex, 6–9μ wide at the base; septa averaging 16μ apart.

On dead branches and leaves, usually associated with and growing over pycnidia and perithecia of other fungi; Guinea, N. America, Sierra Leone, Uganda.

74. CLASTEROSPORIUM

Clasterosporium Schweinitz, 1832, *Trans. Am. phil. Soc.*, N.S., **4**: 300–301.
Cometella Schw. in Fr., 1835, Corpus Florarum Provincialium Sueciae, **1**, Floram Scanicam, Upsala: 361–362.
Hymenopodium Corda, 1837, Icon. Fung., **1**: 7.
Cheiropodium Syd., 1915, *Annls mycol.*, **13**: 42–43.
Sporhelminthium Speg., 1918, *Physis*, **4** (17): 292.

Colonies usually effuse, dark brown to black, often velvety. *Mycelium* superficial. *Stroma* none. *Setae* present or absent, sometimes seen only in old colonies; when present simple, dark, smooth. *Hyphopodia* present. *Conidiophores* macronematous, mononematous, straight or flexuous, unbranched, mid to dark brown, smooth. *Conidiogenous cells* monoblastic, integrated, terminal, determinate or percurrent, cylindrical. *Conidia* solitary, acrogenous, simple, straight or curved, cylindrical or obclavate, sometimes rostrate, transversely septate, mid to dark brown, smooth, rugose or verrucose.

Type species: Clasterosporium caricinum Schw.

Ellis, M. B., *Mycol. Pap.*, **70**: 1–13, 1958; **72**: 71–73, 1959; **93**: 23–24, 1963.

KEY

Conidia strongly constricted at septa on inner curved side only	*anomalum*
Conidia not strongly constricted at septa on inner curved side only . . .	1
1. Hyphopodia lobed	2
Hyphopodia not lobed	5
2. Mature conidia smooth	3
Mature conidia verrucose	4
3. Hyphopodia generally not deeply lobed	*caricinum*
Hyphopodia always very deeply lobed	*flagellatum*
4. Setae present, numerous	*cyperi*
Setae absent	*scleriae*
5. Conidia cylindrical, rounded at apex	*Asterodothis solaris*
Conidia obclavate	6
6. Conidia strongly constricted at septa, often rugose . . .	*cocoicola*
Conidia not strongly constricted at septa, not rugose . . .	7
7. Conidia 18–30×4–6μ	*pistaciae*
Conidia 45–85×7–9μ	*Asterina clasterosporium*

Clasterosporium caricinum Schw., 1832, *Trans. Am. phil. Soc.*, N.S., **4**: 300–301.
For synonymy see *Mycol. Pap.*, 70.

(Fig. 79 A)

Colonies effuse, black, velvety. *Mycelium* a repent network of branched and anastomosing, pale brown hyphae. *Setae* in old colonies up to 460μ long, 7–10μ thick at base, 5–7μ at apex. *Hyphopodia* brown, 14–24 × 10–20μ. *Conidiophores* up to 130μ long, 4–7μ thick. *Conidia* obclavate, 7–12-septate, 70–360μ long, 11–20μ thick in the broadest part, 4–8μ at base.

On leaves of various sedges, including *Carex acutiformis*, *C. riparia* and *C. elata*; usually found on marshes and fens which are subject to periodic flooding. Specimens seen from Canada, Czechoslovakia, Great Britain, U.S.A. Other

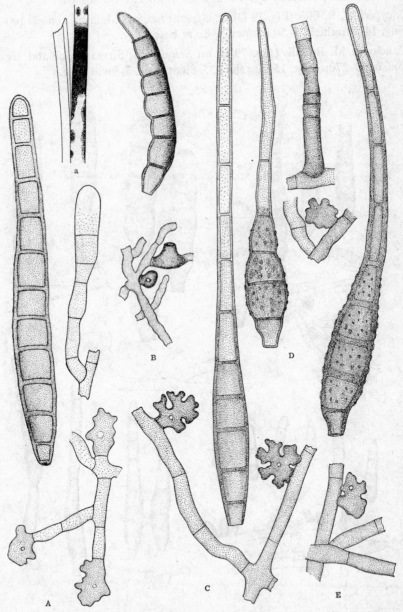

FIG. 79. *Clasterosporium* species (1): A, *caricinum*; B, *anomalum*; C, *flagellatum*; D, *cyperi*; E, *scleriae* (a, habit sketch; other figures × 650).

species keyed out above and more fully described in *Mycological Papers* are figured; substrata, distribution and measurements of conidia are given below.

C. anomalum (Speg.) Hughes (Fig. 79 B) on coriaceous leaves of a dicotyledonous plant; Brazil; *conidia* 65–120μ long, 10–12μ thick, 3–5μ at base.

C. flagellatum (Syd.) M. B. Ellis (Fig. 79 C) on *Carex breviculmis*; Japan; *conidia* 90–330μ long, 16–20μ thick, 6–8μ at apex, 7–8μ at base.

C. cyperi M. B. Ellis (Fig. 79 D) on *Cyperus haspan*; Malaya; *conidia* 100–210μ long, 14–19μ thick, 3–5μ at apex, 6–8μ at base.

C. scleriae M. B. Ellis (Fig. 79 E) on *Scleria* sp.; Sierra Leone and Uganda; *conidia* 60–270μ long, 15–20μ thick, 5–8μ at apex, 7–9μ at base.

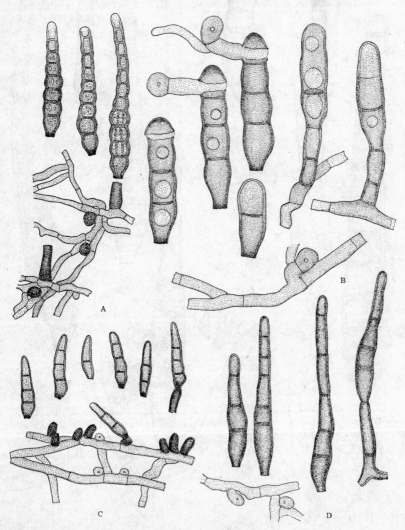

FIG. 80. *Clasterosporium* species (2): A, *cocoicola*; B, state of *Asterodothis solaris*; C, *pistaciae*; D, state of *Asterina clasterosporium* (× 650).

C. state of **Asterodothis solaris** (Kalchbr. & Cooke) Theiss. (Fig. 80 B) on *Olea* spp.; S. Africa, Sudan, Tanzania, Uganda; *conidia* 40–65µ long, 9–13µ thick, 6–8µ at base.

C. cocoicola M. B. Ellis & D. Shaw (Fig. 80 A) on *Cocos nucifera*: New Guinea; *conidia* 35–70µ long, 7–9µ thick, 4–5µ at apex, 4–4·5µ at base.

C. pistaciae M. B. Ellis (Fig. 80 C) on *Pistacia lentiscus*; Cyprus; *conidia* 18–30µ long, 4–6µ thick, 2–3·5µ at apex, 2–3µ at base.

C. state of Asterina clasterosporium Hughes (Fig. 80 D) on *Maba warneckei*; Dahomey, Ghana; *conidia* 45–85µ long, 7–9µ thick, 4–5µ at apex, 3·5–4µ at base.

75. ANNELLOPHORA

Annellophora Hughes emend. M. B. Ellis, 1958, *Mycol. Pap.*, **70**: 84.
  Chaetotrichum Syd. non Rabenh., 1927, *Annls mycol.*, **25**: 150.

Colonies effuse, hairy, thin, brown to black. *Mycelium* superficial or immersed; hyphae subhyaline to brown or olivaceous brown. *Stroma* none. *Setae* and *hyphopodia* absent. *Conidiophores* macronematous, mononematous, simple, brown or dark brown, septate. *Conidiogenous cells* integrated, terminal,

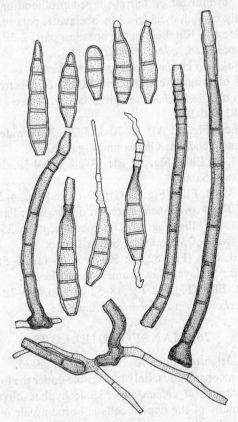

FIG. 81. *Annellophora solani* (× 650).

monoblastic, percurrent. *Primary conidia* terminal, cylindrical, obclavate or fusiform, subhyaline to brown, smooth, transversely septate or pseudoseptate, often germinating in situ at the apex to form a short, percurrent secondary conidiophore which bears shorter secondary conidia one at a time at the apices of successive proliferations.

Type species: Annellophora solani (H. Syd.) Hughes.

Ellis, M. B., *Mycol. Pap.*, **70**: 84–89, 1958; **82**: 44–45, 1961; **103**: 36–37, 1965. Hughes, S. J., *Trans. Br. mycol. Soc.*, **34**: 544–550, 1951.

Annellophora solani (H. Syd.) Hughes, 1951, *Trans. Br. mycol. Soc.*, **34**: 544. *Chaetotrichum solani* H. Syd., 1927, *Annls mycol.*, **25**: 150.

(Fig. 81)

Colonies hypophyllous, dark brown to black, effuse, sometimes covering nearly the whole of the lower leaf surface. *Mycelium* superficial, composed of a repent network of branched and anastomosing, subhyaline to brown, smooth, 2–4μ thick hyphae. *Conidiophores* arising singly or in small groups often from swollen, darker cells, erect, simple, straight or flexuous, mid to dark brown, pale at apex, septate, up to 150μ long, 5–6μ thick, percurrent, with up to 6 successive short, cylindrical or barrel-shaped proliferations. *Primary conidia* straight or slightly curved, cylindrical to obclavate, pale brown to brown, 2–6-septate, 25–50μ long, 8–10μ thick in the broadest part, 2–3μ wide at the truncate base. Up to 7 smaller *secondary conidia* formed.

On leaves of *Solanum erythrotrichum;* Costa Rica.

Measurements of primary conidia and details of substrata and distribution of 6 other species of *Annellophora* are given below. These are fully described in *Mycol. Pap.* **70, 82** and **103**.

A. africana Hughes (Fig. 82 A): 25–70 × 5–7μ, 3–4μ wide at base; on *Anthocleista, Bridelia* and *Toddalia*; Ghana and Uganda.

A. borneoensis M. B. Ellis (Fig. 82 B): 18–28 × 4–4·5μ, 3μ wide at base; on leaves of *Theobroma*; Sabah.

A. dendrographii M. B. Ellis (Fig. 82 C): 30–62 × 8–9μ, 4–5μ wide at base; on conidiophores of *Dendrographium atrum* on dead wood; Paraguay.

A. mussaendae M. B. Ellis (Fig. 82 D): pseudoseptate, 35–70 × 9–11μ, 3–5μ wide at base; on twigs of *Mussaenda*; Sierra Leone.

A. phoenicis M. B. Ellis (Fig. 82 E): 50–70 × 10–11μ, 3–5μ wide at base; on *Cocos* and *Phoenix*; Malaya, New Guinea, Sierra Leone.

A. sydowii M. B. Ellis (Fig. 82 F): 18–32 × 3·4μ, 1·5–2μ wide at base; on *Sporidesmium baccharidis* on *Baccharis*; Ecuador.

76. HANSFORDIELLOPSIS

Hansfordiellopsis Deighton, 1960, *Mycol. Pap.*, **78**: 33–35.

Colonies effuse, inconspicuous, only just visible under the low-power binocular dissecting microscope. *Mycelium* superficial; hyphae olivaceous. Near the distal septum of many of the hyphal cells is borne a pair of small prominent, hemispherical lateral cells or one such cell; these cells resemble hyphopodia

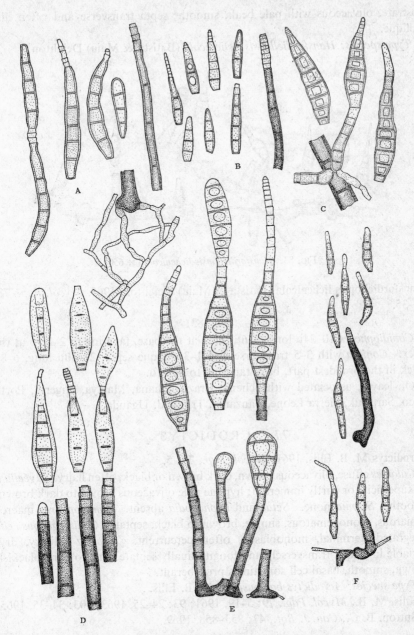

FIG. 82. *Annellophora* species: A, *africana*; B, *borneoensis*; C, *dendrographii*; D, *mussaendae*; E, *phoenicis*; F, *sydowii* (× 650).

but are without haustoria. *Stroma* none. *Setae* and *hyphopodia* absent. *Conidiophores* macronematous, mononematous, straight, erect, unbranched, olivaceous, smooth. *Conidiogenous cells* monoblastic, integrated, terminal, determinate, obclavate or subulate. *Conidia* solitary, dry, acrogenous, simple, obclavate, rostrate, olivaceous with pale beak, smooth; septa transverse and often also oblique.

Type species: Hansfordiellopsis lichenicola (Batista & Maia) Deighton.

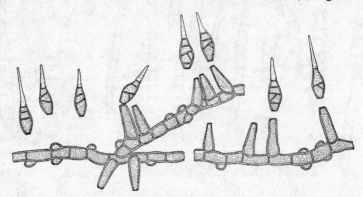

FIG. 83. *Hansfordiellopsis lichenicola* (× 650).

Hansfordiellopsis lichenicola (Batista & Maia) Deighton, 1965, *Mycol. Pap.*, **78**: 34–35.

(Fig. 83)

Conidiophores 10–24μ long, 4–6μ thick at the base, tapering to 2–2·5μ at the apex. *Conidia* with 3–5 transverse and 1–3 oblique septa, 20–50μ long, 6–8μ thick in the broadest part, beak tapering to 0·5–1μ.

On leaves, associated with lichens; Brazil, Ghana, Malaya, Nigeria, Porto Rico, Sarawak, Sierra Leone, Tanzania, Trinidad, Uganda.

77. ACRODICTYS

Acrodictys M. B. Ellis, 1961, *Mycol. Pap.*, **79**: 5.

Colonies effuse, olivaceous brown, dark brown or black, often hairy. *Mycelium* all superficial or partly immersed; hyphae pale olivaceous brown to dark brown, smooth. *Stroma* none. *Setae* and *hyphopodia* absent. *Conidiophores* macronematous, mononematous, simple, brown to black, septate. *Conidiogenous cells* integrated, terminal, monoblastic, often percurrent. *Conidia* solitary, dry, variable in shape, transversely and longitudinally septate, pale to dark blackish brown, smooth, basal cell sometimes protuberant.

Type species: Acrodictys bambusicola M. B. Ellis.

Ellis, M. B., *Mycol. Pap.*, **79**: 5–19, 1961; **93**: 24–25, 1963; **103**: 34–35, 1965.
Sutton, B. C., *Can. J. Bot.*, **47**: 853–858, 1969.

Acrodictys bambusicola M. B. Ellis, 1961, *Mycol. Pap.*, **79**: 6.

(Fig. 84)

Colonies effuse, dark blackish brown to black. *Mycelium* partly superficial, partly immersed in the substratum, composed of a network of branched, septate, pale to mid brown, smooth, 1–5μ thick hyphae. *Conidiophores* arising singly and laterally on the hyphae, erect, straight or flexuous, brown to dark brown, smooth, septate, up to 280μ long, 5–8μ thick, tapering to 2–4μ at the apex, percurrent with up to 5 proliferations. *Conidia* solitary, terminal, broadly clavate or pyriform, pale to dark brown, smooth, with 2–5 transverse and usually 1 or more longitudinal septa, often slightly constricted at the septa, 17–36μ long, 12–18μ thick in the broadest part; basal cell obconical, pale brown, truncate at the base, 2–4μ wide.

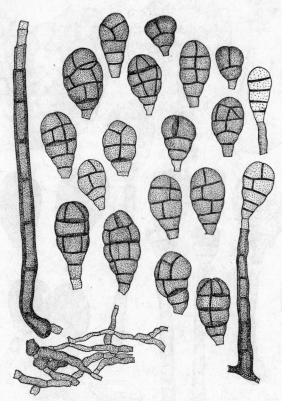

FIG. 84. *Acrodictys bambusicola* (× 650).

On bamboo and *Pennisetum*; Uganda and Venezuela.

Measurements of conidia and details of substrata and distribution of 10 other species of *Acrodictys* are given below; these have been more fully described in *Mycol. Pap.* **79, 93** and **103**.

A. balladynae (Hansf.) M. B. Ellis (Fig. 85 A): 18–25 × 13–19μ, base 2μ wide; on *Balladyna* and *Balladynopsis*; Ghana, Sierra Leone, Uganda.

A. brevicornuta M. B. Ellis (Fig. 85 B): 40–60 × 25–58μ; on bamboo; Venezuela.

A. deightonii M. B. Ellis (Fig. 85 C): 40–86 × 30–55μ, basal cell protruding, 4–6μ wide; on dead branches of *Cassia, Gardenia* and *Rauwolfia*; Sierra Leone.

A. dennisii M. B. Ellis (Fig. 85 D): 26–57 × 19–30μ, basal cell protruding, 3–5μ wide; on bamboo; Venezuela.

A. elaeidicola M. B. Ellis (Fig. 85 E): 17–26 × 11–19μ, base 2–3μ wide; on oil palm; Ghana, Sierra Leone.

A. erecta (Ellis & Everh.) M. B. Ellis (Fig. 85 F): 24–40 × 15–22μ, base 3·5–4·5μ wide; on *Arundo* and maize; U.S.A., Venezuela.

A. fimicola M. B. Ellis & Gunnell (Fig. 85 G): 15–24 × 11–15μ; on elephant dung and bamboo culms; Malaya, Sierra Leone.

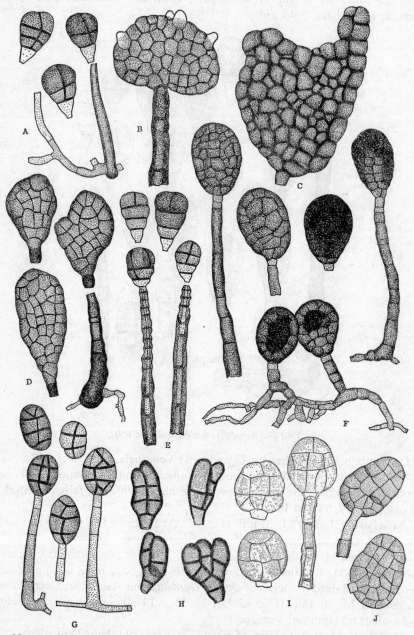

FIG. 85. *Acrodictys* species: A, *balladynae*; B, *brevicornuta*; C, *deightonii*; D, *dennisii*; E, *elaeidicola*; F, *erecta*; G, *fimicola*; H, *furcata*; I, *globulosa*; J, *obliqua* (× 650).

A. furcata M. B. Ellis (Fig. 85 H): 27–37 × 11–21μ, 4–5μ wide at the base; overgrowing *Balladyna* on *Canthium*; Uganda.

A. globulosa (Toth) M. B. Ellis (Fig. 85 I): 22–27 × 17–23μ, basal cell protruding, 5–6μ wide; on *Clematis* and bark of a tree; Hungary, Sierra Leone.

A. obliqua M. B. Ellis (Fig. 85 J): 24–33 × 18–26μ, basal cell protruding, 3·5–5μ wide; on dead wood and perithecia of *Calyculosphaeria*; Ghana.

78. XENOSPORIUM

Xenosporium Penzig & Saccardo, 1902, *Malpighia*, **15**: 248.

Xenosporella Höhnel, 1923, *Zentbl. Bakt. ParasitKde*, Abt. 2, **60**: 17–18.

Colonies punctiform or effuse, grey to black, velvety. *Mycelium* mostly superficial. *Stroma* none. *Setae* and *hyphopodia* absent. *Conidiophores* macro-

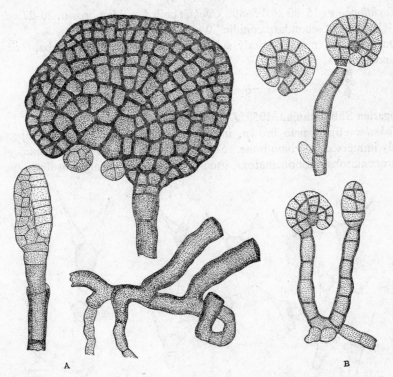

Fig. 86. A, *Xenosporium mirabile*; B, *X. berkeleyi* (× 650).

nematous, mononematous, usually scattered, occasionally caespitose, mostly unbranched but sometimes branched, straight or flexuous, hyaline to dark brown, smooth. *Conidiogenous cells* monoblastic, integrated, terminal, determinate or percurrent, cylindrical. *Conidia* solitary, dry, acrogenous, simple, usually helicoid, curved dorsiventrally and flattened from side to side, pale to dark brown, smooth, muriform, bearing along the inner curved side 1 or more spherical or subspherical, smooth, brown, either 0-septate or muriform ' secondary conidia '.

Type species: Xenosporium mirabile Penz. & Sacc.

Ellis, M. B., *Mycol. Pap.*, **87**: 13–15, 1963.

Pirozynski, K. A., *Mycol. Pap.*, **105**: 21–35, 1966; with key and descriptions of 6 species.

Xenosporium mirabile Penz. & Sacc., *Malpighia*, **15**: 248.
(Fig. 86 A)

Conidiophores up to 110μ long, 7–11μ thick. *Conidia* 60–105μ long, 30–70μ broad, 8–18μ thick; ' secondary conidia ' muriform, 7–20μ diam.

On rotten leaves and stems of *Elettaria*; Java.

Xenosporium berkeleyi (Curtis) Pirozynski, 1966, *Mycol. Pap.*, **105**: 27–29.
(Fig. 86 B)

Conidiophores 15–80 × 4·5–8μ. *Conidia* coiled, golden brown, 20–27μ diam., 6·5–11μ thick; ' secondary conidia ' 0-septate, 3–7μ diam.

On decaying, mostly woody substrata; Bermuda, Brazil, Cuba, Trinidad, Uganda, U.S.A.

79. IYENGARINA

Iyengarina Subramanian, 1958, *J. Indian bot. Soc.*, **37**: 404–407.

Colonies effuse, pale brown, inconspicuous. *Mycelium* partly superficial, partly immersed. *Stroma* none. *Setae* and *hyphopodia* absent. *Conidiophores* macronematous, mononematous, usually unbranched, straight or flexuous, pale

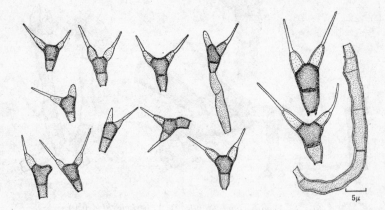

FIG. 87. *Iyengarina elegans* (× 650 except where indicated by the scale).

to dark brown, smooth. *Conidiogenous cells* monoblastic, integrated, terminal, determinate or percurrent, cylindrical. *Conidia* solitary, acrogenous, Y-shaped, each of the two divergent branches arising from the upper part of the conidium ends in a narrow, pointed appendage; body of conidium smooth, unequally coloured, the upper cell usually dark brown, the basal cell pale or mid pale brown.

Type species: Iyengarina elegans Subram.

Iyengarina elegans Subram., 1958, *J. Indian bot. Soc.*, **37**: 404–407.
(Fig. 87)

Conidiophores up to 60μ long, 3–5μ thick. *Conidia*: body 11–15μ long, 6–8μ thick in the broadest part, 2–4μ wide at the base; branches up to 35μ long but usually 10–20μ, 3–4μ thick at the base.

On dead branches; India.

80. TERATOSPERMA

Teratosperma Sydow, 1909, *Annls mycol.*, **7**: 172–173.

Colonies orbicular, 1–3 mm. diam. or effuse, olivaceous brown to black. *Mycelium* superficial and immersed; hyphae in some species thick. *Stroma* none. *Setae* and *hyphopodia* absent. *Conidiophores* macronematous, mononematous, scattered, unbranched, straight or flexuous, pale to mid brown or olivaceous brown, smooth. *Conidiogenous cells* monoblastic, integrated, terminal, percurrent, with well-defined annellations. *Conidia* solitary, dry, acrogenous, appendiculate, mid to dark brown or olivaceous brown, smooth; the main body obclavate, sometimes rostrate, appendages thick, subulate or conical, arising singly, in pairs or threes from the proximal or occasionally the second cell.

Type species: Teratosperma singulare Syd.

Ellis, M. B., *Mycol. Pap.*, **69**: 1–7, 1957.

KEY

Conidia 50–200μ long 1
Conidia not more than 50μ long 3
1. Conidial appendages less than 10μ long . . . . . . *pulchrum*
   Conidial appendages more than 10μ long 2
2. Conidial appendages 5–12μ thick. . . . . *singulare*
   Conidial appendages 2–4μ thick . . . . *cornigerum*
3. Conidia 7–9μ thick . . . . *appendiculatum*
   Conidia 4·5–7μ thick . . . . *anacardii*

Teratosperma singulare Syd., 1909, *Annls mycol.*, **7**: 172–173.
(Fig. 88 A)

Colonies epiphyllous, 1–3 mm. diam., black. *Hyphae* 3–10μ thick. *Conidiophores* pale olivaceous brown, 12–45 × 6–10μ with up to 9 annellations. *Conidia* anchor-like, the main body of the conidium resembling that of *Clasterosporium* but with the proximal or occasionally the second cell bearing 1–3 lateral appendages; body straight, obclavate, truncate at the base, dark olivaceous brown, subhyaline and sometimes rostrate at the apex, 3–10-septate, 60–130μ long, 17–22μ thick in the broadest part, 7–10μ wide at the base; appendages subulate, subhyaline to pale brown, 0–3-septate, up to 55μ long, 5–12μ thick.

On leaves of *Ulmus*, associated with a pycnidial fungus; Japan.

Main measurements of the body of the conidium, together with details of substrata and distribution for 4 other species of *Teratosperma* are given below.

T. pulchrum (Ellis & Everh.) M. B. Ellis (Fig. 88 B): 70–180 × 15–19μ, on bark of *Carpinus* and perithecia of *Nummularia*; N. America. Possibly the same as the next species but the type specimens look different.

T. cornigerum (Ellis & Everh.) M. B. Ellis (Fig. 88 C): 70–200 × 14–17µ, on bark and cortex of *Carpinus* and perithecia; N. America.

T. appendiculatum (Hughes) M. B. Ellis (Fig. 88 D): 17–50 × 7–9µ, on blue-green algae on leaves of *Culcasia*; Ghana.

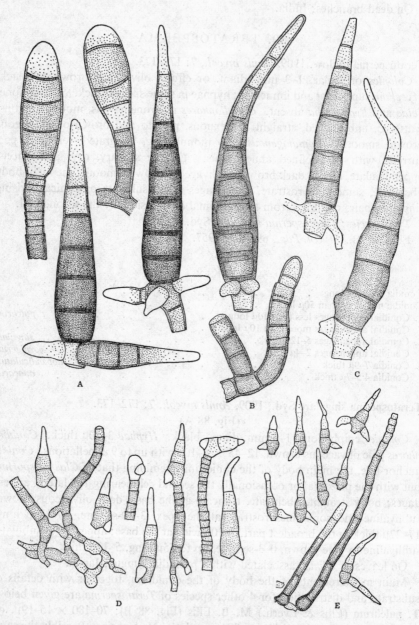

FIG. 88. *Teratosperma* species: A, *singulare*; B, *pulchrum*; C, *cornigerum*; D, *appendiculatum*; E, *anacardii* (× 650).

T. anacardii Hansf. (Fig. 88 E): 17–30 × 4·5–7μ, on foliicolous lichens; Ghana, Sierra Leone, Uganda.

81. CERATOSPORELLA

Ceratosporella Höhnel, 1919, *Ber. dt. bot. Ges.*, **37**: 155.

Colonies effuse, dark brown or black, velvety. *Mycelium* mostly immersed. *Stroma* none or small and composed of just a few cells. *Setae* and *hyphopodia* absent. *Conidiophores* macronematous, mononematous, straight, unbranched, mid to dark brown, smooth. *Conidiogenous cells* integrated, terminal, monoblastic, percurrent, cylindrical. *Conidia* solitary, acrogenous, branched, brown or dark brown, smooth; branches pluriseptate.

Type species: Ceratosporella bicornis (Morgan) Höhnel.

Hughes, S. J., *Mycol. Pap.*, **46**: 22–25, 1951; *Trans. Br. mycol. Soc.*, **35**: 243–247, 1952.

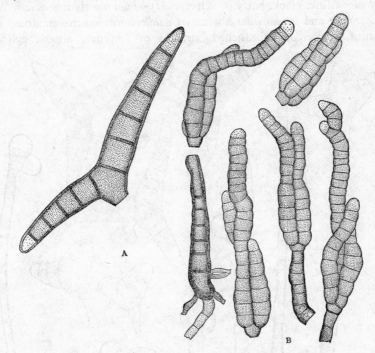

FIG. 89. A, *Ceratosporella bicornis*; B, *C. stipitata* (× 650).

Ceratosporella bicornis (Morgan) Höhnel, 1923, *Zentbl. Bakt. ParasitKde*, Abt. 2, **60**: 5–6.

(Fig. 89 A)

Conidiophores up to 170μ long, 7–8μ thick. *Conidia* with 2 tapering, 5–7-septate branches 50–80μ long, 9–13μ thick at base, splaying out from a central cell with truncate hilum 7–9μ wide.

On old corn (*Zea*) stalks; U.S.A.

Ceratosporella stipitata (Goidànich) Hughes, 1952, *Trans. Brit. mycol. Soc.*, **35**: 243.

(Fig. 89 B)

Hyphae 2–4μ thick. *Conidiophores* subulate, very dark brown, closely septate, up to 110μ long, 7–10μ thick just above the swollen base, tapering to 5–6μ. *Conidia* brown, smooth, composed of a basal cell 5–6 × 5–7μ which bears a whorl of 2–5 (usually 3–4) branches which are often closely adpressed along part of their length; branches up to 130μ long, 6–9μ thick, up to 32-septate, constricted at the septa.

On dead branches; specimens seen on hornbeam (*Carpinus*) and sweet chestnut (*Castanea*) from Great Britain.

82. TRIPOSPORIUM

Triposporium Corda, 1837, Icon. Fung., **1**: 16.

Colonies effuse, black, hairy or velvety. *Mycelium* mostly immersed. *Stroma* none. *Setae* and *hyphopodia* absent. *Conidiophores* macronematous, mononematous, scattered, unbranched, straight or flexuous, almost cylindrical,

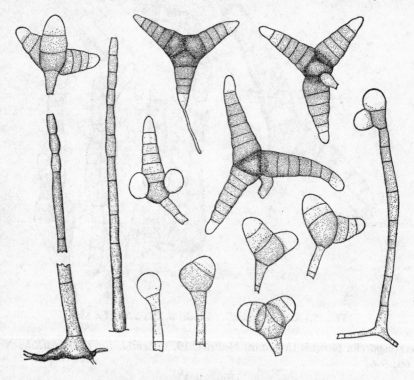

FIG. 90. *Triposporium elegans* (× 650).

broadened at the base to form a flat plate, brown, smooth. *Conidiogenous cells* monoblastic, integrated, terminal, percurrent, cylindrical, doliiform or lageniform. *Conidia* solitary, dry, acrogenous, branched, usually made up of a small, clavate, doliiform or cylindrical stalk cell and 3 or occasionally 4 conical, smooth, septate arms joined by their wide, rounded bases; the arms are dark brown near the centre of the conidium, hyaline or subhyaline at the tips.

Type species: Triposporium elegans Corda.

Ellis, M. B., E. A. & J. P., *Trans. Br. mycol. Soc.*, **34**: 161–163, 1951.

Hughes, S. J., *Mycol. Pap.*, **46**: 1–10, 1951.

Triposporium elegans Corda, 1837, Icon. Fung., **1**: 16.
(Fig. 90)

Conidiophores up to 230μ long, 2·5–5μ thick at the apex, 12–30μ at the base, 5–8μ immediately above the basal plate. *Conidia* with arms 3–9-septate, often slightly constricted at the septa, up to 70μ but mostly 20–40μ long, 9–12μ thick at the base, 3–5μ at the tip.

Common on dead wood and herbaceous stems; hosts include *Alnus, Cladium, Corylus, Epilobium, Fagus, Filipendula, Fraxinus, Ilex, Prunus, Quercus, Rhododendron, Rubus* and *Ulex*; Europe, N. America.

83. ACTINOCLADIUM

Actinocladium Ehrenb. ex Pers.; Ehrenberg, 1819, *Jb. Gewächskde*, **1**: 52; Persoon, 1822, Mycol. eur., **1**: 31; Fries, 1832, Syst. mycol. **3**: 352.

Colonies effuse, hairy, dark brown to black. *Mycelium* all immersed or partly superficial; hyphae pale to mid brown. *Stroma* none. *Setae* and *hyphopodia* absent. *Conidiophores* macronematous, mononematous, simple, brown, septate. *Conidiogenous cells* integrated, terminal, monoblastic, sometimes percurrent. *Conidia* solitary, dry, brown, each composed of a 1–4-celled stalk surmounted by a group of cells from which usually 3, occasionally 2, septate arms radiate upwards and outwards at an angle of 30°–80°. A conidial initial, at first a swelling of the thin-walled apex of the conidiophore, is cut off from the conidiophore by a septum, 1–4 transverse septa dividing off the stalk cells; in the terminal swollen cell an oblique septum arises dividing it unequally and the larger cell becomes transversely septate and originates one of the arms. The other 2 arms develop as lateral swellings from the smaller cell which eventually becomes divided into two by a vertical septum.

Type species: Actinocladium rhodosporum Ehrenb. ex Pers.

Actinocladium rhodosporum Ehrenb. ex Pers.; Ehrenberg, 1819, *Jb. Gewächskde*, **1**: 52; Persoon, 1822, Mycol. eur., **1**: 32; Fries, 1832, Syst. mycol., **3**: 352.

Triposporium cambrense Hughes, 1951, *Mycol. Pap.*, **46**: 6.
(Fig. 91)

Colonies effuse, hairy, dark brown to black. *Mycelium* all immersed or partly superficial; hyphae branched, septate, pale to mid brown, 2–5μ thick. *Conidiophores* arising singly or in small groups terminally and laterally on the hyphae, erect, simple, straight or slightly flexuous, septate, dark brown below, paler above, smooth, up to 130μ long, 6–8μ thick at the base, tapering upwards to 2·5–4μ, sometimes percurrent. *Conidia* formed singly as blown-out ends at the

apex of each conidiophore which, after the first conidium has fallen, may proliferate straight on, forming another conidium at a higher level, pale to dark
brown, each composed of a 1–4-celled, 7–23µ long stalk 3–4µ wide at the base,
surmounted by a group of cells 9–12µ thick, from which 3, or occasionally 2,

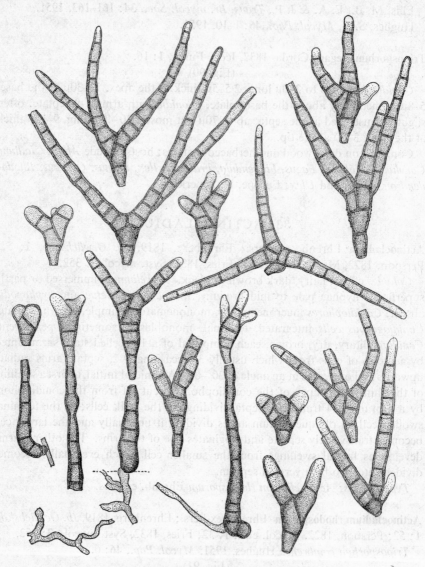

FIG. 91. *Actinocladium rhodosporum* (× 650).

septate arms radiate upwards and outwards at an angle of 30°–80°; arms up
to 140µ long, 7–10µ thick at the base, 2–4·5µ at the apex.

On wood and bark of various trees and shrubs including ash, birch, box,
elm, gorse, hazel, hornbeam, oak, rowan, spruce; Germany, Great Britain,
Portugal; also one collection on *Citrus limetta* from Sierra Leone.

84. ARACHNOPHORA

Arachnophora Hennebert, 1963, *Can. J. Bot.*, **41**: 1165.

Colonies small, brown or blackish. *Mycelium* superficial. *Stroma* none. *Setae* and *hyphopodia* absent. *Conidiophores* macronematous, mononematous, simple, brown, septate. *Conidiogenous cells* integrated, terminal, monoblastic, percurrent. *Conidia* solitary, terminal, complex, with a central brown, often 1-septate body and lateral protuberances bearing incurved, subhyaline, claw- or spine-like processes.

Type species: Arachnophora fagicola Hennebert.

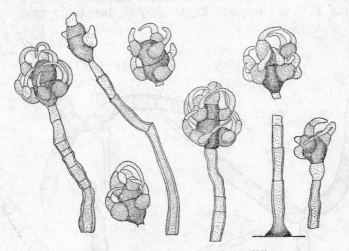

FIG. 92. *Arachnophora fagicola* (× 650).

Arachnophora fagicola Hennebert, 1963, *Can. J. Bot.*, **41**: 1166.
(Fig. 92)

Colonies small, brown or blackish. *Mycelium* superficial; hyphae pale, about 2µ thick. *Conidiophores* arising singly or in groups, usually simple, straight, septate, percurrent, lower part brown or dark brown, upper part paler, up to 150µ long, 4–6µ thick, swollen at the base to 9–13µ and often lobed. *Conidia* solitary, terminal, complex, with a central body which is usually 1-septate, smooth, 16–20 × 10–13µ; each of the central cells bears several pale brown lateral protuberances 4–7µ diam. which themselves each bear 1 or a number of inwardly curved, hyaline, claw- or spine-like processes 10–25µ long and 1–3µ thick.

On cupules of beech; Belgium.

85. GRALLOMYCES

Grallomyces F. L. Stevens, 1918, *Bot. Gaz.*, **65**: 245.
Phialetea Batista & Nascimento, 1960, *Atas Inst. Micol.*, **1**: 264–266.
Ophiopodium Arnaud, 1954, *Bull. trimest. Soc. mycol. Fr.*, **69**: 300.

Colonies effuse, sometimes covering most of the leaf surface, pale to mid olivaceous brown, arachnoid. *Mycelium* superficial, composed of a chain-like series of slightly curved, olivaceous brown, minutely verruculose, multiseptate segments, each with a narrow, smooth, stalked attachment organ at one end which passes down to the leaf surface where it terminates in a vesicle. *Stroma* none. *Setae* and true *hyphopodia* absent. *Conidiophores* macronematous, mononematous, unbranched, straight, olivaceous brown, smooth. *Conidiogenous cells* monoblastic, integrated, terminal, percurrent, cylindrical. *Conidia* solitary, dry, acrogenous, branched, gently curved, brown, minutely verruculose, multiseptate, with a stalked attachment organ at the end of the main axis and each branch.

Type species: Grallomyces portoricensis F. L. Stevens.

Deighton, F. C. & Pirozynski, K., *Mycol. Pap.*, **105**: 10–17, 1966.

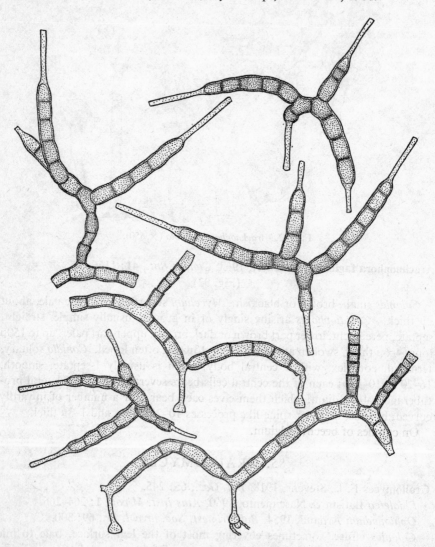

FIG. 93. *Grallomyces portoricensis* (× 650).

Grallomyces portoricensis F. L. Stevens, 1918, *Bot. Gaz.*, **65**: 245–246.
(Fig. 93)

Hyphal segments usually 50–90 × 5–8μ, attachment organs 25–45 × 2–3μ, vesicles 5–7μ diam. *Conidiophores* up to 70μ long but usually about 30μ, 4–5μ thick. *Conidia* 50–75 × 5–8μ, branches 20–55 × 5–8μ.

Very common in tropical countries on leaves of many different kinds of plants, often overgrowing but apparently not parasitic on other micro-fungi.

86. POLLACCIA

Pollaccia Baldacci & Ciferri, 1937, *Atti Ist. bot. Univ. Lab. crittogam. Pavia*, Ser. 4, **10**: 55–172, 1938 (but first appeared as a separate in 1937).

Sporodochia scattered, olivaceous, sometimes confluent. *Mycelium* immersed, forming radiating hyphal plates. *Stroma* present, irregularly erumpent, composed of cells rather loosely aggregated in or below the epidermis. *Setae* and

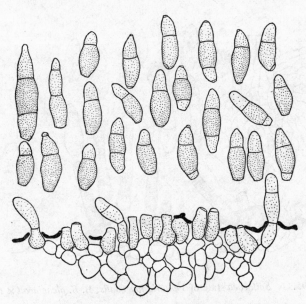

FIG. 94. *Pollaccia radiosa* (× 650).

hyphopodia absent. *Conidiophores* macronematous, mononematous, crowded, unbranched, more or less straight, hyaline to olive, smooth. *Conidiogenous cells* monoblastic, integrated, the whole conidiophore often composed of a single large conidiogenous cell, determinate or percurrent, cylindrical, ampulliform or doliiform. *Conidia* solitary, dry, acrogenous, simple, cylindrical, rounded at the apex and truncate at the base, ellipsoidal or obclavate, hyaline or olivaceous, smooth or verrucose, 1–2-septate, often constricted at the septa.

Type species: Pollaccia radiosa (Lib.) Baldacci & Cif.

Hughes, S. J., *Can. J. Bot.*, **31**: 572–574, 1953.

Pollaccia radiosa (Lib.) Baldacci & Cif., 1937, *Atti Ist. bot. Univ. Lab. crittogam. Pavia*, Ser. 4, **10**: 55–172.

(Fig. 94)

Conidiophores 8–17 × 3–7μ. *Conidia* most commonly 2-septate, hyaline to pale olive or straw-coloured, smooth, 15–42 × 6–11μ, mostly 18–26 × 5–8μ, often with a minute frill at the base.

On living leaves of *Populus*, causing round or irregular, sometimes confluent, buff spots 1–2 cm. diam., each with a blackish purple margin; Europe, N. America.

87. SPILOCAEA

Spilocaea Fries ex Fries; Fries, 1919, Novitiae Florae Sueciae, **5**: 79; 1825, Syst. Orb. veg.: 198.

Cycloconium Castagne, 1845, Cat. Pl. Environs Marseilles: 220.

Napicladium Thümen, 1875, *Hedwigia*, **14**: 3–4.

Basiascum Cavara, 1888, *Atti Ist. bot. Univ. Lab. crittogam. Pavia*, Ser. 2, **1**: 433–434.

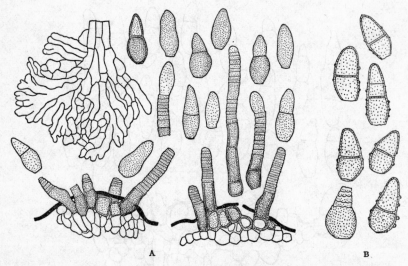

FIG. 95. A, *Spilocaea* state of *Venturia inaequalis*; B, *S. oleaginea* (× 650).

Colonies discrete and orbicular or effuse, olive, olivaceous brown, grey or black, sometimes velvety. *Mycelium* mostly subcuticular or intraepidermal, forming radiating plates. *Stroma* often formed, pseudoparenchymatous. *Setae* and *hyphopodia* absent. *Conidiophores* macronematous, mononematous, mostly unbranched, straight or flexuous, pale to dark brown or olivaceous brown, smooth. *Conidiogenous cells* monoblastic, integrated, terminal, percurrent, often with numerous distinct annellations, cylindrical, ampulliform or lageniform. *Conidia* solitary, dry, acrogenous, simple, usually obpyriform or obclavate, truncate at the base, pale to mid brown or olivaceous brown, smooth or verrucose, mostly 0–1-septate, occasionally with 2 or more septa, sometimes constricted at the septa.

Type species: Spilocaea state of *Venturia inaequalis* (Cooke) Wint. = *S. pomi* Fr. ex Fr.

Hughes, S. J., *Can. J. Bot.*, **31**: 560–565, 1953.

Spilocaea state of **Venturia inaequalis** (Cooke) Wint., 1875, apud Thüm. in *Mycotheca universalis*: 261.

(Fig. 95 A)

Colonies effuse, mid to dark olivaceous brown, velvety. *Conidiophores* cylindrical, pale to mid brown or olivaceous brown, up to 90μ long, 5–6μ thick, sometimes swollen to 10μ at the base. *Conidia* obpyriform or obclavate, pale to mid olivaceous brown, smooth, 0–1 septate, 16–24 (20·5)μ long, 7–10 (8·5)μ thick in the broadest part, truncate base 4–5μ wide.

On living leaves and fruit of *Malus* and *Pyrus*, the common apple scab fungus; cosmopolitan [CMI Distribution Map, 120].

Spilocaea oleaginea (Cast.) Hughes, 1953, *Can. J. Bot.*, **31**: 564–565.

(Fig. 95 B)

Colonies effuse, grey to black. *Conidiophores* ampulliform or lageniform, mid-pale olivaceous brown, smooth or verrucose, up to 30μ long, 8–15μ thick at the base, 5–7μ at the apex. *Conidia* obclavate, mid pale olivaceous brown, verrucose, mostly 1-septate, occasionally 2-septate, 20–30μ long, 10–13μ thick in the broadest part, truncate base 5–7μ wide.

On living leaves of *Olea*; Cyprus, Italy, Malta, Palestine.

88. OEDOTHEA

Oedothea Sydow, 1930, *Annls mycol.*, **28**: 202–204.

Colonies hypophyllous, dark brown, narrow, growing only on the leaf veins. *Mycelium* immersed. *Stroma* present, pseudoparenchymatous, very pale olivaceous brown. *Setae* and *hyphopodia* absent. *Conidiophores* macronematous, mononematous, closely packed together forming elongated sporodochia, unbranched, straight or flexuous, pale to mid olive or olivaceous brown, smooth.

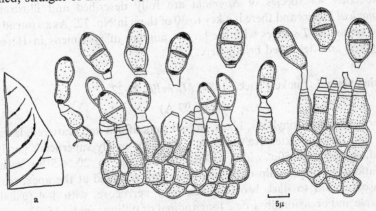

FIG. 96. *Oedothea vismiae* (a, habit sketch; other figures × 650).

Conidiogenous cells monoblastic, integrated, terminal, percurrent, cylindrical or doliiform. *Conidia* solitary, dry, acrogenous; when detached often carrying away a small part of the conidiophore, simple, ellipsoidal, pale olive with a small dark area at the apex, smooth, 1-septate.

Type species: Oedothea vismiae Syd.

Oedothea vismiae Syd., 1930, *Annls mycol.*, **28**: 202–204.
(Fig. 96)

Conidiophores 10–20 × 2–5µ. *Conidia* 9–13 × 5–8µ.
On leaves of *Vismia*; Venezuela.

89. STIGMINA

Stigmina Saccardo, 1880, *Michelia*, **2**: 22.
Thyrostroma Höhnel, 1911, *Sber. Akad. Wiss. Wien*, Abt. 1, **120**: 472–473.
Thyrostromella Sydow, 1924, *Annls mycol.*, **22**: 406–407 (non Höhnel, 1919).
Colonies usually punctiform, brown to black. *Mycelium* mostly immersed. *Stroma* always present, immersed, erumpent or almost all superficial, subhyaline to dark brown. *Setae* and *hyphopodia* absent. *Conidiophores* macronematous, mononematous, usually short and packed closely together forming pulvinate sporodochia, mostly unbranched, straight or flexuous, subhyaline to brown or olivaceous brown, smooth or verrucose. *Conidiogenous cells* monoblastic, integrated, terminal, percurrent, cylindrical, doliiform or lageniform. *Conidia* solitary, dry, acrogenous, simple, clavate, cylindrical, rounded at the apex, truncate at the base, ellipsoidal or obclavate, occasionally rostrate, subhyaline to dark brown, smooth, rugose, verrucose or echinulate, with 1 or more transverse and sometimes 1 or more oblique or longitudinal septa.

Type species: Stigmina platani (Fuckel) Sacc.

Ellis, M. B., *Mycol. Pap.*, **72**: 36–71, 1959; **82**: 47–48, 1961; **87**: 31–39, 1963; **93**: 32–33, 1963.

Hughes, S. J., *Mycol. Pap.*, **49**: 7–14 and 24–25, 1952.

Altogether 43 species of *Stigmina* are fully described and illustrated in *Mycological Papers* and there is a key to 30 of them in No. 72. As an introduction to the genus the 7 species with the largest number of specimens in Herb. IMI are figured and described briefly here.

Stigmina platani (Fuckel) Sacc., 1880, *Michelia*, **2**: 22.
(Fig. 97 A)

Sporodochia hypophyllous, black, at first punctiform and scattered, later confluent and forming extensive colonies. *Stromata* partly superficial, partly immersed, usually 40–50µ wide. *Conidiophores* up to 20 × 5–9µ with 0–5 annellations. *Conidia* ellipsoidal or cylindrical, rounded at the apex, truncate at the base, mid to dark brown, smooth or verrucose, with 1–4 (usually 3) transverse and occasionally 1 or 2 longitudinal or oblique septa, 15–27 × 8–11µ.

On leaves of *Platanus*; Europe, India, Pakistan, N. America.

Stigmina glomerulosa (Sacc.) Hughes, 1958, *Can. J. Bot.*, **36**: 814.
(Fig. 97 B)

Sporodochia dark olivaceous brown to black, punctiform. *Stromata* superficial, 80–250μ wide. *Conidiophores* 4–24 × 3–6μ, with up to 8 annellations.
Conidia straight or curved, cylindrical to fusiform, rounded at the apex, truncate

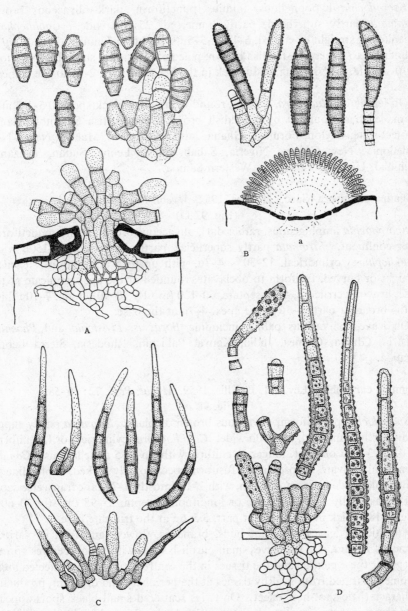

FIG. 97. *Stigmina* species (1): A, *platani*; B, *glomerulosa*; C, *mangiferae*; D, *palmivora* (a, sketch
of section, other figures × 650).

at the base, pale to mid brown, verruculose, almost always 7-septate, 27–50µ long, 6–10µ thick in the broadest part, 3–5µ at the base.

On leaves of *Juniperus communis*; Europe, N. America.

Stigmina mangiferae (Koorders) M. B. Ellis, 1959, *Mycol. Pap.*, **72**: 49–50.

(Fig. 97 C)

Sporodochia hypophyllous, minute, punctiform, dark olivaceous brown. *Stromata* partly superficial, partly immersed, 25–50µ wide. *Conidiophores* lageniform, straight or curved, 6–18 × 3–5µ, with up to 7 annellations. *Conidia* cylindrical to obclavate, reddish brown, pale near the apex, smooth or rugulose, 1–10-septate, 20–80µ long, 4–6µ thick in the broadest part, 2–4µ at the truncate base.

On leaves of *Mangifera*, causing round or angular black spots, 0·5–1·5 mm. diam., each surrounded by a reddish brown zone; British Solomon Islands Protectorate, Ceylon, Formosa, Ghana, Jamaica, Malawi, Malaya, Nepal, New Caledonia, New Guinea, Nigeria, Sabah, Sierra Leone, Sudan, Tanzania, Trinidad, Uganda, Venezuela, Western Samoa.

Stigmina palmivora (Sacc.) Hughes, 1952, *Mycol. Pap.*, **49**: 13.

(Fig. 97 D)

Sporodochia amphigenous, rather dark olivaceous brown, at first punctiform, later confluent. *Stromata* partly superficial, partly immersed, 25–150µ wide. *Conidiophores* cylindrical, 12–30 × 4–7µ, with up to 5 annellations. *Conidia* straight or curved, fusiform to obclavate, rounded at the apex, truncate at the base, brown, verrucose, 4–16-septate, 60–120 (mostly 70–90)µ long, 7–10µ thick in the broadest part, 3–6µ at the apex, 4–7µ at the base.

On leaves of various palms, including *Borassus*, *Livistonia* and *Phoenix*; Austria, Ghana, Guinea, India, Kenya, Pakistan, Rhodesia, Sierra Leone, Uganda.

Stigmina carpophila (Lév.) M. B. Ellis, 1959, *Mycol. Pap.*, **72**; 56–58.

(Fig. 98 A)

Sporodochia punctiform, olivaceous brown or black. *Stromata* partly superficial, partly immersed, 50–250µ wide. *Conidiophores* cylindrical or lageniform, 14–45 × 3–11µ on host, longer in culture, with up to 5 annellations. *Conidia* cylindrical, clavate, ellipsoidal or fusiform, occasionally forked, subhyaline to rather pale golden brown, smooth, with 2–11 (mostly 3–7) dark transverse septa and occasionally 1 or 2 oblique or longitudinal septa, 25–95 (mostly 30–60)µ long, 9–18µ thick in the broadest part, 3·5–6µ at the truncate base.

Parasitic on leaves, dormant buds, branches, flowers and fruit of various species of *Prunus*. On the leaves small purplish spots with yellow centres appear first; later these enlarge and the tissues in the centre turn brown. Infected areas become separated from healthy tissues at the periphery and drop out, producing a characteristic shot-hole effect. On twigs scattered small black spots appear, later these increase in size, the centres become pale and sink; longitudinal cracks may appear in the periderm across these lesions with exudation of gum.

Lesions occur abundantly on peach twigs where they cause severe blighting. Blighting of dormant buds is frequently very severe in the case of apricots, although it occurs also on almond and peach trees. The disease is important in parts of Europe, N. America, Africa, Australia and New Zealand [CMI Distribution Map, 188].

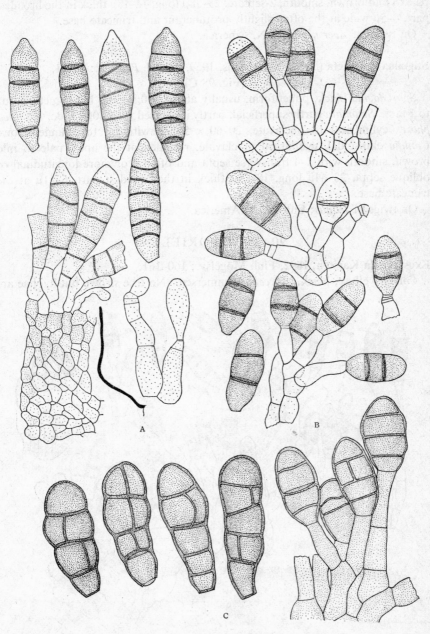

FIG. 98. *Stigmina* species (2): A, *carpophila*; B, *negundinis*; C, *compacta* (× 650).

Stigmina negundinis (Berk. & Curt.) M. B. Ellis, 1959, *Mycol. Pap.*, **72**: 44–45.
(Fig. 98 B)

Sporodochia black, punctiform, at first scattered, later sometimes confluent. *Stromata* mostly immersed, 200–400μ wide. *Conidiophores* cylindrical, doliiform or lageniform, 12–26 × 5–9μ, with up to 5 annellations. *Conidia* ellipsoidal, pale to mid brown, smooth, 2-septate, 25–38μ long, 12–18μ thick in the broadest part, 4–5μ wide at the often slightly protuberant and truncate base.

On twigs of *Acer negundo*; N. America.

Stigmina compacta (Sacc.) M. B. Ellis, 1959, *Mycol. Pap.*, **72**: 40–41.
(Fig. 98 C)

Sporodochia black, punctiform, usually about 500μ diam. but sometimes up to 1 mm. *Stromata* partly superficial, partly immersed, 200–800μ wide. *Conidiophores* cylindrical to obclavate, 20–60 × 5–12μ with up to 3 annellations. *Conidia* ellipsoidal, cylindrical or clavate, rounded at the apex, pale to mid brown, smooth with 2–4 transverse septa and often 1 or more longitudinal or oblique septa, 28–64μ long, 18–25μ thick in the broadest part, 6–11μ at the truncate base.

On twigs of *Ulmus*; Europe, N. America.

90. EXOSPORIELLA

Exosporiella Karsten, 1892, Finl. Mögelsv.: 160–161.

Colonies effuse, black. *Mycelium* immersed. No true *stroma* but hyphae are

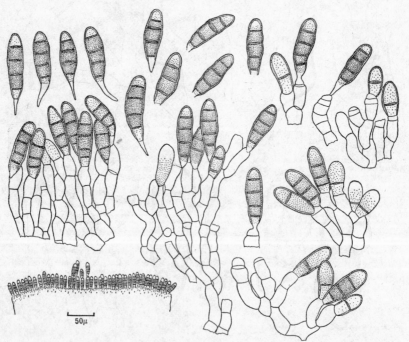

FIG. 99. *Exosporiella fungorum* (× 650 except where indicated by the scale).

closely interwoven to form a pseudostroma. *Setae* and *hyphopodia* absent. *Conidio-phores* close together forming a palisade, macronematous or semi-macronematous, mononematous, flexuous, branched, hyaline, smooth. *Conidiogenous cells* mono-blastic, integrated and terminal or discrete, percurrent, cylindrical or doliiform. *Conidia* solitary, dry, acrogenous, simple, straight or slightly curved, ellipsoidal or cylindrical, rounded at the apex, caudate at the base, usually 4-septate, smooth, the cells at each end hyaline or pale brown, intermediate cells brown or dark brown.

Type species: Exosporiella fungorum (Fr.) Karst.

Exosporiella fungorum (Fr.) Karst., 1892, Finl. Mögelsv.: 160–161.

(Fig. 99)

Conidiophores branched, variable in length, 4–7µ thick. *Conidia* 23–35 × 7–9µ including the caudate appendage.

On *Corticium*; Europe.

91. TRICHODOCHIUM

Trichodochium H. Sydow, 1927, *Annls mycol.*, **25**: 159–160.

Colonies compact or effuse, dark blackish brown, individual sporodochia appearing as black dots. *Mycelium* superficial. *Sporodochia* superficial, orbicular or broadly elliptical, dark, blackish brown to black. *Stroma* present, pseudo-parenchymatous, dark, thick in the centre, thin, flattened and tending to be radiate at the margin which often bears long, simple, smooth, brown *setae*. Each sporodochium is usually fixed to the substratum by a small *hypostroma* located in a stoma. *Hyphopodia* absent. *Conidiophores* macronematous, mono-nematous, crowded, short, unbranched, straight, brown, smooth. *Conidiogenous cells* monoblastic, integrated, terminal, percurrent, with distinct annellations, cylindrical. *Conidia* in unbranched basipetal chains, dry, acrogenous, simple, oblong to ellipsoidal, pale to mid golden brown, longitudinally striate, mostly 1-septate, rarely 2-septate.

Type species: Trichodochium disseminatum Syd.
Ellis, M. B., *Mycol. Pap.*, **111**: 36–39, 1967.

Trichodochium disseminatum H. Syd., 1927, *Annls mycol.*, **25**: 159–160.

(Fig. 100)

Sporodochia 100–220µ diam. *Setae* up to 300 × 3–7µ. *Conidiophores* up to 10 × 2·5–4µ with up to 8 annellations. *Conidia* 8–14 (10·8) × 3–5·5 (4·4)µ.

On leaves of *Rapanea*; Costa Rica, Venezuela.

T. pirozynskii M. B. Ellis found also on *Rapanea* in India and Venezuela has conidia 14–23 (18·3) × 6–9 (7·5)µ.

92. DICTYODESMIUM

Dictyodesmium Hughes, 1951, *Mycol. Pap.*, **36**: 27–29.

Sporodochia epiphyllous, erumpent, pulvinate, olivaceous brown. *Mycelium* immersed forming hyphal cushions at the point of origin of the conidiophores but no definite stroma. *Setae* and *hyphopodia* absent. *Conidiophores* macronematous, mononematous, caespitose, crowded, straight or flexuous, unbranched, pale brown, smooth. *Conidiogenous cells* monoblastic, integrated,

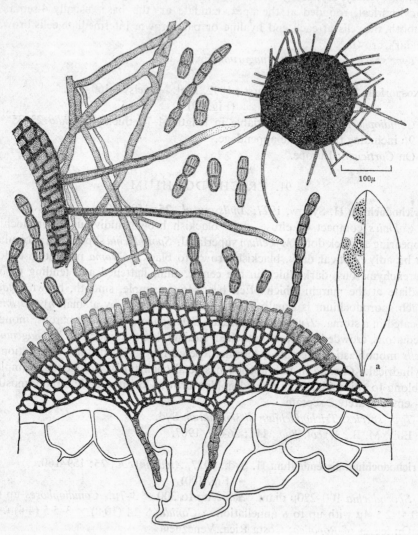

FIG. 100. *Trichodochium disseminatum* (a, habit sketch; other figures × 650 except where indicated by the scale).

terminal, determinate, cylindrical. *Conidia* solitary, acrogenous, simple, fusiform to obclavate, rostrate, truncate at the base, rather pale olivaceous brown, palest at the ends, smooth, with transverse septa throughout and longitudinal and oblique septa in the central 6–9 cells.

   Type species: Dictyodesmium ulmicola (Ellis & Kellerman) Hughes.

Dictyodesmium ulmicola (Ellis & Kellerman) Hughes, 1951, *Mycol. Pap.*, **36**: 27–29.

Ceratophorum ulmicola Ellis & Kellerman, 1887, *J. Mycol.*, **3**: 127.

(Fig. 101)

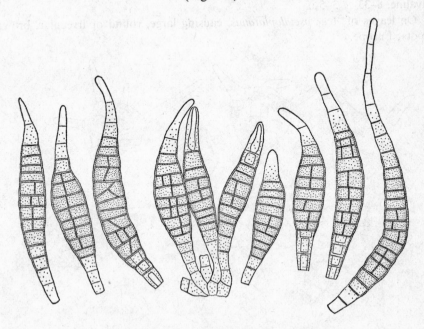

FIG. 101. *Dictyodesmium ulmicola* (× 650).

Sporodochia 40–80μ wide. *Conidiophores* 10–15 × 4–6μ. *Conidia* up to 140μ long, 10–16μ thick in the broadest part, 3–4μ wide at the apex, 4·5–6μ at the base.

On living leaves of *Ulmus fulva*; U.S.A.; causing brown spots with white centres.

93. PETRAKIA

Petrakia Sydow, 1913, *Annls mycol.*, **11**: 406.

Echinosporium Woronichin, 1913, *Vêst. tiflis. bot. Sada*, **28**: 25.

Sporodochia minute, erumpent. *Mycelium* immersed. *Stroma* present, pale brown, pseudoparenchymatous. *Setae* and *hyphopodia* absent. *Conidiophores* macronematous, mononematous, unbranched, straight or curved, short, pale brown, smooth, forming pulvinate sporodochia. *Conidiogenous cells* monoblastic, integrated, terminal, determinate, cylindrical. *Conidia* solitary, dry, acrogenous, corniculate, subspherical, ellipsoidal or irregular in shape, with a projecting stalk cell, pale to mid brown or olivaceous brown, smooth, muriform; when detached often carrying away a part of the conidiophore.

Type species: Petrakia echinata (Peglion) Syd.

Petrak, F., *Sydowia*, **20**: 186–189, 1968.

Petrakia echinata (Peglion) Syd., 1913, *Annls mycol.*, **11**: 407.
(Fig. 102)

Stromata up to 150μ wide. *Conidiophores* 10–30 × 4–6μ. *Conidia* 20–50 ×
15–40μ, pale to mid brown or olivaceous brown; horn-like projections subhyaline, 8–33 × 3–5μ.

On leaves of *Acer pseudoplatanus*, causing large, round or irregular, brown
spots; Europe.

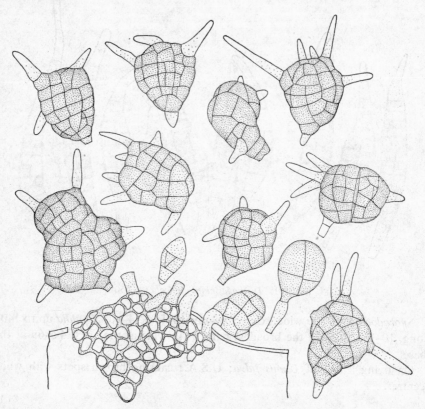

FIG. 102. *Petrakia echinata* (× 650).

94. ENDOPHRAGMIA

Endophragmia Duvernoy & Maire, 1920, *Bull. trimest. Soc. mycol. Fr.*, **36**:
86–89.

Phragmocephala Mason & Hughes, 1951, *Naturalist*, Hull, 1951: 97–105.

Colonies effuse, dark brown to black. *Mycelium* all immersed or partly
superficial. *Stroma* present or absent. *Setae* and *hyphopodia* absent. *Conidiophores* macronematous, mononematous sometimes caespitose, or synnematous,
unbranched, straight or flexuous, brown, smooth; the wall of the lower part
of a conidium may remain attached to the apex of the conidiophore forming
a cup and the conidiophore then proliferates straight on through the cup.
Conidiogenous cells monoblastic, integrated, terminal, occasionally determinate

but more frequently percurrent, calyciform or cylindrical. *Conidia* mostly solitary, in a few species catenate, dry, acrogenous, simple, ellipsoidal, pyriform, obovoid or clavate, pale or dark brown, cells often unequally coloured, 1–5-septate, frequently with very dark brown or black bands at the septa, smooth

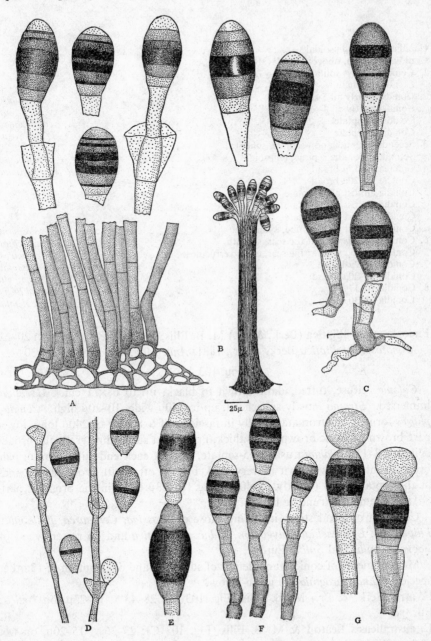

FIG. 103. *Endophragmia* species (1): A, *elliptica*; B, *atra*; C, *australiensis*; D, *taxi*; E, *catenulata*; F, *stemphylioides*; G, *prolifera* (× 650 except where indicated by the scale).

or verrucose, generally becoming detached through a break or tear in the conidiophore wall below the base of the conidium.

Type species: Endophragmia mirabilis Duvernoy & Maire.

Ellis, M. B., *Mycol. Pap.*, **72**: 19–36, 1959; **82**: 48–50, 1961; **106**: 54–55, 1966.

KEY

Conidia mostly ellipsoidal	1
Conidia pyriform, obovoid or clavate	5
1. Conidia always solitary	2
Conidia catenate	3
2. Conidia nearly all 5-septate	*elliptica*
Conidia nearly all 4-septate	*atra*
Conidia 2-septate	*australiensis*
Conidia 1-septate	*taxi*
3. Secondary conidia non-septate, spherical	4
Secondary conidia septate, ellipsoidal	*catenulata*
4. Conidia 10–15µ broad	*stemphylioides*
Conidia 16–20µ broad	*prolifera*
5. Conidia almost all 1-septate	6
Conidia almost all 2-septate	7
Conidia almost all 3-septate	8
Conidia almost all 4-septate, sybhyaline	*hyalosperma*
6. Conidia 2–3.5µ wide at the scar, smooth	*uniseptata*
Conidia 4–5.5µ wide at the scar, often verruculose	*verruculosa*
7. Conidia 8–12µ broad	*biseptata*
Conidia 12–20µ broad	*nannfeldtii*
8. Conidia 10–17µ long	*glanduliformis*
Conidia 22–30µ long	*boothii*

Endophragmia elliptica (Berk. & Br.) M. B. Ellis, 1959, *Mycol. Pap.*, **72**: 20–22. *Monotospora elliptica* Berk. & Br., 1881, *Ann. Mag. nat. Hist.*, 5, **7**: 30.

(Fig. 103 A)

Colonies effuse, tufted, dark brown to black, up to 6 × 1 cm. *Mycelium* immersed. *Stroma* mostly superficial, up to 350µ wide, 10–20µ high. *Conidiophores* formed on stromata usually in fascicles of 6–80, up to 240µ long, lower part brown or dark brown, 4–5µ thick, upper part subhyaline gradually broadening to 8–11µ. *Conidia* usually 5-septate, cells at each end subhyaline or pale brown, central cells brown to very dark brown, often with broad black bands at the septa, 27–54 (mostly 35–40)µ long, 16–23µ thick in the broadest part, 8–11µ wide at the truncate base.

Common on dead stems including those of *Brassica*, *Centaurea*, *Epilobium*, *Filipendula*, *Lysimachia*, *Polygonum*, *Rubus* and *Urtica* and on rotten wood of *Fagus*, *Populus* and *Salix*; Europe.

Measurements of conidia and details of substrata and distribution of 13 other species of *Endophragmia* are given below.

E. atra (Berk. & Br.) M. B. Ellis (Fig. 103 B): 28–43 × 15–23µ, on *Urtica*; Europe.

E. australiensis Beaton & M. B. Ellis (Fig. 103 C): 23–36 × 15–20µ, on wet timber; Australia.

E. taxi M. B. Ellis (Fig. 103 D): 12–21 × 7–13µ, on *Taxus*; Europe, N. America.

E. catenulata M. B. Ellis (Fig. 103 E): 25–45 × 14–19µ, on *Fagus* cupules; Great Britain.

E. stemphylioides (Corda) M. B. Ellis (Fig. 103 F): 15–32 × 10–15µ, on *Eucalyptus*, *Taxus* and *Vitis*; Europe, U.S.A.

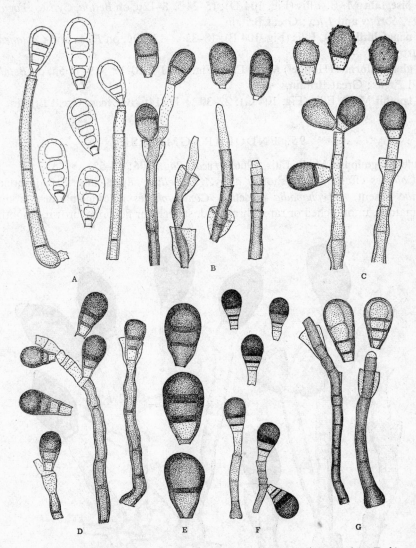

FIG. 104. *Endophragmia* species (2): A, *hyalosperma*; B, *uniseptata*; C, *verruculosa*; D, *biseptata*; E, *nannfeldtii*; F, *glanduliformis*; G, *boothii* (× 650).

E. prolifera (Sacc., Rouss. & Bomm.) M. B. Ellis (Fig. 103 G): 28–34 × 16–20µ, on *Filipendula* and *Urtica*; Europe.

E. hyalosperma (Corda) Morgan-Jones & Cole, 1964, *Trans. Br. mycol. Soc.*, 47: 489–495 (Fig. 104 A): 20–29 × 10–13µ, on *Bambusa*, *Betula*, *Brassica*, *Carex*, *Cytisus*, *Filipendula*, *Quercus*, *Rubus*, *Sambucus* and *Tilia*; Europe.

E. uniseptata M. B. Ellis (Fig. 104 B): 13–27 × 9–13μ, on *Castanea* and *Fagus*; Great Britain.

E. verruculosa M. B. Ellis (Fig. 104 C): 16–21 × 10–13μ, on *Fagus*; Great Britain.

E. biseptata M. B. Ellis (Fig. 104 D): 15–24 × 8–12μ, on *Betula*, *Cytisus*, *Fagus*, *Ilex*, *Sorbus* and *Ulex*; Great Britain.

E. nannfeldtii M. B. Ellis (Fig. 104 E): 18–33 × 12–20μ, on *Picea* and *Castanea*; Europe.

E. glanduliformis (Höhnel) M. B. Ellis (Fig. 104 F): 10–17 × 9–11·5μ, on *Betula* and *Fagus*; Great Britain.

E. boothii M. B. Ellis (Fig. 104 G): 22–30 × 11–14μ, on *Acer*; Great Britain.

95. ENDOPHRAGMIOPSIS

Endophragmiopsis M. B. Ellis, 1966, *Mycol. Pap.*, **106**: 55–57.

Colonies effuse, black, shortly hairy. *Mycelium* superficial. *Stroma* none. *Setae* absent. *Hyphopodia* present. *Conidiophores* macronematous, mononematous, unbranched or rarely branched, straight or flexuous, brown, smooth.

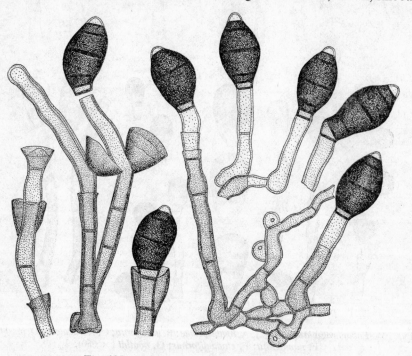

FIG. 105. *Endophragmiopsis pirozynskii* (× 650).

Frequently when a conidium is shed a section of the wall of the lower part of the conidium remains attached to the top of the conidiophore forming a cup; the conidiophore sometimes proliferates straight on through the cup, forming another conidium at a higher level but often grows out laterally just below the cup, pushing it over to one side. *Conidiogenous cells* monoblastic or polyblastic, integrated, terminal becoming intercalary, percurrent or sympodial, calyciform or cylindrical. *Conidia* solitary, dry, acropleurogenous, simple, ellipsoidal, septate, smooth, the cell at each end subhyaline or pale brown, intermediate cells dark brown or blackish brown.

Type species: Endophragmiopsis pirozynskii M. B. Ellis.

Endophragmiopsis pirozynskii M. B. Ellis, *Mycol. Pap.*, **106**: 55–57.

(Fig. 105)

Hyphopodia 5–8 × 7–9μ. *Conidiophores* up to 190μ long, 6–9μ thick. *Conidia* usually 5-septate, 27–38 × 16–19μ, 6–9μ wide at the base.

On *Grewia*; India.

96. BRACHYSPORIELLA

Brachysporiella Batista, 1952, *Bolm Secr. Agric. Ind. Com. Est. Pernambuco,* **19**: 108.

Monosporella Hughes, 1953, *Can. J. Bot.*, **31**: 654 (non Keilin, 1920).

Monotosporella Hughes, 1958, *Can. J. Bot.*, **36**: 787.

Colonies effuse, brown to black, hairy. *Mycelium* all immersed or partly superficial. *Stroma* none. *Setae* and *hyphopodia* absent. *Conidiophores* macronematous, mononematous, straight or flexuous, usually with one or a number of short branches near the apex, brown or dark brown, smooth. *Conidiogenous cells* integrated, terminal, monoblastic, percurrent, cylindrical, doliiform or lageniform. *Conidia* solitary, acrogenous, simple, 2- or pluri-septate, brown, smooth, in some species cells unequally coloured with very dark bands at the septa, clavate, pyriform or turbinate. Conidiogenous cells often remain attached to the base of conidia when these fall.

Type species: Brachysporiella gayana Batista.

Ellis, M. B., *Mycol. Pap.*, **72**: 15–19, 1959.

KEY

Conidia nearly all 3-septate	*gayana*
Conidia nearly all 2-septate	*setosa*
Conidia often 4-septate	*turbinata*

Brachysporiella gayana Batista, 1952, *Bolm Secr. Agric. Ind. Com. Est. Pernambuco*, **19**: 109.

(Fig. 106 A)

Conidiophores up to 250μ long, 8–12μ thick at the base, 5–7μ in the middle, 3–4μ at the apex. *Conidia* 20–38μ long, 12–20μ thick in the broadest part, 3–4μ wide at the base.

On date (*Phoenix dactylifera*) and oil palms (*Elaeis guineensis*), *Pennisetum purpureum* and wood; Brazil, Ghana, Sierra Leone, U.S.A.

Brachysporiella setosa (Berk. & Curt.) M. B. Ellis, 1959, *Mycol. Pap.*, **72**: 17.
(Fig. 106 B)

Conidiophores up to 450µ long, 9–13µ thick at the swollen base, 6–8µ in the middle, 4–6µ at the apex. *Conidia* 20–40µ long, 15–25µ thick in the broadest part, 4–6µ wide at the base.

On rotten wood; U.S.A.

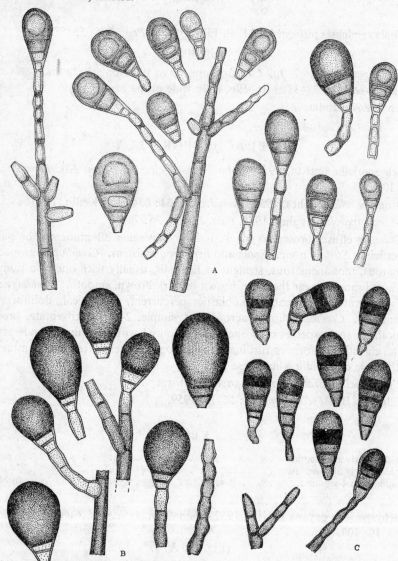

FIG. 106. *Brachysporiella* species: A, *gayana*; B, *setosa*; C, *turbinata* (× 650).

Brachysporiella turbinata (Cooke & Harkn.) M. B. Ellis, 1959, *Mycol. Pap.*, **72**: 18.

(Fig. 106 C)

Conidiophores up to 45μ long, 6–8μ thick at the base, 4–5μ in the middle, 1·5–3μ at the apex. *Conidia* 20–35μ long, 10–16μ thick in the broadest part, 1·5–3μ wide at the base.

On wood of *Sequoia*; U.S.A.

97. CIRCINOCONIS

Circinoconis Boedijn, 1942, *150th Anniversary Vol. R. bot. Gdn Calcutta*: 209–211.

Colonies effuse, dark blackish brown, hairy. *Mycelium* partly superficial, partly immersed. *Stroma* none. *Setae* and *hyphopodia* absent. *Conidiophores* macronematous, mononematous, straight or flexuous, dark blackish brown to

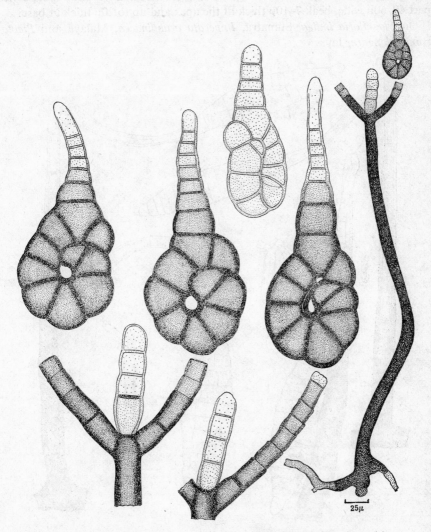

Fig. 107. *Circinoconis paradoxa* (× 650 except where indicated by the scale).

black, smooth, trichotomously branched at the apex, the middle branch remaining fairly short and sterile whilst the branches each side proliferate percurrently and bear terminal conidia. *Conidiogenous cells* monoblastic, integrated, terminating branches, percurrent, cylindrical. *Conidia* solitary, acrogenous on branches, simple, circinate, rostrate, multiseptate, mid-dark brown, paler at apex, smooth.

Type species: Circinoconis paradoxa Boedijn.

Circinoconis paradoxa Boedijn, 1942 (ref. as for genus).
(Fig. 107)

Conidiophores up to 500μ long, 10–24μ thick at the base, 8–15μ thick below the apex; branches up to 130μ long, 8–14μ thick. *Conidia* 70–120μ long, coiled part 40–60μ wide, beak 7–10μ thick at the apex and up to 20μ thick at base.

On *Flagellaria indica*, Sumatra, *Imperata arundinacea*, Malaya and *Plectocomia elongata*, Java.

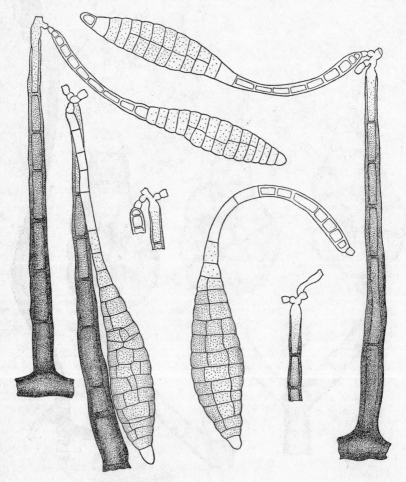

FIG. 108. *Pendulispora venezuelanica* (× 650).

98. PENDULISPORA

Pendulispora M. B. Ellis, 1961, *Mycol. Pap.*, **82**: 41–42.

Colonies effuse, dark blackish brown, hairy. *Mycelium* partly superficial, partly immersed. *Stroma* none. *Setae* and *hyphopodia* absent. *Conidiophores* macronematous, mononematous, with a few very small branches at the apex, straight, subulate, erect, brown, smooth. *Conidiogenous cells* monoblastic, integrated, terminal on stipe and branches, determinate, short, cylindrical. *Conidia* solitary, dry, pendulous, acrogenous, simple, clavate or fusiform, caudate, subhyaline to brown, smooth, muriform.

Type species: Pendulispora venezuelanica M. B. Ellis.

Pendulispora venezuelanica M. B. Ellis, 1961, *Mycol. Pap.*, **82**: 41–42.
(Fig. 108)

Conidiophores up to 200μ long, 9–12μ thick at the base, tapering to 4–5μ at the apex; branches 3–15 × 2–4μ. *Conidia* 112–150μ (130)μ long, 15–22 (18)μ thick in the broadest part, 4–6μ thick at the apex, 2–5μ at the base.

On dead wood of *Fagus* and other trees; Japan, Venezuela.

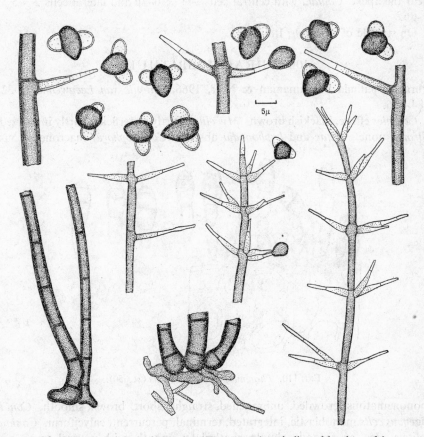

5μ

FIG. 109. *Physalidium elegans* (× 650 except where indicated by the scale).

D.H.—6

99. PHYSALIDIUM

Physalidium Mosca, 1965, *Allionia*, **11**: 73–79.

Colonies effuse, thin, arachnoid or velvety, greyish olive. *Mycelium* partly superficial, partly immersed. *Stroma* none. *Setae* and *hyphopodia* absent. *Conidiophores* macronematous, mononematous, with short branches arranged in verticils; stipe straight or flexuous, subulate, sometimes swollen at the base, brown, smooth, branches pale brown. *Conidiogenous cells* monoblastic, integrated, terminal on stipe and branches, determinate or occasionally percurrent, subulate or cylindrical. *Conidia* solitary, dry, acrogenous, smooth, complex, each composed of an obovoid or ellipsoidal, mid to dark brown central cell and two smaller hemispherical or subspherical, hyaline or very pale brown lateral cells.

Type species: Physalidium elegans Mosca.

Physalidium elegans Mosca, 1965, *Allionia*, **11**: 73–79.
(Fig. 109)

Conidiophores up to 600μ long, 5–9μ thick near the base, tapering to 2–2·5μ near the apex. *Conidia* with central cell 7–11 × 6–7μ and lateral cells 2–4·5 × 4–6μ.

On stubble of *Triticum*; Italy.

100. PHRAGMOSPATHULA

Phragmospathula Subramanian & Nair, 1966, *Antonie van Leeuwenhoek*, **32**: 384–386.

Colonies effuse, blackish brown. *Mycelium* partly superficial, partly immersed. *Stroma* none. *Setae* and *hyphopodia* absent. *Conidiophores* macronematous,

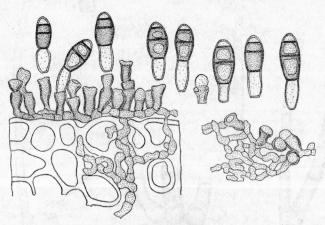

FIG. 110. *Phragmospathula phoenicis* (× 650).

mononematous, crowded, unbranched, straight, short, brown, smooth. *Conidiogenous cells* monoblastic, integrated, terminal, percurrent, calyciform. *Conidia* solitary, dry, acrogenous, simple, spathulate, smooth, with several transverse septa, unequally coloured, the long cell at the base and the short one at the apex hyaline or very pale, intermediate cells golden brown.

Type species: Phragmospathula phoenicis Subram. & Nair.

Phragmospathula phoenicis Subram. & Nair, 1966, *Antonie van Leeuwenhoek*, **32**: 384–386.

(Fig. 110)

Conidiophores 5–15 × 3–7μ. *Conidia* usually 3–septate, 25–35μ long, 7–10μ thick in the broadest part, basal cell 4·5–6μ thick.

On dead leaf rachis of *Phoenix*; India.

101. DEIGHTONIELLA

Deightoniella Hughes, 1952, *Mycol. Pap.*, **48**: 27; emend. D. Shaw, 1959. *Papua New Guin. agric. J.*, **11** (3): 77.

Colonies effuse, grey, brown or black, often hairy. *Mycelium* immersed in most species, occasionally superficial. *Stroma* none. *Setae* and *hyphopodia* absent. *Conidiophores* macronematous, mononematous, torsive or flexuous, usually unbranched but in some species occasionally dichotomously branched, brown, smooth. *Conidiogenous cells* monoblastic, integrated, terminal, percurrent, cylindrical, doliiform, spherical or subspherical. After the first conidium has fallen the thin upper part of the conidiogenous cell tends to collapse and growth recommences inside and usually near its base. *Conidia* solitary, acrogenous, simple, clavate, cylindrical, doliiform, broadly ellipsoidal, obclavate, obpyriform, obturbinate, ovoid, or subspherical, pale to mid brown or golden brown, smooth or verruculose, with 0, 1, 2, 3 or more transverse septa or pseudosepta.

Type species: Deightoniella africana Hughes.
Ellis, M. B., *Mycol. Pap.*, **66**: 1–12, 1957.

KEY

Conidia 0-septate . . . . *papuana*
Conidia 1-septate . . . . *africana*
Conidia 2-septate . . . . *arundinacea*
Conidia mostly 3-septate 1
Conidia frequently with more than 3 septa or pseudosepta 2
1. Conidia obpyriform or obclavate, on *Irenina* . . . . . *leonensis*
   Conidia broadly ellipsoidal or ovoid, on *Arthraxon* . . . *bhopalensis*
   Conidia broadly ellipsoidal or clavate, on *Euphorbia* . . . *jabalpurensis*
2. Conidia obpyriform or obclavate, on *Musa* . . . . . . *torulosa*
   Conidia cylindrical or doliiform, on *Fimbristylis* . . . . *fimbristylidis*

With the exception of *Deightoniella bhopalensis* Satya [*Curr. Sci.*, **34**: 122–123, 1965], *D. jabalpurensis* Agarwal & Hasija [*J. Indian bot. Soc.*, **40**: 542–543, 1962] and *D. papuana* D. Shaw [*Papua New Guin. agric. J.*, **11** (3): 77–81, 1959], the species keyed out above are fully described and illustrated in *Mycol. Pap.*, **66**. Their substrata, geographical distribution and conidial measurements in μ are given below.

Deightoniella africana Hughes (Fig. 111 A) on *Imperata*; W. Africa (20–29 × 11–16).

D. arundinacea (Corda) Hughes (Fig. 111 B) on *Phragmites*; Australia, Japan, Europe, N. America (20–49 × 11–18).

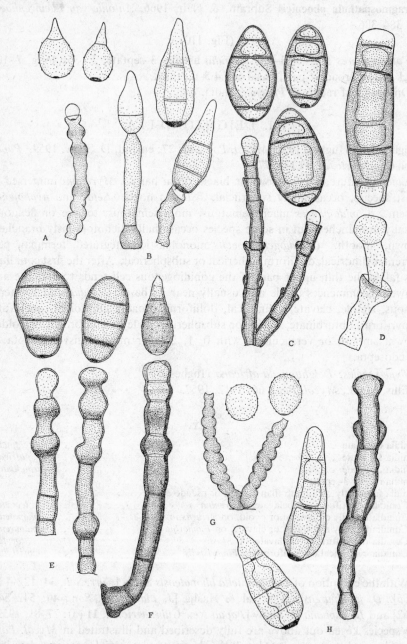

FIG. 111. *Deightoniella* species: A, *africana*; B, *arundinacea*; C, *bhopalensis*; D, *fimbristylidis*; E, *jabalpurensis*; F, *leonensis*; G, *papuana*; H, *torulosa* (× 650).

D. bhopalensis Satya (Fig. 111 C) on *Arthraxon*; India (30–50 × 17–30).

D. fimbristylidis Sawada ex M. B. Ellis (Fig. 111 D) on *Fimbristylis*; Formosa (47–75 × 15–22).

D. jabalpurensis Agarwal & Hasija (Fig. 111 E) on *Euphorbia*; India (24–42 × 15–28).

D. leonensis M. B. Ellis (Fig. 111 F) on *Asteridiella* on *Tetracera*; Sierra Leone (26–34 × 12–14).

D. papuana D. Shaw (Fig. 111 G) on *Saccharum*; Papua and New Guinea (15–20 × 15–18).

D. torulosa (Syd.) M. B. Ellis (Fig. 111 H) on *Musa*; common in the tropics, see CMI Distribution Map, 175 (usually 35–70 × 13–25 but sometimes up to 150μ long). Causes black tip disease of bananas, also leaf spots and blotches, pseudostem rot and leaf spot of *Musa textilis*; *CMI Descr. Path. Fung. Bact.* No. 165. See also D. S. Meredith in *Trans. Br. mycol. Soc.*, **44**: 95–104, 265–284, 391–405, 1961 and *Ann. appl. Biol.*, **49**: 488–496, 1961.

102. STEMPHYLIUM

Stemphylium Wallroth, 1833, Fl. crypt. Ger., Pars. post.: 300.

Colonies effuse, grey, brown, olivaceous brown or black, velvety or cottony. *Mycelium* immersed or partly superficial. *Stroma* sometimes present. *Setae* and *hyphopodia* absent. *Conidiophores* macronematous, mononematous, scattered or caespitose, unbranched or occasionally loosely branched, straight or flexuous, usually nodose with a number of vesicular swellings, pale to mid brown or olivaceous brown, smooth or in part verruculose. *Conidiogenous cells* monoblastic, integrated, terminal, percurrent, at first clavate or subspherical with the wall at the apex thin, later often becoming calyciform by invagination. *Conidia* solitary, dry, acrogenous, oblong, rounded at the ends, ellipsoidal, obclavate or subspherical, some species with a pointed conical apex and one with lateral conical protrusions, pale to mid dark or olivaceous brown, smooth, verrucose or echinulate, muriform, often constricted at one or more of the septa, cicatrized at the base.

Type species: Stemphylium state of *Pleospora herbarum* (Pers. ex Fr.) Rabenh. = *S. botryosum* Wallr.

Hannon, C. I. & Weber, G. F., *Phytopathology*, **45**: 11–16, 1955.

Graham, J. H. & Zeiders, K. E., *Phytopathology*, **50**: 757–760, 1960.

Simmons, E. G., *Mycologia*, **59**: 67–92, 1967; **61**: 1–26, 1969.

Wiltshire, S. P., *Trans. Br. mycol. Soc.*, **21**: 211–239, 1938.

KEY

Conidia rounded at the apex	1
Conidia with pointed, conical apex	3
1. Conidia smooth; on *Trifolium*	*sarciniforme*
Conidia verrucose or verruculose; on dead herbaceous stems, etc. . . .	2
2. Conidia mostly constricted at the median septum only . . .	*Pleospora herbarum*
Conidia frequently constricted at the 3 major transverse septa . .	*. vesicarium*
3. Conidia constricted at median septum, l/b ratio not more than 2:1 . . .	*solani*
Conidia constricted at 3 major transverse septa, l/b ratio 3:1 or more .	*. lycopersici*

Stemphylium sarciniforme (Cav.) Wiltshire, 1938, *Trans. Br. mycol. Soc.*, **21**: 224–228.

(Fig. 112 A)

Conidiophores pale to mid golden brown, up to 50μ long, 6–10μ thick, vesicular swellings 11–14μ diam. *Conidia* subspherical or very broadly ellipsoidal, golden

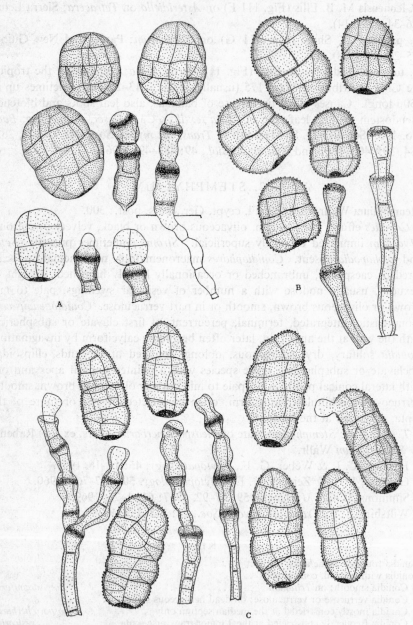

Fig. 112. *Stemphylium* species (1): A, *sarciniforme*; B, state of *Pleospora herbarum*; C, *vesicarium* (× 650).

brown, smooth with usually 3 transverse and several longitudinal septa, sometimes constricted at the median transverse septum, 30–50 × 22–33μ.

On leaves of *Trifolium* causing dark brown, zonate spots 1–8 mm. diam.; Africa, Australasia, Europe, N. & S. America [CMI Distribution Map, 139]. Another species which is reported by Simmons to be common on leguminous hosts is **S. globuliferum** (Vestergren) Simmons; it has conidia similar in shape and septation to those of *S. sarciniforme* but smaller (28–30 × 25–28μ) and verrucose.

Stemphylium state of **Pleospora herbarum** (Pers. ex Fr.) Rabenh., 1855, *Herb. mycol. Ed. nov.*, **2**: 547.
S. botryosum Wallr., 1833, Fl. crypt. Germ., Pars. post.: 300.
(Fig. 112 B)

Conidiophores often caespitose and arising from a submerged stroma, up to 80μ long on natural substrata, sometimes longer in culture, 4–7μ thick, pale to mid brown or olivaceous brown: terminal swellings which become through percurrent proliferation intercalary 7–11μ diam., each with a broad, dark verrucose band a little way below the apex. *Conidia* oblong rounded at the ends, broadly ellipsoidal or subspherical, with usually 3 transverse and 1–3 longitudinal septa, constricted at the median transverse septum, pale to mid dark brown or olivaceous brown, minutely verruculose or echinulate, 27–42 × 24–30μ.

Common on dead herbaceous plants and occasionally on wood; isolated from air, paper, soil, etc.; cosmopolitan.

Stemphylium vesicarium (Wallr.) Simmons, 1969, *Mycologia*, **61**: 9–12.
(Fig. 112 C)

Conidiophores often caespitose, up to 70μ long, 3–8μ thick, pale to mid brown; swellings 8–11 diam., with dark band smooth or minutely verruculose. *Conidia* oblong rounded at the ends or broadly ellipsoidal, pale to mid brown or olivaceous brown, verrucose, with up to 6 transverse and several longitudinal septa, often constricted at the 3 major transverse septa, 20–50 × 15–26μ.

On dead herbaceous stems; Africa, Europe, N. & S. America.

Stemphylium solani Weber, 1930, *Mycologia*, **20**: 516–518.
(Fig. 113 A)

Conidiophores pale to mid brown, up to 200μ long, 4–7μ thick, vesicular swellings 8–10μ diam. *Conidia* pointed at the apex, with 3–6 transverse and several longitudinal septa, mostly constricted at the median septum only, pale to mid golden brown, smooth or minutely verruculose, 35–55 × 18–28μ (1/b ratio 2:1).

On living leaves of *Lycopersicon* and *Solanum*, causing very small spots; Africa, Asia, Australasia, Europe, N., Central and S. America [CMI Distribution Map, 333].

Stemphylium lycopersici (Enjoji) Yamamoto, 1960, *Trans. mycol. Soc. Japan*, **2** (5): 93.
(Fig. 113 B)

Conidiophores pale to mid brown, up to 140 × 6–7μ, vesicular swellings 8–10μ diam. *Conidia* pointed, conical at the apex, with 1–8 transverse and several longitudinal septa, usually constricted at the 3 major transverse septa, pale to mid brown, smooth or minutely verruculose, mostly 50–74 × 16–23μ (l/b ratio 3:1 or more).

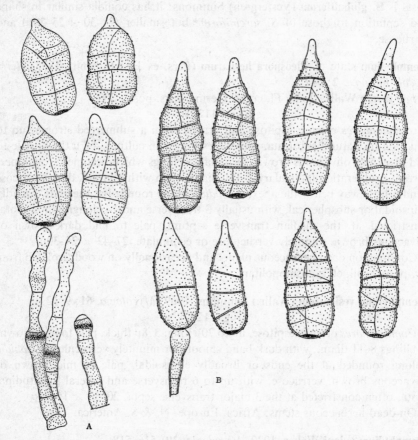

Fig. 113. *Stemphylium* species (2): A, *solani*; B, *lycopersici* (× 650).

Most records on *Lycopersicon* but found also on other plants including *Allium*, *Carthamus* and *Gladiolus*; Cuba, Hong Kong, Japan, Saudi Arabia and Venezuela.

103. HANSFORDIELLA

Hansfordiella Hughes, 1951, *Mycol. Pap.*, **47**: 10–15.

Colonies effuse, inconspicuous, overgrowing other fungi. *Mycelium* superficial. *Stroma* none. *Setae* and *hyphopodia* absent. *Conidiophores* macronematous, mononematous, solitary or in small clusters, unbranched, straight or flexuous, pale to mid brown, smooth. *Conidiogenous cells* monoblastic, integrated, terminal, determinate, cylindrical, cicatrized; the characteristic scars are large, dark brown, cupulate with an inner inverted cone. *Conidia* solitary, dry, acrogenous, simple, obclavate, rostrate, truncate at the base, brown, smooth, with several transverse and sometimes oblique or longitudinal septa.

Type species: Hansfordiella asterinarum Hughes.

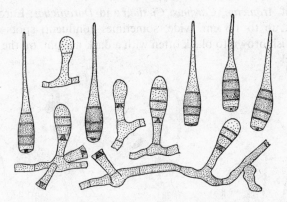

FIG. 114. *Hansfordiella asterinarum* (× 650).

Hansfordiella asterinarum Hughes, 1951, *Mycol. Pap.*, **47**: 11–12.

(Fig. 114)

Conidiophores cylindrical, pale brown, 10–14 × 3–4μ, expanded at the dark apical scar to 5–6μ. *Conidia* 40–55 × 7·5–8·5μ, beak tapering to 0·5–1μ.

On *Asterina* and *Asterolibertia*; Ghana, Sierra Leone.

104. FUSICLADIELLA

Fusicladiella Höhnel, 1919, *Ber. dt. bot. Ges.*, **37**: 155.

Colonies suborbicular or angular. *Mycelium* immersed. *Stroma* sometimes present in the host cuticle. *Setae* and *hyphopodia* absent. *Conidiophores* macronematous, mononematous, crowded, unbranched, at first erect, straight or slightly curved, cylindrical, almost colourless, later strongly curved, brown or olivaceous brown, pale and thin-walled on one side, dark and thick-walled on the other, the curvature always taking place towards the thin-walled side, smooth or sometimes finely verruculose near the apex. *Conidiogenous cells* monoblastic, integrated, terminal, determinate, cylindrical, cicatrized, the single apical scar broad and flat. *Conidia* solitary, dry, acrogenous, straight or slightly curved, often cylindrical, rounded at the apex, truncate with a thin scar at the base, but sometimes clavate, ellipsoidal or obclavate, colourless to pale olive, smooth to finely verruculose, almost always 1-septate, rarely 2-septate.

Type species: Fusicladiella melaena (Fuckel) Hughes = *F. aronici* (Sacc.) Höhnel.

Deighton, F. C., *Mycol. Pap.*, **101**: 23–28, 1965.

Hughes, S. J., *Mycol. Pap.*, **49**: 20–24, 1952.

Fusicladiella melaena (Fuckel) Hughes, 1952, *Mycol. Pap.*, **49**: 21–24.

(Fig. 115)

Conidiophores up to 120 × 4–8μ. *Conidia* cylindrical, rounded at the apex, truncate at the base, 28–66 × 8–13μ.

On leaves of *Aronicum, Carduus, Cirsium* and *Doronicum*; Europe. Colonies hypophyllous, up to 1·5 cm. wide, sometimes confluent, spots on the upper surface yellowish brown to black often with a dark margin, on the lower surface brown to black.

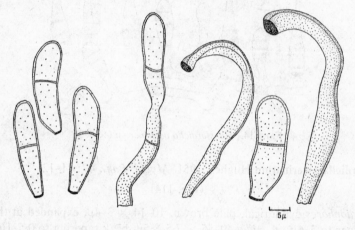

FIG. 115. *Fusicladiella melaena* (× 650 except where indicated by the scale).

Deighton (1965) made three transfers to the genus *Fusicladiella*. **F. pimpinellae** (Vestergren) Deighton on *Pimpinella* from Sweden resembles *F. melaena* but has very short conidiophores (13–25μ long); the conidia measure 39–43 × 6·5–10μ. **F. cousiniae** (Petrak) Deighton and **F. phaeopappi** (Petrak) Deighton, both from Iran have broadly ellipsoidal or short clavate conidia; the former on *Cousinia* has conidiophores up to 110μ long and conidia 18–26 × 9·5–13·5μ, the latter on *Phaeopappus* conidiophores 18–35μ long and conidia 16–34 × 8–11μ.

105. HERPOSIRA

Herposira Sydow, 1938, *Annls mycol.*, **36**: 312–313.

Colonies elongated, black, velvety. *Mycelium* immersed. *Stroma* present, immersed, pale brown. *Setae* numerous, unbranched, subulate, acutely pointed, thick-walled, mid to dark brown. *Hyphopodia* absent. *Conidiophores* macronematous, mononematous, closely packed together, short, irregularly branched, straight or flexuous, pale to mid brown, verrucose. *Conidiogenous cells* polyblastic, integrated, terminal and intercalary, determinate, broadly ellipsoidal, coarsely verrucose. *Conidia* solitary, dry, acropleurogenous, simple, ellipsoidal, brown, smooth or verrucose, 0-septate.

Type species: Herposira velutina Syd.

Herposira velutina Syd., 1938, *Annls mycol.*, **36**: 312–313.

(Fig. 116)

Colonies up to 5 × 2 mm. *Setae* up to 400μ long, 8–11μ thick in the broadest part. *Conidiophores* up to 50 × 5–7μ. *Conidia* 11–18 × 7–8μ.

On dead leaves of *Xanthorrhoea*; Australia.

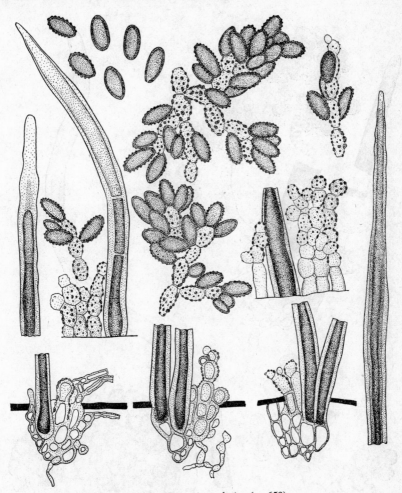

FIG. 116. *Herposira velutina* (× 650).

106. LACELLINA

Lacellina Saccardo, 1913, *Annls mycol.*, **11**: 418–419.

Colonies discrete, round or elliptical, black, velvety or hairy. *Mycelium* immersed. *Stroma* present, partly superficial, partly immersed, subhyaline to brown or olivaceous brown. *Setae* simple, straight, flexuous or vermiform, grey, olivaceous grey, dark brown or black, smooth. *Hyphopodia* absent. *Conidiophores* macronematous, mononematous, caespitose, surrounding bases of setae, unbranched or loosely branched, flexuous, subhyaline to pale brown, verrucose. *Conidiogenous cells* polyblastic, integrated, terminal and intercalary, determinate, cylindrical, verrucose. *Conidia* dry, acropleurogenous, catenate, simple, spherical or subspherical, often somewhat flattened in one plane, pale to mid brown, smooth to verrucose, 0-septate.

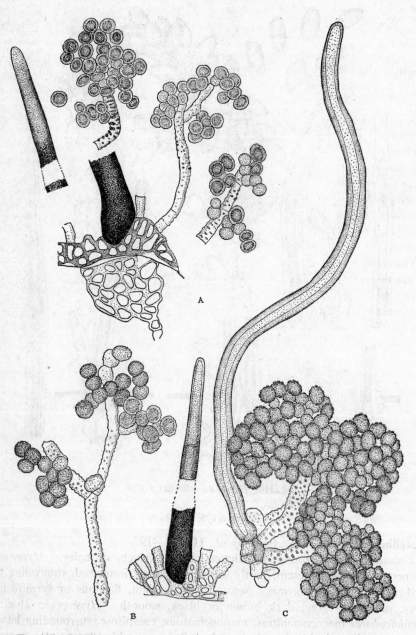

Fig. 117. *Lacellina* species: A, *graminicola*; B, *macrospora*; C, *leonensis* (× 650).

Type species: Lacellina graminicola (Berk. & Br.) Petch = *L. libyca* Sacc. & Trott.

Ellis, M. B., *Mycol. Pap.*, **67**: 10–15, 1957.

KEY

Setae straight or slightly flexuous dark brown to black   1
Setae strongly flexuous, vermiform, pale grey or olivaceous grey .   .   .   *leonensis*
1. Conidia 4–7 (rarely 8)µ diam. .   .   .   .   *graminicola*
   Conidia 7–10µ diam. or up to 12×9µ .   .   .   .   .   .   *macrospora*

Lacellina graminicola (Berk. & Br.) Petch, 1924, *Ann. R. bot. Gdns Peradeniya*, **9**: 171. For synonymy see *Mycol. Pap.*, **67**.
(Fig. 117 A)

Setae usually many in each colony but formed singly or in groups of 2–3 on each stroma, nearly black below the middle, paler above, 300–1200µ long, 11–17µ thick at the base, tapering to 4–6µ near the apex. *Conidiophores* 40–120µ long, 3–5µ thick. *Conidia* mostly 4–7µ diam., somewhat flattened.

On *Andropogon, Aristida, Borassus, Erianthus, Gynerium, Hyparrhenia, Miscanthidium, Phoenix, Pogonarthria, Saccharum, Typha* and *Vetiveria*; Australia, Barbados, Brazil, Ceylon, Cuba, Ghana, India, Jamaica, Nigeria, Pakistan, Puerto Rico, Rhodesia, S. Africa, Sudan, Tripoli, U.S.A., Venezuela, Zambia.

Two other species, both collected on *Borassus* in West Africa, are figured: **L. macrospora** M. B. Ellis (Fig. 117 B) with straight or flexuous, dark blackish brown setae up to 1500µ long and conidia 7–10µ diam. or up to 12 × 9µ; and **L. leonensis** M. B. Ellis (Fig. 117 C) with vermiform setae 200–400µ long and conidia 5–9µ diam. or up to 10 × 8µ.

107. GONATOBOTRYUM

Gonatobotryum Saccardo, 1880, *Michelia*, **2**: 24–25.

Colonies compact or effuse, brown to dark brown. *Mycelium* mostly superficial. *Stroma* none. *Setae* and *hyphopodia* absent. *Conidiophores* macronematous, mononematous, unbranched, straight or flexuous, nodose with terminal and intercalary conidiogenous ampullae, often also swollen at the base, pale to dark brown, smooth. *Conidiogenous cells* polyblastic, integrated, terminal becoming intercalary, percurrent, spherical to subspherical, cicatrized, scars slightly raised. *Conidia* catenate, dry, acropleurogenous, simple, ellipsoidal, limoniform or oblong rounded at the ends, pale to mid brown, smooth, 0-septate.

Type species: Gonatobotryum fuscum (Sacc.) Sacc.

Kendrick, Cole & Bhatt, *Can. J. Bot.*, **46**: 591–596, 1968.

Shigo, A. L., *Mycologia*, **52**: 584–598, 1961.

KEY

Conidia ellipsoidal or oblong rounded at the ends, 10–25×6–13µ .   .   .   *fuscum*
Conidia mostly limoniform, 5–12×2·5–6µ .   .   .   .   .   .   *apiculatum*

Gonatobotryum fuscum (Sacc.) Sacc., 1886, Syll. Fung., **4**: 278.

(Fig. 118 A)

Conidiophores up to 700µ long, 12–15µ thick, ampullae up to 28µ diam. *Conidia* in chains of usually 2, 10–25 × 6–13µ.

On rotten wood of *Quercus* and *Fagus* associated with and parasitic on *Ceratocystis* and other fungi; Europe, N. America.

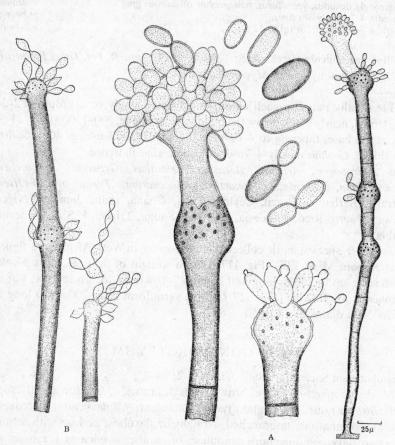

Fig. 118. A, *Gonatobotryum fuscum*; B, *G. apiculatum* (× 650 except where indicated by the scale).

Gonatobotryum apiculatum (Peck) Hughes, 1953, *Can. J. Bot.*, **31**: 594, with full synonymy.

(Fig. 118 B)

Conidiophores up to 550µ long, but mostly 200–300µ, 6–9µ thick, ampullae, 9–15µ diam. *Conidia* in chains of up to 12, sometimes branched, 5–12 × 2·5–6µ.

Parasitic on leaves of *Hamamelis*, causing orbicular, sometimes confluent reddish brown spots up to 4 mm. diam., also recorded on *Rhus* and isolated from soil; N. America.

108. OEDEMIUM

Oedemium Link, 1824, in Linné's Sp. Pl., Ed. 4 (Willdenow's), **6** (1): 42–43.
  Diplosporium Link, 1824, in Linné's Sp. Pl., Ed. 4 (Willdenow's), **6** (1): 64.
  Dimera Fries, 1825, Syst. Orb. veg.: 183.
  Cladotrichum Corda, 1831, in Sturm's Deut. Fl., III, **3** (12): 39–40.
  Gongylocladium Wallroth, 1833, Fl. crypt. Germ., Pars Post.: 160.

  Colonies effuse or pulvinate, very dark brown to black, felted or velvety.
Mycelium partly immersed, partly superficial; hyphae thick, brown or dark
brown. *Stroma* none. *Setae* present in some species, unbranched or occasionally branched, subulate, acute at the apex, dark brown, smooth. *Hyphopodia*

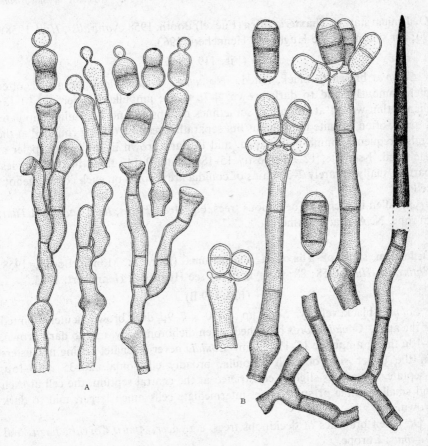

FIG. 119. A, *Oedemium* state of *Thaxteria fusca*; B, *O.* state of *Thaxteria phaeostroma* (× 650).

absent. *Conidiophores* macronematous, mononematous, branched, often
dichotomously or sometimes unbranched, flexuous, distinctly nodose with terminal and intercalary conidiogenous ampullae, mid to dark brown, smooth.
Conidiogenous cells polyblastic, integrated, terminal becoming intercalary,
percurrent, spherical, subspherical or pyriform, cicatrized; scars small,

slightly raised. *Conidia* solitary or catenate, dry, acrogenous or acropleurogenous, simple, ellipsoidal, oblong rounded at the ends or dumb-bell shaped, pale to dark brown, cells sometimes unequally coloured, smooth, 1–3-septate.

Type species: Oedemium state of *Thaxteria fusca* (Fuckel) Booth = *O. atrum* Link.

Hughes, S. J. & Hennebert, G. L., *Can. J. Bot.*, **41**: 773–809, 1963.

KEY

Conidia 1-septate . . . . *Thaxteria fusca*
Conidia almost always 3-septate . . . . *Thaxteria phaeostroma*

Oedemium state of **Thaxteria fusca** (Fuckel) Booth, 1958, *Naturalist, Hull*, 1958: 90; for synonymy see Hughes & Hennebert, 1963.

(Fig. 119 A)

Colonies brown to black, felted. No *setae*. *Conidiophores* branched, often dichotomously, mid to dark brown, 4–7μ thick; ampullae numerous, 10–13μ diam., thin-walled at the apex, sometimes collapsing and becoming cupulate. *Conidia* often in quite long chains but separating readily, oblong rounded at the ends, frequently dumb-bell-shaped, mid to dark brown in the centre, paler at each end, 1-septate, 12–20 (mostly 15–18)μ long, 9–14μ thick in the broadest part. Usually 1, rarely 2–3 chains of conidia are formed on each conidiogenous cell.

On fallen branches of deciduous trees, e.g. *Acer, Fagus, Ilex, Salix* and *Tilia*; Europe, N. Africa, N. America.

Oedemium state of **Thaxteria phaeostroma** (Dur. & Mont.) Booth, 1958, *Naturalist, Hull*, 1958: 88; for synonymy see Hughes & Hennebert, 1963.

(Fig. 119 B)

Colonies black, velvety. *Setae* up to 450 × 7–9μ, dark brown, acutely pointed at the apex. *Conidiophores* branched, often dichotomously, mid to dark brown, 7–10μ thick; ampullae 13–17μ diam. *Conidia* never catenate, arising in clusters on the upper side of each ampulla, broadly ellipsoidal, 20–35 × 10–15μ, 3-septate, sometimes slightly constricted at the central septum, the cell at each end small, hyaline or subhyaline, intermediate cells much larger, mid to dark brown.

On fallen branches of deciduous trees, e.g. *Acer, Alnus, Corylus, Fagus* and *Fraxinus*; Europe.

109. CEPHALIOPHORA

Cephaliophora Thaxter, 1903, *Bot. Gaz.*, **35**: 157–158.
　　Cephalomyces Bainier, 1907, *Bull. trimest. Soc. mycol. Fr.*, **23**: 109–110.
　　Colonies effuse, pink, buff or reddish brown. *Mycelium* immersed and superficial. *Stroma* none. *Setae* and *hyphopodia* absent. *Conidiophores* semi-

macronematous or macronematous, mononematous, straight or flexuous, un-branched, with terminal ampulla, colourless or pale brown, smooth. *Conidiogenous cells* integrated, terminal, polyblastic, determinate, clavate, spherical or subspherical. *Conidia* solitary, acropleurogenous, formed simultaneously over the curved surface of the conidiogenous cell, simple, clavate, cuneiform, cylindrical, pyriform or turbinate, sometimes lobed, with a protuberant hilum, at first colourless, later pale to mid brown, often darker at the septa, end cells frequently paler than central ones, smooth, 1- to pluri-septate.

Lectotype species: Cephaliophora tropica Thaxter.

Chabelska-Frydman, Chaja, *Bull. Res. Coun. Israel*, **7** (D): 81–84, 1959.

Crook, F. M. & Hindson, W. R., *Trans. Br. mycol. Soc.*, **38**: 218–220, 1955.

Goos, R. D., *Mycologia*, **56**: 133–136, 1964.

KEY

Conidia 1–2-septate, sometimes lobed . . . . *irregularis*
Conidia 3–7-septate . . . . *tropica*

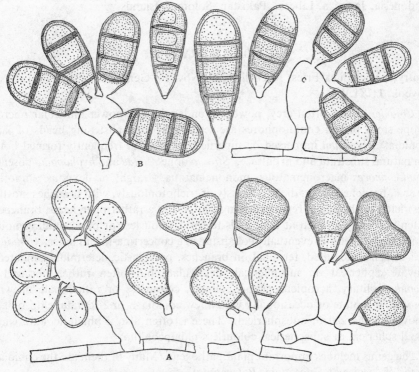

FIG. 120. A, *Cephaliophora irregularis*; B, *C. tropica* (× 650).

Cephaliophora irregularis Thaxter, 1903, *Bot. Gaz.*, **35**: 158.
(Fig. 120 A)

Colonies pink to reddish brown. *Hyphae* 2–9μ thick. *Conidiophores* clavate, up to 120μ long, 6–10μ thick near the base; swollen conidiogenous cell 20–60 × 15–35μ. *Conidia* very variable in shape, often pyriform or turbinate, sometimes lobed, colourless to rather pale brown, reddish brown in mass when mature, 1- to 2-septate, 20–44μ long, 13–33μ thick in the broadest part, protuberant hilum 1·5–4μ wide.

On cocoa beans, dung, wood, soil etc.; Germany, Great Britain, India, Japan, Pakistan, Puerto Rico, U.S.A.

Cephaliophora tropica Thaxter, 1903, *Bot. Gaz.*, **35**: 158.

(Fig. 120 B)

Colonies buff or cinnamon. *Hyphae* 3–6μ thick. *Conidiophores* up to 75μ long, 3–5μ thick near base; swollen conidiogenous cell 27–35 × 13–27μ. *Conidia* cylindrical to clavate, 27–60 × 14–25μ, protuberant hilum 1·5–3μ wide, bright yellowish or cinnamon brown when mature, end cells paler than central ones, 3–7- (usually 3–5-) septate, darker at septa.

On cacao, dung, *Piper betle*, shoe leather, soil etc.; Australia, China, Ghana, Indonesia, Jamaica, Liberia, Pakistan, Solomon Islands.

110. BOTRYTIS

Botrytis Micheli ex Fries; Micheli, 1729, Nov. Pl. Gen.: 212; Fries, 1821, Syst. mycol., **1**: XLV.

Colonies effuse, often grey, powdery; under the low-power binocular microscope stout brown conidiophores are seen supporting glistening heads of pale conidia. *Mycelium* immersed or superficial. *Sclerotia* frequently formed both on natural substrata and in culture. *Stroma* none. *Setae* and *hyphopodia* absent. *Conidiophores* macronematous, mononematous, straight or flexuous, smooth, brown, branched, often dichotomously or trichotomously, with branches mostly restricted to the apical region forming a stipe and a rather open head; branches often markedly swollen at their ends to form colourless or pale conidiogenous ampullae, sometimes eventually collapsing in a concertina-like manner. *Conidiogenous cells* integrated, terminal on branches, polyblastic, determinate, inflated, clavate, spherical or subspherical, denticulate but often rather obscurely. *Conidia* solitary, acropleurogenous, simple, colourless or rather pale brown, smooth, 0-septate or occasionally with a few conidia 1- or 2-septate, ellipsoidal, obovoid, spherical or subspherical. There is often also a phialidic state with small spherical or subspherical colourless phialoconidia.

The genus includes important plant pathogens. Many of them are the conidial states of species of *Sclerotinia* (*Botryotinia*).

Lectotype species: Botrytis cinerea Pers. ex Pers., conidial *Sclerotinia fuckeliana* (de Bary) Fuckel.

Hennebert, G. L., Meded. LandbHoogesch. OpzoekStns Gent, **28**: 851–876, 1963; *Parasitica*, **20**: 138–153, 1964.

Host plants	B. = *Botrytis* S. = *Sclerotinia*	*Conidia–usual* *dimensions in* μ
Plurivorous	*B. cinerea*	8–14 × 6–9
Allium	*B. allii*	7–11 × 5–6
„	*B. byssoidea*	10–14 × 6–9
„	*S. porri*	11–14 × 7–10
A. cepa	*S. squamosa*	15–21 × 13–16
A. triquetrum	*S. sphaerosperma*	20–26 diam.
A. ursinum	*S. globosa*	12–18 diam.
Gladiolus	*S. draytonii*	10–22 × 8–13
Iris	*S. convoluta*	10–12 × 8–10
Lilium	*B. elliptica*	20–30 × 13–18
Narcissus	*S. narcissicola*	8–16 × 7·5–12
„	*S. polyblastis*	30–50 diam.
Paeonia	*B. paeoniae*	12–18 × 8–10
Ricinus (Other Euphorbiaceae)	*S. ricini*	7–10 diam.
Trifolium pratense (anthers)	*B. anthophila*	11–16 × 4–5
T. repens (seeds)	*S. spermophila*	10–18 × 6–10
Tulipa (Other Liliaceae)	*B. tulipae*	16–20 × 10–13
Vicia faba	*B. fabae*	16–25 × 13–16

Botrytis state of **Sclerotinia fuckeliana** (de Bary) Fuckel, 1870, Symb. mycol.: 330.

B. cinerea Pers. ex Pers.; Persoon, 1801, Syn. method. Fung.: 690; 1822, Mycol. eur., **1**: 32.

(Fig. 121 A)

Colonies grey or greyish brown. Production, size and shape of sclerotia on natural substrata and in culture extremely variable; in culture some strains form no sclerotia, in others they are abundant. *Sclerotia* black, usually smaller and thinner than those of *Sclerotinia sclerotiorum*. *Conidiophores* frequently 2 mm. or more long, mostly 16–30μ thick, branched, often with a stipe and a rather open head of branches, smooth, clear brown below, paler near the apex, with the ends of the branches often quite colourless. *Conidia* ellipsoidal or obovoid, often with a slightly protuberant hilum, colourless to pale brown, smooth, 6–18 × 4–11 (mostly 8–14 × 6–9)μ [l/b ratio 1·35–1·5].

The cosmopolitan ' grey mould ' damages flowers, leaves, stems, fruit and other parts of all sorts of plants (more than 140 represented in Herb. IMI), including many of economic importance such as bean, lettuce, gooseberry, tomato and vine.

Botrytis allii Munn, 1917, *Bull. N.Y. St. agric. Exp. Stn*, **437**: 396.
(Fig. 121 B)

Sclerotia 1–5 mm. diam., aggregated, frequently found on natural substrata but less so in cultures. *Conidiophores* abundant, rather short, usually about 1 mm. *Conidia* narrowly ellipsoidal, distinctly narrower than those of other species on *Allium*, sometimes pyriform or cuneiform, occasionally septate in culture, 5–10 × 3–8 (mostly 7–11 × 5–6)μ [l/b ratio 1·7–2·5].

On various species of *Allium*, cosmopolitan [CMI Distribution Map 169]; causes extensive damage to onions (neck rot) and shallots in storage.

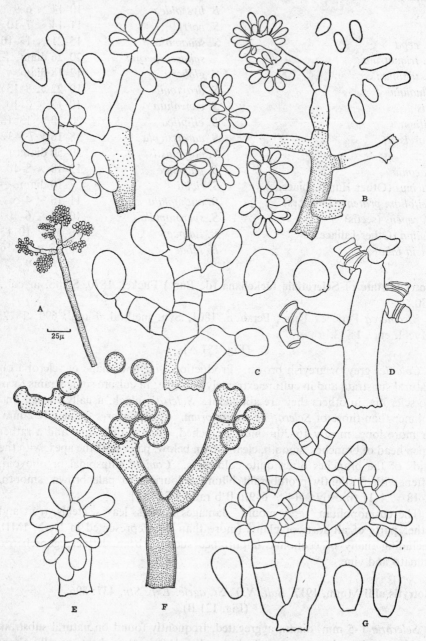

FIG. 121. *Botrytis* species: A, *cinerea*; B, *allii*; C, *squamosa*; D, *globosa*; E, *paeoniae*; F, *ricini*; G, *fabae* (× 650 except where indicated by the scale).

Botrytis byssoidea Walker, 1925, *Phytopathology*, **15**: 709.

Colonies densely cottony on the surface of bulbs and in culture; *conidiophores* few, *sclerotia* rare or absent. *Conidia* 8–19 × 5–11 (mostly 10–14 × 6–9)μ [l/b ratio 1·5–1·65].

Causes mycelial neck rot of onion (*Allium cepa*); recorded also on leek (*Allium porrum*) and other species of *Allium* from Australia, Europe, Japan and N. America [CMI Distribution Map 165].

Botrytis state of **Sclerotinia porri** v. Beyma, 1927, *Meded. phytopath. Lab. Willie Commelin Scholten*, **10**: 46.

Characterised by the very large, irregular, often cerebriform *sclerotia* formed in culture on p.d.a. etc. *Conidia* 9–18 × 6–13 (mostly 11–14 × 7–10)μ [l/b ratio 1·35–1·5].

On leek (*Allium porrum*) and occasionally on other species of *Allium*; Belgium, Brazil, Bulgaria, Great Britain, Netherlands, Norway etc.

Botrytis state of **Sclerotinia squamosa** (Viennot-Bourgin) Dennis, 1956, *Mycol. Pap.*, **62**: 157.

Botrytis squamosa Walker, 1925, *Phytopathology*, **15**; 710.

(Fig. 121 C)

Sclerotia on natural substrata and in culture flat or globular, oval or rounded, 1–2 mm. diam., discrete. Concertina-like collapse of *conidiophore* branches very marked. *Conidia* 10–26 × 10–18 (mostly 15–21 × 13–16)μ [l/b ratio 1·25–1·45].

Causes tip and leaf blight of onion (*Allium cepa*) and forms numerous small black sclerotia on bulb scales. Recorded from Australasia, Europe, Japan, Mauritius, N. & S. America [CMI Distribution Map 164].

Botrytis state of **Sclerotinia sphaerosperma** Gregory, 1941, *Trans. Br. mycol. Soc.*, **25**: 37.

Sclerotia 1·5–2·5 × 0·5–1 mm. *Conidia* spherical, 18–28 (mostly 20–26)μ diam.

On *Allium triquetrum*; Isles of Scilly.

Botrytis state of **Sclerotinia globosa** (Buchwald) Webster, 1954, *Trans. Br. mycol. Soc.*, **37**: 168.

Botrytis globosa Raabe, 1938, *Hedwigia*, **78**: 71.

(Fig. 121 D)

Sclerotia on leaves linear, 3–15 × 1–2 mm. formed along the midrib, in culture discrete, oval or reniform, 2–10 mm. diam. Concertina-like collapse of *conidiophore* branches often well marked. *Conidia* spherical or subspherical, 9–23 (mostly 12–18)μ diam.

On leaves of garlic (*Allium ursinum*); Belgium, Germany, Great Britain. Fully described by Webster & Jarvis, *Trans. Br. mycol. Soc.*, **34**: 187–189, 1951 and G. L. Hennebert, *Bull. Jard. bot. État Brux.*, **28**: 193–207, 1958.

Botrytis state of **Sclerotinia draytonii** Buddin & Wakefield, 1946, *Trans. Br. mycol. Soc.*, **29**: 150–151.

Botrytis gladiolorum Timmermans, 1941, *Meded. Inst. Phytopath. Lab. BloembollOnderz-Lisse*, **67**: 15.

Sclerotia abundant, aggregated in large masses. *Conidia* 10–22 × 8–13µ.

Causes corm rot, leaf spots and water-soaked, oval lesions on flowers of *Gladiolus*. There are two types of leaf spot, one oblong, brown with red margin, the other small, round, rust-coloured. Specimens seen from Europe, N. America and E. Africa.

Botrytis state of **Sclerotinia convoluta** Drayton, 1937, *Mycologia*, **29**: 314.

Botrytis convoluta Whetzel & Drayton, 1932, *Mycologia*, **24**: 475.

Sclerotia characteristically convoluted, often confluent, shiny black, both on rhizomes and in culture. *Conidia* 7–18 × 5–13 (commonly 10–12 × 8–10)µ.

Parasitic on rhizomes of *Iris*; Canada, France, Germany, Great Britain, Netherlands, U.S.A., etc.

Botrytis elliptica (Berk.) Cooke, 1902, *Jl. R. Hort. Soc.*, **26**: CXXIX.

Characterised by its large ellipsoidal *conidia*, 16–35 × 10–24 (mostly 20–30 × 13–18)µ.

On many lilies including *Lilium candidum*; at first causing brown spots on leaves, later attacking flowering shoots. Recorded from N. America and many countries in Europe including Great Britain.

Botrytis state of **Sclerotinia narcissicola** Gregory, 1941, *Trans. Br. mycol. Soc.*, **25**: 36.

Botrytis narcissicola Kleb., 1906, *Jb. hamb. wiss. Anst.*, **24**, Beih. 3: 43.

Sclerotia black, smooth, spherical, 1–1·5 mm. diam. *Conidia* pale brown, 8–16 × 7·5–12µ.

On *Narcissus*, sometimes rotting the leaves at ground level, also bulbs; disease referred to as smoulder or grey mould. Distribution Australasia, Europe including Great Britain and N. America [CMI Distribution Map 315].

Botrytis state of **Sclerotinia polyblastis** Gregory, 1938, *Trans. Br. mycol. Soc.*, **22**: 202.

Botrytis polyblastis Dowson, 1928, *Trans. Br. mycol. Soc.*, **13**: 95–102.

Sclerotia elongated, flattened, 3–8 × 1·5–3 mm., immersed. *Conidia* when young pyriform, when mature almost spherical with a protuberant hilum, colourless to pale brown, 30–50µ diam.; on germination they form numerous germ tubes.

On living leaves of *Narcissus* causing a yellow leaf-blotch disease frequently called ' fire '; it also causes brown spots on the flowers. Recorded from Great Britain and North America.

Botrytis paeoniae Oudem., 1898, *Hedwigia*, **37**: 182.
(Fig. 121 E)

Sclerotia small, usually about 1 mm. diam. *Conidiophores* often swollen at the base. *Conidia* 10–22 × 7–11 (mostly 12–18 × 8–10)µ.

On peony (*Paeonia*) causing wilt of young shoots in the spring, mature leaves also later becoming infected; widely distributed in Europe, also recorded from N. & S. America.

Botrytis state of **Sclerotinia ricini** Godfrey, 1919, *Phytopathology*, **9**: 565–567.
(Fig. 121 F)

Sclerotia often elongated, usually 3–9 mm. long, but sometimes up to 2·5 cm. *Conidiophores* much branched, dichotomously. *Conidia* spherical, 6–12 (mostly 7–10)µ diam.

Causes an important disease of castor-oil plants (*Ricinus communis*); the inflorescence is the main part affected. Most records are on *Ricinus* in U.S.A.; it also occurs on species of *Acalypha, Euphorbia,* and *Phyllanthus* as well as *Ricinus* in other countries including Brazil, Bulgaria, Ghana, Jamaica, Malawi, Nigeria, Rhodesia and Sierra Leone. A comprehensive account is given by G. H. Godfrey in *J. agric. Res.*, **23**: 679–715, 1923.

Botrytis anthophila Bondarzew, 1913, *Bolez. Rast.*, **7**: 3.

Conidia 9·5–20 × 3·5–7 (mostly 11–16 × 4–5)µ.

Grows systemically in shoots of red clover (*Trifolium pratense*) and sporulates in the anthers making them appear grey. It is recorded for Australasia, Europe and U.S.A. [CMI Distribution Map 167]. An interesting account of this fungus is that of R. A. Silow, *Trans. Br. mycol. Soc.*, **18**: 239–248, 1923.

Botrytis state of **Sclerotinia spermophila** Noble, 1948, *Trans. Br. mycol. Soc.*, **30**: 84–91.

Sclerotia on seeds and in culture small, irregular, 0·5–1·5 mm. diam. *Conidia* mostly 10–18 × 6–10µ.

A parasite in seeds of white clover (*Trifolium repens*); Great Britain, New Zealand; affected seeds fail to germinate.

Botrytis tulipae Lind, 1913, Dan. Fung., Errata at end of book, no page number.

Sclerotia discrete or confluent, commonly 1–2 mm. diam., abundant in cultures where few conidia are formed. *Conidia* 12–22 × 8–15 (mostly 16–20 × 10–13)µ.

Most commonly recorded on tulip, causing 'fire', but also found accasionally on other plants such as asparagus; Australasia, many countries in Europe including Great Britain, N. & S. America [CMI Distribution Map 170]. A good account is that of Beaumont, Dillon Weston & Wallace in *Ann. appl. Biol.*, **23**: 57–88, 1936.

Botrytis fabae Sardiña, 1929, *Mems R. Soc. esp. Hist. nat.*, **15**: 291.
(Fig. 121 G)

Sclerotia abundant in cultures, discrete or sometimes confluent, mostly 1–1·7 mm. diam. *Conidia* 14–29 × 11–20 (mostly 16–25 × 13–16)μ.

On stems and leaves of broad bean (*Vicia faba*), the cause of chocolate spot disease. Recorded from Africa, Asia, Australasia, Europe including Great Britain, S. America [CMI Distribution Map 162].

111. PACHNOCYBE

Pachnocybe Berkeley, 1836, in J. E. Smith's Engl. Fl., **5** (2): 333–335.

Colonies effuse, rust-coloured; scattered, rather dark reddish brown synnemata each capped by a pale head easily seen under a low-power dissecting microscope. *Mycelium* mostly immersed. *Stroma* none. *Setae* and *hyphopodia* absent. *Conidiophores* macronematous, synnematous; synnemata mid to dark reddish brown; individual threads branched penicillately at the apex, stipes straight or flexuous, pale burnt sienna, smooth. *Conidiogenous cells* polyblastic, integrated and terminal or discrete, determinate, clavate. *Conidia* solitary, dry, arising close together from the upper part of the conidiogenous cell, simple, ellipsoidal, hyaline to straw-coloured, smooth, 0-septate.

Lectotype species: Pachnocybe ferruginea (Sow. ex Fr.) Berk.

Mason, E. W. & Ellis, M. B., *Mycol. Pap.*, **56**: 41–47, 1953.

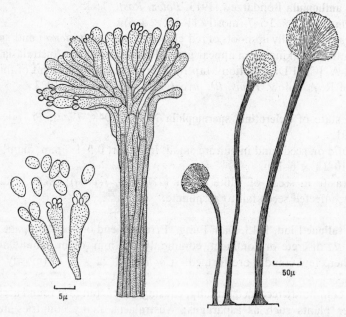

Fig. 122. *Pachnocybe ferruginea* (× 650 except where indicated by scales).

Pachnocybe ferruginea (Sow. ex Fr.) Berk., 1836, in J. E. Smith's Engl. Fl., **5** (2): 334.

(Fig. 122)

Synnemata up to 1 mm. high, 10–20μ thick, splaying out in the head to

80–120μ; individual threads 1–3μ thick. Swollen part of *conidiogenous cells* 4–5μ thick. *Conidia* 4–7 × 2–3·5μ.

On floor boards and other timber; Europe.

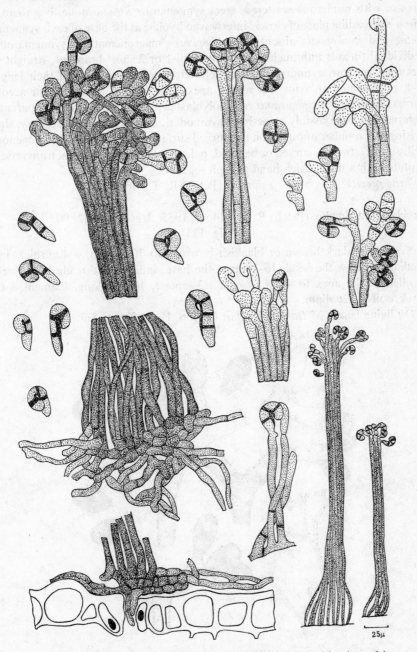

FIG. 123. *Trochophora simplex* (× 650 except where indicated by the scale).

112. TROCHOPHORA

Trochophora R. T. Moore, 1955, *Mycologia*, **47**: 90.

Colonies hypophyllous, small, often square or oblong, mid to dark olivaceous brown, with numerous scattered, erect synnemata. *Stroma* none, but there is often a spreading plate of closely interwoven hyphae at the base of each synnema. *Setae* and *hyphopodia* absent. *Conidiophores* macronematous, synnematous; individual threads unbranched or rarely with 1 or 2 short branches, straight or flexuous, narrow, cylindrical and closely adpressed along most of their length but swollen and splaying out at the apex, pale to mid brown or olivaceous brown, smooth. *Conidiogenous cells* polyblastic, integrated, terminal, apparently determinate but possibly sometimes sympodial, clavate. *Conidia* solitary, dry, arising at a number of points on the curved surface of the swollen conidiogenous cell, simple, strongly curved or helicoid, pale to mid brown, smooth, transversely septate with a dark, thick band at each septum.

Type species: Trochophora simplex (Petch) R. T. Moore.

Trochophora simplex (Petch) R. T. Moore, 1955, *Mycologia*, **47**: 90.
(Fig. 123)

Synnemata dark brown or blackish brown, 200–700µ high, 9–40µ thick immediately below the head, 16–100µ at the base; individual threads 2–4µ thick, swollen at the apex to 4–8µ. *Conidia* 3-septate, 10–26µ long, filament 3–6µ thick, coil 9–12µ diam.

On living leaves of *Daphniphyllum*; Ceylon, Hong Kong, Taiwan.

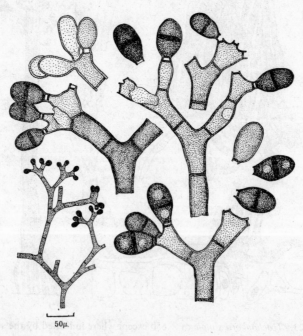

50µ

FIG. 124. *Balanium stygium* (× 650 except where indicated by the scale).

113. BALANIUM

Balanium Wallroth, 1833, *Fl. crypt. Ger.*, **2**: 159.

Colonies effuse, black. *Mycelium* partly superficial, partly immersed. *Stroma* none. *Setae* and *hyphopodia* absent. *Conidiophores* macronematous, mononematous, dichotomously and trichotomously branched. *Conidiogenous cells* integrated, terminal, polyblastic, determinate, cylindrical, denticulate, with broad, cylindrical separating cells. *Conidia* solitary, acrogenous, simple, dark, 1-septate, smooth, ellipsoidal, base slightly protruding, frilled.

Type species: Balanium stygium Wallroth.

Ellis, M. B., *Mycol. Pap.*, **73**: 19–20, 1961.

Hughes, S. J. & Hennebert, G. L., *Can. J. Bot.*, **39**: 1505–1508, 1961.

Balanium stygium Wallr., 1833, Fl. crypt. Ger., **2**: 160.
(Fig. 124)

Colonies effuse, black, velvety. *Mycelium* partly superficial, partly immersed; hyphae pale to dark brown, smooth, 4–15μ thick. *Conidiophores* repeatedly dichotomously or trichotomously branched, brown to dark brown, end branches paler, smooth, septate, 5–14μ thick; terminal branches at point of abscission of conidia 3–7μ thick. *Conidia* solitary at tip of each terminal branch, more or

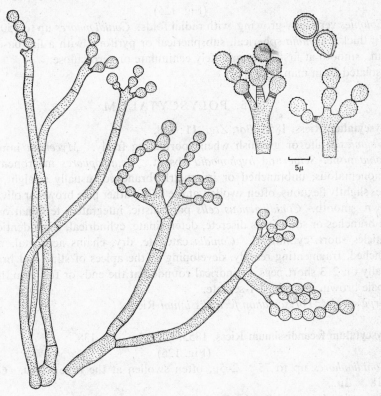

Fig. 125. *Polypaecilum insolitum* (× 650 except where indicated by the scale).

less oval, very dark brown to black, smooth, 1-septate, 13–23μ long, 8–14μ
thick, with a slightly protruding and truncate base, 3–4 (occasionally up to 7)μ
wide.

On dead wood of elder, *Hippophaë rhamnoides*, etc.; Belgium, Germany, Great
Britain.

114. POLYPAECILUM

Polypaecilum G. Smith, 1961, *Trans. Br. mycol. Soc.*, **44**: 437–440.
Colonies slow-growing, effuse, brownish grey. *Mycelium* partly superficial,
partly immersed; thick-walled *chlamydospores* sometimes formed in old cultures.
Stroma none. *Setae* and *hyphopodia* absent. *Conidiophores* macronematous,
mononematous, branched, often dichotomously or trichotomously, straight or
flexuous, pale brown, smooth. *Conidiogenous cells* polyblastic, integrated,
terminal, clavate or cylindrical, denticulate. Each broad cylindrical denticle
bears a succession of conidia and proliferates percurrently. *Conidia* catenate,
acrogenous, simple, spherical, subspherical, pyriform or ellipsoidal, pale brown,
smooth, finely echinulate or verruculose, 0-septate.
Type species: Polypaecilum insolitum G. Smith.

Polypaecilum insolitum G. Smith, 1961, *Trans. Br. mycol. Soc.*, **44**: 437–439.
(Fig. 125)
Colonies very slow-growing, with radial folds. *Conidiophores* up to 150μ long,
3–7μ thick. *Conidia* spherical, subspherical or pyriform with a flat base, 4–7μ
diam., smooth at first, later minutely echinulate or verruculose.
Isolated from man; Europe.

115. POLYSCYTALUM

Polyscytalum Riess, 1853, *Bot. Ztg.*, **11**: 138.
Colonies white or greenish when sporulating freely. *Mycelium* immersed.
Stroma none. *Setae* and *hyphopodia* absent. *Conidiophores* macronematous,
mononematous, unbranched or irregularly branched, usually straight, sometimes slightly flexuous, often swollen at the base, rather pale brown or olivaceous
brown, smooth. *Conidiogenous cells* polyblastic, integrated, terminal on stipe
and branches or sometimes discrete, determinate, cylindrical, often denticulate;
denticles short, cylindrical. *Conidia* catenate, dry, chains acropetal, usually
branched, fragmenting readily, developing at the apices of stipe and branches
usually on 2–3 short pegs, cylindrical rounded at the ends or fusiform, hyaline
or pale brown, smooth, 0–1-septate.
Type species: Polyscytalum fecundissimum Riess.

Polyscytalum fecundissimum Riess, 1853, *Bot. Ztg.*, **11**: 138.
(Fig. 126)
Conidiophores up to 75 × 2–5μ, often swollen at the base to 7μ. *Conidia*
13–18 × 2μ.
On rotting leaves of *Betula*, *Fagus*, *Quercus* and *Salix*; Europe.

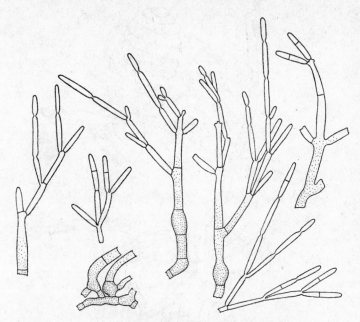

FIG. 126. *Polyscytalum fecundissimum* (× 650).

116. HELICOON

Helicoon Morgan, 1892, *J. Cincinn. Soc. nat. Hist.*, **15**: 49–51.

Colonies effuse, yellow, grey, olivaceous or brown, velvety or loosely cottony. *Mycelium* partly superficial, partly immersed. *Stroma* none. *Setae* and *hyphopodia* absent. *Conidiophores* macronematous or micronematous, mononematous, unbranched or branched, straight or flexuous, hyaline to brown, smooth. *Conidiogenous cells* polyblastic or monoblastic, integrated, terminal and intercalary, sympodial or determinate, denticulate; denticles cylindrical. *Conidia* solitary, dry, acropleurogenous or pleurogenous, simple, coiled in 3 planes to form an ellipsoidal or sometimes a cylindrical spore body, colourless to brown; filaments smooth, multiseptate.

Lectotype species: Helicoon sessile Morgan.

Linder, D. H., *Ann. Mo. Bot. Gdn.*, **16**: 322–329, 1929.

Helicoon ellipticum (Peck) Morgan, 1892, *J. Cincinn. Soc. nat. Hist.*, **15**: 50–51.

(Fig. 127)

Colonies olivaceous brown, cottony or velvety. *Conidiophores* micronematous, branched and anastomosing, mid pale olivaceous brown, 3·5–5μ thick. *Conidia* ellipsoidal, 32–40 × 22–30μ, composed of 7–9 coils, pale straw coloured; filaments multiseptate, 3–5μ thick.

On decaying coniferous wood in moist places; U.S.A.

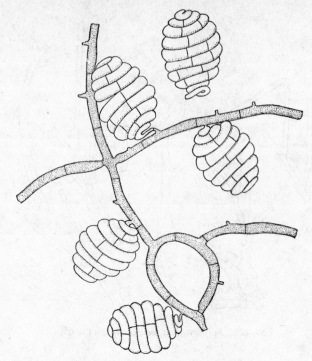

FIG. 127.   *Helicoon ellipticum* (× 650).

117. HELICOMA

Helicoma Corda, 1837, Icon. Fung., **1**: 15.
  Helicocoryne Corda, 1854, Icon. Fung., **6**: 9.
  Colonies effuse, often hairy or velvety, variable in colour but frequently buff, greyish brown, brown or olivaceous. *Mycelium* partly immersed, partly superficial. *Stroma* none. *Setae* and *hyphopodia* absent. *Conidiophores* macronematous, mononematous, branched or unbranched, apex sometimes setiform, straight or flexuous, pale to dark brown or olivaceous brown, smooth. *Conidiogenous cells* polyblastic or monoblastic, integrated, terminal and intercalary, determinate or sympodial, cylindrical, denticulate; denticles cylindrical. *Conidia* solitary, dry, acropleurogenous or pleurogenous, simple, helicoid, colourless to brown or olivaceous brown, smooth, multiseptate.
  Type species: Helicoma state of *Lasiosphaeria pezicula* (Berk. & Curt.) Sacc. = *H. muelleri* Corda.
  Linder, D. H., *Ann. Mo. bot. Gdns*, **16**: 295–317, 1929.

Helicoma state of **Lasiosphaeria pezicula** (Berk. & Curt.) Sacc., 1883, Syll. Fung., **2**: 195. For synonymy see Hughes, *Can. J. Bot.*, **36**: 772, 1958.
(Fig. 128)
  Colonies effuse, hairy or velvety, buff, olivaceous brown or brown. *Conidiophores* simple or occasionally branched, apex sometimes setiform, up to 180μ

long, 5–8μ thick, brown or olivaceous brown, paler towards the apex. *Conidia* mostly pleurogenous, borne on cylindrical pegs, $1\frac{1}{2}$–$1\frac{3}{4}$ times coiled, colourless to rather pale olivaceous brown, smooth, 17–21μ diam.; filaments rounded at the apex truncate at the base, 5–12-septate, 3·5–6μ thick.

On decaying wood and bark of deciduous trees; Europe, N. America.

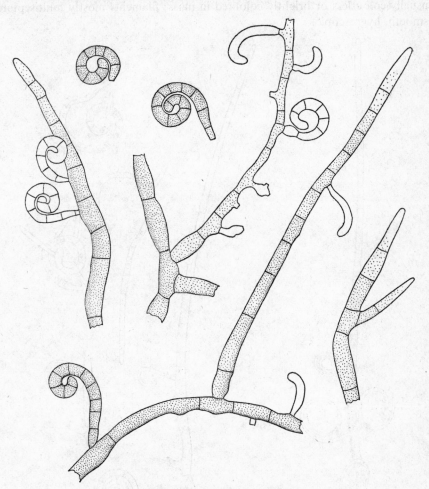

FIG. 128. *Helicoma* state of *Lasiosphaeria pezicula* (× 650).

118. HELICOSPORIUM

Helicosporium C. G. Nees, 1816, Syst. Pilze Schwämme; 68.

Helicotrichum C. G. & T. F. L. Nees, 1818, *Nova Acta Acad. Caesar. Leop. Carol.*, **9**: 246–247.

Colonies effuse, brightly coloured or fuscous, hairy or cottony. *Mycelium* partly superficial, partly immersed. *Stroma* none. Separate *setae* not formed but the upper part of the conidiophore is often sterile and setiform. *Hyphopodia* absent. *Conidiophores* macronematous, mononematous, unbranched or loosely branched, branches when present sometimes anastomosing, straight or flexuous, apex often setiform. *Conidiogenous cells* polyblastic or monoblastic, integrated, intercalary and sometimes terminal, occasionally small and discrete, sympodial or determinate, cylindrical, denticulate; denticles cylindrical, often narrow. *Conidia* solitary, dry, pleurogenous or acropleurogenous, simple, helicoid, usually colourless or brightly coloured in mass; filaments mostly multiseptate, smooth, hygroscopic.

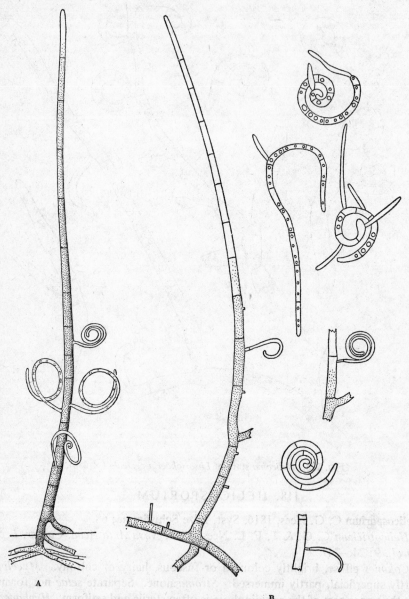

FIG. 129. A, *Helicosporium* state of *Tubeufia cerea*; B, *H.* state of *Tubeufia helicomyces* (× 650).

Type species: Helicosporium state of *Tubeufia cerea* (Berk. & Curt.) Booth = *H. vegetum* C. G. Nees.

Linder, D. H., *Ann. Mo. bot. Gdn*, **16**: 275–294, 1929: with key to 14 species.

Helicosporium state of **Tubeufia cerea** (Berk. & Curt.) Booth, 1964, *Mycol. Pap.*, **94**: 11–13. For additional synonyms see Hughes, *Can. J. Bot.*, **36**: 773–774, 1958.

(Fig. 129 A)

Colonies hairy, bright lemon or greenish yellow. *Conidiophores* usually un-branched, subulate, up to 350μ long, lower part mid to dark brown, 4–5μ thick, fertile, upper part pale, tapering to a point, sterile. *Conidia* helically coiled 2–3 times in one plane, hyaline, greenish yellow in mass, 10–20μ diam.; filaments about 1μ thick, septa if present indistinct.

Common on dead wood and bark lying on the ground; Europe, N. America.

Helicosporium state of **Tubeufia helicomyces** Höhnel, 1909, *Sber. Akad. Wiss. Wien*, **118**: 1477.

H. phragmitis Höhnel, 1905, *Annls mycol.*, **3**: 338.

(Fig. 129 B)

Colonies raised, cottony pale grey to pale brown, tinged with pink when sporing freely. *Conidiophores* unbranched or loosely branched, branches some-times anastomosing, often forked close to the base, upper part frequently sterile, colourless to pale brown, up to 370μ long, 4–5μ thick at the base, tapering to 2–3·5μ at the apex. *Conidia* helically coiled 2–4 times in one plane, colourless, pink in mass, 14–21 (18)μ diam.; filaments 1·5–2·5μ thick, 7–12-septate.

Common on decaying culms of *Glyceria* lying in the very damp basal leaf mat, also found on *Arrhenatherum*, *Phalaris*, *Phragmites* and occasionally other substrata, always in damp situations; Europe. The connection between conidial and perfect state was demonstrated by J. Webster in *Trans. Br. mycol. Soc.*, **34**: 304–305, 1951.

119. HIOSPIRA

Hiospira R. T. Moore, 1962, *Trans. Br. mycol. Soc.*, **45**: 143–146.

Colonies effuse, brown, usually hypophyllous. *Mycelium* superficial. *Stroma* none. *Setae* absent but the upper part of the conidiophore is often sterile and sometimes the whole of what looks just like a conidiophore remains sterile. *Hyphopodia* absent. *Conidiophores* macronematous, mononematous, unbranched, straight or flexuous, olive to rather pale brown, narrow and smooth near the base but soon expanded into a broader cylinder with coarsely reticulate walls, upper part often sterile, tapered to a rounded apex. *Conidiogenous cells* mono-blastic or polyblastic, integrated, intercalary, determinate, cylindrical, denticu-late; denticles cylindrical or slightly tapered. *Conidia* solitary, dry, pleurogenous, simple, helicoid, olive to rather pale brown, filaments multiseptate with coarsely reticulate walls.

Type species: Hiospira state of *Brooksia tropicalis* Hansf. = *H. hendrickxii* (Hansf.) R. T. Moore.

Deighton, F. C. & Pirozynski, K. A., *Mycol. Pap.*, **105**: 2–10, 1966.

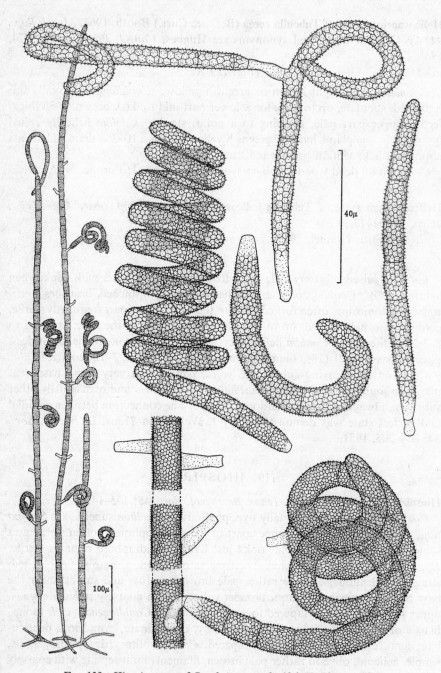

40μ

100μ

FIG. 130. *Hiospira* state of *Brooksia tropicalis* (del. K. Pirozynski).

Hiospira state of **Brooksia tropicalis** Hansf., 1956, *Proc. Linn. Soc. N.S.W.*, **81**: 33.

(Fig. 130)

Conidiophores up to 3 mm. long, 6·5–10μ thick except near the base where they are narrower. *Conidia* much coiled, up to 90 × 40μ; filaments 7–10μ thick.

On leaves of many different plants, often associated with other fungi; very common in the moist tropics.

120. VIRGARIELLA

Virgariella Hughes, 1953, *Can. J. Bot.*, **31**: 653–654.

Colonies effuse, dark brown to black, often hairy. *Mycelium* partly superficial but mostly immersed. *Stroma* none. *Setae* and *hyphopodia* absent. *Conidiophores* macronematous, mononematous, scattered, erect, unbranched, straight or flexuous, dark brown, smooth, thick-walled. *Conidiogenous cells* polyblastic, integrated, terminal, sympodial, cylindrical. *Conidia* solitary, dry,

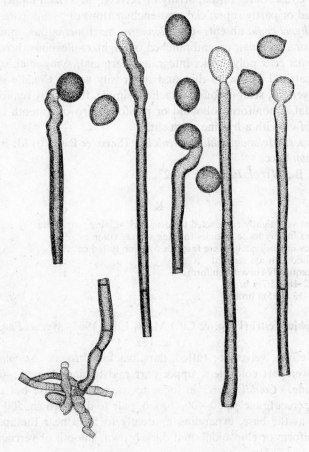

Fig. 131. *Virgariella atra* (× 650).

acropleurogenous, simple, broadly ellipsoidal, subspherical or spherical, dark brown, smooth, thick-walled, 0-septate.

Type species: Virgariella globigera (Sacc. & Ellis) Hughes.

Virgariella atra Hughes, 1953, *Can. J. Bot.*, **31**: 654.
(Fig. 131)

Colonies effuse, black, hairy. *Conidiophores* up to 200 × 4–7μ. *Conidia* 11–15 × 10–13μ.

On rotten wood of *Fagus, Fraxinus* and *Quercus*; Great Britain.

121. MELANOGRAPHIUM

Melanographium Saccardo, 1913, *Annls mycol.*, **11**: 557–558.

Sporostachys Saccardo, 1919, *Atti Accad. scient. veneto-trent.-istriana*, Ser. 3, **10**: 92 [1917–1919, published 1919].

Pseudocamptoum Fragoso & Ciferri, 1925, *Boln R. Soc. esp. Hist. nat.*, **25**: 453–455.

Colonies effuse, dark, tufted, bristly or velvety. *Mycelium* immersed. *Stroma* all immersed or partly superficial, prosenchymatous or pseudoparenchymatous. *Setae* and *hyphopodia* absent. *Conidiophores* macronematous, mononematous, caespitose or synnematous, unbranched, straight or flexuous, brown, smooth. *Conidiogenous cells* polyblastic, integrated, terminal, sympodial, cylindrical or clavate, cicatrized but scars thin and not easily seen. *Conidia* solitary, dry, acropleurogenous, seldom found attached, simple, frequently reniform but may be ellipsoidal, limoniform, obovoid or pyriform, brown, smooth or verrucose, 0-septate, often with a hyaline germ slit.

Type species: Melanographium selenioides (Sacc. & Paoletti) M. B. Ellis = *M. spleniosporum* Sacc.

Ellis, M. B., *Mycol. Pap.*, **93**: 14–23, 1963.

KEY

Conidiophores very tightly compacted to form dark shining synnemata .	*selenioides*
Conidiophores in very loose fascicles forming velvety colonies . . .	*spinulosum*
Conidiophores in moderately dense fascicles forming tufted colonies	1
1. Conidia almost always straight	*fasciculatum*
Conidia frequently curved, reniform	2
2. Conidia 12–18 (15·7)μ broad	*cookei*
Conidia 8–13 (10·5)μ broad	*citri*

Melanographium citri (Frag. & Cif.) M. B. Ellis, 1963, *Mycol. Pap.*, **93**: 21–23.
(Fig. 132 A)

Colonies effuse, velvety or tufted, dark blackish brown. *Stroma* mostly immersed, lower part colourless, upper part mid to dark brown, 40–120μ high, 30–300μ wide. *Conidiophores* in loose to moderately dense but never tightly compacted, erect fascicles, 30–400μ broad, pale to dark brown, 500–1200μ long, 3–5μ thick at the base, expanding gradually to 5–7μ near the apex. *Conidia* curved, reniform or ellipsoidal, mid dark brown, smooth or verruculose, 14–19 (15·8) × 8–13 (10·5)μ.

On dead wood and bark of various trees including *Aleurites*, *Averrhoa*, *Bombax*, *Borassus*, *Citrus*, *Coccothrinax*, *Cocos*, *Elaeis*, *Hippocratea*, *Loranthus*, *Olax*, *Phoenix*, *Sabal*, *Theobroma* and *Trachycarpus*; Dominican Republic, Ghana, India, Japan, Nigeria, Sierra Leone, Sudan, Sumatra and Venezuela.

Conidial measurements, substratum range and geographical distribution of 4 other species of *Melanographium* are given below.

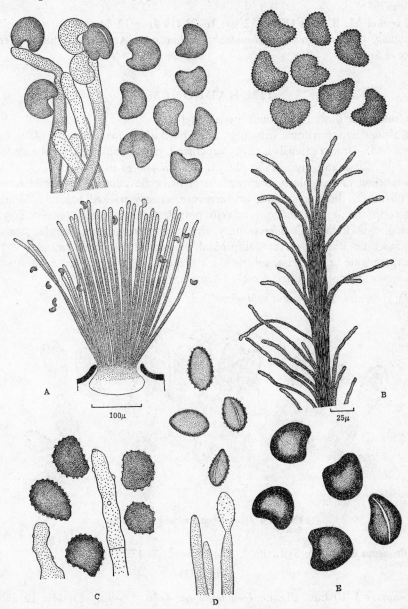

FIG. 132. *Melanographium* species: A, *citri*; B, *selenioides*; C, *spinulosum*; D, *fasciculatum*; E, *cookei* (× 650 except where indicated by scales).

M. selenioides (Sacc. & Paoletti) M. B. Ellis (Fig. 132 B): 15–23 (17) × 8–14 (11·4)μ, on *Arenga* and unknown palm; Java, Malacca, Philippines.

M. spinulosum (Speg.) Hughes (Fig. 132 C): 17–23 (19) × 11–15 (13·5)μ, on unknown wood; Paraquay.

M. fasciculatum Hughes (Fig. 132 D): 13–20 (16·3) × 8–11 (9·5)μ, on *Capparis*; Philippines.

M. cookei M. B. Ellis (Fig. 132 E): 16–23 (19·9) × 12–18 (15·7)μ, on *Acacia, Bauhinia, Cassia, Citrus, Dichrostachys, Mansonia*; Australia, Ghana, Sierra Leone, Sudan.

122. HADRONEMA

Hadronema Sydow, 1909, *Annls mycol.*, **7**: 172.

Colonies hypophyllous, orbicular, dark blackish brown, velvety. *Mycelium* immersed. *Stroma* extensive, partly superficial, partly immersed, brown or dark brown. *Setae* and *hyphopodia* absent. *Conidiophores* macronematous, short, unbranched or irregularly branched, straight or flexuous, pale to mid brown or olivaceous brown, smooth or verrucose, closely packed together forming flat sporodochia. *Conidiogenous cells* polyblastic, integrated, terminal, sympodial, cylindrical. *Conidia* solitary, dry, acropleurogenous, simple, straight or sometimes curved, clavate, ellipsoidal, cylindrical or oblong rounded at the apex, truncate at the base, mid to dark brown, coarsely verrucose often with the warts arranged in lines, mostly 1-septate.

Type species: Hadronema orbiculare Syd.

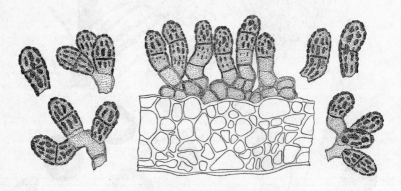

FIG. 133. *Hadronema orbiculare* (× 650).

Hadronema orbiculare Syd., 1909, *Annls mycol.*, **7**: 172.

(Fig. 133)

Colonies 1–10 mm. diam. *Conidiophores* 8–30 × 6–9μ. *Conidia* 19–27 × 8–15μ.

On living leaves of *Quercus glauca*; Japan.

123. IDRIELLA

Idriella Nelson & Wilhelm, 1956, *Mycologia*, **48**: 547–551.

Colonies effuse, grey to blackish brown. *Mycelium* partly superficial, partly immersed. *Stroma* none. *Setae* and *hyphopodia* absent. *Conidiophores* macronematous, mononematous, unbranched, straight or flexuous, subulate, usually swollen at the base, geniculate towards the apex, pale brown, smooth. *Conidiogenous cells* polyblastic, integrated, terminal, sympodial, subulate, denticulate; denticles short, conical. *Conidia* solitary, dry, acropleurogenous, simple, falcate or lunate, hyaline, smooth, 0-septate. There is also a *Trichocladium* state with brown, septate conidia.

Type species: Idriella lunata Nelson & Wilhelm.

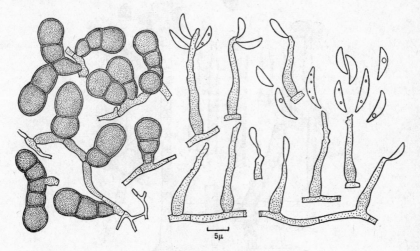

FIG. 134. *Idriella lunata.*

Idriella lunata Nelson & Wilhelm, 1956, *Mycologia*, **48**; 547–551.

(Fig. 134)

Conidiophores up to 35μ long, 3–4μ thick at the base, tapering to 1μ at the apex. *Conidia* 7–13 × 2–2·5μ. *Trichocladium* conidia golden brown, 1–4-septate, smooth, 10–25 × 7–9μ.

Isolated from soil and strawberry (*Fragaria*) roots; N. America. On strawberry it causes a root rot with black, sunken lesions.

124. CORDANA

Cordana Preuss, 1851, *Linnaea*, **24**: 129–130.

Preussiaster O. Kunze, 1891, Revis. Gen. Pl., **2**: 867.

Colonies effuse, brown, greyish brown or black, often thinly hairy. *Mycelium* mostly immersed. *Stroma* none. *Setae* and *hyphopodia* absent. *Conidiophores* macronematous, mononematous, straight or flexuous, unbranched, with terminal and usually also intercalary swellings, brown, smooth. *Conidiogenous cells* polyblastic, integrated, terminal and becoming also intercalary, sympodial, doliiform, spherical or subspherical, denticulate; denticles small, cyclindrical. *Conidia* solitary, acropleurogenous, simple, ellipsoidal, obovoid, ovoid or pyriform, pale to dark brown, smooth, 1-septate, in some species with a thick dark band at the septum, often with a protuberant hilum.

Lectotype species: Cordana pauciseptata Preuss.

Hughes, S. J., *Can. J. Bot.*, **33**: 259–263, 1955.

Meredith, D. S., *Ann. Bot.*, **26**: 233–241, 1962.

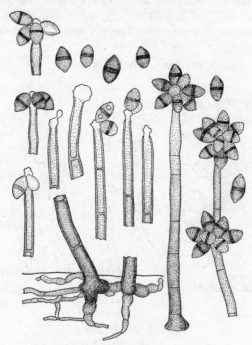

FIG. 135. *Cordana pauciseptata* (× 650).

Cordana pauciseptata Preuss, 1851, *Linnaea*, **24**: 129.
(Fig. 135)

Colonies effuse, almost black. *Conidiophores* subulate, pale to dark brown, up to 170μ long, 3–6μ thick, often swollen at the base to 7–12μ; conidium-bearing terminal and intercalary swellings 4–8μ diam. *Conidia* borne on short denticles, broadly ellipsoidal or ovoid, mid to dark brown, often with a thick, very dark band at the septum, smooth, 8–12 × 5–7μ.

On wood and bark of deciduous trees and conifers; Europe and N. America. Grows and sporulates well in culture.

Cordana musae (Zimm.) Höhnel, 1923, *Zentbl. Bakt. ParasitKde*, Abt. 2, **60**: 7.
Scolecotrichum musae Zimm., 1902, *Zentbl. Bakt. ParasitKde*, Abt. 2, **8**: 220.
(Fig. 136)

Colonies hypophyllous, effuse, greyish brown, hairy. *Conidiophores* straight or flexuous, often nodose, pale to mid brown, smooth, up to 220μ long, 4–6μ thick, usually swollen at the base to 9–11μ, terminal and intercalary swellings

6–8μ diam. *Conidia* solitary on small pegs arising from terminal swellings which later become intercalary, obovoid or pyriform, 1-septate, sometimes slightly constricted at the septum, subhyaline to pale brown, smooth, 11–18μ long, 7–10μ thick in the broadest part; hilum protuberant.

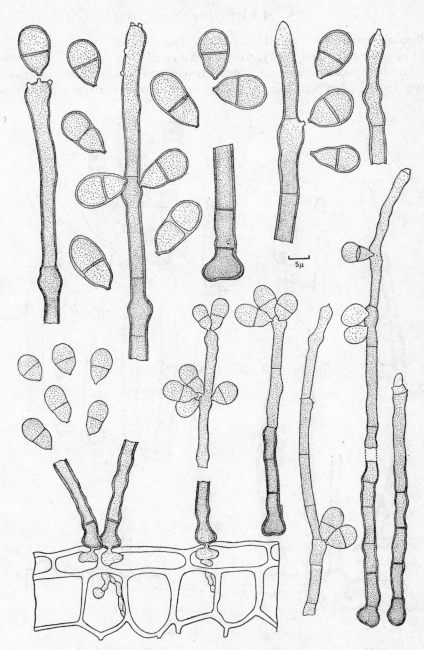

FIG. 136. *Cordana musae* (× 650 except where indicated by the scale).

On *Musa* spp., causing large oval, pale brown spots which are sometimes zonate with a dark border, each surrounded when fresh by a yellow or orange halo; occurs throughout the tropics wherever bananas are grown [CMI Distribution Map 168].

125. PLEUROPHRAGMIUM

Pleurophragmium Costantin, 1888, Mucéd. simpl.: 100.

Colonies effuse, thinly hairy, pale brown to dark greyish brown. *Mycelium* mostly immersed. *Stroma* often present, partly or wholly immersed, mid or dark brown, often plate-like. *Conidiophores* macronematous, mononematous,

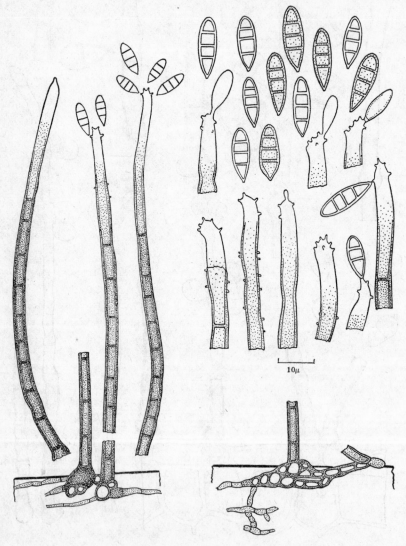

FIG. 137. *Pleurophragmium simplex* (× 650 except where indicated by the scale).

unbranched, straight or flexuous, mid to dark brown, usually paler at the apex, smooth. *Conidiogenous cells* polyblastic, integrated, terminal, sympodial, cylindrical, denticulate; denticles usually tapered to a point. *Conidia* solitary, dry, acropleurogenous, simple, narrowly ellipsoidal to subclavate, rounded at the

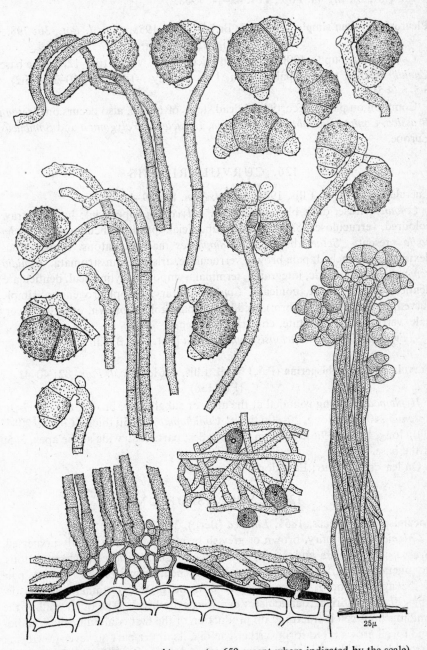

Fig. 138. *Curvulariopsis cymbisperma* (× 650 except where indicated by the scale).

apex, tapered to a point at the base which has no flat scar, hyaline to brown, smooth or verruculose, with 0, 1 or a few septa.

Type species: Pleurophragmium simplex (Berk. & Br.) Hughes = *P. bicolor* Cost.

Ellis, M. B., *Mycol. Pap.*, **114**: 42–44, 1968.

Pleurophragmium simplex (Berk. & Br.) Hughes, 1958, *Can. J. Bot.*, **36**: 798.
(Fig. 137)

Conidiophores up to 400μ long, 4–6μ thick, often swollen to 7–12μ at the base. *Conidia* hyaline to very pale brown, 1–4- (mostly 3-) septate, 10–21 (15·2) × 3·5–6 (4·5)μ.

Common on partly decorticated dead stems of *Urtica*, also occurs on *Arctium, Brassica, Conium, Epilobium, Filipendula, Heracleum, Polygonum* and *Sambucus*; Europe.

126. CURVULARIOPSIS

Curvulariopsis M. B. Ellis, 1961, *Mycol. Pap.*, **82**: 39–41.

Colonies effuse, dark blackish brown. *Mycelium* superficial; hyphae straw-coloured, verruculose. *Stroma* partly superficial, partly subcuticular. *Hypho-podia* present. *Setae* absent. *Conidiophores* macronematous, synnematous, flexuous, unbranched, pale brown, verruculose, arising from stromata. *Conidio-genous cells* polyblastic, integrated, terminal, sympodial, cylindrical, denticulate; denticles broad, short, conical. *Conidia* solitary, acropleurogenous, simple, curved, navicular (cymbiform), middle cell pale golden brown, other cells very pale, verrucose, 2–5-septate, constricted at septa.

Type species: Curvulariopsis cymbisperma (Pat.) M. B. Ellis.

Curvulariopsis cymbisperma (Pat.) M. B. Ellis, 1961, *Mycol. Pap.*, **82**: 40–41.
(Fig. 138)

Hyphopodia oblong rounded at the apex or subglobose, brown, 8–11 × 7–9μ. *Coremia* 240–400μ long, 20–40μ thick. *Conidiophores* 3–5μ thick. *Conidia* 20–45 (38)μ long, 15–24 (19·6)μ thick in the broadest part, 4–8μ wide at the apex, 3–5μ at the base.

On leaves of *Rubus*; Equador.

127. CACUMISPORIUM

Cacumisporium Preuss, 1851, *Linnaea* (Berl.), **24**: 130.

Colonies effuse, hairy, brown or greyish brown. *Mycelium* mostly immersed. *Stroma* none. *Setae* and *hyphopodia* absent. *Conidiophores* macronematous, mononematous, straight or flexuous, unbranched, brown or dark brown, pale at the apex, smooth. *Conidiogenous cells* integrated, terminal, polyblastic, sympodial, cylindrical, denticulate; denticles cylindrical. The outer wall of the conidiogenous cell is fractured on production of the first conidium; the conidio-genous cell grows on to form further conidia, its upper part however now being surrounded only by the inner wall. *Conidia* aggregated in slimy heads although formed singly at the ends of denticles, acropleurogenous, simple, septate, at first colourless, later brown, the end cells often remaining paler than the central ones, smooth, allantoid, ellipsoidal or oblong rounded at the ends.

Type species: Cacumisporium tenebrosum Preuss? = *C. capitulatum* fide Hughes.

Goos, R. D., *Mycologia*, **61**: 52–56, 1969.

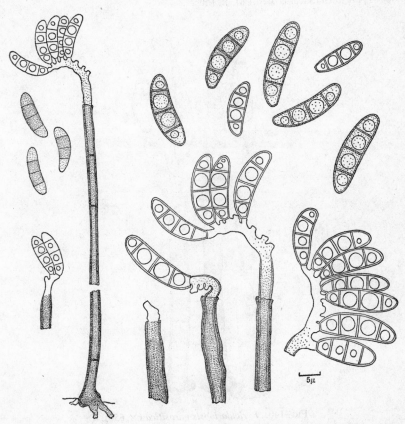

FIG. 139. *Cacumisporium capitulatum* (× 650 except where indicated by the scale).

Cacumisporium capitulatum (Corda) Hughes, 1958, *Can. J. Bot.*, **36**: 743 (with full synonymy).

(Fig. 139)

Conidiophores up to 260μ long, 4–7μ thick. *Conidia* often allantoid, 3-septate, at first colourless, later brown, the central cells darker than the end ones, 16–27 (20) × 5–7 (5·7)μ.

On dead wood and bark of beech, birch, lime, oak, sycamore etc.; Europe including Great Britain, N. America.

128. PYRICULARIOPSIS

Pyriculariopsis M. B. Ellis gen. nov.

Coloniae effusae, griseae vel atrae. *Mycelium* immersum ex hyphis ramosis, septatis, hyalinis vel atro-brunneis, laevibus compositum; cellulae myceliales in hypodermide epidermideque inflatae. *Conidiophora* singula, ex cellulis inflatis oriunda, erecta, simplicia, recta vel flexuosa, saepe basi inflata, laevia, atro-brunnea, crasse tunicata, apicem versus pallidiora, denticulis conico-truncatis vel cylindricis numerosis praedita. *Conidia* singula ex denticulis oriunda, sicca, acropleurogena, recta vel curvata, obclavata, rostrata, laevia, plerumque 3-septata, cellulis extimis hyalinis vel subhyalinis, cellulis intermediis pallide brunneis. Species typica: *Pyriculariopsis parasitica* (Sacc. & Berl.) M. B. Ellis.

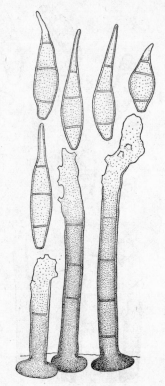

FIG. 140. *Pyriculariopsis parasitica* (× 650).

Colonies effuse, grey to black. *Mycelium* immersed. *Stroma* none but mycelial cells in hypodermis and epidermis swollen. *Setae* and *hyphopodia* absent. *Conidiophores* macronematous, mononematous, scattered, unbranched, straight or flexuous, often swollen at the base, smooth, thick-walled, dark brown, paler towards the apex. *Conidiogenous cells* polyblastic, integrated, terminal becoming intercalary, sympodial, cylindrical to clavate, denticulate; denticles large, conico-truncate or cylindrical, not cut off by septa to form separating cells as they are in *Pyricularia*. *Conidia* solitary at ends of denticles, dry, acropleurogenous, simple, straight or curved, obclavate, rostrate, smooth, mostly 3-septate, cell at each end hyaline or subhyaline, intermediate cells rather pale brown.

Pyriculariopsis parasitica (Sacc. & Berl.) M. B. Ellis comb. nov.

Helminthosporium parasiticum Sacc. & Berl., 1889, *Revue mycol.*, **11**: 204.

Pyricularia musae Hughes, 1958, *Can. J. Bot.*, **36**: 800.

(Fig. 140)

Conidiophores up to 350μ long, 6–10μ thick, basal cell up to 20μ diam.
Conidia 30–55μ long, 9–12μ thick in the broadest part.

On leaves and stems of *Musa*; Cameroons, Ghana, Guinea, Jamaica, Sabah,
San Thomé, Sierra Leone, Taiwan.

129. TOMENTICOLA

Tomenticola Deighton, 1969, *Mycol. Pap.*, **117**: 20–25.

Colonies effuse, dark brown, velvety. *Mycelium* superficial, sometimes ascending leaf hairs. *Stroma* none. *Setae* and *hyphopodia* absent. *Conidiophores*

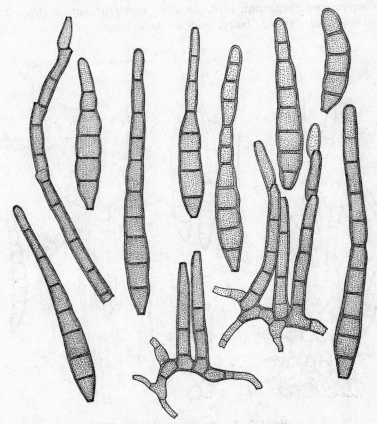

FIG. 141. *Tomenticola trematis* (× 650).

macronematous, mononematous, solitary or caespitose, unbranched, straight
or flexuous, moderately dark brown, smooth. *Conidiogenous cells* monoblastic,
integrated, terminal becoming intercalary, sympodial, cylindrical, conico-
truncate at the apex, successive apices becoming lateral and then appearing as

short, broad, conical denticles; cicatrized, scars very thin with a slight rim. *Conidia* solitary, dry, acropleurogenous, simple, straight or slightly curved, usually obclavate, sometimes rostrate, conico-truncate at the base, brown or olivaceous brown, smooth, multiseptate.

Type species: Tomenticola trematis Deighton.

Tomenticola trematis Deighton, 1969, *Mycol. Pap.*, **117**: 20–25.
(Fig. 141)

Conidiophores up to 450 × 4–7μ. *Conidia* 1–24-septate, 23–360μ long, 9–15μ thick in the broadest part, tapering to 3–6μ at the apex, 2·5–4μ wide at the scar.

On living leaves of *Trema*; Gabon, Ghana, Nigeria, Sierra Leone, Uganda.

130. PSEUDOCERCOSPORA

Pseudocercospora Spegazzini, 1910, *An. Mus. nac. Hist. nat. B. Aires*, **20**: 437.
Colonies effuse, tufted, hairy, mid to dark brown or olivaceous brown.

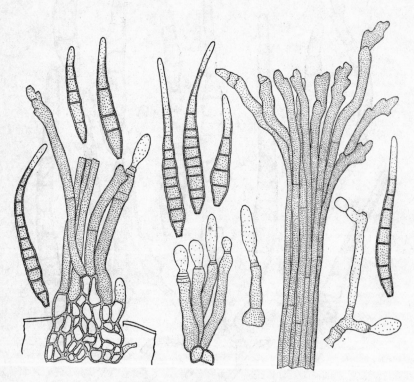

FIG. 142. *Pseudocercospora vitis* (× 650).

Mycelium immersed. *Stroma* present. *Setae* and *hyphopodia* absent. *Conidiophores* macronematous, synnematous or mononematous caespitose; individual threads unbranched, flexuous, often narrow, cylindrical and closely adpressed near the base, splaying out and somewhat swollen towards the apex, pale to mid brown or olivaceous brown, smooth. *Conidiogenous cells* integrated, terminal, often monoblastic and percurrent whilst the conidiophores are young, later polyblastic, sympodial and denticulate, with short, broad conical denticles and no scars. *Conidia* solitary, dry, acrogenous on young conidiophores, later acropleurogenous, simple, mostly obclavate, often rostrate, conico-truncate at the base, pale to mid brown, smooth or rugulose, with numerous transverse and occasionally one or two longitudinal or oblique septa.

Type species: Pseudocercospora vitis (Lév.) Speg.

Pseudocercospora vitis (Lév.) Speg., 1910, *An. Mus. nac. Hist. nat. B. Aires*, **20**: 438.

   Septonema vitis Lév., 1848, *Annls Sci. nat.*, 3 Sér., **9**: 261.
   Cercospora vitis (Lév.) Sacc., 1876, *Nuovo G. bot. ital.*, **8**: 188.
                              (Fig. 142)
   Colonies amphigenous. *Conidiophores* up to 500μ long, 2–7μ thick. *Conidia* 5–14-septate, 35–95μ long, 6–8μ thick in the broadest part, 2–3μ at apex and base.

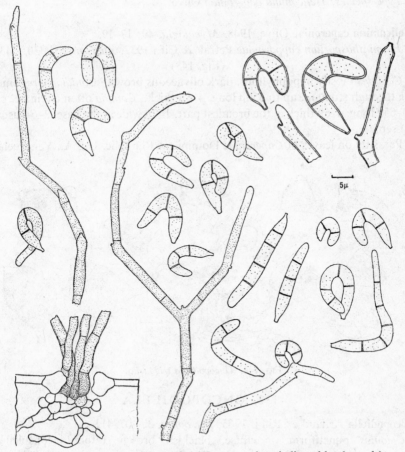

FIG. 143. *Helicomina caperoniae* (× 650 except where indicated by the scale).

On leaves of *Vitis*, causing large, round or irregular, sometimes confluent, brown, purplish brown or purple spots; Europe, India, Japan, Malaya, Mauritius, N. America, Pakistan, Rhodesia, Saudi Arabia, Tanzania, Trinidad, Zambia.

131. HELICOMINA

Helicomina Olive, 1948, *Mycologia*, **40**: 16–19.

Colonies effuse, velvety, mid to dark olivaceous brown. *Mycelium* immersed. *Stroma*, if present, composed of only a few cells in the substomatal cavity. *Setae* and *hyphopodia* absent. *Conidiophores* macronematous, mononematous, caespitose, unbranched or loosely branched, straight or flexuous, pale to mid brown or olivaceous brown, smooth. *Conidiogenous cells* polyblastic or monoblastic, integrated, terminal becoming intercalary, sympodial, cylindrical, denticulate; denticles cylindrical or conical and truncate. *Conidia* solitary, dry, acropleurogenous, simple, mostly curved, frequently circinate, hyaline to pale brown or olivaceous brown, smooth, with several transverse septa.

Type species: Helicomina caperoniae Olive.

Helicomina caperoniae Olive, 1948, *Mycologia*, **40**: 17–19.
   Helminthosporium capparoniae Petrak & Cif., 1932, *Annls mycol.*, **30**: 343.
(Fig. 143)

Colonies mostly hypophyllous, dark olivaceous brown. *Conidiophores* emerging through stomata, up to 350µ long, 4–5µ thick. *Conidia* often circinate, pale, 20–65µ long, 4–6µ thick in the broadest part, 1–2µ wide at the base, 2–6- (usually 3-) septate.

Parasitic on leaves of *Caperonia*; Dominican Republic, U.S.A., Venezuela.

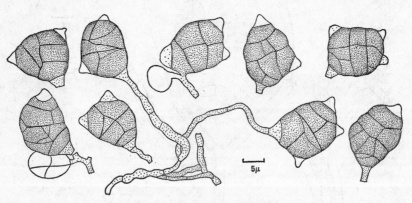

Fig. 144. *Oncopodiella trigonella*.

132. ONCOPODIELLA

Oncopodiella Arnaud ex Rifai, 1965, *Persoonia*, **3**: 407–411.

Colonies punctiform or effuse, blackish brown, rather inconspicuous. *Mycelium* partly superficial but mostly immersed. *Stroma* none. *Setae* and *hyphopodia* absent. *Conidiophores* macronematous, mononematous, slender, unbranched, flexuous, pale brown, smooth. *Conidiogenous cells* polyblastic, integrated, terminal, sympodial, cylindrical, denticulate; denticles cylindrical. *Conidia* solitary, dry, acropleurogenous, corniculate, trigonous, pyriform, subspherical or turbinate, mid pale to dark reddish brown, smooth, muriform, with a protuberant hilum; the 2–4 short, conical horns remain hyaline or subhyaline.

Type species: Oncopodiella trigonella (Sacc.) Rifai = *O. tetraedrica* Arnaud.

Oncopodiella trigonella (Sacc.) Rifai, 1965, *Persoonia*, **3**: 407–411.
(Fig. 144)
Conidiophores up to 30µ long, 1–3µ thick. *Conidia* 14–19 × 12–16µ, horns 2–3µ long.

On dead wood and bark of *Ulmus* and other trees; Europe.

133. VIRGARIA

Virgaria Nees ex S. F. Gray; Nees, 1816, Syst. Pilze Schwämme: 54; S. F. Gray, 1821, Nat. Arr. Br. Pl., **1**: 552.
Colonies effuse, dark olive to brown or almost black, often thick and felt-like.

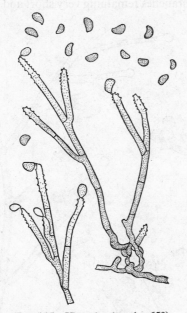

Fig. 145. *Virgaria nigra* (× 650).

Mycelium partly superficial, partly immersed. *Stroma* none. *Setae* and *hyphopodia* absent. *Conidiophores* macronematous, mononematous, erect or ascending, extensively branched, straight or flexuous, pale to mid brown, smooth. *Conidiogenous cells* polyblastic, integrated, terminal on branches, sympodial, cylindrical, denticulate; denticles cylindrical, numerous. *Conidia* solitary, dry, acropleurogenous, simple, reniform, often obliquely attenuated at the base, pale to mid brown, smooth, 0-septate.

Type species: Virgaria nigra (Link) Nees ex S. F. Gray.

Virgaria nigra (Link) Nees ex S. F. Gray, 1821, Nat. Arr. Br. Pl., **1**: 553.
(Fig. 145)

Conidiophores 60–250µ long, 2–3µ thick. *Conidia* 4–6 × 2·5–4µ.

On wood and bark of various trees including *Acer, Betula, Carpodinus, Combretum, Corylus, Diplorrhynchus, Fagus, Fraxinus, Heeria, Juglans, Pittosporum, Quercus* and *Sorbus*; Europe, New Zealand, N. America, Sierra Leone, Zambia.

134. DICYMA

Dicyma Boulanger, 1897, *Revue gén. Bot.*, **9**; 17–20.

Colonies effuse, dark greenish grey becoming black, velvety. *Mycelium* immersed and superficial. *Stroma* none. *Setae* and *hyphopodia* absent. *Conidiophores* macronematous or semi-macronematous, mononematous, straight or flexuous, dark brown or olivaceous brown near the base becoming paler upwards, smooth, sympodially and sometimes dichotomously or trichotomously branched, some of the branches remaining very short and non-septate as clavate,

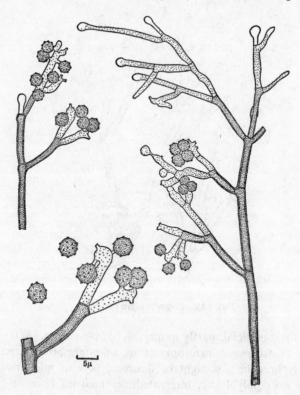

FIG. 146. **Dicyma** state of *Ascotricha chartarum* (× 650 except where indicated by the scale).

hyaline, sterile processes. *Conidiogenous cells* on branches polyblastic, integrated, terminal, or discrete, sympodial, cylindrical to clavate, denticulate, denticles cylindrical. *Conidia* solitary, dry, acropleurogenous, simple, obovoid, ellipsoidal or subspherical, 0-septate, olivaceous brown, smooth or verruculose.

Type species: Dicyma state of *Ascotricha chartarum* Berk. = *D. ampullifera* Boulanger.

Dicyma state of **Ascotricha chartarum** Berk., 1838, *Ann. Mag. nat. Hist.*, I, **1**: 257.

(Fig. 146)

Conidiophores much branched 3–5μ thick. *Conidia* obovate or subspherical, verruculose, 5–7 × 4–6μ.

On dead stems and leaves, paper, cardboard, cotton wool, etc.; Europe, India, New Zealand, N. & S. America.

135. PHAEOISARIA

Phaeoisaria Höhnel, 1909, *Sber. Akad. Wiss. Wien*, Abt. 1, **118**: 329–330.

Graphiopsis Bainier, 1907, *Bull. trimest. Soc. mycol. Fr.*, **23**: 19–20 (non Trail, 1889).

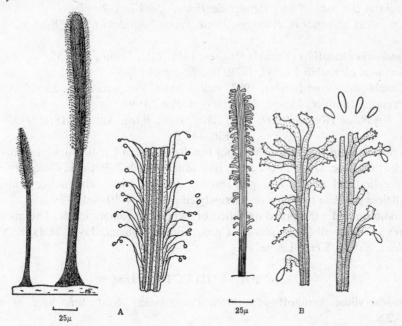

FIG. 147. A, *Phaeoisaria clavulata*; B, *P. clematidis* (× 650 except where indicated by scales).

Colonies effuse, black, hairy; under a dissecting microscope the upper half or two thirds of each synnema is seen to be covered with a white or pale grey powdery mass of conidia. *Mycelium* mostly immersed. *Stroma* usually none. *Setae* and *hyphopodia* absent. *Conidiophores* macronematous, synnematous;

individual threads narrow, branched towards their apices, pale to mid brown, smooth, straight or flexuous, splaying out at the apex and along the sides of the upper half or two thirds of each synnema. *Conidiogenous cells* polyblastic, integrated and terminal or discrete, sympodial, subulate, cylindrical or clavate, denticulate; denticles cylindrical. *Conidia* solitary, dry, acropleurogenous, simple, fusiform, ellipsoidal or subspherical, hyaline or subhyaline, smooth, 0-septate.

Type species: Phaeoisaria clematidis (Fuckel) Hughes = *P. bambusae* Höhnel.

KEY

Conidia subspherical or broadly ellipsoidal 1–2µ diam. . . . . *clavulata*
Conidia fusiform or narrowly ellipsoidal, 4–10 × 1·5–2·5µ . . . . . *clematidis*

Phaeoisaria clavulata (Grove) Mason & Hughes, 1953, apud Mason & M. B. Ellis, *Mycol. Pap.*, **56**: 42–44.

Pachnocybe clavulata Grove, 1885, *J. Bot., Lond.*, **30**: 168.
Graphium grovei Sacc., 1886, Syll. Fung., **4**: 613.
(Fig. 147 A)
Synnemata up to 450µ high, 10–20µ thick; individual threads about 2µ thick. *Conidiogenous cells* very pale, subulate, tapering to a fine point and bearing the conidia at the ends of very slender denticles. *Conidia* 1–2µ diam.

On fallen branches of *Fraxinus, Pinus, Prunus, Sambucus*, etc.; Europe.

Phaeoisaria clematidis (Fuckel) Hughes, 1958, *Can. J. Bot.*, **36**: 795.

Stysanus clematidis Fuckel, 1870, Symb. mycol.: 365.
Graphiopsis cornui Bainier, 1907, *Bull. trimest. Soc. mycol. Fr.*, **23**: 19–20.
P. cornui (Bainier) Mason, 1937, *Mycol. Pap.*, **4**: 94.
P. bambusae Höhnel, 1909, *Sber. Akad. Wiss. Wien*, Abt. 1, **118**: 329–330.
(Fig. 147 B)
Synnemata up to 1·5 mm. high but usually less than 1 mm., subulate, 20–80µ thick at the base, 8–25µ at the apex, individual threads 2–3µ thick. *Conidiogenous cells* cylindrical or clavate, pale brown, usually with numerous cylindrical denticles. *Conidia* fusiform or narrowly ellipsoidal, 4–10 × 1·5–2·5µ.

Common and widespread on fallen branches and sometimes dead stems and leaves of many different plants; Cuba, Europe, Ghana, Java, Malaya, New Guinea, Sabah, Sierra Leone.

136. PHAEODACTYLIUM

Phaeodactylium Agnihothrudu, 1968, *Proc. Indian Acad. Sci.*, Sect. B, **62**: 206–209.

Colonies effuse, pale, whitish when sporing freely. *Mycelium* partly superficial, partly immersed. *Stroma* none. *Setae* and *hyphopodia* absent. *Conidiophores* macronematous, mononematous, branching above the middle often dichotomously or trichotomously; stipe straight or flexuous, pale brown, smooth, branches often hyaline. *Conidiogenous cells* polyblastic, integrated and terminal or discrete, sympodial, cylindrical or clavate, denticulate; denticles tapered to a point. *Conidia* solitary, dry, acropleurogenous, simple, narrowly ellipsoidal or clavate, tapered to a fine point at the base, colourless or subhyaline, smooth, transversely septate.

Type species: P. alpiniae (Sawada) M. B. Ellis = *P. venkatesanum* Agnihothrudu.

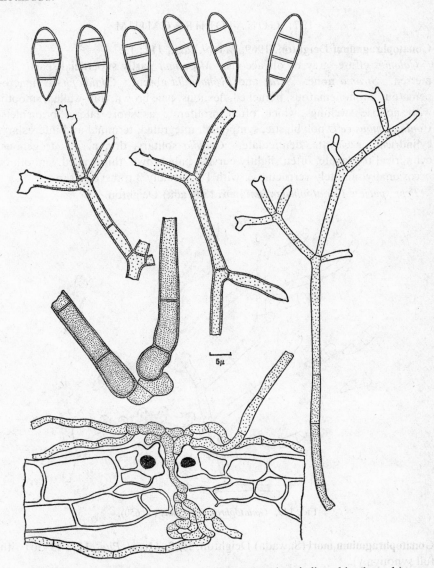

FIG. 148. *Phaeodactylium alpiniae* (× 650 except where indicated by the scale).

Phaeodactylium alpiniae (Sawada) M. B. Ellis comb. nov.

Dactylium alpiniae Sawada, 1928, Descriptive catalogue of the Formosan fungi, **4**: 102–103.

(Fig. 148)

Colonies hypophyllous. *Conidiophores* up to 350μ long, 3–5μ thick, swollen to 7–9μ at the base. *Conidia* almost always 3-septate, 16–25μ long, 6–9μ thick in the broadest part.

On leaves of *Alpinia* and *Elettaria* causing water-soaked lesions; India, Taiwan.

137. GONATOPHRAGMIUM

Gonatophragmium Deighton, 1969, *Mycol. Pap.*, **117**: 13.

Colonies effuse, grey or olivaceous. *Mycelium* partly superficial, partly immersed. *Stroma* none. *Setae* and *hyphopodia* absent. *Conidiophores* macronematous, mononematous, branched, flexuous, pale brown, thin-walled, smooth, with nodose swellings which often proliferate as short lateral branchlets. *Conidiogenous cells* polyblastic, sympodial, integrated, terminal and intercalary, cylindrical or clavate, denticulate. *Conidia* solitary, dry, acropleurogenous, cylindrical to clavate, often slightly curved, pale brown, thin-walled, smooth or occasionally minutely verruculose, with 1–7, usually 3 transverse septa.

Type species: Gonatophragmium mori (Sawada) Deighton.

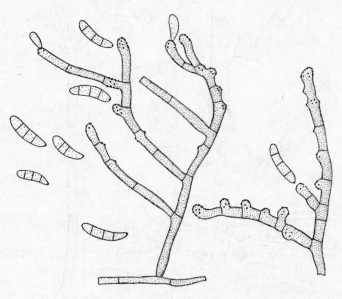

FIG. 149. *Gonatophragmium mori* (× 650).

Gonatophragmium mori (Sawada) Deighton, 1969, *Mycol. Pap.*, **117**: 13–30 (with full synonymy).

(Fig. 149)

Description as for genus. *Conidiophores* up to 500μ long, 3–4·5μ thick. *Conidia* 9–45 × 3·5–5·5 (mostly 13–26 × 4·5)μ.

Widespread in tropical countries on host plants belonging to many different families, often found on *Ficus*; it causes large, distinctive, zonate leaf spots.

138. HELICORHOIDION

Helicorhoidion Hughes, 1958, *Can. J. Bot.*, **36**: 773–774.

Colonies effuse, olivaceous brown to black. *Mycelium* partly superficial, partly immersed. *Stroma* none. *Setae* and *hyphopodia* absent. *Conidiophores* macronematous or semi-macronematous, branched or unbranched, flexuous, brown, smooth. *Conidiogenous cells* polyblastic or sometimes monoblastic, integrated, terminal becoming intercalary, sympodial, cylindrical, denticulate; denticles cylindrical. *Conidia* solitary, dry, acropleurogenous, helicoid, very tightly coiled, simulating dictyospores, filaments broad, multiseptate and often constricted at the septa, mid to dark golden brown, smooth or verruculose.

Type species: Helicorhoidion botryoideum (Cooke) Hughes.

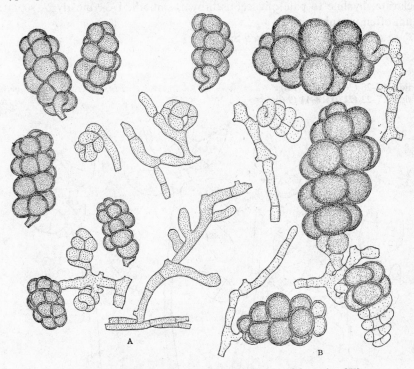

Fig. 150. A, *Helicorhoidion botryoideum*; B, *H. pulchrum* (× 650).

Helicorhoidion botryoideum (Cooke) Hughes, 1958, *Can. J. Bot.*, **36**: 773.
(Fig. 150 A)
Conidiophores up to 80 × 4–6μ. *Conidia* 25–40 × 15–22μ.
On logs; N. America.

Helicorhoidion pulchrum (Berk. & Corda) Hughes, 1958, *Can. J. Bot.*, **36**: 773–774.
(Fig. 150 B)
Conidiophores up to 60 × 3–5μ. *Conidia* 40–60 × 23–38μ.
On dead wood; Australia.

139. PYRICULARIA

Pyricularia Saccardo, 1880, *Michelia,* **2**: 20.

Colonies effuse, thinly hairy, grey, greyish brown or olivaceous brown. *Mycelium* immersed, *chlamydospores* sometimes formed in culture. *Stroma* none. *Setae* and *hyphopodia* absent. *Conidiophores* macronematous, mononematous, slender, thin-walled, usually emerging singly or in small groups through stomata, mostly unbranched, straight or flexuous, geniculate towards the apex, pale brown, smooth. *Conidiogenous cells* polyblastic, integrated, terminal, sympodial, cylindrical, geniculate, denticulate; each denticle cylindrical, thin-walled, cut off as a rule by a septum to form a separating cell. *Conidia* solitary, dry, acropleurogenous, simple, obpyriform, obturbinate or obclavate, hyaline to pale olivaceous brown, smooth, 1–3- (mostly 2-) septate; hilum often protuberant.

Type species: Pyricularia grisea Sacc.

KEY

Conidia 17–28 (20·9) × 6–9 (7·6)μ . . . . *grisea*
Conidia 17–23 (21·2) × 8–11 (9·6)μ . . . . *oryzae*

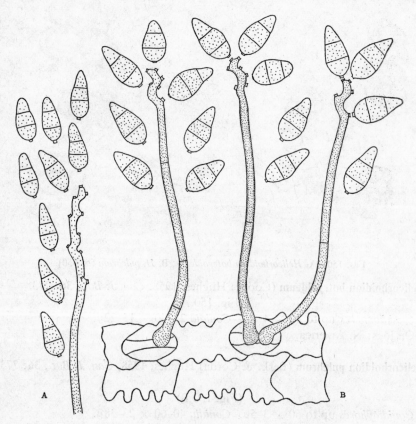

FIG. 151. A, *Pyricularia grisea;* B, *P. oryzae* (× 650).

Pyricularia grisea Sacc., 1880, *Michelia*, **2**: 20.
(Fig. 151 A)

Conidiophores up to 150μ long, 2·5–4·5 (usually 3–4)μ thick. *Conidia* mostly 17–28 (20·9) × 6–9 (7·6)μ.

Common on leaves of many different grasses and sometimes other plants, causing round or elliptical, pale tan or brown spots often with a purple or dark brown border, hosts include *Bambusa, Brachiaria, Commelina, Cynodon, Digitaria, Echinochloa, Eleusine, Eragrostis, Nicotiana, Panicum, Pennisetum, Secale, Setaria, Sorghum* and *Zea*; Cuba, Ethiopia, Europe, Fiji, Guinea, Hong Kong, India, Japan, Kenya, Malawi, Mauritius, Nepal, Nigeria, N. America, Papua, Paraguay, Rhodesia, Sierra Leone, Sudan, Tanzania, Trinidad, Uganda, Venezuela, Zambia.

Pyricularia oryzae Cav., 1891, Fungi Longob. exsicc., 49 (with description); *Atti ist. bot. Univ. Pavia*, Ser. 2, **2**: 280, 1892.
(Fig. 151 B)

Conidiophores up to 130μ long, 3–4μ thick. *Conidia* 17–23 (21·2) × 8–11 (9·6)μ.

On *Oryza* and occasionally on other grasses, the cause of rice blast disease; widely distributed [CMI Distribution Map 51].

140. NAKATAEA

Nakataea Hara, 1939, Diseases of the rice plant, Ed. 2: 185.

Vakrabeeja Subramanian, 1956, *J. Indian bot. Soc.*, **35**: 465–466.

Colonies effuse, black. *Mycelium* partly immersed, partly superficial. Spherical or subspherical black *sclerotia* are formed on natural substrata and in culture. *Stroma* none. *Setae* and *hyphopodia* absent. *Conidiophores* macronematous, mononematous, unbranched or rarely branched, brown, smooth. *Conidiogenous cells* polyblastic, integrated, terminal becoming intercalary, sympodial, cylindrical, sometimes geniculate, denticulate; denticles thin-walled, cylindrical or broadly conical, each cut off by a septum to form a separating cell. *Conidia* solitary, dry, acropleurogenous, becoming detached by a break across the thin separating cell wall, simple, usually falcate, often sigmoid, smooth, almost always 3-septate, cells unequally coloured, the cell at each end hyaline or very pale brown, intermediate cells pale to mid-pale brown.

Type species: Nakataea state of *Leptosphaeria salvinii* Catt. = *N. sigmoidea* Hara.

Nakataea state of **Leptosphaeria salvinii** Cattaneo, 1879, *Archo bot. Lab. Critt. R. Univ. Pavia*, **2–3**: 115–128.

N. sigmoidea Hara, 1939, Diseases of the rice plant, Ed. 2: 185.

Helminthosporium sigmoideum Cav., 1889, *Revue mycol.*, **11**: 185.

Vakrabeeja sigmoidea (Cav.) Subram., 1956, *J. Indian bot. Soc.*, **35**: 465–466.

Curvularia sigmoidea (Cav.) Hara, 1959, A monograph of rice diseases: 42.
(Fig. 152)

Sclerotia mostly 200–300μ diam. *Conidiophores* up to 200μ long, 4–6μ thick. *Conidia* 40–83 (mostly 45–55)μ long, 11–14μ thick in the broadest part tapering rather abruptly at the ends.

On *Oryza* causing stem rot, a serious disease; Ceylon, Egypt, Europe, Fiji, India, Japan, Kenya, Malaya, Nigeria, Sabah, U.S.A. A good account of the

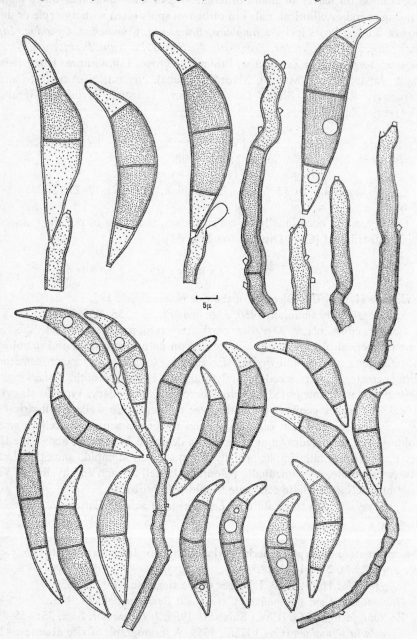

FIG. 152. *Nakataea* state of *Leptosphaeria salvinii* (× 650 except where indicated by the scale).

life history of this fungus is given by E. C. Tullis in *J. agric. Res.*, **47**: 675–687, 1933.

141. BRACHYSPORIUM

Brachysporium Saccardo, 1880, *Michelia*, **2**: 28; emend. Mason & Hughes in Hughes, 1951, *Naturalist, Hull*, **1951**: 45.

Colonies effuse, brown or dark brown, hairy. *Mycelium* mostly immersed. *Stroma* none. *Setae* and *hyphopodia* absent. *Conidiophores* macronematous, mononematous, straight or flexuous, unbranched, often swollen at the base, brown or dark brown, smooth. *Conidiogenous cells* integrated, terminal, polyblastic, sympodial, cylindrical, denticulate, the conidia being borne at the ends of long, sometimes twisted, narrow, cylindrical pedicels or separating cells. *Conidia* solitary, usually pendulous, acropleurogenous, simple, clavate, ellipsoidal, fusiform, limoniform, obovoid or pyriform, septate, brown, often with one or more cells paler than the others, smooth except in one species in which they are sometimes verrucose.

Lectotype species: Brachysporium obovatum (Berk.) Sacc.

Ellis, M. B., *Mycol. Pap.*, **106**: 43–54, 1966.

Hughes, S. J., *Naturalist, Hull*, **1951**: 45–48, 1951; *Can. J. Bot.*, **33**: 263–266, 1953; *N.Z. Jl Bot.*, **3**: 27–30, 1965.

KEY

Conidia 2-septate	**1**
Conidia 3-septate	**2**
Conidia 4-septate	**4**
Conidia 5-septate	*novae-zelandiae*
1. Conidia rather pale brown	*obovatum*
Conidia dark brown	*britannicum*
2. Conidia ellipsoidal (oval)	*nigrum*
Conidia obovoid or clavate.	**3**
3. Conidial measurements averaging 29 × 11·7μ	*bloxami*
Conidial measurements averaging 23·5 × 13·3μ	*dingleyae*
4. Conidia broadly fusiform to ellipsoidal	*masonii*
Conidia limoniform, pendulous	**5**
5. Conidia 28–39 × 12–17μ	*pendulisporum*
Conidia 41–51 × 20–25μ	*pulchrum*

Brachysporium obovatum (Berk.) Sacc., 1886, Syll, Fung., **4**: 427.
(Fig. 153 A)

Conidiophores up to 380μ long, often swollen at base to 9–13μ, 6–8μ just above the base, tapering to 2·5–4μ. *Conidia* 2-septate, 19–28 (23) × 11–14 (12·3)μ, basal cell very small, subhyaline, other cells much larger and rather pale brown.

Common on rotten wood and bark of various trees and shrubs including beech, blackthorn, elder, oak, poplar, *Prunus avium* and sycamore; Germany, Great Britain.

Brachysporium britannicum Hughes, 1951, *Naturalist, Hull*, **1951**: 48.
(Fig. 153 B)

Conidiophores up to 500μ long, 10–14μ thick at base, 7–8μ just above the base tapering to 4–5μ. *Conidia* 2–septate, 18–24 (21) × 11–14 (13)μ, basal cell small, subhyaline, other cells dark brown or very dark brown.

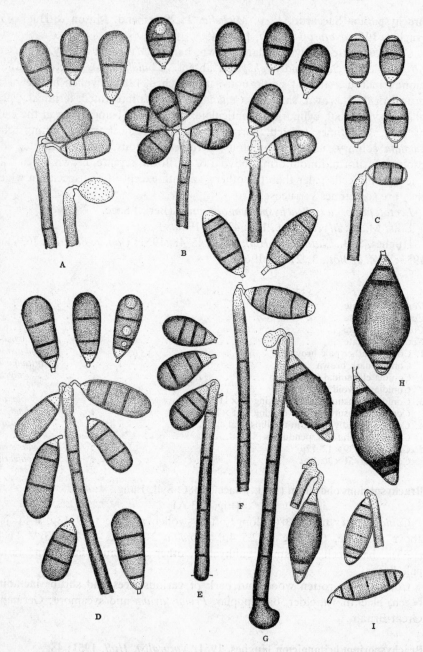

FIG. 153. *Brachysporium* species: A, *obovatum*; B, *brittanicum*; C, *nigrum*; D, *bloxami*; E, *dingleyae*; F, *masonii*; G, *pendulisporum*; H, *pulchrum*; I, *novae-zelandiae* (× 650).

On wood and bark of ash, beech, birch, box, chestnut, elder, oak, etc.; common in Great Britain.

Brachysporium nigrum (Link) Hughes, 1958, *Can. J. Bot.*, **36**: 742.
(Fig. 153 C)

Conidiophores up to 300μ long, sometimes swollen at base to 9–13μ, 6–8μ thick just above the base, tapering to 3–5μ. *Conidia* 3-septate, 17–24 (19·8) × 8–11 (9·6)μ, two middle cells brown, cell at each end hyaline or very pale brown.

Very common on rotten wood and bark of various trees, including ash, beech, birch, chestnut, lime and oak; Belgium, Canada, Germany, Great Britain.

Brachysporium bloxami (Cooke) Sacc., 1886, Syll. Fung., **4**: 426.
(Fig. 153 D)

Conidiophores up to 380μ long, 10–12μ thick at base, 6–8μ just above the base, tapering to 3–5μ. *Conidia* 3-septate, 22–37 (29) × 9–14 (11·7)μ, basal cell small and very pale, other cells much larger and mid-pale to rather dark brown.

Common on rotten wood and bark of alder, ash, beech, birch, chestnut, oak, pine, *Prunus avium*, etc.; Belgium, Great Britain.

Brachysporium dingleyae Hughes, 1965, *N.Z. Jl Bot.*, **3**: 27
(Fig. 153 E)

Conidiophores up to 250μ long, 8–10μ thick at base, 5–7μ just above the base, tapering to 3–4μ. *Conidia* 3-septate, 22–25 (23·5) × 11–15 (13·3)μ, basal cell small, pale, other cells larger, brown or dark brown.

On rotten wood; New Zealand.

Brachysporium masonii Hughes, 1951, *Naturalist, Hull*, **1951**: 46.
(Fig. 153 F)

Conidiophores up to 400μ long, 10–12μ thick at base, 6–8μ just above the base, tapering to 2·5–4μ. *Conidia* nearly all 4-septate, 26–34 (29·5) × 10–14 (12·2)μ, cell at each end hyaline or pale, middle cells brown.

On rotten wood and bark of beech, chestnut, oak, *Prunus avium*, etc.; Great Britain.

Brachysporium pendulisporum Hughes, 1955, *Can. J. Bot.*, **33**: 266.
(Fig. 153 G)

Conidiophores up to 250μ long, 9–14μ thick at base, 6–7μ just above the base, tapering to 3–5μ. *Conidia* pendulous, nearly all 4–septate, 28–39 (35) × 12–17 (15·3)μ, often verrucose, central cell very large and dark brown, other cells small and progressively paler towards the ends.

On rotten wood; Canada.

Brachysporium pulchrum M. B. Ellis, 1966, *Mycol. Pap.*, **106**: 52.
(Fig. 153 H)

Conidiophores up to 360μ long, 12–15μ thick at base, 6–8μ just above the base, tapering to 4–5μ. *Conidia* pendulous, mostly 4-septate, 40–51 (46) ×

20–25 (21·5)μ, central cell very large, dark brown or dark blackish brown, other cells pale and smaller.

On *Phylica*; Tristan da Cunha.

Brachysporium novae-zelandiae Hughes, 1965, *N.Z. Jl Bot.*, **3**: 29.
(Fig. 153 I)

Conidiophores up to 280μ long, 10–11μ thick at base, 6–8μ just above the base, tapering to 3–4μ. *Conidia* 5-septate, the middle septum usually much thinner than the others, 26–33 (29·5) × 10–12 (11)μ, cell at each end subhyaline, other cells rather pale golden brown.

On rotten wood; New Zealand.

142. SCOLECOBASIDIUM

Scolecobasidium Abbot, 1927, *Mycologia*, **19**: 29–31.

Colonies effuse, usually slow-growing, flat or raised in the centre, sometimes funiculose, grey, brown, olivaceous brown or blackish olive. *Mycelium* superficial or partly immersed; *chlamydospores* formed in some species. *Stroma* none. *Setae* and *hyphopodia* absent. *Conidiophores* macronematous, mononematous, often short, unbranched, straight or flexuous, brown or olivaceous, smooth. *Conidiogenous cells* polyblastic, integrated, terminal sometimes becoming intercalary, sympodial, cylindrical, clavate or cuneiform, denticulate; denticles long, narrow cylindrical, thread-like, often breaking across the middle leaving part attached to the conidium and part to the conidiogenous cell. *Conidia* solitary, dry, acropleurogenous, simple, ellipsoidal, oblong or cylindrical rounded at the ends, fusiform, T- or Y- shaped, pale to mid brown or olivaceous brown, smooth, verruculose or echinulate, 0–3-septate.

Type species: Scolecobasidium terreum Abbott.

Barron, G. L. & Busch, L. V., *Can. J. Bot.*, **40**: 77–84, 1962.

KEY

Conidia almost always 1-septate 1
Conidia frequently 3-septate . . . . *variabile*
1. Conidia Y- or T-shaped . . . . *terreum*
　Conidia ellipsoidal or oblong to cylindrical rounded at the ends . . . 2
2. Conidia not or rarely constricted at the septum . . . . . *humicola*
　Conidia mostly constricted at the septum 3
3. Conidia minutely verruculose or finely echinulate, average dimensions 8·5 × 2·5μ
　　　　　　　　　　　　　　　　　　　　　　　　　　constrictum
　Conidia with prominent spines, average dimensions 7·5 × 4μ . . . *verruculosum*

Scolecobasidium variabile Barron & Busch, 1962, *Can. J. Bot.*, **40**: 83–84.
(Fig. 154 A)

Conidiophores pale olivaceous, 4–25 × 1·5–3μ. *Conidia* cylindrical rounded at the ends or ellipsoidal, 1–3- (most frequently 3-) septate, sometimes constricted at the septa, pale olivaceous, verruculose or finely echinulate, 7–18 × 2·5–4·5 (15 × 3·3)μ.

Isolated from soil, decaying leaves, latex and lentils; Congo, Egypt, India, N. America.

Scolecobasidium terreum Abbott, 1927, *Mycologia*, **19**: 29–31.
(Fig. 154 B)

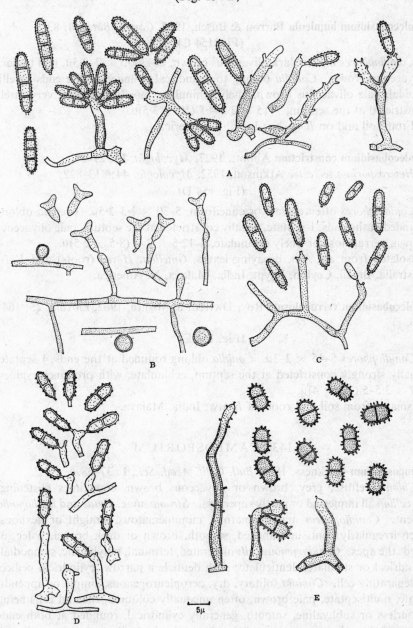

FIG. 154. *Scolecobasidium* species: A, *variabile*; B, *terreum*; C, *humicola*; D, *constrictum*; E, *verruculosum*.

D.H.—8

Chlamydospores 4–6·5μ diam. *Conidiophores* often clavate, pale olivaceous, 3–15 × 1·5–3·5μ. *Conidia* Y- or T-shaped, hyaline or pale olive, smooth or minutely verruculose, 4–12μ long, 1·5–2·5μ thick, 5–7μ between the tips of the arms, 1-septate.

Isolated from soil; India, N. America.

Scolecobasidium humicola Barron & Busch, 1962, *Can. J. Bot.*, **40**: 83.

(Fig. 154 C)

Conidiophores mid to dark olivaceous brown, 20–300μ × 2–2·5μ, mid to dark olivaceous brown. *Conidia* oblong to cylindrical rounded at the ends or ellipsoidal, pale olivaceous brown, finely echinulate, mostly 1-septate, very rarely constricted at the septum, 7–15 × 2·5–4·5 (10 × 3·5)μ.

From soil and on *Borassus*; India, N. America.

Scolecobasidium constrictum Abbott, 1927, *Mycologia*, **19**: 29–31.
Heterosporium terrestre Atkinson, 1952, *Mycologia*, **44**: 813–822.

(Fig. 154 D)

Conidiophores often clavate or cuneiform, 5–30 × 1·5–2·5μ. *Conidia* oblong rounded at the ends, 1-septate, usually constricted at the septum, pale olivaceous brown, verruculose or finely echinulate, 5–12·5 × 2–4 (8·5 × 2·5)μ.

Isolated from air, soil, decaying leaves, *Gmelina*, *Hevea* (roots) and *Vitis*; Australia, Brazil, Ceylon, Egypt, India, Malaya, N. America.

Scolecobasidium verruculosum Roy, Dwivedi & Mishra, 1962, *Lloydia*, **25**: 164–166.

(Fig. 154 E)

Conidiophores 5–45 × 2–3μ. *Conidia* oblong rounded at the ends, 1-septate, usually strongly constricted at the septum, echinulate, with prominent spines, 6–9 × 3·5–5 (7·5 × 4)μ.

Isolated from soil and roots of *Hevea*; India, Malaya.

143. CAMPOSPORIUM

Camposporium Harkness, 1884, *Bull. Calif. Acad. Sci.*, **1**: 37–38.

Colonies effuse, grey, brown or olivaceous brown, sometimes glistening. *Mycelium* all immersed or partly superficial. *Stroma* none. *Setae* and *hyphopodia* absent. *Conidiophores* macronematous, mononematous, straight or flexuous, often irregularly bent, unbranched, smooth, brown or dark brown, paler towards the apex. *Conidiogenous cells* integrated, terminal, polyblastic, sympodial, cylindrical or subulate, denticulate; each denticle a narrow, cylindrical pedicel or separating cell. *Conidia* solitary, dry, acropleurogenous, simple or appendiculate, multiseptate, pale brown, often unequally coloured, the end cells being colourless or subhyaline, smooth, generally cylindrical, rounded at both ends or rounded at the apex, conico-truncate at the base, but sometimes tapered towards the apex, rostrate.

Type species: Camposporium antennatum Harkness.
Hughes, S. J., *Mycol Pap.*, **36**: 3–16, 1951.
Rao, P. R. and Dev Rao, *Antonie van Leeuwenhoek*, **30**: 60–64, 1964.

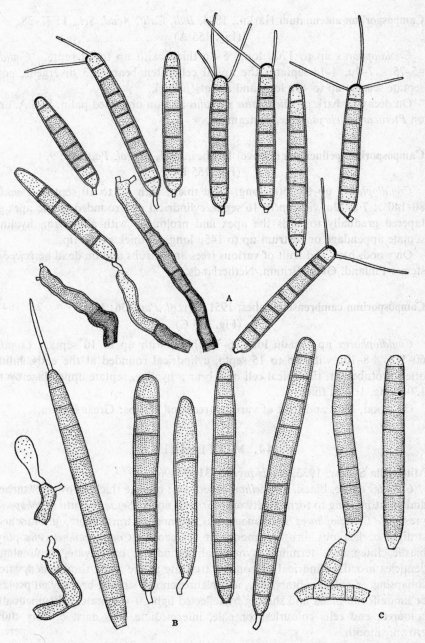

FIG. 155. *Camposporium* species: A, *antennatum*; B, *pellucidum*; C, *cambrense* (× 650).

KEY

Conidial appendages when present 1–3 and not septate	*antennatum*
Conidial appendages when present 1 only and septate	1
1. Conidia often tapered towards the apex	*pellucidum*
2. Conidia nearly always cylindrical, rounded at apex	*cambrense*

Camposporium antennatum Harkn., 1884, *Bull. Calif. Acad. Sci.*, **1**: 37–38.
(Fig. 155 A)

Conidiophores up to 170μ long, 5–8μ thick, with up to 12 septa. *Conidia* 45–75 × 7–9μ, 4–14-septate; the apical cell often bears 1–3 divergent, non-septate setulae up to 40μ long and about 1μ thick.

On decaying bark of *Eucalyptus globulus* and on unnamed palm, U.S.A. and on *Pterocarpus chrysothrix*, Tanzania.

Camposporium pellucidum (Grove) Hughes, 1951, *Mycol. Pap.*, **36**: 9.
(Fig. 155 B)

Conidiophores up to 150μ long, 5–8μ thick, with up to 10 septa. *Conidia* 80–140 × 7·5–12μ with up to 16 septa, cylindrical and rounded at the apex or tapered gradually towards the apex and prolonged with a filiform, hyaline, septate appendage or rostrum up to 145μ long, 2μ thick at the tip.

On wood, bark and fruit of various trees and shrubs and on dead herbaceous stems; Finland, Great Britain, Netherlands.

Camposporium cambrense Hughes, 1951, *Mycol. Pap.*, **36**: 11.
(Fig. 155 C)

Conidiophores up to 80μ long, 6–7μ thick, with up to 10 septa. *Conidia* 60–115 × 8–10μ with up to 15 septa, cylindrical rounded at the ends, hilum often protuberant, the apical cell may bear a hyaline, septate appendage up to 130μ long, 1·5–2μ thick.

On wood, bark and fruit of various trees and shrubs; Great Britain.

144. MITTERIELLA

Mitteriella Sydow, 1933, *Annls mycol.*, **31**: 95–96.

Colonies effuse, black. *Mycelium* superficial; hyphae thick, brown, branched and anastomosing to form a network. *Stroma* none. *Setae* absent. *Hyphopodia* present. *Conidiophores* macronematous, mononematous, short, unbranched, straight or flexuous, brown, smooth or verrucose. *Conidiogenous cells* poly-blastic, integrated, terminal, sympodial, cylindrical or clavate, denticulate; denticles broad, cylindrical or conical, truncate at the apex, thin-walled, often collapsing. *Conidia* solitary, dry, acropleurogenous, simple, broadly ellipsoidal or limoniform, black and shining by reflected light, 0–4-septate, cells unequally coloured, end cells colourless or pale, intermediate cells dark or very dark brown, smooth.

Type species: Mitteriella ziziphina Syd.

Mitteriella ziziphina Syd., 1933, *Annls mycol.*, **31**: 95–96 [as ' *zizyphina* '].
(Fig. 156)

Colonies epiphyllous. *Hyphopodia* hemispherical or oval, 7–10µ diam. or 11×8µ. *Conidiophores* 12–27 × 5–10µ, mid-pale brown. *Conidia* 30–40µ long, 17–21µ thick in the broadest part, central cell large and usually very dark, other cells shorter, end cells very pale.

On leaves of *Ziziphus*; India, Pakistan, Sudan, Uganda, Zambia.

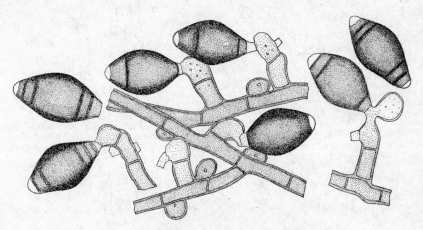

FIG. 156. *Mitteriella ziziphina* (× 650).

145. GENICULOSPORIUM

Geniculosporium Chesters & Greenhalgh, 1964, *Trans. Br. mycol. Soc.*, **47**: 393–401.

Colonies effuse, grey, brown or dark blackish brown, hairy or velvety. *Mycelium* partly superficial, partly immersed. *Stroma* none. *Setae* and *hyphopodia* absent. *Conidiophores* macronematous, mononematous, flexuous, irregularly or sometimes dichotomously or trichotomously branched, very pale to mid brown, smooth. *Conidiogenous cells* polyblastic, integrated and terminal on branches or discrete, sympodial, geniculate, denticulate with very short, thin-walled separating cells which break across the middle leaving a minute frill or collar at each geniculation which corresponds to a frill at the base of each conidium. *Conidia* solitary, dry, acropleurogenous, simple, ellipsoidal or obovoid, colourless to pale brown or olivaceous brown, smooth, 0-septate.

Type species: Geniculosporium state of *Hypoxylon serpens* (Pers. ex Fr.) Kickx = *G. serpens* Chesters & Greenhalgh.

Geniculosporium state of **Hypoxylon serpens** (Pers. ex Fr.) Kickx, 1835, *Flore Crypt. Louvain*: 115.
(Fig. 157)

Colonies effuse, grey. *Conidiophores* up to 300µ long or occasionally longer, 2–3µ thick. *Conidia* colourless or very pale brown, 2·5–4·5 × 2–3µ.

Very common on dead wood and bark; Australia, Europe, New Zealand, N. America, Sierra Leone, S. Africa. In herbaria specimens are often incorrectly labelled ' *Haplaria grisea* '.

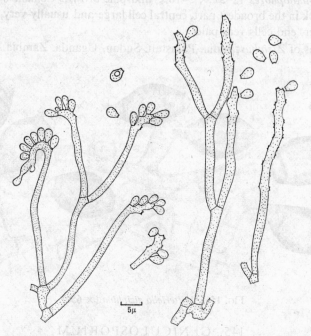

FIG. 157. *Geniculosporium* state of *Hypoxylon serpens.*

146. DEMATOPHORA

Dematophora Hartig, 1883, *Unters. forstbot. Inst. München*, **3**: 95 and 122–126.

Colonies effuse, dark chocolate to blackish brown, with numerous erect, dark brown synnemata clearly visible to the naked eye. *Mycelium* partly superficial, partly immersed. *Stroma* none. *Setae* and *hyphopodia* absent. *Conidiophores* macronematous, synnematous; component threads flexuous, intertwined, repeatedly branched, often dichotomously towards the apex, splaying out to form a head, pale to mid brown or olivaceous brown. *Conidiogenous cells* polyblastic, integrated and terminal on branches or discrete, sympodial, geniculate, denticulate with short, thin-walled separating cells which break across the middle leaving a minute frill or collar at each geniculation which corresponds to a frill at the base of each conidium. *Conidia* solitary, dry, acropleurogenous, simple, ellipsoidal or obovoid, colourless to pale olivaceous brown, 0-septate, smooth.

Type species: Dematophora necatrix Hartig.

Dematophora necatrix Hartig, 1883, *Unters. forstbot. Inst. München*, **3**: 122–126 [= **Dematophora** state of **Rosellinia necatrix** (Hartig) Berl. ex Prill.].

(Fig. 158)

Synnemata up to 1½ mm. high. *Stipe* 40–300μ thick. *Branches* 2–3·5μ thick. *Conidia* 3–4·5 × 2–2·5μ.

On roots and stumps of trees, including apple and pear, also on herbaceous plants, e.g. *Narcissus*; the cause of 'white root rot' in its mycelial state. Reported from Africa, Asia, Australasia, Europe, N. & S. America [CMI Distribution Map 306].

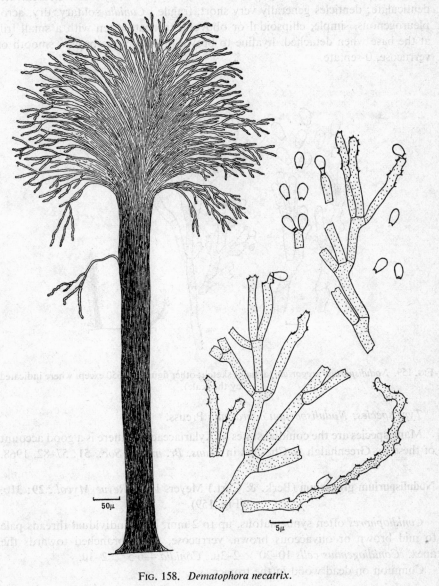

50μ

5μ

FIG. 158. *Dematophora necatrix.*

147. NODULISPORIUM

Nodulisporium Preuss, 1851, *Linnaea*, **24**: 120–121.

Colonies effuse, grey, brown, olivaceous or blackish brown, often velvety. *Mycelium* partly immersed, partly superficial. *Stroma* none seen. *Setae* and *hyphopodia* absent. *Conidiophores* macronematous, mononematous or synnematous, individual threads often much branched especially towards the apex, flexuous, pale to dark brown or olivaceous brown, smooth or verrucose. *Conidiogenous cells* polyblastic, integrated and terminal becoming intercalary, or discrete, solitary or arranged penicillately, sympodial, cylindrical to clavate, denticulate; denticles generally very short, fragile. *Conidia* solitary, dry, acropleurogenous, simple, ellipsoidal or obovoid, truncate often with a small frill at the base when detached, hyaline to brown or olivaceous brown, smooth or verrucose, 0-septate.

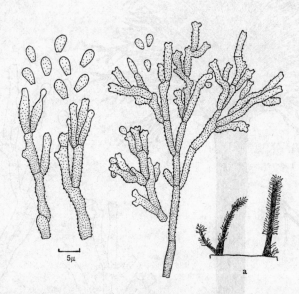

Fig. 159. *Nodulisporium gregarium* (a, habit sketch; other figures × 650 except where indicated by the scale).

Type species: Nodulisporium ochraceum Preuss.

Many species are the conidial states of Xylariaceae and there is a good account of these by Greenhalgh & Chesters in *Trans. Br. mycol. Soc.*, **51**: 57–82, 1968.

Nodulisporium gregarium (Berk. & Curt.) Meyer, 1965, *Revue Mycol.*, **29**: 310. (Fig. 159)

Conidiophores often synnematous, up to 2 mm. high, individual threads pale to mid brown or olivaceous brown, verrucose, much branched towards the apex. *Conidiogenous cells* 10–20 × 2–3μ. *Conidia* 4–6·5 × 2–3μ.

Common on dead wood in the tropics.

148. HANSFORDIA

Hansfordia Hughes, 1951, *Mycol. Pap.*, **43**: 15–24.

Colonies effuse, pale to dark olivaceous grey or greyish brown, hairy or velvety. *Mycelium* superficial or immersed. *Stroma* none. Separate *setae* absent

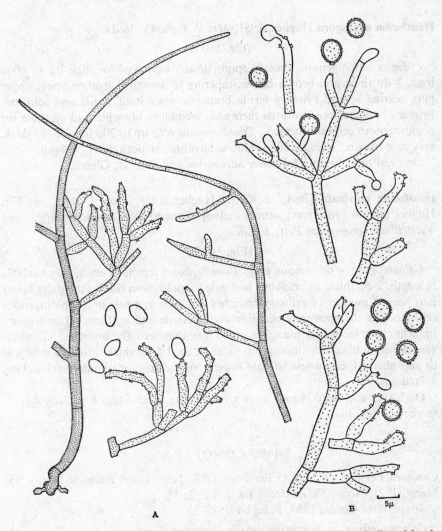

A   B

Fig. 160. A, *Hansfordia ovalispora*; B, *H. pulvinata* (× 650 except where indicated by the scale).

but in some species the upper part of the conidiophore is sterile and setiform or there are setiform branches. *Hyphopodia* absent. *Conidiophores* macronematous, mononematous, erect or repent, branched, straight or flexuous, hyaline to brown, smooth, apex and branches sometimes setiform. *Conidiogenous cells* polyblastic, integrated and terminal, or discrete, sympodial, cylindrical or clavate, denticulate, each denticle a thin-walled separating cell. *Conidia* solitary, dry, acropleurogenous, liberated by a break across the separating cell, simple, ellipsoidal, fusiform, spherical or subspherical, hyaline to pale brown, smooth or verruculose, 0-septate.

Type species: Hansfordia ovalispora Hughes.

Hansfordia ovalispora Hughes, 1951, *Mycol. Pap.*, **43**: 16–18.

(Fig. 160 A)

Colonies amphigenous, mostly epiphyllous. *Conidiophore* stipe up to 600μ long, 3–4μ thick and brown below, tapering to a narrow hyaline apex; upper part bearing several primary fertile branches, apex itself sterile and setiform. Primary branches bear towards their ends secondary branches and on these the conidiogenous cells are formed. *Conidiogenous cells* up to 30μ long, 2–4μ thick, very pale brown. *Conidia* ellipsoidal, subhyaline, smooth, 8–11 × 4–6μ.

On dead branches of *Heliconia*, *Saccharum* and *Setaria*; Ghana.

Hansfordia pulvinata (Berk. & Curt.) Hughes, 1958, *Can. J. Bot.*, **36**: 771. Hughes cites 6 synonyms; others include *Sporotrichum canescens* Speg. and *Verticillium cercosporae* Petr. & Cif.

(Fig. 160 B)

Colonies grey or olivaceous grey. *Conidiophores* repent or erect, very variable in length, 2–5μ thick, unbranched and pale to mid brown below; the paler upper part bears a number of primary branches and ends as a rule in a conidiogenous cell although the apex may occasionally be sterile and setiform. The primary branches bear secondary branches which may themselves be branched. *Conidiogenous cells* subhyaline, smooth, up to 20μ long, 1·5–4μ thick. *Conidia* spherical or subspherical, colourless to pale brown, verruculose or minutely echinulate, 4–7μ diam.

On leaves and dead wood, often overgrowing other fungi especially *Cercospora*; cosmopolitan.

149. CONOPLEA

Conoplea Persoon ex Mérat; Persoon, 1797, Tent. Disp. method. Fung.: 55; Mérat, 1821, Nouv. Fl. Environs Paris, Ed. 2: 16.

Streptothrix Corda, 1839, Pracht-Fl.: 27–28.

Colonies effuse or pulvinate, brown, sometimes velvety. *Mycelium* mostly immersed. *Stroma* usually present. *Setae* and *hyphopodia* absent. *Conidiophores* macronematous, mononematous, straight or flexuous, torsive in some species, branched, pale to dark brown, mostly finely and densely echinulate. *Conidiogenous cells* polyblastic, integrated and terminal or discrete, sympodial, more or less cylindrical, denticulate but often rather indistinctly so, with cylindrical, thin, disk-like separating cells. *Conidia* solitary, dry, acropleurogenous, simple, ellipsoidal, obovoid, pyriform or subspherical, pale to dark brown, often finely and densely echinulate, rarely smooth, 0-septate, germ slits or pores frequently seen.

Lectotype species: Conoplea olivacea Fr. = *C. sphaerica* Pers.

Hughes, S. J., *Can J. Bot.*, **38**; 659–696, 1960; good descriptions, historical survey, synonyms.

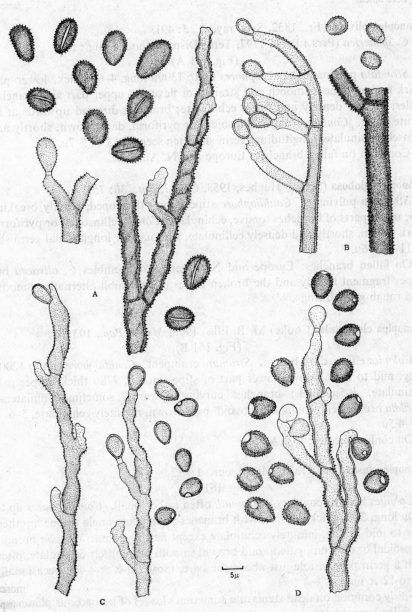

Fig. 161. *Conoplea* species: A, *olivacea*; B, *elegantula*; C, *fusca*; D, *juniperi* var. *juniperi*.

KEY

Conidia with longitudinal germ slit 1
Conidia without germ slit; germ pore sometimes present 2
1. Conidiophores up to 1300μ long with well defined stipe not breaking up . . *olivacea*
   Conidiophores up to 650μ long, stipe not well defined, often breaking up . *globosa*
2. Conidiophores torsive 3
   Conidiophores not torsive . . . . *elegantula*
3. Pore lateral, usually just above the base . . . . *fusca*
   Pore apical . . . . *juniperi*

Conoplea olivacea Fr., 1832, Syst. mycol., 3: 491.

C. sphaerica (Pers.) Pers., 1797, Tent. Disp. method. Fung.: 55.

(Fig. 161 A)

Stromata pulvinate. *Conidiophores* up to 1300μ long, 4–6μ thick, lower part dark brown, almost smooth and straight or flexuous, upper part and branches paler, torsive, densely and finely echinulate; branches directed upwards at an acute angle. *Conidia* broadly ellipsoidal or pyriform, dark brown, shortly and densely echinulate, longitudinal germ slit often seen, 7–13 × 5–7μ.

Common on fallen branches; Europe and N. America.

Conoplea globosa (Schw.) Hughes, 1958, *Can. J. Bot.*, 36: 755.

Stromata pulvinate. *Conidiophore* stipe poorly developed, readily breaking up; upper parts of branches torsive, echinulate. *Conidia* ellipsoidal or pyriform, dark brown, shortly and densely echinulate, with narrow longitudinal germ slit, 7–11 × 5–6μ.

On fallen branches; Europe and N. America. Resembles *C. olivacea* but stipes fragment readily and the broken pieces often exhibit alternating smooth and rough growth rings.

Conoplea elegantula (Cooke) M. B. Ellis, 1965, *Mycol. Pap.*, 103: 38.

(Fig. 161 B)

Colonies effuse, dark brown. *Stromata* erumpent. *Conidiophores* up to 1200μ long, mid to dark brown, lower part of stipe smooth, 7–8μ thick, upper part echinulate, 4–5μ thick; branches curving upwards, sometimes unilateral. *Conidia* broadly ellipsoidal or obovoid, pale brown, minutely echinulate, 5–6 × 3·5–4·5μ.

On conifer needles; U.S.A.

Conoplea fusca Pers., 1822, Mycol. eur., 1: 12.

(Fig. 161 C)

Colonies brown, powdery. *Stromata* often rather small. *Conidiophores* up to 400μ long, 3–5μ thick, torsive, with branches at an acute angle to one another, pale to mid brown, minutely echinulate except near the base. *Conidia* broadly ellipsoidal or obovoid, pale to mid brown, smooth or minutely echinulate, often with a germ pore visible just above the base, mostly 6–8 × 4–5μ, occasionally up to 12 × 6μ.

Fairly common on dead stems and sometimes leaves of herbaceous plants and on fallen branches of shrubs and trees; Europe and N. America.

Conoplea juniperi Hughes var. **juniperi**, 1958, *Can. J. Bot.*, **36**: 755.

(Fig. 161 D)

Colonies pulvinate, reddish brown. *Stromata* subglobose. *Conidiophores* up to 500µ long, 3–6µ thick, torsive except near the base, pale to dark brown, finely but densely echinulate, with branches directed upwards at an acute angle. *Conidia* subglobose, broadly ellipsoidal, obovoid or pyriform, mid to dark brown, finely but densely echinulate, with germ pore usually at or near the apex, 6–8 × 5–7µ.

On wood and bark of *Juniperus virginiana*; U.S.A.

150. BELTRANIA

Beltrania O. Penzig, 1882, *Nuovo G. bot. ital.*, **14**: 72.

Colonies effuse, velutinous, brown to black. *Mycelium* all immersed or partly superficial. *Stroma* usually present, often confined to epidermis. *Setae* simple, dark, smooth or verrucose, thick-walled, arising from flat, radially lobed basal cells. *Hyphopodia* absent. *Conidiophores* macronematous, mononematous, usually simple, straight or flexuous, pale olive to brown, smooth, septate, arising from basal cells of setae or from separate radially lobed cells. *Conidiogenous cells* integrated, terminal, polyblastic, sympodial, clavate or cylindrical, denticulate (denticles cylindrical); separating cells when present swollen. *Conidia* solitary, acropleurogenous, biconic, appendiculate, the free end being usually spicate or apiculate, 0-septate, smooth, pale olive to dark reddish brown with a distinct hyaline transverse band immediately above the widest part of the conidium.

Type species: Beltrania rhombica O. Penzig.

Hughes, S. J., *Mycol. Pap.*, **47**: 1–5, 1951.

Pirozynski, K. A., *Mycol Pap.*, **90**: 4–16, 1963.

KEY

Conidia spicate .	1
Conidia apiculate 36–48 × 18–22µ	*africana*
1. Conidia 15–30 × 7–14µ (without appendage) .	2
Conidia 30–38 × 22–25µ (without appendage)	*malaiensis*
2. Conidia biconic symmetrical, proximal end V-shaped	*rhombica*
Conidia biconic mostly asymmetrical, proximal end U-shaped	*querna*

Beltrania rhombica O. Penzig, 1882, *Nuovo G. bot. ital.*, **14**: 72–75 [syn. *B. indica* Subram. and *B. multispora* Swart, fide Pirozynski].

(Fig. 162 A)

Setae smooth, usually less than 200µ long but occasionally up to 300µ, 4–6µ thick near the base. *Conidiophores* up to 130µ long, 4–8µ thick. *Separating cells* ellipsoidal or obovoid, pale, 6–15 × 3–8µ. *Conidia* 15–30 × 7–14µ, appendage 3–20µ long, 2µ wide at the base, tapering to a point.

On dead leaves of many tropical plants including lime, pineapple and tea and isolated from air, soil, seeds and stems; Brazil, Congo, England, India, Japan, Java, Malaya, Mozambique, Sicily, Sierra Leone and Tanzania. It sporulates in culture on oat agar but rarely produces normal setae.

Beltrania querna Harkn. (Fig. 162 B) has smooth *setae* up to 400μ long, *conidiophores* up to 200 × 3–7μ, *separating cells* 8–12 × 4–7μ, *conidia* 18–30 × 7–10μ with appendage 2–5μ long. On various species of *Quercus*, mostly from America.

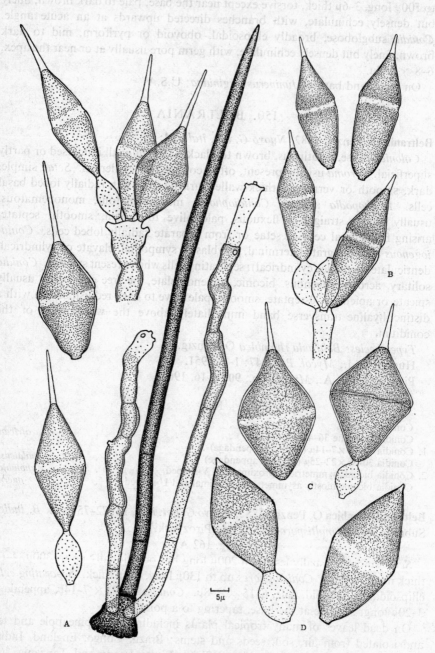

FIG. 162. *Beltrania* species: A, *rhombica*; B, *querna*; C, *malaiensis*; D, *africana* (del. K. Pirozynski).

Beltrania malaiensis Wakefield (Fig. 162 C) has smooth *setae* up to 160μ long, *conidiophores* up to 250 × 4–6μ, *separating cells* ellipsoidal or subspherical 10–12 × 7–9μ, *conidia* 30–38 × 22–25μ, appendage 9–14μ long. Isolated from gutta percha; Malaya.

Beltrania africana Hughes (Fig. 162 D) has verrucose *setae* up to 230μ long, *conidiophores* 20–80 × 5–7μ, *separating cells* ellipsoidal, 14–20 × 8–12μ, *conidia* 35–45 × 17–20μ, including short apiculus. On dead branches of *Averrhoa carambola* and cacao, Ghana.

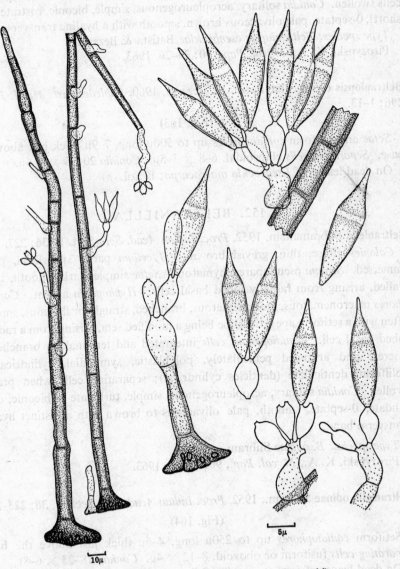

FIG. 163. *Beltraniopsis esenbeckiae* (del. K. Pirozynski).

151. BELTRANIOPSIS

Beltraniopsis Batista & Bezerra, 1960, *Publções Inst. Micol. Recife*, **296**: 1–13.
   Colonies effuse, velutinous, fuscous. *Mycelium* partly superficial, partly immersed. *Stroma* pseudoparenchymatous. *Setae* simple, olivaceous brown, smooth, septate, arising from radially lobed basal cells. *Conidiophores* macronematous, mononematous, branched, straight or flexuous, smooth, often with a setiform apex, the stipe being a modified seta, arising from a radially lobed basal cell. *Conidiogenous cells* integrated, terminal on banches, polyblastic, sympodial, cylindrical, doliiform, denticulate (denticles cylindrical); separating cells swollen. *Conidia* solitary, acropleurogenous, simple, biconic, rostrate (beak short), 0-septate, pale olivaceous brown, smooth with a hyaline transverse band.
   Type species: Beltraniopsis esenbeckiae Batista & Bezerra.
   Pirozynski, K. A., *Mycol. Pap.*, **90**: 24–26, 1963.

Beltraniopsis esenbeckiae Batista & Bezerra, 1960, *Publções Inst. Micol. Recife*, **296**: 1–13.

(Fig. 163)

Setae and setiform *conidiophores* up to 500μ long, 7–9μ thick just above the base. *Separating cells* ellipsoidal, 6–8 × 3–5μ. *Conidia* 20–25 × 4–6μ.
   On dead leaves of *Esenbeckia macrocarpa*; Brazil.

152. BELTRANIELLA

Beltraniella Subramanian, 1952, *Proc. Indian Acad. Sci.*, Sect. B, **36**: 227.
   Colonies effuse, thin, greyish-brown. *Mycelium* partly superficial, partly immersed. *Stroma* pseudoparenchymatous. *Setae* simple, dark, smooth, thick-walled, arising from radially lobed basal cells. *Hyphopodia* absent. *Conidiophores* macronematous, mononematous, branched, straight or flexuous, smooth, often with a setiform apex, the stipe being a modified seta, arising from a radially lobed basal cell. *Conidiogenous cells* integrated and terminal on branches or discrete and arranged penicillately, polyblastic, sympodial, cylindrical or doliiform, denticulate (denticles cylindrical); separating cells when present swollen. *Conidia* solitary, acropleurogenous, simple, turbinate or biconic, often caudate, 0-septate, smooth, pale olivaceous to brown with a distinct hyaline transverse band.

Type species: B. *odinae* Subram.
   Pirozynski, K. A., *Mycol. Pap.*, **90**: 26–28, 1963.

Beltraniella odinae Subram., 1952, *Proc. Indian Acad. Sci.*, Sect. B, **36**: 223–228.
(Fig. 164)
Setiform *conidiophores* up to 250μ long, 4–6μ thick just above the base. *Separating cells* fusiform or obovoid, 8–12 × 4μ. *Conidia* 18–25 × 6–8μ.
   On dead leaves of *Lannea grandis* (*Odina woodier*); India.

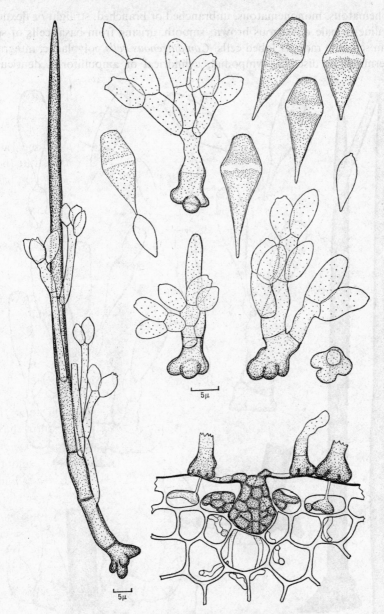

FIG. 164. *Beltraniella odinae* (del. K. Pirozynski).

153. ELLISIOPSIS

Ellisiopsis Batista, 1956, *Anais Soc. Biol. Pernamb.*, **14**: 20–22.

Colonies punctiform to effuse, velutinous, brown to black. *Mycelium* immersed or partly superficial. *Stroma* usually present, confined to the epidermis. *Setae* simple, mid to dark olivaceous brown or brown, smooth to verrucose, arising from flat, radially lobed basal cells. *Hyphopodia* absent. *Conidiophores* macronematous, mononematous, unbranched or branched, straight or flexuous, subhyaline to pale olivaceous brown, smooth, arising from basal cells of setae or from separate radially lobed cells. *Conidiogenous cells* polyblastic, integrated and terminal or discrete, sympodial, cylindrical or ampulliform, denticulate

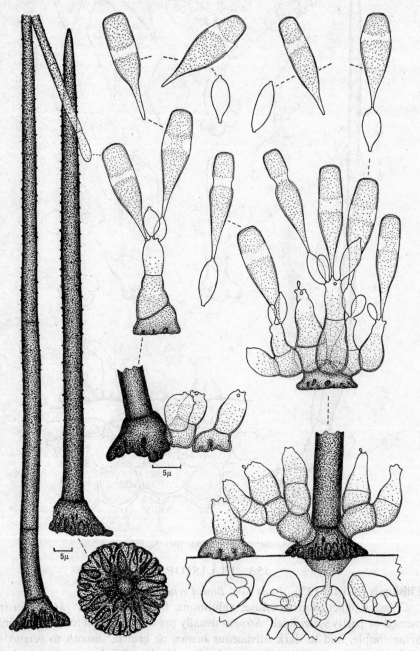

FIG. 165. *Ellisiopsis gallesiae* (del. K. Pirozynski).

(denticles cylindrical); separating cells when present swollen. *Conidia* solitary, acropleurogenous, simple, straight, turbinate, the base drawn out to a fine point, pale olivaceous with a hyaline transverse band just above the centre, smooth, 0-septate.

Type species: Ellisiopsis gallesiae Batista & Nascimento.
Pirozynski, K. A., *Mycol. Pap.*, **90**: 16–22, 1963.

KEY

Conidiophores much branched, crowded, not longer than 25µ . . . . *gallesiae*
Conidiophores unbranched, scattered, up to 40µ long . . . . *portoricensis*

Ellisiopsis gallesiae Batista & Nascimento, 1956, *Anais Soc. Biol. Pernamb.*, **14**: 21.

(Fig. 165)

Setae up to 400µ (mostly 100–150µ) long, about 6µ thick at the base. *Conidiophores* up to 25 × 5–7µ. *Separating cells* limoniform or fusiform, 7–14 × 3–4µ. *Conidia* 18–27 × 5–8·5µ.

On leaves of *Artocarpus, Esenbeckia, Euonymus, Gallesia, Lithocarpus* and fallen fruit of *Chrysobalanus*; Brazil, Japan, Pakistan and W. Africa.

Ellisiopsis portoricensis (F. L. Stevens) Pirozynski has *setae* up to 450 (usually 100–200)µ long, 8µ thick at the base, *conidiophores* up to 50 × 4–5µ, *separating cells* limoniform, 9–12 × 4µ and *conidia* 20–23 × 6·5–8µ.

On dead leaves of *Clusia rosea*; Puerto Rico, Venezuela.

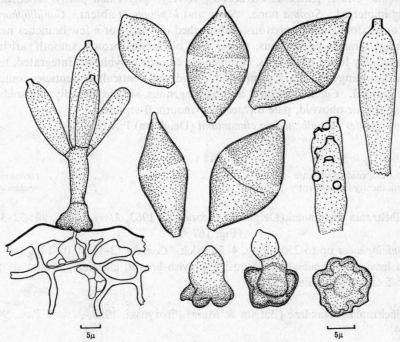

5µ 5µ

FIG. 166. *Pseudobeltrania cedrelae* (del. K. Pirozynski).

154. PSEUDOBELTRANIA

Pseudobeltrania P. Hennings, 1902, *Hedwigia*, **41**: 310.

Colonies effuse, olivaceous or greyish green, velvety. *Mycelium* partly superficial, partly immersed. *Stroma* none. *Setae* and *hyphopodia* absent. *Conidiophores* macronematous, mononematous; stipe unbranched, straight or flexuous, pale brown, smooth, arising from a radially lobed basal cell. *Conidiogenous cells* polyblastic or sometimes monoblastic, integrated and terminal or discrete in a whorl at the apex of the conidiophore stipe, sympodial, cylindrical or clavate, denticulate; denticles cylindrical. *Conidia* solitary, dry, acropleurogenous or acrogenous, simple, biconic, apiculate, olivaceous brown, smooth, 0-septate but with a median transverse hyaline band.

Type species: Pseudobeltrania cedrelae P. Hennings.
Pirozynski, K. A., *Mycol. Pap.*, **90**: 28–30, 1963.

Pseudobeltrania cedrelae P. Hennings, 1902, *Hedwigia*, **41**: 310.
(Fig. 166)

Conidiophores up to 60 × 3–4µ. *Conidiogenous cells* often inflated up to 7–8µ, each with 1–10 large denticles. *Conidia* 20–25 × 10–13µ.
On living leaves of *Cedrela*; Brazil.

155. HEMIBELTRANIA

Hemibeltrania Pirozynski, 1963, *Mycol. Pap.*, **90**: 30–34.

Colonies effuse, parasitic, olivaceous, velvety. *Mycelium* partly superficial, partly immersed. *Stroma* none. *Setae* and *hyphopodia* absent. *Conidiophores* macronematous, mononematous, unbranched or with 1 or a few branches near the apex, straight or flexuous, rather pale olivaceous brown, smooth, arising from radially lobed basal cells. *Conidiogenous cells* polyblastic, integrated, terminal becoming intercalary, sympodial, cylindrical, denticulate; denticles conical to cylindrical. *Conidia* solitary, acropleurogenous, simple, broadly ellipsoidal, limoniform or obovoid, pale olivaceous, smooth, 0-septate.

Type species: Hemibeltrania cinnamomi (Deighton) Pirozynski.

KEY

Conidia ellipsoid to obovate . . . . *cinnamomi*
Conidia mostly limoniform . . . . *nectandrae*

Hemibeltrania cinnamomi (Deighton) Pirozynski, 1963, *Mycol. Pap.*, **90**: 32–33.
(Fig. 167 A)

Conidiophores up to 250µ long, 4–5µ thick. *Conidia* 12–16 × 10–12µ.
On leaves of *Cinnamomum*, causing greyish-brown, elliptical, zonate spots up to 5 cm. across; Sierra Leone.

Hemibeltrania nectandrae (Batista & Maia) Pirozynski, 1963, *Mycol. Pap.*, **90**: 33–34.

(Fig. 167 B)

Conidiophores up to 250μ long, 4–6μ thick. *Conidia* 15–23 × 10–13μ.
On leaves of *Nectandra*; Brazil.

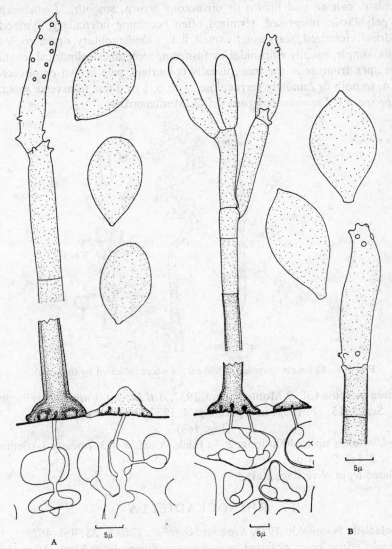

Fig. 167. A, *Hemibeltrania cinnamomi*; B, *H. nectandrae* (del. K. Pirozynski).

156. VERONAEA

Veronaea Ciferri & Montemartini, 1957, *Atti Ist. bot. Univ. Lab. crittogam.*
Pavia, Ser. 5, **15**: 68, 1958; in separate p. 4, 1957.
Sympodina Subramanian & Lodha, 1964, *Antonie van Leeuwenhoek*, **30**: 317–
330.

Colonies effuse, brown, greyish brown or blackish brown, cottony, hairy or
velvety. *Mycelium* partly immersed, partly superficial. *Stroma* none. *Setae*

and *hyphopodia* absent. *Conidiophores* macronematous, monomenatous, un-branched or occasionally loosely branched, straight or flexuous, sometimes geniculate, pale to mid brown or olivaceous brown, smooth. *Conidiogenous cells* polyblastic, integrated, terminal often becoming intercalary, sympodial, cylindrical, cicatrized; scars usually small, flat. *Conidia* solitary, dry, acropleuro-genous, simple, usually ellipsoidal or fusiform, sometimes cylindrical, rounded at the apex truncate at the base, usually colourless, pale brown or olivaceous brown, smooth or minutely verruculose, with 0, 1 or a few transverse septa.

Type species: Veronaea botryosa Cif. & Montemartini.

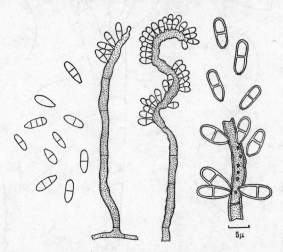

FIG. 168. *Veronaea botryosa* (× 650 except where indicated by the scale).

Veronaea botryosa Cif. & Montemartini, 1957, *Atti Ist. bot. Univ. Lab. crittogam. Pavia*, Ser. 5, **15**: 68, 1958; in separate p. 4, 1957.

(Fig. 168)

Conidiophores up to 400μ long, 2–4μ thick. *Conidia* 0–3- (mostly 1-) septate, 5–12 × 2–4 (8·5 × 3)μ.

Isolated from olive slag; Italy.

157. RHINOCLADIELLA

Rhinocladiella Nannfeldt, 1934, *Svenska skogsFör. Tidskr.*, **32**: 461–462.

Colonies effuse, hairy or felted, grey, brown, olivaceous or black, often slow-growing. *Mycelium* partly immersed, partly superficial. *Stroma* none. *Setae* and *hyphopodia* absent. *Conidiophores* macronematous or semi-macronematous, mononematous, unbranched or loosely branched, straight or flexuous, pale to mid dark brown or olivaceous brown, smooth, verruculose or echinulate. *Conidiogenous cells* polyblastic, integrated, terminal sometimes becoming inter-calary, sympodial, cylindrical, cicatrized, scars small; *ramo-conidia* often present. *Conidia* solitary, dry, acropleurogenous, simple, ellipsoidal, cylindrical, clavate or fusiform, colourless to pale brown or olivaceous brown, smooth or verru-culose, almost always 0-septate, 1 or more septa very rarely formed.

Type species: Rhinocladiella state of *Dictyotrichiella mansonii* Schol-Schwarz
= *R. atrovirens* Nannf.

Schol-Schwarz, M.B., *Antonie van Leeuwenhoek*, **34**: 119–152, 1968.

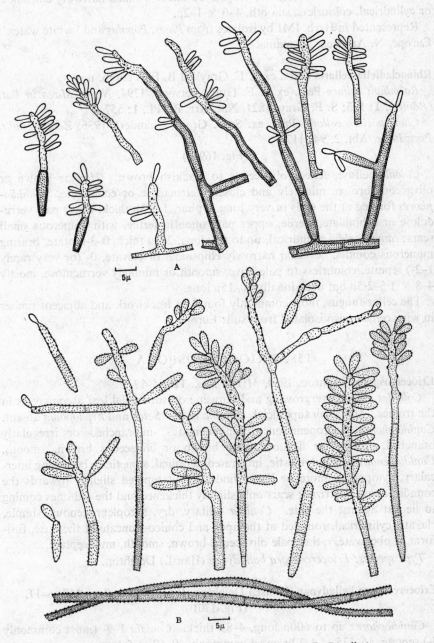

FIG. 169. A, *Rhinocladiella* state of *Dictyotrichiella mansonii*; B, *R. cellaris*.

Rhinocladiella state of **Dictyotrichiella mansonii** Schol-Schwarz, 1968, *Antonie van Leeuwenhoek*, **34**: 122–135, with full synonymy.

(Fig. 169 A)

Colonies slow-growing, olivaceous or blackish green. *Conidiophores* pale to mid olivaceous brown, up to 60μ long, 2–3μ thick. *Conidia* narrowly ellipsoidal or cylindrical, colourless, smooth, 4–6 × 1–2μ.

Represented in Herb. IMI by isolates from *Pinus*, *Populus* and ' white water '; Europe, N. America, Tasmania.

Rhinocladiella cellaris (Pers. ex S. F. Gray) M.B. Ellis comb. nov.

Racodium cellare Pers. ex S. F. Gray; Persoon, 1794, *Neues Magazin Bot.* (*Römer*), **1**: 123; S. F. Gray, 1821, Nat. Arr. Br. Pl., **1**: 557.

Cladosporium cellare (Pers. ex. S. F. Gray) Schanderl, 1936, *Zentbl. Bakt. Parasitkde*, Abt. 2, **94**: 117.

(Fig. 169 B)

Colonies effuse, felted, olivaceous to blackish brown. *Hyphae* brown or olivaceous brown, minutely and closely verruculose or echinulate. *Conidiophores* formed at the ends of very long hyphae, 1·5–3μ thick, lower part verruculose or echinulate, sterile, upper part smooth, fertile, with numerous small scars; *ramo-conidia* cylindrical, up to 65μ long, 2–3μ thick, 0–3-septate, bearing numerous conidia. *Conidia* narrowly ellipsoidal or clavate, 0- (or very rarely 1–2-) septate, colourless to pale olive, smooth or minutely verruculose, mostly 4–8 × 1·5–2·5μ but occasionally 10–15μ long.

The cellar fungus, most commonly found on brickwork and adjacent timber in wine cellars, also isolated from soil; Europe.

158. ERIOCERCOSPORA

Eriocercospora Deighton, 1969, *Mycol. Pap.*, **118**: 5–17.

Colonies effuse, overgrowing and parasitic on superficial leaf ascomycetes in the tropics. *Mycelium* superficial. *Stroma* none. *Setae* and *hyphopodia* absent. *Conidiophores* macronematous, mononematous, unbranched or irregularly branched, straight or flexuous, pale brown or olivaceous brown, smooth. *Conidiogenous cells* polyblastic, integrated, terminal, sometimes becoming intercalary, sympodial, more or less cylindrical but tapered slightly towards the rounded apex, cicatrized; scars only slightly thickened and the old ones coming to lie flat against the side. *Conidia* solitary, dry, acropleurogenous, simple, clavate, cylindrical, rounded at the apex and conico-truncate at the base, fusiform or obclavate, rather pale olivaceous brown, smooth, multiseptate.

Type species: Eriocercospora balladynae (Hansf.) Deighton.

Eriocercospora balladynae (Hansf.) Deighton, 1969, *Mycol. Pap.*, **118**: 6–17.

(Fig. 170)

Conidiophores up to 600μ long, 4–8μ thick. *Conidia* 1–9- (most commonly 3-) septate, 16–135 × 4–12μ, most commonly 40–60 × 5–8μ.

Overgrowing and parasitic on *Asterina, Balladyna, Balladynopsis, Clypeolella* and *Schiffnerula*; Costa Rica, Ghana, Malaysia, Panama, Sierra Leone, S. Africa, Uganda.

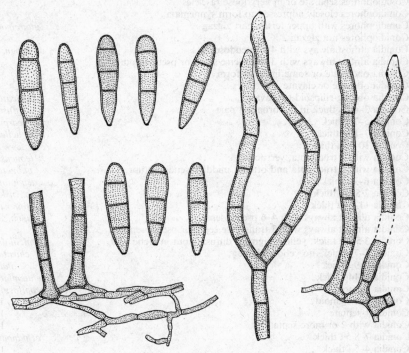

FIG. 170. *Eriocercospora balladynae* (× 650).

159. SPIROPES

Spiropes Ciferri, 1955, *Sydowia*, **9**: 302–303.

Nascimentoa Ciferri & Batista, 1956, *Publções Inst. Micol. Recife*, **44**: 4.

Colonies usually effuse, hairy or velvety, pale olive to brown or black, often overgrowing and apparently parasitic on Meliolineae and other tropical leaf ascomycetes. *Mycelium* mostly superficial. *Stroma* none. *Setae* and *hyphopodia* absent. *Conidiophores* macronematous, almost always unbranched, straight or flexuous, sometimes geniculate, arising from hyphae singly, in loose fascicles or closely adpressed to form synnemata, pale olivaceous brown to dark blackish brown, smooth, often thick-walled. *Conidiogenous cells* polyblastic, integrated, terminal becoming intercalary, sympodial, cylindrical, cicatrized, scars often numerous and conspicuous. *Conidia* solitary, dry, acropleurogenous, simple, most commonly obclavate but sometimes clavate, ellipsoidal, fusiform or oblong rounded at the apex truncate at the base, straight, curved or sigmoid, subhyaline to olivaceous brown or dark brown, smooth, rugose or verruculose, with 1–9 transverse septa or pseudosepta.

Type species: Spiropes guareicola (Stev.) Cif.

Ellis, M.B., *Mycol. Pap.*, **114**: 2–42, 1968.

KEY

Overgrowing and apparently parasitic on Meliolineae 1
Overgrowing fungi other than Meliolineae or on leaves and stems 10
  1. Conidiophores separate or in very loose fascicles 2
     Conidiophores closely adpressed to form synnemata 8
  2. Conidiophores with upper fertile part zigzag . . . . . *guareicola*
     Conidiophores not zigzag 3
  3. Conidia almost always with 4–6 pseudosepta . . . . *capensis*
     Conidia almost always with 3 transverse septa or pseudosepta . . . 4
  4. Conidia obclavate or sometimes fusiform 5
     Conidia obovate or clavate 7
     Conidia oblong-ellipsoid to clavate . . . . . . *intricatus*
  5. Conidia 3–4·5μ thick in the broadest part . . . . . *effusus*
     Conidia 5–7μ thick . . . . *dorycarpus*
     Conidia 7–10μ thick . . . . *helleri*
     Conidia 10–15μ thick 6
  6. Conidia with 3 true septa, verrucose . . . . . *leonensis*
     Conidia with 2 true septa and often a middle pseudoseptum . . . *palmetto*
  7. Conidia 6–7μ thick . . . . *deightonii*
     Conidia 7·5–9μ thick . . . . *pirozynskii*
     Conidia 11–16μ thick . . . . *fumosus*
  8. Conidia almost always with 4–6 pseudosepta . . . . *japonicus*
     Conidia almost always with 3 transverse septa or pseudosepta . . . 9
  9. Conidia 3·5–5μ thick, yellow pigment diffusing out in lactic acid . . *penicillium*
     Conidia 5–7μ thick, no yellow pigment . . . . . *clavatus*
     Conidia 7–9μ thick . . . . *dialii*
     Conidia 9–14μ thick . . . . *melanoplaca*
 10. Conidia sigmoid . . . . *scopiformis*
     Conidia not sigmoid 11
 11. Conidia 1-septate 12
     Conidia with 2 or more septa 14
 12. Conidia 7–8·5μ thick . . . . *echidnodis*
     Conidia 4–5μ thick 13
 13. Conidia subhyaline, smooth . . . . *bakeri*
     Conidia brown, often verruculose . . . . . . *lembosiae*
 14. Conidia 3·5–4·5μ thick . . . . *balladynae*
     Conidia 5–6μ thick . . . . *nothofagi*
     Conidia 6–8μ thick 15
     Conidia 8–13μ thick 16
 15. Conidia with 1–3 septa . . . . *shoreae*
     Conidia with 3–8 septa . . . . *davillae*
 16. Conidia almost always with 2 septa . . . . . *harunganae*
     Conidia with 5–9 transverse septa . . . . . *caaguazuensis*

Spiropes guareicola (Stev.) Cif., 1955, *Sydowia*, **9**: 303.
(Fig. 171 A)

Colonies effuse, dark blackish brown to black, hairy. *Mycelium* superficial, composed of a network of branched and anastomosing, rather pale olivaceous brown, smooth, 2–4μ thick hyphae. *Conidiophores* arising singly or in groups as lateral branches on the hyphae, erect, sterile lower part straight or flexuous, fertile upper part zigzag, sometimes with several fertile zigzag regions separated by sterile areas, mid to dark brown, paler near the apex, with numerous dark conidial scars, up to 400μ long, 6–9μ thick. *Conidia* broadly fusiform, pale to dark brown or olivaceous brown, smooth, with 3–5 (usually 3) pseudosepta, 25–52 (35)μ long, 10–13 (11·8)μ thick in the broadest part, 3·5–5μ wide at the truncate base.

Overgrowing colonies of *Asteridiella, Irenopsis* and *Meliola* on many different flowering plants; Assam, Bougainville, Ghana, India, Malaya, Netherlands New Guinea, Philippines, Puerto Rico, Sabah, Sarawak, Sierra Leone, Solomon Islands, Uganda.

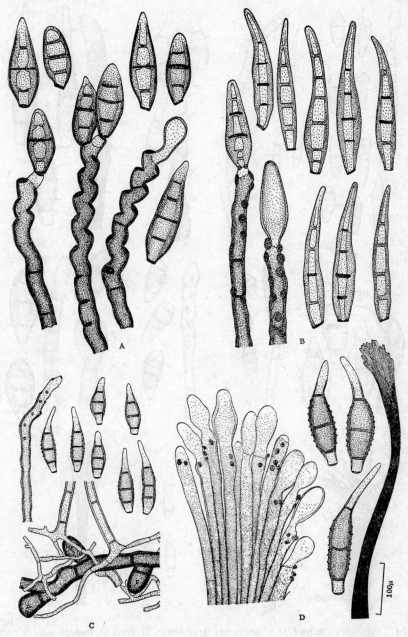

FIG. 171. *Spiropes* species (1): A, *guareicola*; B, *capensis*; C, *dorycarpus*; D, *melanoplaca* (× 650 except where indicated by the scale).

Spiropes capensis (Thüm.) M. B. Ellis, 1968, *Mycol. Pap.*, **114**: 5–8.
(Fig. 171 B)

Colonies dark blackish brown to black, hairy. *Conidiophores* arising singly or in groups, sometimes in very large groups of 100 or more, straight or flexuous,

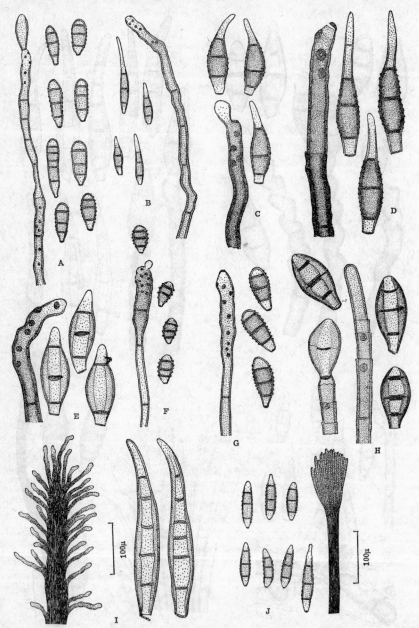

FIG. 172. *Spiropes* species (2): A, *intricatus*; B, *effusus*; C, *helleri*; D, *leonensis*; E, *palmetto*; F, *deightonii*; G, *pirozynskii*; H, *fumosus*; I, *japonicus*; J, *penicillium* (× 650 except where indicated by scales).

brown to dark brown, paler near the apex, with usually rather pale but quite well-defined conidial scars, up to 500μ long, 6–8μ thick. *Conidia* straight or curved, fusiform to obclavate, subhyaline to brown, smooth, with 3–6 (usually 4 or 5) transverse pseudosepta, 38–67 (50)μ long, 6–11 (8·4)μ thick in the broadest part, tapering to 1·5–4μ at the apex, 3–5μ wide at the base.

Very common on colonies of *Appendiculella, Asteridiella, Irenopsis* and *Meliola* on many different plants; Amboina, Bolivia, Brazil, Cameroons, Congo, Dominican Republic, Ghana, India, Jamaica, Malaya, Peru, Philippines, Puerto Rico, Sabah, Sarawak, Sierra Leone, S. Africa, Tanzania, Uganda, Venezuela.

Spiropes dorycarpus (Mont.) M. B. Ellis, 1968, *Mycol. Pap.*, **114**: 11–14.
(Fig. 171 C)

Colonies effuse, pale olive, brown or dark brown, hairy or velvety. *Conidiophores* arising singly or in groups, straight or flexuous, pale to mid brown, with scars towards the apex, up to 700μ long, 3–7 (4)μ thick. *Conidia* straight or curved, variable in shape but generally obclavate to fusiform, conico-truncate at the base, sometimes strongly attenuated towards the apex, nearly always 3-septate but occasionally with 4 or 5 septa, rather pale brown, the central cells often slightly darker than the end ones, smooth, rugose or verruculose, 17–38 (23)μ long, 5–7 (6·1)μ thick in the broadest part.

Very common on colonies of *Appendiculella, Asteridiella, Clypeolella, Irenopsis, Meliola* and *Schiffnerula* on many different kinds of flowering plants; Australia, Brazil, Chile, Congo, Cuba, Dominican Republic, Ghana, Guyana, India, Indo-China, Malaya, Nigeria, Puerto Rico, Sarawak, Sierra Leone, S. Africa, Taiwan, Tanzania, Uganda.

Spiropes melanoplaca (Berk. & Curt.) M. B. Ellis, 1968, *Mycol. Pap.*, **114**: 28–30.
(Fig. 171 D)

Colonies effuse, dark blackish brown to black, hairy, with large, erect, dark synnemata clearly visible under a low-power dissecting microscope. *Conidiophore* threads very tightly packed together to form erect, dark blackish brown to black synnemata up to 1·5 mm. long, 20–80μ thick, splaying out at apex and base, individually brown or dark brown, smooth, cylindrical and 2–4μ thick along most of their length, paler and thickening to 5–8μ near the apex, with numerous scars which frequently lie at an acute angle to the wall and overlap like scales. *Conidia* straight or curved, fusiform to obclavate, often rostrate, 3-septate, the two middle cells golden brown or brown, smooth or verruculose, the cells at each end very pale and usually smooth, 30–67 (50)μ long, 9–14 (10·8)μ thick in the broadest part, tapering to 2–3μ at the apex, 4–5μ wide at the truncate base.

Overgrowing and apparently parasitic on Meliolines, especially *Meliola* species on various plants; Cuba, Dominican Republic, Ghana, Guadalcanal, Malaya, Peru, Philippines, Sarawak, Sierra Leone, Tanzania, Trinidad, Uganda.

Conidial measurements, substrata and geographical distribution of 22 other species of *Spiropes* are given below.

S. intricatus (Sacc.) M. B. Ellis (Fig. 172 A): 12–25 (17) × 4·5–7 (6)μ, on Melioline on unknown tree; Philippines.

S. effusus (Pat.) M. B. Ellis (Fig. 172 B): 15–36 (20) × 3–4·5 (3·9)μ, on Meliolines especially *Amazonia* and on *Asterina* on various plants; Ghana, Puerto Rico, Sierra Leone, Uganda, Venezuela.

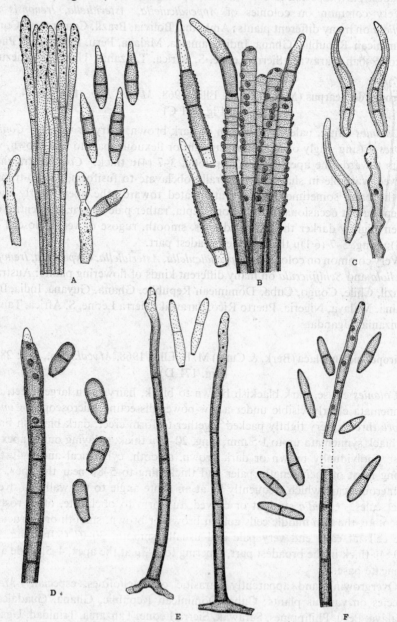

FIG. 173. *Spiropes* species (3): A, *clavatus*; B, *dialii*; C, *scopiformis*; D, *echidnodis*: E. *bakeri*; F, *lembosiae* (× 650).

S. helleri (Stev.) M. B. Ellis (Fig. 172 C): 30–50 (37) × 7–10 (8·4)μ, on *Asteridiella*, *Irenopsis* and *Meliola* on various plants; Ghana, Malaya, New Caledonia, Philippines, Puerto Rico, Sabah, Sierra Leone, Uganda.

S. leonensis M. B. Ellis (Fig. 172 D): 40–63 (54) × 10–13 (11)μ, on *Meliola garciniae* on *Pentadesma*; Sierra Leone.

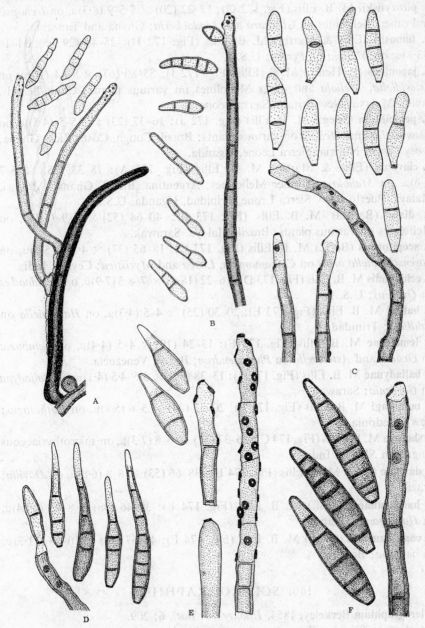

FIG. 174. *Spiropes* species (4): A, *balladynae*; B, *nothofagi*; C, *shoreae*; D, *davillae*; E, *harunganae*; F, *caaguazuensis* (× 650).

S. palmetto (Gerard) M. B. Ellis (Fig. 172 E): 29–46 (38) × 10–15 (12·4)µ, on *Irenopsis* and *Meliola* on various plants; Ghana, New Zealand, Puerto Rico, Sarawak, Sierra Leone, U.S.A.

S. deightonii M. B. Ellis (Fig. 172 F): 10–14 (12·5) × 6–7 (6·8)µ, on *Meliola borneensis* on *Uvaria*; Sierra Leone.

S. pirozynskii M. B. Ellis (Fig. 172 G): 17–22 (20) × 7·5–9 (8·3)µ, on *Irenopsis* and other Meliolines on *Glyphaea* and *Lindackeria*; Ghana and Tanzania.

S. fumosus (Ellis & Martin) M. B. Ellis (Fig. 172 H): 25–40 (29·7) × 11–16 (13·8)µ, on *Meliola* on *Persea*; U.S.A.

S. japonicus (P. Henn.) M. B. Ellis (Fig. 172 I): 55–80 (67) × 8–14 (10)µ, on *Asteridiella*, *Meliola* and other Meliolines on various plants; Cook Islands, Japan, Malaya, New Britain, Sierra Leone.

S. penicillium (Speg.) M. B. Ellis (Fig. 172 J): 16–37 (23) × 3·5–5 (4·5)µ, on *Asteridiella* and *Meliola* on various plants; Brazil, Congo, Costa Rica, Ghana, Ivory Coast, Nigeria, Sierra Leone, Uganda.

S. clavatus (Ellis & Martin) M. B. Ellis (Fig. 173 A): 18–33 (25) × 5–7 (5·6)µ, on *Meliola* and other Meliolines; Argentina, Brazil, Ghana, Malawi, Malaya, Puerto Rico, Sierra Leone, Trinidad, Uganda, U.S.A.

S. dialii (Batista) M. B. Ellis (Fig. 173 B): 40–64 (52) × 7–9 (8·1)µ, on Meliolines on various plants; Brazil, Malaya, Sarawak.

S. scopiformis (Berk.) M. B. Ellis (Fig. 173 C): 18–65 (37) × 4·5–8 (5·6)µ, on *Rosenscheldiella orbis* on *Cinnamomum*, *Litsea* and *Myristica*; Ceylon, India.

S. echidnodis M. B. Ellis (Fig. 173 D): 16–22 (18·4) × 7–8·5 (7·9)µ, on *Echidnodes* on *Quercus*; U.S.A.

S. bakeri M. B. Ellis (Fig. 173 E): 20–30 (25) × 4–5 (4·3)µ, on *Halbaniella* on *Paullinia*; Trinidad.

S. lembosiae M. B. Ellis (Fig. 173 F): 13–24 (19) × 4–5 (4·4)µ, on *Lembosia* on *Ocotea* and *Asterinella* on *Phoradendron*; Brazil, Venezuela.

S. balladynae M. B. Ellis (Fig. 174 A): 13–38 (32) × 3·5–4·5 (4·1)µ, on *Balladyna* on *Gardenia*; Sarawak.

S. nothofagi M. B. Ellis (Fig. 174 B): 20–32 (24) × 5–6 (5·8)µ, on *Nothofagus*; New Caledonia.

S. shoreae M. B. Ellis (Fig. 174 C): 26–37 (31) × 6–8 (7·3)µ, on microthyriaceous fungus on *Shorea*; India.

S. davillae (Syd.) M. B. Ellis (Fig. 174 D): 38–66 (53) × 6–8 (6·8)µ, on *Davilla*; Brazil.

S. harunganae (Hansf.) M. B. Ellis (Fig. 174 E): 30–46 (38) × 8–12 (10·4)µ, on *Harungana*; Uganda.

S. caaguazuensis (Speg.) M. B. Ellis (Fig. 174 F): 42–67 (58) × 10–13 (11·3)µ, on bamboo; Brazil.

160. SCLEROGRAPHIUM

Sclerographium Berkeley, 1854, *Hooker's J. Bot.*, **6**: 209.

Colonies effuse, with scattered, erect, black or dark brown synnemata. *Mycelium* superficial. *Stroma* none. *Setae* and *hyphopodia* absent. *Conidio-*

phores macronematous, synnematous, subulate; individual threads mostly un-branched, rarely with a few short branches near the apex, straight or flexuous, brown or olivaceous brown, narrow and smooth except near the apex, where they are swollen and usually verruculose. *Conidiogenous cells* polyblastic, integrated, terminal sometimes becoming intercalary, sympodial, clavate, cicatrized; scars often slightly raised. *Conidia* solitary, dry, acropleurogenous, simple, cylindrical with rounded ends, ellipsoidal or fusiform, mid to dark brown or olivaceous brown, verrucose or echinulate, with numerous transverse and a few longitudinal septa.

Type species: Sclerographium aterrimum Berk.

Hughes, S. J., *Indian Phytopath.*, **4**: 5–6, 1951.

KEY

Conidia less than 40μ long, on *Indigofera* . . . . *aterrimum*
Conidia more than 40μ long, on *Phyllanthus* . . . . *phyllanthicola*

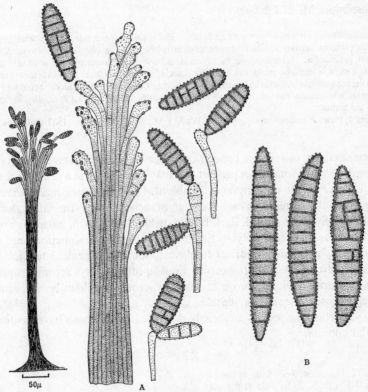

FIG. 175. A, *Sclerographium aterrimum*; B, *S. phyllanthicola* (× 650 except where indicated by the scale).

Sclerographium aterrimum Berk., 1854, *Hooker's J. Bot.*, **6**: 209.

(Fig. 175 A)

Colonies hypophyllous. *Synnemata* black, up to 1 mm. high, 20–30μ thick at the base, tapering to 11–20μ; individual threads brown, 2–3μ thick, swollen near the apex to 4–6µ. *Conidia* cylindrical rounded at the ends or ellipsoidal, mid to dark brown, verrucose 30–38 × 9–12µ, with usually 7 transverse and a few longitudinal septa.

On leaves of *Indigofera*; India.

Sclerographium phyllanthicola Deighton, 1960, *Mycol. Pap.*, **78**: 25–28.
(Fig. 175 B)

Colonies hypophyllous. *Synnemata* dark brown, up to 1300µ high, 16–28µ thick at the base, tapering to 10–12µ; individual threads olivaceous brown, 1·5–3µ thick, swollen near the apex to 9µ. *Conidia* fusiform, olivaceous brown, echinulate or verrucose, 40–85 × 10–15µ, with up to 16 transverse and 1–6 longitudinal septa.

On leaves of *Phyllanthus*; Sierra Leone.

161. PSEUDOSPIROPES

Pseudospiropes M. B. Ellis gen. nov.

Coloniae effusae, olivaceo-brunneae, atro-brunneae vel atrae, velutinae. *Mycelium* immersum, ex hyphis ramosis, septatis, pallide brunneis vel atro-brunneis, laevibus compositum. *Stromata* interdum praesentia. *Conidiophora* ex apice lateribusque hypharum et ex stromate quoque oriunda, erecta, simplicia, recta vel flexuosa, laevia, crasse tunicata, atro-brunnea vel fusca, apicem versus pallidiora, cicatricibus magnis praedita. *Conidia* solitaria, sicca, acropleurogena, fusiformia, navicularia vel obclavata, olivaceo-brunnea vel aureo-brunnea, laevia, pseudo-septata vel septata.

Species typica: *Pseudospiropes nodosus* (Wallr.) M. B. Ellis (Typus: IMI 68038 ex Herb. STR).

Colonies effuse, olivaceous brown, dark brown or black, velvety. *Mycelium* immersed. *Stroma* sometimes present and may consist of just a few cells forming a small plate. *Setae* and *hyphopodia* absent. *Conidiophores* macronematous, mononematous, scattered or caespitose, unbranched, erect, straight or flexuous, smooth, usually thick-walled, dark brown or blackish brown, paler towards the apex. *Conidiogenous cells* polyblastic, integrated, terminal, sometimes becoming intercalary, sympodial, cylindrical to clavate, sometimes geniculate, cicatrized; scars large, often dark and prominent. *Conidia* solitary, dry, acropleurogenous, simple, fusiform, navicular or obclavate, olivaceous or golden brown, smooth, transversely pseudoseptate or septate.

Two very common species are briefly described. These both have fusiform or navicular conidia.

KEY

Conidia 12–18µ thick, 5–7µ wide at the scar   .   .   .   . *nodosus*
Conidia 9–13µ thick, 2–3µ wide at the scar   .   .   .   . *simplex*

Pseudospiropes nodosus (Wallr.) M. B. Ellis comb. nov.

Helminthosporium nodosum Wallr., 1833, Fl. crypt. Ger., **2**: 165. For other synonyms see Hughes in *Can. J. Bot.*, **36**: 798, 1958, under *Pleurophragmium nodosum*.

(Fig. 176 A)

Colonies black. *Conidiophores* dark blackish brown except near the apex where they are pale, 100–350μ long, mostly 8–10μ thick but sometimes swollen in the upper part to 15–20μ; scars large, dark, prominent. *Conidia* broadly fusiform to navicular, truncate at the base, pale to dark golden brown, 6–10-

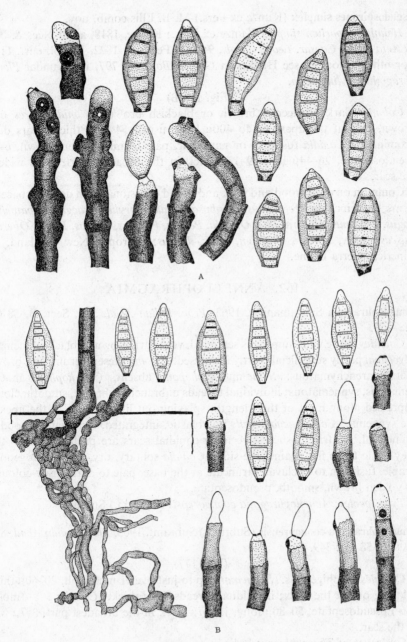

Fig. 176. A, *Pseudospiropes nodosus*; B, *P. simplex* (× 650).

pseudoseptate, 32–50µ long, 12–18µ thick in the broadest part, 5–7µ wide at the scar.

Common on dead wood and bark and found occasionally on dead herbaceous stems, hosts include *Acer, Berberis, Betula, Corylus, Eupatorium, Fagus, Fraxinus, Hedera, Populus, Rubus, Salix* and *Sorbus*; Europe, N. America.

Pseudospiropes simplex (Kunze ex Pers.) M. B. Ellis comb. nov.

Helminthosporium simplex Kunze ex Pers.; Kunze, 1818, apud Nees & Nees in *Acta Acad. Caesar. Leop. Carol.*, **9**: 241; Persoon, 1822, Mycol. eur., **1**: 18. For other synonyms see Hughes in *Can. J. Bot.*, **36**: 797, 1958, under *Pleurophragmium cylindricum*.

(Fig. 176 B)

Colonies dark olivaceous brown or blackish brown. *Conidiophores* dark brown, pale at the apex, up to 400µ long, usually 4·5–6µ thick, scars dark, prominent. *Conidia* fusiform or navicular, pale to mid golden brown, 6–11-pseudoseptate, 26–44µ long, 9–13µ thick in the broadest part, 2–3µ wide at the scar.

Common on dead wood and bark and found occasionally on dead herbaceous stems, hosts include *Acer, Alnus, Bambusa, Betula, Corylus, Craegus, Eupatorium, Fagus, Fraxinus, Hedera, Ilex, Olearia, Parkia, Populus, Prunus, Pyrus, Quercus, Ruscus, Salix, Sorbus, Triticum, Ulex, Ulmus*; Europe, New Zealand, N. America, Sierra Leone.

162. ANNELLOPHRAGMIA

Annellophragmia Subramanian, 1963, *Proc. Indian Acad. Sci.*, Sect. B, **58** (6): 348–350.

Colonies effuse with numerous scattered, very dark brown to black synnemata. *Mycelium* partly superficial, partly immersed. *Stroma* present, erumpent, brown, pseudoparenchymatous. *Setae* and *hyphopodia* absent. *Conidiophores* macronematous, synnematous; individual threads unbranched, brown, smooth, closely adpressed along most of their length, splaying out like a brush at the apex of the synnema. *Conidiogenous cells* polyblastic, integrated, terminal, sympodial, cylindrical, cicatrized; successive large conidial scars are pushed over so that they come to lie flat against the side. *Conidia* solitary, dry, acropleurogenous, simple, fusiform to obclavate, truncate at the base, pale to dark straw-coloured or golden brown, smooth, pseudoseptate.

Type species: *Annellophragmia coonoorensis* (Subram.) Subram.

Annellophragmia coonoorensis (Subram.) Subram., 1963, *Proc. Indian Acad. Sci.*, Sect. B, **58** (6): 349.

(Fig. 177)

Colonies amphigenous. *Synnemata* up to just over 1 mm. high, 20–60µ thick, splaying out at the apex; individual threads 5–9µ thick. *Conidia* 3–8- (mostly 4–6-) pseudoseptate, 50–80µ long, 12–17µ thick in the broadest part, 5–7µ wide at the scar.

On leaves of *Thysanolaena*; India.

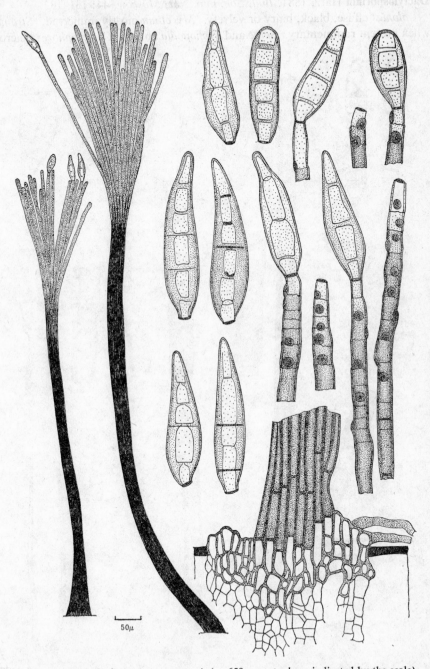

Fig. 177. *Annellophragmia coonoorensis* (× 650 except where indicated by the scale).

163. DACTYLOSPORIUM

Dactylosporium Harz, 1881, *Bull. Soc. imp. Nat. Moscow*, **44**: 131–132.

Colonies effuse, black, hairy or velvety. *Mycelium* mostly immersed. *Stroma* when present rudimentary. *Setae* and *hyphopodia* absent. *Conidiophores* macro-

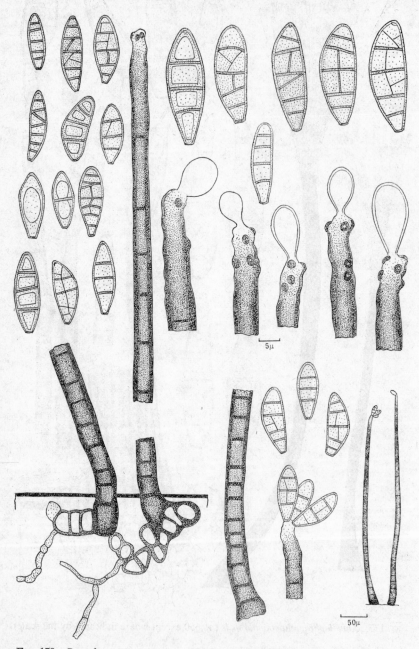

FIG. 178. *Dactylosporium macropus* (× 650 except where indicated by scales).

nematous, mononematous, straight or flexuous, unbranched, subulate, closely septate near the base, dark brown except towards the apex where they are much paler, smooth. *Conidiogenous cells* polyblastic, integrated, terminal, sympodial, cylindrical, cicatrized; scars prominent. *Conidia* solitary, acropleurogenous, simple, clavate, ellipsoidal or navicular, subhyaline to brown, smooth, with transverse and also usually longitudinal and oblique septa, with a truncate base and often a small, thickened, refractive area of wall at the apex.

Type species: Dactylosporium macropus (Corda) Harz.
Hughes, S. J., *Naturalist, Hull*, **1952**: 63–64, 1952.

Dactylosporium macropus (Corda) Harz, 1871, *Bull. Soc. imp. Nat. Moscow*, **44**: 131–132.
Mystrosporium macropus Corda, 1839, Icon. Fung., **3**: 10.
Helminthosporium tingens Cooke, 1883, *Grevillea*, **72**: 37.
Brachysporium tingens (Cooke) Sacc., 1886, Syll. Fung., **4**: 427.

(Fig. 178)

Conidiophores up to 1 mm. long, 9–12μ thick just above the base, tapering to 4–7μ at the apex. *Conidia* 18–36 × 7–13μ.
On fallen branches and occasionally on dead herbaceous stems; Europe.

164. PLEIOCHAETA

Pleiochaeta (Sacc.) Hughes, 1951, *Mycol. Pap.*, **36**: 39.
Colonies effuse, grey, olivaceous brown or black. *Mycelium* mostly immersed. *Stroma* none. *Setae* and *hyphopodia* absent. *Conidiophores* macronematous, mononematous, usually unbranched, flexuous, frequently geniculate, hyaline or pale olivaceous, smooth. *Conidiogenous cells* polyblastic, integrated, terminal becoming intercalary, sympodial, cylindrical, cicatrized; scars broad, flat, thin. *Conidia* solitary, dry, acropleurogenous, appendiculate, roughly cylindrical, narrowed at the apex, truncate at the base, colourless or with the cell at each end hyaline or subhyaline and intermediate cells straw-coloured to golden brown, smooth, multiseptate; the apical cell bears several long hyaline, subulate appendages which are sometimes branched.

Type species: Pleiochaeta setosa (Kirchn.) Hughes.

Pleiochaeta setosa (Kirchn.) Hughes, 1951, *Mycol. Pap.*, **36**: 32–39.
Ceratophorum setosum Kirchn., 1892, *Z. PflKrankh. PflPath. PflSchutz*, **2**: 324–327.

(Fig. 179)

Conidiophores up to 150μ long, 8–13μ thick. *Conidia* 4–8- (most commonly 5-) septate, 60–90 × 14–22μ, scar 7–11μ wide, appendages up to 100μ long, 3–4μ thick at the base, tapering to about 1μ.
Parasitic on *Crotalaria*, *Cytisus*, *Lupinus* and *Ornithopus*; Africa, Asia, Australasia, Europe, N. & S. America [CMI Distribution Map 243].

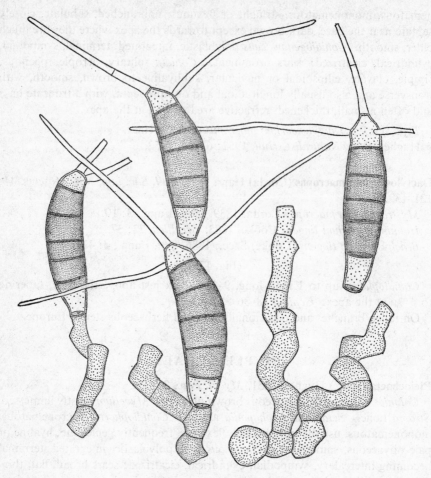

FIG. 179. *Pleiochaeta setosa* (× 650).

165. CENTROSPORA

Centrospora Neergaard, 1942, *Zentbl. Bakt. ParasitKde*, Abt. 2, **104**: 408–411.

Ansatospora Newhall, 1944, *Phytopathology*, **34**: 98.

Colonies effuse, at first hyaline, later variously coloured, finally almost black. *Mycelium* superficial and immersed; with torulose groups of swollen, mid to dark brown cells. *Stroma* none. *Setae* and *hyphopodia* absent. *Conidiophores* macronematous, mononematous, straight or flexuous, geniculate in the upper part, usually unbranched, colourless, smooth. *Conidiogenous cells* integrated, terminal, polyblastic, sympodial, clavate or cylindrical with broad, flat scars. *Conidia* solitary, acropleurogenous, often appendiculate with a septate lateral appendage from the basal cell or sometimes simple, obclavate, rostrate, frequently strongly curved, truncate at the base, colourless or with the broader cells rather pale brown, smooth, multiseptate.

Type species: Centrospora acerina (Hartig) Newhall = *C. ohlsenii* Neergaard.
Channon, A. G., *Ann. appl. Biol.*, **56**: 119–128, 1965.

Centrospora acerina (Hartig) Newhall, 1946, *Phytopathology*, **36**: 893–896 [with full synonymy].

(Fig. 180)

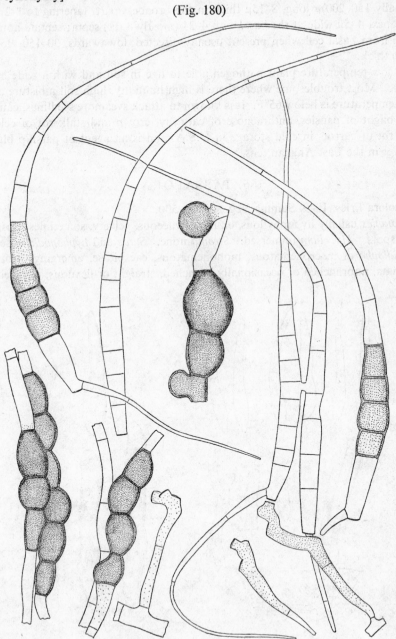

FIG. 180. *Centrospora acerina* (× 650).

Colonies at first hyaline, later green, grey or reddish purple, finally almost black. *Hyphae* mostly cylindrical, colourless, 4–8μ thick, but torulose groups of mid to dark brown cells swollen to 17–30μ are often formed especially near the centre of colonies. *Conidiophores* usually about 50μ long, 5–7μ thick, markedly geniculate towards the apex with broad, flat scars. *Conidia* 60–250 (usually 150–200)μ long, 8–15μ thick in the broadest part, tapering to 1–2μ at the apex, 4–5μ wide at the base, with 4–24 (mostly 8–11) septa; septate appendage from basal cell when present usually directed downwards, 30–150μ long, 2–3μ thick.

A low temperature plant pathogen able to live in soil and with a wide host range. Most troublesome where there is high humidity, high soil moisture and the temperature is below 65°F. It is known to attack sycamore seedlings, causes leaf blight of pansies, anthracnose of caraway, crown and stalk rot of celery and rot of carrots in cold storage in U.S.A. and is an agent of parsnip black canker in the East Anglian fens.

166. PASSALORA

Passalora Fries, 1849, Summa Veg. scand.: 500.

Colonies usually hypophyllous, effuse, olivaceous, velvety, sometimes causing leaf spots. *Mycelium* immersed. *Stroma* none. *Setae* and *hyphopodia* absent. *Conidiophores* macronematous, mononematous, caespitose, emerging through stomata, unbranched or occasionally branched, straight or flexuous, olivaceous

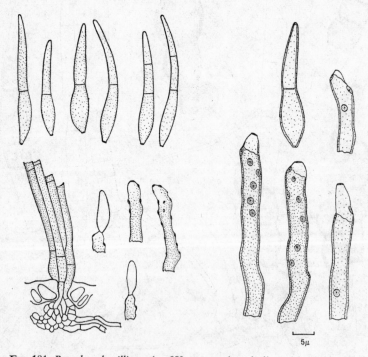

5μ

FIG. 181. *Passalora bacilligera* (× 650 except where indicated by the scale).

or brown, smooth. *Conidiogenous cells* polyblastic, sympodial, integrated, terminal becoming intercalary, cicatrized; scars slightly but distinctly thickened, not or very slightly prominent. *Conidia* solitary, dry, acropleurogenous, obclavate, pale olivaceous brown, smooth, mostly 1-septate, the proximal cell swollen and long ellipsoidal, the distal cell narrow, subcylindrical to very long ellipsoidal, rarely 0- or 2–3-septate.

Type species: Passalora bacilligera (Mont. & Fr.) Mont. & Fr.
Deighton, F. C., *Mycol. Pap.*, **112**: 3–16, 1967.

Passalora bacilligera (Mont. & Fr.) Mont. & Fr., 1856, in Montagne's Syll. Gen. Sp. crypt.: 305.

(Fig. 181)

Conidiophores often in fascicles of about 12, up to 180μ long, 3–5μ thick at the base, swelling towards the apex to 4–6·5μ. *Conidia* 21–68 (mostly 35–45)μ long, proximal cell 4·5–8·5μ thick, distal cell 3·5–5μ thick.

On *Alnus glutinosa*, often on pale, yellowish-green angular areas 1–2 mm. wide, limited by small leaf veins and without definite margin; Europe.

F. C. Deighton recognises two other species of *Passalora* on *Alnus*: **P. alni** (Chupp & Greene) Deighton on *A. crispa* in N. America, *A. alnobetula* and *A. viridis* in Europe with simple, rather rigid conidiophores not more than 110μ long, very pale conidia 26–87 × 4–7μ and more conspicuous scars, and **P. microsperma** Fuckel on *A. incana* in Europe with conidiophores up to 360μ long and conidia 13–24 × 5·5–9μ.

167. STENELLOPSIS

Stenellopsis Huguenin, 1966, *Bull. trimest. Soc. mycol. Fr.*, **81**: 693–696.

Colonies effuse, greyish olive, hairy. *Mycelium* immersed. *Stroma* rudimentary or prosenchymatous, immersed. *Setae* and *hyphopodia* absent. *Conidiophores* macronematous, mononematous, caespitose, unbranched, usually rather short, straight or flexuous, olivaceous, smooth or verruculose. *Conidiogenous cells* polyblastic, integrated, terminal, sympodial, cylindrical, cicatrized; scars broad, flat. *Conidia* solitary, dry, acropleurogenous, simple, cylindrical to obclavate, rounded at the apex, truncate at the base, pale olivaceous brown, verrucose, multiseptate.

Type species: Stenellopsis fagraeae Huguenin.

Stenellopsis fagraeae Huguenin, 1966, *Bull. trimest. Soc. mycol. Fr.*, **81**: 693–696.

(Fig. 182)

Colonies amphigenous. *Conidiophores* mostly 20–30 (rarely 50)μ long, 5–7μ thick. *Conidia* 40–130 × 4–7μ, 3–14-septate.

Parasitic on leaves of *Fragraea*, causing large, round, pale brown spots, each with a broad, dark brown border; New Caledonia.

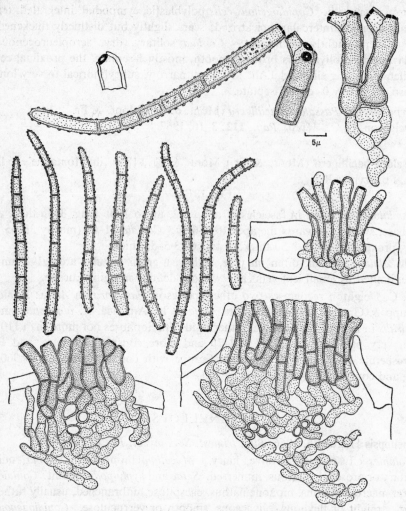

FIG. 182. *Stenellopsis fagraeae* (× 650 except where indicated by the scale).

168. PHAEOISARIOPSIS

Phaeoisariopsis Ferraris, 1909, *Annls mycol.*, **7**: 280.

Colonies effuse, olivaceous brown, cottony or hairy. *Mycelium* immersed. *Stroma* present, prosenchymatous, brown or olivaceous brown. *Setae* and *hyphopodia* absent. *Conidiophores* macronematous, mononematous and caespitose or synnematous, individual threads unbranched, straight or flexuous, pale to mid brown or olivaceous brown, smooth. *Conidiogenous cells* polyblastic, integrated, terminal, sympodial, cylindrical or clavate, cicatrized; scars thin but visible, flattened against the side of the conidiogenous cell. *Conidia* solitary, dry, acropleurogenous, simple, obclavate or cylindrical, pale olive, olivaceous brown or brown, smooth or verruculose, mostly with 3 or more transverse septa.

Lectotype species: Phaeoisariopsis griseola (Sacc.) Ferraris.

Phaeoisariopsis griseola (Sacc.) Ferraris, 1909, *Annls mycol.*, **7**: 280.

Isariopsis griseola Sacc., 1878, *Michelia*, **1**: 273.

(Fig. 183)

Conidiophores caespitose or forming synnemata, up to 500μ long, threads 2–4μ thick near the base, swelling to 5–6μ near the apex. *Conidia* mostly obclavate, conico-truncate at the base, very pale olive or olivaceous brown, smooth, 3–6-septate, 30–70μ long, 5–8μ thick in the broadest part, 1·5–2μ wide at the base.

Common on leaves and pods of *Phaseolus*; cosmopolitan [CMI Distribution Map 328]. D. Hocking in *Pl. Dis. Reptr.*, **51**: (4) 276–278, 1967, described a new virulent form causing circular leaf spot of french beans.

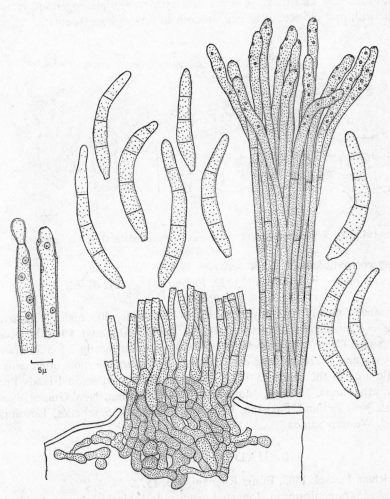

FIG. 183. *Phaeoisariopsis griseola* (× 650 except where indicated by the scale).

169. PSEUDOEPICOCCUM

Pseudoepicoccum M. B. Ellis gen. nov.

Sporodochia punctiformia, pulvinata, fusca. *Mycelium* partim superficiale, partim immersum. *Stromata* superficialia, hemisphaerica, brunnea, pseudoparenchymatica. *Conidiophora* simplicia, cylindrica vel clavata, brevia, recta vel leniter flexuosa, pallide olivaceo-brunnea, laevia, cicatricibus numerosis praedita. *Conidia* solitaria, sicca, acropleurogena, subsphaerica, pallide olivaceo-brunnea, laevia vel verruculosa, 0-septata.

Species typica: *Pseudoepicoccum cocos* (F. L. Stevens) M. B. Ellis.

Sporodochia punctiform, pulvinate, dark blackish brown. *Mycelium* partly superficial, partly immersed. *Stromata* superficial, hemispherical, brown, pseudoparenchymatous. *Setae* and *hyphopodia* absent. *Conidiophores* unbranched, cylindrical or clavate, short, straight or slightly flexuous, pale olivaceous brown, smooth. *Conidiogenous cells* polyblastic, integrated, terminal, sympodial, cylindrical, cicatrized. *Conidia* solitary, dry, acropleurogenous, simple, subspherical, pale olivaceous brown, smooth to verruculose, 0-septate.

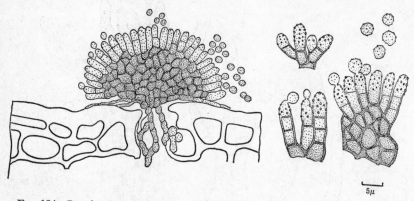

FIG. 184. *Pseudoepicoccum cocos* (× 650 except where indicated by the scale).

Pseudoepicoccum cocos (F. L. Stevens) M. B. Ellis comb. nov.
Epicoccum cocos F. L. Stevens, 1932, *Philipp. Agric.*, **21**: 81–82.
(Fig. 184)

Sporodochia mostly hypophyllous, up to 120µ wide, 60µ high; lying over stomata. *Hyphae* pale olivaceous, 1–2·5µ thick. *Stromata* 40–100µ wide, 20–50µ high. *Conidiophores* 8–14 × 2·5–3·5µ. *Conidia* mostly 2·5–3·5µ diam.

On *Areca* and *Cocos*, causing oval, reddish brown, zonate spots often with a very pale area in the centre, 5–18 × 2–6 mm.; British Solomon Islands Protectorate, Indonesia, Jamaica, Mauritius, New Caledonia, New Guinea, New Hebrides, New Zealand, Pakistan, Philippines, Sarawak, Seychelles, Tanzania, Thailand, Western Samoa.

170. HADROTRICHUM

Hadrotrichum Fuckel, 1865, Fungi Rhenani, No. 1522.

Sporodochia punctiform, in elongated groups, dark blackish brown to black, sometimes mistaken for rust pustules. *Mycelium* immersed. *Stroma* well

developed, erumpent; lower part hyaline or subhyaline, upper part brown.
Setae and *hyphopodia* absent. *Conidiophores* macronematous, mononematous,
closely packed together, unbranched, straight or flexuous, colourless to very
pale brown, smooth. *Conidiogenous cells* polyblastic, integrated, terminal,
sympodial, cylindrical or clavate, cicatrized. *Conidia* solitary, dry, acropleurogenous, simple, spherical or subspherical, at first colourless, later pale to mid pale
brown, verruculose or finely echinulate, 0-septate.

Type species: Hadrotrichum phragmitis Fuckel.

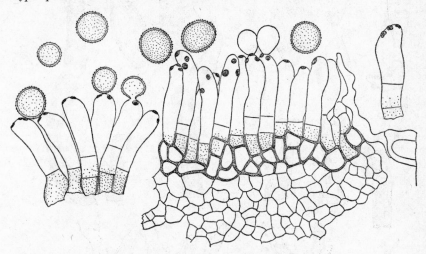

Fig. 185. *Hadrotrichum phragmitis* (× 650).

Hadrotrichum phragmitis Fuckel, 1865, Fungi Rhenani, No. 1522.
(Fig. 185)
Conidiophores up to 50μ long, 6–12μ thick. *Conidia* 10–16μ diam.
On *Phragmites* and other grasses; Europe, Iraq, Pakistan, Papua, Sudan,
Zambia.

171. FUSICLADIUM

Fusicladium Bonorden, 1851, Hndb. allg. Mykol.: 80.
Mycelium immersed, sometimes subcuticular. *Stroma* often present, sometimes subcuticular. *Setae* absent. *Hyphopodia* absent. *Conidiophores* macronematous, mononematous, simple or occasionally once branched, often
olivaceous brown, septate, usually fasciculate, bursting through the cuticle of
the host plant. *Conidiogenous cells* integrated, terminal, polyblastic, sympodial,
cicatrized; old conidial scars usually thickened, conspicuous and prominent,
sometimes situated at the end of short lateral projections, numerous and often
crowded, giving the conidiophore a nodular appearance. *Conidia* solitary or
occasionally in short chains, dry, variable in shape but often tending to be
broadly fusiform, truncate at the base and pointed at the apex, 0–3- (often 0- or
1-) septate, pale to mid olive or olivaceous brown, frequently minutely verruculose.

Type species: Fusicladium state of *Venturia pirina* Aderh. = *F. virescens* Bon.
Deighton, F. C., *Mycol. Pap.*, **112**: 16–27, 1967. Species of *Fusicladium* on
Euphorbia.
Hughes, S. J., *Can. J. Bot.*, **31**: 565–569, 1953.

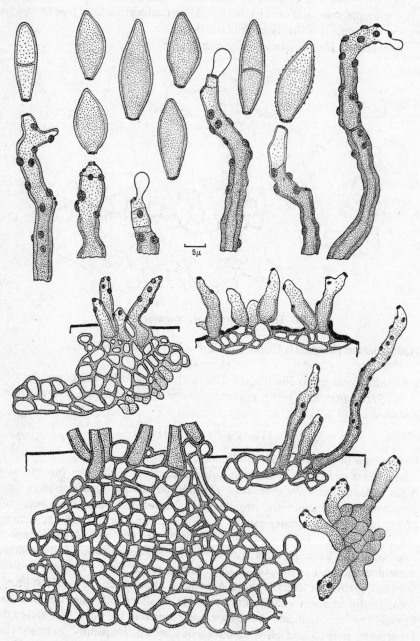

FIG. 186. *Fusicladium* state of *Venturia pirina* (× 650 except where indicated by the scale).

Fusicladium state of **Venturia pirina** Aderh., 1896, *Landw. Jbr*, **25**: 875. For synonymy see Hughes, *Can. J. Bot.*, **36**: 768, 1958.

(Fig. 186)

Colonies compact or effuse, dark brown or olivaceous brown, velvety. *Stroma* sometimes composed of only a few cells, often, however, well-developed and on fruit may be many cells thick. *Conidiophores* up to 90μ long but usually shorter, 4–9μ thick, mid to dark brown or olivaceous brown, pale at the apex, scars very conspicuous. *Conidia* broadly fusiform, pointed at the apex, truncate at the base, mid pale olivaceous brown, smooth, rugulose or verruculose, 0–1-septate, 17–28 × 8–10μ.

On leaves, shoots, bud scales, flowers and fruit of pear (*Pyrus communis*), causing scab; cosmopolitan [CMI Distribution Map 367].

172. ASPERISPORIUM

Asperisporium Maublanc, 1913, *Lavoura*, **16**: 212.

Sporodochia punctiform, pulvinate, brown, olivaceous brown or black. *Mycelium* immersed. *Stroma* usually well-developed, erumpent. *Setae* and *hyphopodia* absent. *Conidiophores* macronematous, mononematous, closely packed together forming sporodochia, usually rather short, unbranched or occasionally branched, straight or flexuous, hyaline to olivaceous brown, smooth.

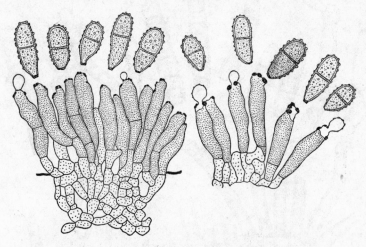

FIG. 187. *Asperisporium caricae* (× 650).

Conidiogenous cells polyblastic, integrated, terminal, sympodial, cylindrical or clavate, cicatrized; scars prominent. *Conidia* solitary, dry, acropleurogenous, ellipsoidal, fusiform, obovoid, pyriform, clavate or obclavate, hyaline to brown or olivaceous brown, smooth or verrucose, with 0–3 transverse and sometimes 1 or more longitudinal or oblique septa.

Type species: Asperisporium caricae (Speg.) Maubl.

Asperisporium caricae (Speg.) Maubl., 1913, *Lavoura*, **16**: 212.
(Fig. 187)

Sporodochia hypophyllous, dark blackish brown to black. *Conidiophores* up to 45 × 6–9μ. *Conidia* ellipsoidal, pyriform or clavate, almost always 1-septate, hyaline to mid pale brown, verrucose, 14–26 × 7–10μ.

On living leaves of *Carica papaya*; Brazil, Costa Rica, Cuba, Dominican Republic, Jamaica, Venezuela.

173. VERRUCISPORA

Verrucispora Shaw & Alcorn, 1967, *Proc. Linn. Soc. N.S.W.*, **92** (2): 171–173.

Colonies round or angular, dark blackish brown to black, hairy. *Mycelium* mostly immersed. *Stroma* well-developed in substomatal region, pseudoparenchymatous, brown. *Setae* and *hyphopodia* absent. *Conidiophores* macronematous, mononematous, caespitose, emerging through stomata, unbranched, straight or flexuous, sometimes geniculate, mid to dark reddish or olivaceous brown, paler at the often slightly thickened apex, smooth. *Conidiogenous cells* polyblastic, integrated, terminal becoming intercalary, sympodial, cylindrical or clavate, cicatrized; scars conspicuous. *Conidia* solitary, dry, acropleurogenous, simple, straight or curved, cylindrical, rounded at the apex, truncate at the base,

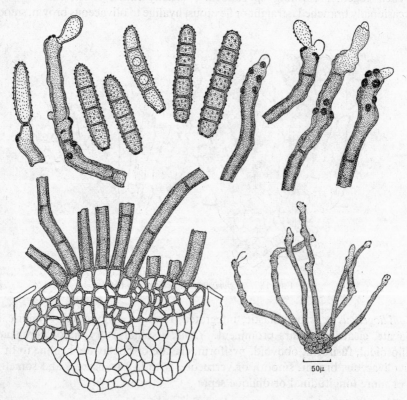

FIG. 188. *Verrucispora proteacearum* (× 650 except where indicated by the scale).

multiseptate, sometimes constricted at the septa, olivaceous or reddish brown, end cells often paler than intermediate ones, verrucose.

Type species: Verrucispora proteacearum Shaw & Alcorn.

Verrucispora proteacearum Shaw & Alcorn, 1967, *Proc. Linn. Soc. N.S.W.*, **92** (2): 171–173.

(Fig. 188)

Conidiophores in fascicles of up to 40, up to 300×5–9μ. *Conidia* 1–7-septate, 25–50×7–10μ.

On leaves of *Finschia* and *Hakea*; Australia and New Guinea.

174. CERCOSPORA

Cercospora Fresenius, 1863, Beitr. Mykol., **3**: 91–93.

Virgasporium Cooke, 1875, *Grevillea*, **3**: 182.

Colonies effuse, greyish, tufted. *Mycelium* mostly immersed. *Stroma* often present but not large. *Setae* and *hyphopodia* absent. *Conidiophores* macronematous, mononematous, caespitose, straight or flexuous, sometimes geniculate, unbranched or rarely branched, olivaceous brown or brown, paler towards the apex, smooth. *Conidiogenous cells* integrated, terminal, polyblastic, sympodial, cylindrical, cicatrized, scars usually conspicuous. *Conidia* solitary, acropleurogenous, simple, obclavate or subulate, colourless or pale, pluriseptate, smooth.

Lectotype species: Cercospora apii Fres.

Chupp, C., A Monograph of the Fungus Genus Cercospora, 667 pp., Ithaca, New York, 1953.

Deighton, F. C., *Mycol. Pap.*, **71**: 1–23, 1959.

Johnson, E. M. and Valleau, W. D., *Phytopathology*, **39**: 763–770, 1949.

Jones, J. P., *Phytopathology*, **49**: 430–432, 1959.

Katsuki, S., *Trans. mycol. Soc. Japan*, Extra Issue **1**: 1–100, 1965.

Vasudeva, R. S., Indian Cercosporae, 245 pp., New Delhi, 1963.

Vestal, E. F., *Res. Bull. Iowa agric. Exp. Stn*, **168**: 44–72, 1933.

Welles, C. G., *Am. J. Bot.*, **12**: 195–218, 1925.

More than 2000 species names have been attributed to *Cercospora* and it has been customary for plant pathologists and mycologists to describe as new any *Cercospora* found on a host plant for the first time. This practice still continues in spite of evidence provided by Johnson and Valleau, Vestal, Welles and others that the host range of *Cercospora* species, especially those belonging to the *C. apii* group, is undoubtedly very wide and is not by any means restricted even to host families. Johnson and Valleau isolated *Cercospora* from 28 different host plants in 16 families. Morphologically and culturally they all appeared to belong to a single species. Numerous successful cross-inoculations were carried out. Welles and others have demonstrated that great variation in the size of conidiophores and conidia is induced by change in environmental conditions, especially humidity. *Cercospora* species are often weak parasites on dead, dying or physiologically diseased plant tissues with occasional serious injury to healthy plants.

Inside the genus *Cercospora* as treated by Chupp in his most useful monograph are many species which can be suitably accommodated in such genera as *Cercoseptoria*, *Mycovellosiella* and *Pseudocercospora* and undoubtedly new generic names are needed for others. Only the lectotype species is being described here as in addition to Chupp's monograph there are available monographic treatments on a regional basis by Katsuki, Vasudeva and others.

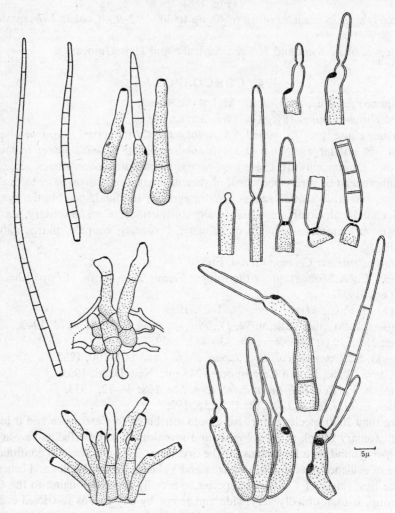

FIG. 189. *Cercospora apii* (× 650 except where indicated by the scale).

Cercospora apii Fres., 1863, Beitr. Mykol., **3**: 91.

(Fig. 189)

Colonies grey, with small, scattered tufts of conidiophores. *Mycelium* immersed, hyphae colourless to pale olive, 2–4μ diam., thickening to 8μ and often forming hyphal knots or pseudostromata in the substomatal cavities. *Conidiophores* caespitose in groups of up to 30, usually emerging through stomata, unbranched, straight or flexuous, sometimes geniculate, often swollen slightly at the base and just below the apex, lower part olivaceous brown or brown, upper part pale, with conspicuous widely spaced scars, usually 30–70μ long but sometimes much longer, 3–4μ thick near the apex, 5–9μ at the base. *Conidia* straight or curved, narrow subulate, truncate at the base, 3·5–5μ thick below

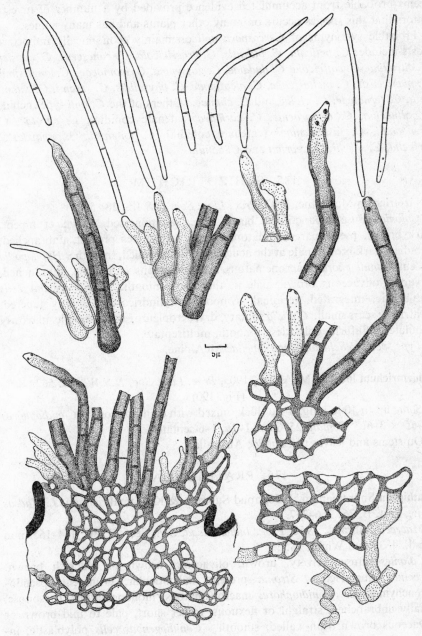

FIG. 190. *Schizotrichum lobeliae* (× 650 except where indicated by the scale).

the middle, tapering towards the apex to 1–2μ, colourless, smooth, 9–17-septate, usually 60–200μ long but sometimes longer; hilum conspicuous, refractive, appearing dark, 2·5–3·5μ wide.

On living leaves of *Apium graveolens*, causing pale brown, usually round spots 4–12 mm. diam. which sometimes have a distinct narrow, raised margin. It seems probable from accumulated evidence provided by a number of investigators that this species occurs on many other plants and has many names.

Probable synonyms of *Cercospora apii* on mainly economically important hosts include: *C. beticola*, *C. bidentis*, *C. brassicicola*, *C. canescens*, *C. capsici*, *C. citrullina*, *C. coffeicola*, *C. duddiae*, *C. gossypina*, *C. ipomoeae*, *C. longissima*, *C. malayensis*, *C. melongenae*, *C. nicotianae*, *C. physalidis*, *C. sesami*, *C. solanicola*, *C. ternateae*, *C. violae* and *C. zinniae*. Others of the *C. apii* type include *C. althaeina*, *C. armoraciae*, *C. caribaea* (catenate conidia), *C. carotae*, *C. chenopodii*, *C. fusimaculans* (catenate conidia), *C. longipes*, *C. papayae*, *C. ricinella*, *C. sorghi*, *C. zebrina* and *C. zonata*.

175. SCHIZOTRICHUM

Schizotrichum McAlpine, 1903, *Proc. Linn. Soc. N.S.W.*, **28**: 562.

Colonies effuse or punctiform, black. *Mycelium* immersed. *Stroma* erumpent, dark brown, pseudoparenchymatous. *Setae* sometimes present, unbranched, subulate, dark brown, pale at the acute apex, thick-walled, smooth. *Hyphopodia* absent. *Conidiophores* macronematous, mononematous, caespitose, unbranched, flexuous, olivaceous brown, pale at the apex, smooth. *Conidiogenous cells* polyblastic, integrated, terminal, sympodial, cylindrical or slightly tapered, cicatrized; scars small. *Conidia* solitary, dry, acropleurogenous, simple, narrowly subulate or filiform, colourless, smooth, multiseptate.

Type species: Schizotrichum lobeliae McAlpine.

Schizotrichum lobeliae McAlpine, 1903, *Proc. Linn. Soc. N.S.W.*, **28**: 562.
(Fig. 190)

Setae up to 100μ long, 4–6μ thick, mixed with conidiophores. *Conidiophores* 10–65 × 3–6μ. *Conidia* 25–60 × 1–2μ, 1–6-septate.

On stems and leaves of *Lobelia*; Australia.

176. PRATHIGADA

Prathigada Subramanian, 1956, apud Subram. & K. Ramakrishnan, *J. Madras Univ.*, Sect. B, **26**: 366–367.

Macraea Subram., 1952, *Proc. Indian Acad. Sci.*, Sect. B, **36**: 164–165, non Lindl., 1830 and Wight, 1852.

Colonies effuse, velvety, brown, olivaceous brown or blackish brown. *Mycelium* immersed. *Stroma* present, subcuticular, erumpent, pseudoparenchymatous. *Conidiophores* macronematous, mononematous, caespitose, usually unbranched, straight or flexuous, rather short, pale to mid brown or olivaceous brown, thin-walled, smooth. *Conidiogenous cells* polyblastic, integrated, terminal, sympodial, cylindrical or clavate, cicatrized; scars few, thin, flat. *Conidia* solitary, dry, acropleurogenous, simple, often obclavate, sometimes rostrate, pale to mid brown or olivaceous brown with cells sometimes unequally coloured, smooth or rugulose, multiseptate, septa often rather thick and dark.

Type species: *Prathigada crataevae* (Syd.) Subram.

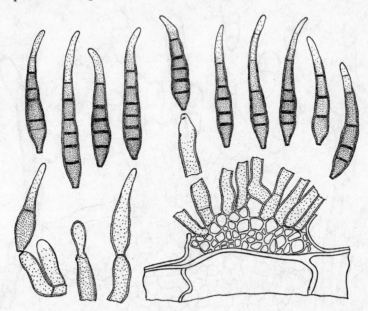

FIG. 191. *Prathigada crataevae* (× 650).

Prathigada crataevae (Syd.) Subram., 1956, apud Subram. & K. Ramakrishnan, *J. Madras Univ.*, Sect. B, **26**: 366–367.

Napicladium crataevae Syd., 1913, *Annls mycol.*, **11**: 329.

(Fig. 191)

Colonies usually irregular, up to 1 cm. long, amphigenous but mostly hypophyllous. *Stroma* 1–3 cells thick. *Conidiophores* rather pale olivaceous brown, up to 35 × 5–9μ. *Conidia* obclavate, often rostrate, 3–6-septate, distal cells usually much paler than the proximal ones, 35–70μ long, 7–9μ thick in the broadest part, 2–3μ at base and apex.

On living leaves of *Crataeva*; Hong Kong, India.

177. CERCOSPORIDIUM

Cercosporidium Earle, 1901, *Muhlenbergia*, **1** (2): 16.

Berteromyces Ciferri, 1954, *Sydowia* **8**: 267.

Mycelium immersed. *Stroma* present, usually well-developed. *Setae* and *hyphopodia* absent. *Conidiophores* macronematous, mononematous, usually simple, rarely branched, brown, septate or continuous, geniculate or not, densely fasciculate, the fascicles in many species incurved, particularly when old as a result of the formation of a thickened band along one side of the conidiophore. *Conidiogenous cells* integrated, terminal, polyblastic, sympodial, cicatrized, conidial scars always thickened and conspicuous, usually prominent, the old scars situated on rounded shoulders or on short peg-like protrusions, but in some species lying more or less flat against the side of the conidiophore. *Conidia*

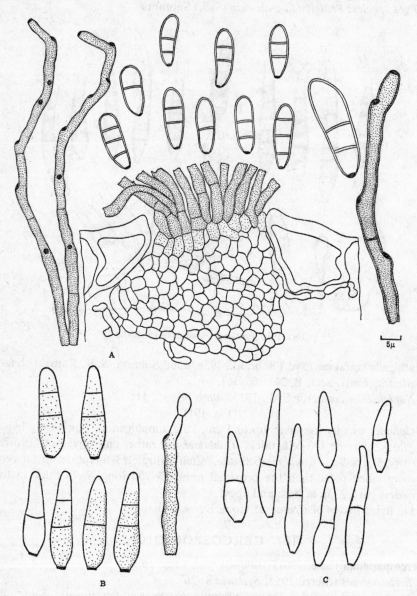

FIG. 192. *Cercosporidium* species (1): A, *chaetomium*; B, *graminis*; C, *depressum* (× 650 except where indicated by the scale).

solitary, dry, clavate, cylindrical, obclavate or broadly fusiform, with a conspicuous thickened hilum, more or less colourless or relatively pale brown, smooth to verrucose, 1–7- most frequently 1–3-septate.

Type species: Cercosporidium chaetomium (Cooke) Deighton = *Scolecotrichum euphorbiae* Tracy & Earle.

Deighton, F. C., *Mycol. Pap.*, **112**: 27–77, 1967.

Cercosporidium chaetomium (Cooke) Deighton, 1967, *Mycol. Pap.*, **112**: 27.

Cladosporium chaetomium Cooke, 1889, *Grevillea*, **17**: 66.

(Fig. 192 A)

Colonies usually hypophyllous, occasionally epiphyllous or on pedicels or ovaries, round or elliptical, olivaceous brown. *Stroma* well developed, substomatal. *Conidiophores* very numerous, geniculate, olivaceous, up to 260μ long, mostly 4–5μ thick; conidial scars conspicuous, thickened, prominent. *Conidia* straight or slightly curved, clavate or cylindric-clavate, very pale greenish, smooth, with thickened hilum; 0–3 septate, 14–29 × 5–8·5μ.

On *Euphorbia* spp.; Cuba, U.S.A.

Deighton (1967) described 16 other species on 7 host families. Measurements of conidia, substratum range and geographical distribution of these are given below; also his very useful key to 5 species on Umbelliferae.

ARACEAE

C. caladii (Stev.) Deighton (Fig. 193 G): 25–50 × 6–9·5μ; on *Caladium bicolor*; Puerto Rico and Trinidad.

COMPOSITAE

C. scariolae (Syd.) Deighton (Fig. 193 A): 11–24 × 5–9μ; on *Lactuca*; Bavaria, Iran, Kashmir.

GRAMINEAE

C. compactum (Berk. & Curt.) Deighton (Fig. 194 G): 22–53 × 5·5–8μ; on *Arundinaria*; U.S.A.

C. graminis (Fuckel) Deighton (Fig. 192 B): 21–44 × 10–12μ; on grasses; Asia, Europe, N. and S. America.

LEGUMINOSAE

C. state of Mycosphaerella berkeleyi W. A. Jenkins, syn. *C. personatum* (Berk. & Curt.) Deighton (Fig. 193 C): 20–77 × 4–9 (mostly 30–50 × 6·5–7·5)μ; on *Arachis hypogaea* (groundnut) throughout the tropics and warmer temperate regions.

C. caracasanum (Syd.) Deighton (Fig. 194 B): 20–30 × 9–15·5μ; on *Parosela barbata*; Venezuela.

C. cassiae (P. Henn.) Deighton (Fig. 194 E): 27–65 × 8–11·5μ; on *Cassia*; E. Africa, U.S.A., West Indies.

C. desmanthi (Ellis & Kellerman) Dearness (Fig. 194 A): 18–34 × 5–7·5μ; on *Desmanthus*; Jamaica, U.S.A.

C. pulchellum (T. S. Ramakrishnan) Deighton (Fig. 193 B): 18–52 × 7–10·5μ; on *Indigofera*; Comoro Is., India, Malaya, Philippines.

PAPAVERACEAE

C. guanicense (Stev.) Deighton (Fig. 194 D): 14·5–32·5 × 6–9·5µ; on *Argemone mexicana*; Brazil, Cuba, Dominican Republic, Puerto Rico, Venezuela.

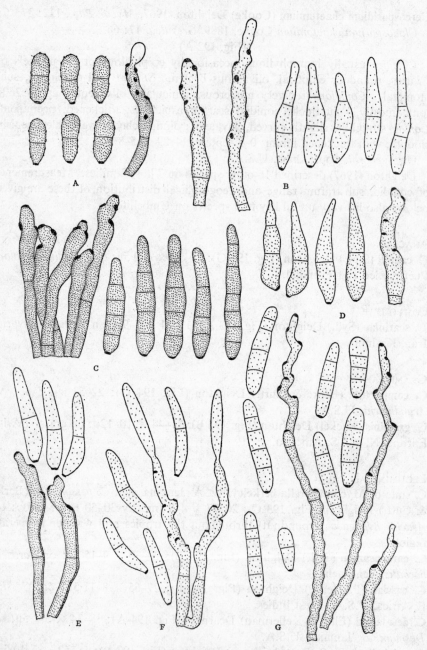

FIG. 193. *Cercosporidium* species (2): A, *scariolae*; B, *pulchellum*; C, state of *Mycosphaerella berkeleyi*; D, *angelicae*; E, *punctum*; F, *sii*; G, *caladii* (× 650).

Rubiaceae
C. galii (Ellis & Holway) Deighton (Fig. 194 C): 15–44 × 2·5–4μ; on *Galium*; Canada, Europe, U.S.A.

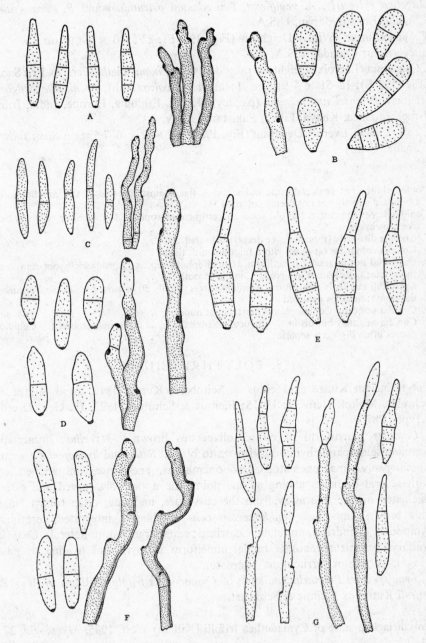

Fig. 194. *Cercosporidium* species (3): A, *desmanthi*; B, *caracasanum*; C, *galii*; D, *guanicense*; E, *cassiae*; F, *punctiforme*; G, *compactum* (× 650).

UMBELLIFERAE

C. angelicae (Ellis & Everh.) Deighton (Fig. 193 D): 25–62 × 6·5–10·5µ, on *Angelica*; Canada, Europe, U.S.A.

C. depressum (Berk. & Br.) Deighton (Fig. 192 C): 20–78 × 6·5–11µ; on *Angelica sylvestris*, *A. genuflexa*, *Peucedanum ostruthium* and *P. decursivum*; Canada, Europe, Japan, U.S.A.

C. punctiforme (Wint.) Deighton (Fig. 194 F): 22·5–40 × 5–8µ; on *Taenidia integerrima*; Canada, U.S.A.

C. punctum (Lacroix) Deighton, pycnidial state *Phoma anethi* (Pers. ex Fr.) Sacc. (Fig. 193 E): 18–51 × 4–9µ; on *Anethum graveolens* (dill), *Foeniculum vulgare* (fennel), *Petroselinum crispum* (parsley); Egypt, Ethiopia, Europe, India, Iran, Israel, Jamaica, Kenya, Libya, Pakistan, U.S.A.

C. sii (Ellis & Everh.) Deighton (Fig. 193 F): 20–52 × 6–7·5µ; on *Sium suave*; Canada, U.S.A.

KEY

Conidiophores not geniculate, the old scars ± flat against the side of the conidiophore.
  Conidia distinctly but not deeply coloured . . . . *depressum*
Conidiophores geniculate, the old scars on conspicuous rounded shoulders or on short peg
  like protrusions 1
1. Conidia distinctly (though never deeply) coloured . . . . . . *angelicae*
  Conidia colourless (at most with a faint greenish tinge) 2
2. Caespituli not markedly punctiform. Conidiophores up to 25 per fascicle, not rupturing
  the stoma; fascicles not incurved. Conidia smooth . . . . . . *sii*
  Caespituli markedly punctiform. Conidiophores usually 50 or more per fascicle, rupturing
  the stoma; fascicles incurved 3
3. Conidia smooth. Conidiophores mostly continuous . . . . . *punctum*
  Conidia minutely but distinctly verruculose over the basal cell (rarely smooth). Conidiophores often distinctly septate . . . . *punctiforme*

178. POLYTHRINCIUM

Polythrincium Kunze ex Ficinus & Schubert; Kunze, 1817, apud Kunze & Schmidt, Mykol. Hefte, **1**: 13–15; Ficinus & Schubert, 1823, Fl. Geg. Dresd. Krypt.: 287.

Colonies punctiform or effuse, olivaceous brown. *Mycelium* immersed. *Stroma* pseudoparenchymatous, brown to black. *Setae* and *hyphopodia* absent. *Conidiophores* macronematous, mononematous, caespitose, unbranched or with several branches arising at one point, the upper part curved and often thickened on the side away from the curvature, undulate, often torsive, mid pale brown, smooth. *Conidiogenous cells* polyblastic, integrated, terminal, sympodial, cylindrical, undulate, cicatrized; scars large, flat, unilateral. *Conidia* solitary, acropleurogenous, simple, cuneiform or pyriform, hyaline to pale brown, smooth or verruculose, 1-septate.

Type species: Polythrincium state of *Cymadothea trifolii* (Killian) Wolf = *P. trifolii* Kunze ex Ficinus & Schubert.

Polythrincium state of **Cymadothea trifolii** (Killian) Wolf, 1935, *Mycologia*, **27**: 71.

(Fig. 195)

Conidiophores up to 100 × 6–9μ. *Conidia* 17–24μ long, 13–24μ thick in the broadest part, 4–5μ at the base.

On leaves of *Trifolium*, causing sooty blotch disease; Ethiopia, Europe, India, Iraq, Israel, Jamaica, Kenya, New Zealand, N. America, Pakistan.

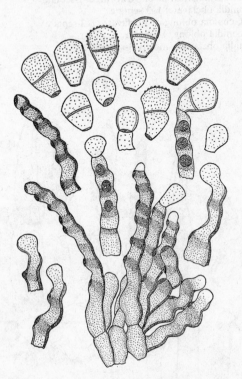

Fig. 195. *Polythrincium* state of *Cymadothea trifolii* (× 650).

179. CAMPTOMERIS

Camptomeris H. Sydow, 1927, *Annls mycol.*, **25**: 142–144.

Sporodochia mostly hypophyllous, pulvinate, punctiform, dark olivaceous brown to black. *Mycelium* immersed. *Stroma* present with either one or several of the cells greatly swollen; each of these vesicle like cells bears a number of conidiophores. *Setae* and *hyphopodia* absent. *Conidiophores* macronematous, often curved inwards, unbranched, pale brown to brown, smooth. *Conidiogenous cells* integrated, terminal, polyblastic, sympodial, cylindrical, cicatrized, with fairly prominent scars. *Conidia* solitary, acropleurogenous, simple, obclavate or oblong rounded at the ends, pale olivaceous brown or brown, 0–3-septate, sometimes smooth but usually verruculose.

Type species: Camptomeris calliandrae Syd.

Bessey, E. A., *Mycologia*, **45**: 364–390, 1953; *Pap. Mich. Acad. Sci.*, **40**: 3–6, 1955.

Hughes, S. J., *Mycol. Pap.*, **49**: 14–19, 1952.

KEY

Swollen stroma cells in each sporodochium 1 (or few)   1
Swollen stroma cells in each sporodochium several or many   2

1. On *Acacia* and *Albizia*, conidia obclavate, 1–3 septate .   .   .   *albiziae*
   On *Astragalus*, conidia spherical or obpyriform, 0–1-septate .   .   *astragali*
   On *Calliandra*, conidia obclavate, 1–3 septate .   .   .   *calliandrae*
   On *Pithecolobium*, conidia oblong or obclavate, 1–3-septate .   .   *floridana*
2. On *Desmanthus*, conidia oblong, 1-septate .   .   .   .   *desmanthi*
   On *Leucaena*, conidia obclavate, mostly 3-septate .   .   .   *leucaenae*

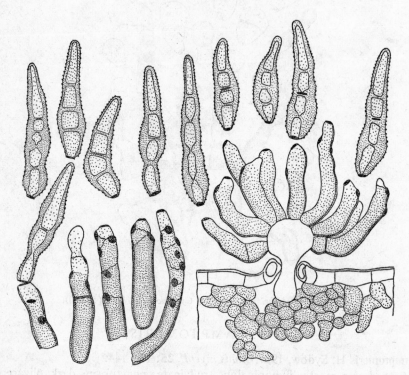

FIG. 196. *Camptomeris albiziae* (× 650).

Conidial measurements in μ *and geographical distribution*

C. albiziae (Petch) Mason, 30–70 × 8–12 (47 × 10); Ceylon, Ghana, India, Pakistan, San Domingo, Sierra Leone, Sudan, Uganda (Fig. 196).

C. astragali Bessey, 12–32 × 8–14; Japan.

C. calliandrae Syd., 25–50 × 8–16 (41 × 12); Costa Rica.

C. desmanthi Cif., 29–48 × 9–11 (41 × 10); San Domingo, U.S.A. (Fig. 197).

C. floridana Bessey, 34–48 × 7–11·5 (42 × 10); Cuba, Jamaica, U.S.A.

C. leucaenae (Stev. & Dalbey) Syd., 30–60 × 8–11 (47 × 9·5); Jamaica, Puerto Rico, San Domingo, Venezuela.

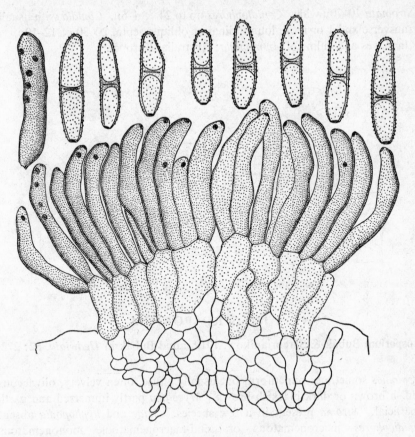

FIG. 197. *Camptomeris desmanthi* (× 650).

180. THYROSTROMELLA

Thyrostromella Höhnel, 1919, *Ber. dt. bot. Ges.*, **37**: 157.

Sporodochia punctiform, black. *Mycelium* immersed. *Stroma* present, erumpent. *Setae* and *hyphopodia* absent. *Conidiophores* macronematous, mono-nematous, short, closely packed together forming sporodochia, unbranched, straight or flexuous, brown or olivaceous brown, smooth. *Conidiogenous cells* polyblastic or monoblastic, integrated, terminal, sympodial when polyblastic, cylindrical or clavate, cicatrized. *Conidia* solitary, dry, acrogenous or acro-pleurogenous, simple, broadly ellipsoidal or somewhat irregular, brown or olivaceous brown, smooth or verruculose, with several transverse and one or more longitudinal or oblique septa.

Type species: Thyrostromella myriana (Desm.) Höhnel.

Hughes, S. J., *Can. J. Bot.*, **33**: 341–343, 1955.

Thyrostromella myriana (Desm.) Höhnel, 1919, *Ber. dt. bot. Ges.*, **37**: 157.

(Fig. 198)

Stromata 40–70μ wide. *Conidiophores* up to 24 × 4–6μ. *Conidia* with usually 3 transverse and 1 or more longitudinal or oblique septa, 20–30 × 12–19μ. On leaves and culms of *Ammophila*; Australia, Europe.

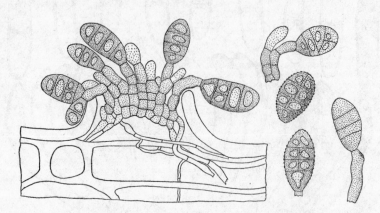

FIG. 198. *Thyrostromella myriana* (× 650).

181. SIROSPORIUM

Sirosporium Bubák & Serebrianikow, 1912, apud Bubák in *Hedwigia*, **52**: 272–273.

Colonies sometimes punctiform but usually effuse, often velvety, olivaceous, reddish brown or dark blackish brown. *Mycelium* partly immersed and partly superficial. *Stroma* present in a few species. *Setae* and *hyphopodia* absent. *Conidiophores* macronematous or semi-macronematous, mononematous, branched or unbranched, straight or flexuous, pale to mid brown or olivaceous brown, smooth or verrucose. *Conidiogenous cells* polyblastic, integrated, terminal on stipe and branches, sometimes becoming intercalary, sympodial, cylindrical or clavate, cicatrized. *Conidia* solitary, dry, acropleurogenous, simple, straight, flexuous or coiled, cylindrical with rounded ends, ellipsoidal or obclavate, subhyaline to olivaceous or golden brown, smooth, rugose or verrucose, with transverse and often also longitudinal or oblique septa; hilum sometimes protuberant.

Type species: Sirosporium antenniforme (Berk. & Curt.) Bubák & Serebrianikow.

Ellis, M. B., *Mycol. Pap.*, **87**: 2–11, 1963.

KEY

Sirosporium antenniforme (Berk. & Curt.) Bubák & Serebrianikow, 1912, apud Bubák in *Hedwigia*, **52**: 272.

(Fig. 199 A)

Colonies hypophyllous, effuse, blackish brown, velvety. *Conidiophores* erect or ascending, stipe and branches pale to mid brown, smooth, 5–9μ thick.

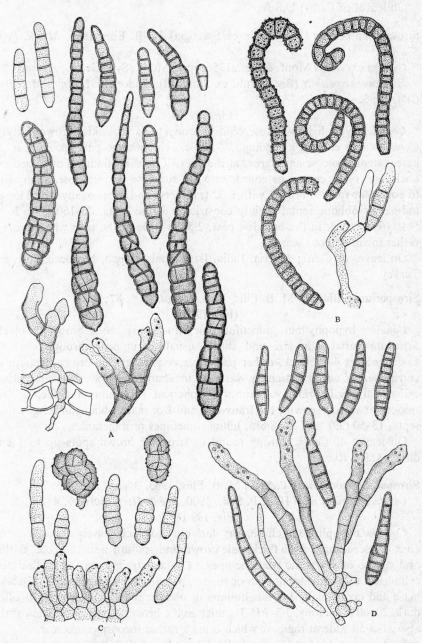

FIG. 199. *Sirosporium* species: A, *antenniforme*; B, *celtidis*; C, *celtidicola*; D, *mori* (× 650).

Conidia straight, slightly curved or flexuous, mostly obclavate and smooth, pale to mid brown, with 1–18 transverse and frequently numerous longitudinal and oblique septa, usually constricted at the septa, 10–96 (47)μ long, 5–14 (8·4)μ thick in the broadest part, tapering to 3–6μ near the apex, hilum almost always protruding, peg-like, 1–2μ thick.

On leaves of *Celtis*; U.S.A.

Sirosporium celtidis (Biv.-Bernh. ex Sprengel) M. B. Ellis, 1963, *Mycol. Pap.*, **87**: 4–5.

Gyroceras celtidis Mont. & Ces., 1956, apud Mont., Syll. Gen. Sp. crypt.: 308.

Helicoceras celtidis (Biv.-Bernh. ex Sprengel) Linder, 1931, *Ann. Mo. Bot. Gdn*, **18**: 3.

(Fig. 199 B)

Colonies hypophyllous, effuse, reddish brown to dark blackish brown, velvety. *Conidiophores* erect or ascending, 3–7μ thick, smooth and pale brown near the base, often verrucose and darker at the apex. *Conidia* cylindrical or sometimes obclavate, often curved or coiled, smooth, rugulose or verrucose, subhyaline to golden or reddish brown, with 1–32 transverse and occasionally 1 or 2 longitudinal or oblique septa, slightly constricted at the septa, 23–160 (70)μ long, 5–10 (6·8)μ thick in the broadest part, 2·5–5μ wide at the base which bears a rather inconspicuous scar.

On leaves of *Celtis*; Algeria, India, Israel, Italy, Japan, Morocco, Portugal, Turkey.

Sirosporium celtidicola M. B. Ellis, 1963, *Mycol. Pap.*, **87**: 5–7.

(Fig. 199 C)

Colonies hypophyllous, punctiform or effuse, olivaceous brown to black. *Stromata* variable in size and shape, usually erumpent through stomata. *Conidiophores* 4–8μ thick, rather pale olivaceous or golden brown, smooth or verrucose. *Conidia* extremely variable in shape, narrow obclavate, broad cylindrical or ellipsoidal, sometimes subspherical, subhyaline to golden brown, smooth or verrucose, with 1–8 transverse and 1 or many oblique or longitudinal septa, 13–50 (29) × 4–27 (8·6)μ; hilum sometimes protuberant.

On leaves of *Celtis*, causing round or irregular brown spots up to 1 cm. diam.; Italy, Russia.

Sirosporium mori (H. & P. Syd.) M. B. Ellis, 1963, *Mycol. Pap.*, **87**: 7–8.

Clasterosporium mori H. & P. Syd., 1900, *Mém. Herb. Boissier*, **4**: 6.

(Fig. 199 D)

Colonies hypophyllous, effuse, very dark olive to black, velvety. *Conidiophores* erect or ascending, 3·5–8μ thick, pale brown and smooth near the base, darker and rugose or verrucose near the apex. *Conidia* straight, curved or flexuous, cylindrical to obclavate, smooth or rugose, pale to mid brown, with 3–11 transverse and occasionally 1–2 longitudinal or oblique septa, walls and septa often dark, 20–60 (38)μ long, 3·5–7 (5·7)μ thick in the broadest part, 1–5μ thick at the apex, 1–3·5μ wide at the base which bears a rather inconspicuous scar.

On leaves of *Morus*; Cyprus, India, Israel, Japan, Pakistan.

182. PODOSPORIELLA

Podosporiella Ellis & Everhart, 1895, *Proc. Acad. nat. Sci. Philad.*, 1894: 385.

Sporodochia scattered on leaf spots, cylindrical or broadly conical, dark brown to black. *Mycelium* immersed. *Stroma* partly immersed but with a large columnar superficial part, spongy, made up of a number of branched and anastomosing hyphae. *Setae* and *hyphopodia* absent. *Conidiophores* macronematous, mononematous, simple or branched, straight or flexuous, brown, smooth, growing

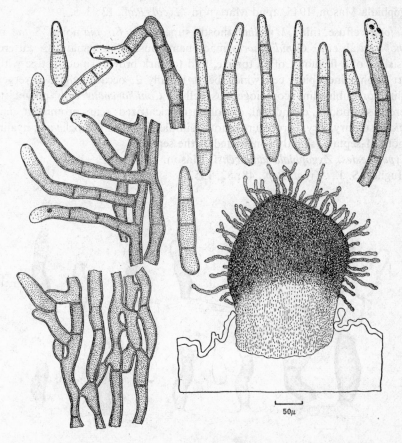

FIG. 200. *Podosporiella glomerata* (× 650 except where indicated by the scale).

out from all sides of the superficial stromatic column. *Conidiogenous cells* polyblastic, integrated, terminal, sympodial, cylindrical, cicatrized; scars small. *Conidia* solitary, dry, acropleurogenous, simple, mostly obclavate, mid pale brown, smooth, rugulose or verruculose, multiseptate.

Type species: Podosporiella glomerata (Harkn.) Bonar = *P. humilis* Ellis & Everh.

Podosporiella glomerata (Harkn.) Bonar, 1965, *Mycologia*, **57**: 395–396.
(Fig. 200)

Sporodochia up to 1 mm. high and 500µ thick. *Conidiophores* up to 60µ long, 4–8µ thick. *Conidia* up to 7-septate, 30–65µ long, 6–8µ thick in the broadest part.

On leaves of *Garrya*, causing round, brown spots 2–4 mm. diam., each with a dark margin; U.S.A.

183. ZYGOPHIALA

Zygophiala Mason, 1945, apud Martyn in *Mycol. Pap.*, **13**: 3–5.

Colonies effuse, thin. *Mycelium* mostly superficial. *Stroma* none. *Setae* and *hyphopodia* absent. *Conidiophores* macronematous, mononematous, scattered, consisting of a flexuous, often torsive, mid to dark brown, smooth stipe with a short subhyaline apical cell which bears usually 2, occasionally 3 divergent, hyaline or subhyaline conidiogenous cells. *Conidiogenous cells* polyblastic, discrete, sympodial, cylindrical, or doliiform, cicatrized; scars prominent, dark. *Conidia* solitary, dry, acropleurogenous, simple, fusiform to obclavate, hyaline, smooth, 1-septate, slightly constricted at the septum.

Type species: Zygophiala jamaicensis Mason.

Hughes, S. J., *Mycol. Pap.*, **48**: 82, 1952.

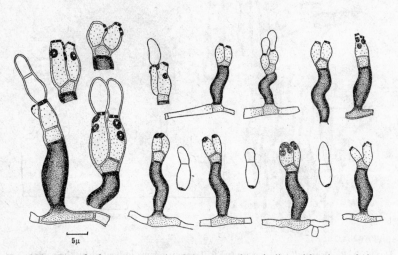

FIG. 201. *Zygophiala jamaicensis* (× 650 except where indicated by the scales).

Zygophiala jamaicensis Mason, 1945, apud Martyn in *Mycol. Pap.*, **13**: 3–5.
(Fig. 201)

Conidiophores up to 35 × 4–8µ. *Conidiogenous cells* 6–15 × 4–6µ. *Conidia* 13–20µ long, 5–6µ thick in the broadest part.

Originally described on leaves of *Musa* it has since been found on many other plants including *Antiaris*, *Barbacenia*, *Calcophyllum*, *Desmodium*, *Dianthus*, *Hymenocardia*, *Psidium* and *Tinospora*; Brazil, Cuba, Europe (in glasshouses), Ghana, Jamaica, Tanzania.

184. HAPLARIOPSIS

Haplariopsis Oudemans, 1903, *Ned. kruidk. Archf*, Ser. 3, **2** (4): 902–903.

Colonies effuse, velvety, grey or olivaceous grey. *Mycelium* partly superficial, partly immersed. *Stroma* none. *Setae* and *hyphopodia* absent. *Conidiophores* macronematous, mononematous, erect or ascending, branched, straight or flexuous, smooth, lower part olivaceous, upper part hyaline or subhyaline. *Conidiogenous cells* polyblastic, integrated, terminal becoming intercalary on stipe and branches, or discrete, sympodial, cylindrical, cicatrized, scars small. *Conidia* solitary, dry, acropleurogenous, simple, cylindrical rounded at the ends, hyaline, smooth, 1-septate.

Type species: Haplariopsis fagicola Oud.

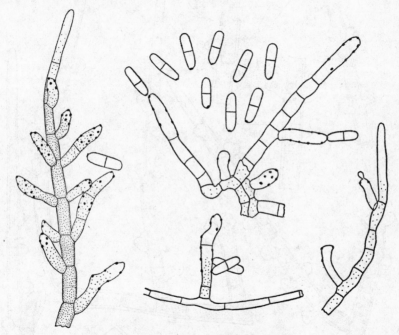

Fig. 202. *Haplariopsis fagicola* (× 650).

Haplariopsis fagicola Oud., 1903, *Ned. kruidk. Archf*, Ser. 3, **2** (4): 902–903.
    Trichothecium cupulicola Lind, 1913, Dan. Fung.: 503.
                                    (Fig. 202)
Conidiophores up to 100µ long, 5–6µ thick. *Conidia* 14–20 × 3·5–5µ.

On dead leaves and fallen cupules of *Fagus*, also on must of *Aesculus* and dead branches of *Quercus*; Europe.

185. SPONDYLOCLADIOPSIS

Spondylocladiopsis M. B. Ellis, 1963, *Mycol. Pap.*, **87**: 15–17.

Colonies effuse, velvety, olivaceous grey. *Mycelium* immersed. *Stroma* none. *Setae* absent but apices of stipes and branches sterile and setiform. *Hyphopodia* absent. *Conidiophores* macronematous, mononematous, unbranched or loosely branched, straight or flexuous, subulate, for the most part subhyaline but reddish brown towards the base, smooth, upper part of stipe and branches sterile and setiform. *Conidiogenous cells* polyblastic, discrete, solitary or in pairs or verticils,

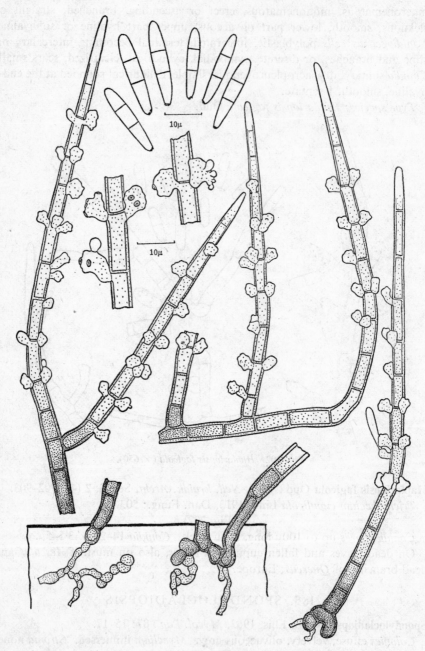

FIG. 203. *Spondylocladiopsis cupulicola* (× 650 except where indicated by scales).

determinate or sympodial, broadly clavate, cicatrized; scars small. *Conidia* solitary, dry, acropleurogenous, simple, cylindrical rounded at the ends or fusiform, hyaline, smooth, transversely septate.

Type species: Spondylocladiopsis cupulicola M. B. Ellis.

Spondylocladiopsis cupulicola M. B. Ellis, 1963, *Mycol. Pap.*, **87**: 15–17.
(Fig. 203)

Conidiophores up to 310μ long, 6–8μ thick at the base, tapering to 2–4μ at the apex; branches 30–200μ long, 6–7μ thick at the base. *Conidiogenous cells* hyaline or pale olivaceous brown, 5–10μ long, 2–3·5μ thick at the base, 3·5–8μ at the apex. *Conidia* almost always 2-septate, 18–23 (20·6) × 3–4 (3·7)μ.

On fallen cupules of *Fagus*; Europe.

186. PERICONIELLA

Periconiella Saccardo, 1885, apud Sacc. & Berlese in *Atti Ist. veneto Sci.*, Ser. 6, **3**: 727.

Acrodesmis H. Sydow, 1926, *Annls mycol.*, **24**: 424.

Ramichloridium Stahel, 1937, *Trop. Agric. Trin.*, **14**: 44.

Colonies occasionally compact but usually effuse, hairy or velvety, sometimes difficult to see with the naked eye. *Mycelium* all superficial or partly immersed, hyphae smooth or verruculose. *Stroma* none. *Setae* and *hyphopodia* absent. *Conidiophores* macronematous, each composed of an erect, straight or flexuous, brown to dark blackish brown, smooth or verruculose stipe and a more or less complex head of branches bearing conidia; in one species the conidiophores are closely grouped together to form coremia. *Conidiogenous cells* polyblastic, integrated and terminal on stipe and branches or discrete, sympodial, cylindrical, cicatrized, scars often numerous and pronounced. *Conidia* solitary or occasionally in very short chains, acropleurogenous, simple, cylindrical rounded at the apex and truncate at the base, ellipsoidal, obclavate or obovoid, hyaline or rather pale olive or olivaceous brown, smooth or verruculose, without septa or with one or a few transverse septa.

Type species: Periconiella velutina (Wint.) Sacc.

Ellis, M. B., *Mycol. Pap.*, **111**: 2–35, 1967.

KEY

Conidia not or very rarely in short chains 1
Conidia frequently in short chains 12
  1. Stipes not aggregated to form coremia 2
     Stipes often forming quite large coremia . . . . *cocoes*
  2. Conidia 0-septate (rarely or never 1-septate) 3
     Conidia frequently with one or more septa 7
  3. Conidia 1·5–2·5μ thick 4
     Conidia 2·5–4μ thick 5
     Conidia 5–7μ thick 6
  4. Conidia 3–4μ long . . . . *mucunae*
     Conidia 3·5–5·5μ long . . . . *daphniphylli*
     Conidia 5–9·5μ long . . . . *musae*

5. Branches erect, pressed together . . . . *anisophylleae*
   Branches spreading . . . . *velutina*
6. Branches simple, inflated at base . . . . *ellisii*
   Primary branches bearing branchlets . . . . *angusiana*
7. Conidia cylindrical or ellipsoidal, never obclavate 8
   Conidia usually obclavate 9
8. Usually not more than one branch at the apex of the conidiophore . . *phormii*
   Several branches at the apex of the conidiophore . . . . *telopeae*
9. Conidia 2·5–3·5μ thick 10
   Conidia thicker than 3·5μ 11
10. Branches 2–4μ thick . . . . *lomateae*
    Branches 4–6μ thick . . . . *smeathmanniae*
11. Conidia 22–34 (28)μ long, smooth or verruculose . . . . *cyatheae*
    Conidia 20–60 (44)μ long, verruculose . . . . . *santaloidis*
    Conidia 20–98 (56)μ long, smooth . . . . . . *rapaneae*
12. Conidia 0-septate (rarely or never 1-septate) 13
    Conidia frequently 1–2-septate 16
13. Conidia 2–3·5 (2·5)μ thick 14
    Conidia 3–6 (4·3)μ thick 15
    Conidia 7–9·5 (8·2)μ thick . . . . *portoricensis*
14. Conidia 5–9μ long . . . . *geonomae*
    Conidia 9–13μ long . . . . *leonensis*
15. Conidia smooth . . . . *cestri*
    Conidia verruculose . . . . *smilacis*
16. Conidia 8–18 (13·9)μ long . . . . *leptoderridis*
    Conidia usually over 20μ long 17
17. Conidia cylindrical 3–6 (4·2)μ thick . . . . . . *deinbolliae*
    Conidia cylindrical-obclavate, 4·5–6·5 (5·4)μ thick . . . . *heveae*

Periconiella velutina (Wint.) Sacc., 1885, apud Sacc. & Berl. in *Atti Ist. veneto Sci.*, Ser. 6, **3**: 727.

(Fig. 204 A)

Colonies hypophyllous, at first small, orbicular, later effuse, dark blackish brown to black, velvety. *Conidiophore* stipes up to 320 × 4–7μ. *Conidia* ellipsoidal, mostly 0-septate, occasionally 1-septate, pale or very pale olive, smooth or verruculose, 7–12 (9)μ long, 2·5–4 (3·4)μ thick in the broadest part.

On leaves of *Brabejum*; S. Africa.

Periconiella musae M. B. Ellis, 1967, *Mycol. Pap.*, **111**: 5–7.

(Fig. 204 B)

Ramichloridium musae Stahel, 1937, *Trop. Agric. Trin.*, **14**: 44 (name not validly published).

Colonies hypophyllous, effuse, grey or greyish brown, velvety. *Conidiophore* stipes up to 500 × 2–3μ; sometimes swollen at the base to 4×7μ. *Conidia* ellipsoidal to obovate, 0-septate, hyaline to subhyaline, smooth or minutely verruculose, 5–9·5 (6·5)μ long, 1·5–2·5 (2·1)μ thick in the broadest part.

On leaves of *Musa* causing a speckling disease; British Solomon Islands Protectorate, Nigeria, Papua, Rhodesia, Sabah, Sierra Leone, West Irian, Western Samoa.

Conidial measurements, host range and geographical distribution of the 21 other species of *Periconiella* in the key are given below; these are fully described in *Mycol. Pap.*, **111**.

P. mucunae M. B. Ellis (Fig. 204 C): 3–4 (3·8) × 1·5–2·5 (2)μ, on *Mucuna*; Sarawak.

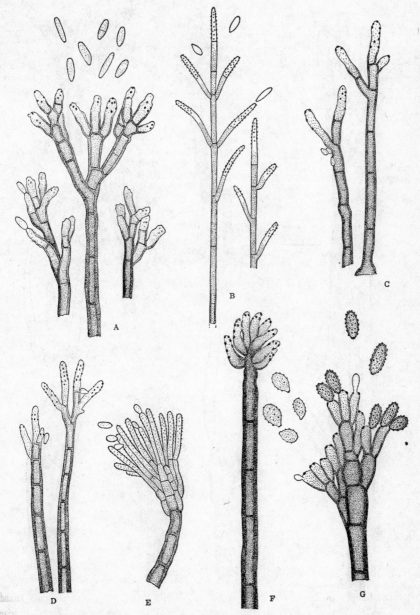

FIG. 204. *Periconiella* species (1): A, *velutina*; B, *musae*; C, *mucunae*; D, *daphniphylli*; E, *anisophylleae*; F, *ellisii*; G, *angusiana* (× 650).

P. daphniphylli M. B. Ellis (Fig. 204 D): 3·5–5·5 (4·6) × 2–2·5 (2·2)μ, on *Daphniphyllum*; Ceylon.

P. anisophylleae M. B. Ellis (Fig. 204 E): 5–8 (6·5) × 2·5–4 (3·2)μ, on *Aniso-phyllea*; Sierra Leone.

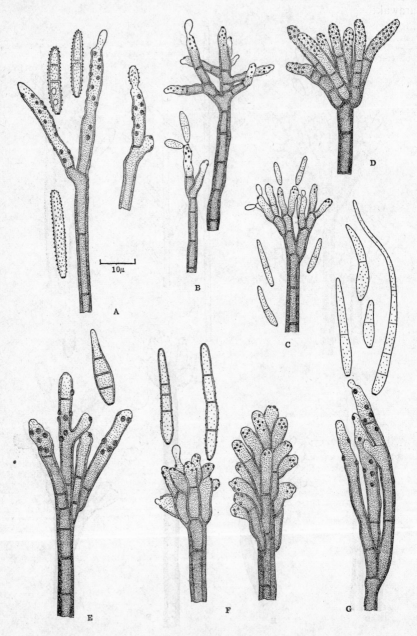

Fig. 205. *Periconiella* species (2): A, *phormii*; B, *telopeae*; C, *lomatiae*; D, *smeathmanniae*; E, *cyatheae*; F, *santaloidis*; G, *rapaneae* (× 650 except where indicated by the scale).

P. ellisii Merny & Huguenin ex M. B. Ellis (Fig. 204 F): 8–12 (9·6) × 5–7 (5·6)μ, hyperparasitic on *Asteridiella*; Congo, Ivory Coast.

P. angusiana M. B. Ellis (Fig. 204 G): 10–16 (12·5) × 5–7 (6)μ; on *Baphia* and *Hymenocardia*; Barotseland, Zambia.

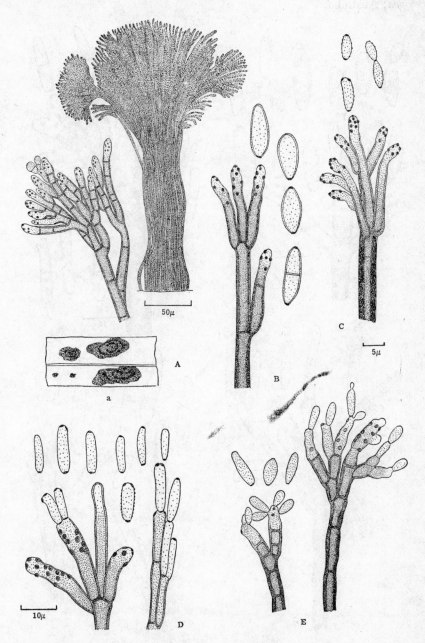

FIG. 206. *Periconiella* species (3): A, *cocoes*; B, *portoricensis*; C, *geonomae*; D, *leonensis*; E, *cestri* (× 650 except where indicated by scales).

P. phormii M. B. Ellis (Fig. 205 A): 12–27 (21) × 3–3·5 (3·2)μ on *Phormium*, causing large, reddish purple blotches on the lower surface of the leaves; New Zealand.

P. telopeae (Hansf.) M. B. Ellis (Fig. 205 B): 7–19 (10·6) × 2·5–4 (3·4)μ, on *Telopea*; Australia.

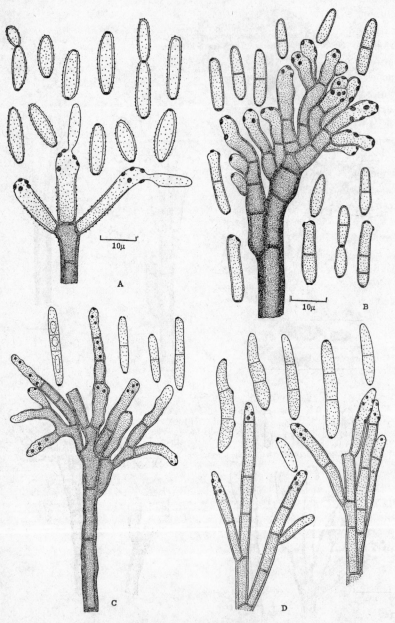

FIG. 207. *Periconiella* species (4): A, *smilacis*; B, *leptoderridis*; C, *deinbolliae*; D, *heveae* (× 650 except where indicated by scales).

P. lomatiae M. B. Ellis (Fig. 205 C): 6–23 (16·7) × 2–3 (2·6)μ, on *Lomatia*; Australia.

P. smeathmanniae M. B. Ellis (Fig. 205 D): 12–26 (20) × 2·5–3·5 (3·1)μ, on *Smeathmannia*; Guinea.

P. cyatheae M. B. Ellis (Fig. 205 E): 22–34 (28) × 6–8 (7·1)μ, on *Cyathea*; New Guinea.

P. santaloidis M. B. Ellis (Fig. 205 F): 20–60 (44) × 4–7 (5·6)μ, on *Santaloides*; Sierra Leone.

P. rapaneae M. B. Ellis (Fig. 205 G): 20–98 (56·1) × 3·5–6 (4·8)μ, on *Rapanea*; Equador.

P. cocoes M. B. Ellis (Fig. 206 A): 5–9 (6·1) × 3–4 (3·4)μ, on *Cocos*, with dark brown, zonate colonies; Malaya, Sarawak.

P. portoricensis (Stev. & Dalbey) M. B. Ellis (Fig. 206 B): 19–27 (22) × 7–9·5 (8·2)μ, on *Canna*; Equador, Guyana, Puerto Rico.

P. geonomae M. B. Ellis (Fig. 206 C): 5–9 (7·2) × 2–3 (2·5)μ, on *Geonoma*; Trinidad.

P. leonensis M. B. Ellis (Fig. 206 D): 9–13 (10·7) × 2–3·5 (2·5)μ on *Smilax* and *Cassia*; Sierra Leone.

P. cestri (H. Syd.) M. B. Ellis (Fig. 206 E): 8–17 (13·6) × 3–6 (4·5)μ, on *Cestrum*; Costa Rica.

P. smilacis M. B. Ellis (Fig. 207 A): 12–17 (14·3) × 3·5–5·5 (4·3)μ, on *Smilax*; India.

P. leptoderridis M. B. Ellis (Fig. 207 B): 8–18 (13·9) × 3–5 (3·6)μ, on *Leptoderris*; Sierra Leone.

P. deinbolliae M. B. Ellis (Fig. 207 C): 21–42 (31) × 3–6 (4·2)μ, on *Deinbollia*; Tanzania.

P. heveae M. B. Ellis (Fig. 207 D): 17–52 (35) × 4·5–6·5 (5·5)μ, on *Hevea*; Sabah.

187. MYSTROSPORIELLA

Mystrosporiella Munjal & Kulshrestha, 1969, *Mycopath. Mycol. appl.*, **39**: 355–357.

Colonies effuse, greyish brown, hairy. *Mycelium* mostly immersed. *Stroma* none. *Setae* and *hyphopodia* absent. *Conidiophores* macronematous, mononematous, branched near the apex forming a stipe and head; stipe straight or slightly flexuous, mid to dark brown, smooth. *Conidiogenous cells* polyblastic, integrated and terminal on branches or discrete, sympodial, cylindrical or doliiform, cicatrized. *Conidia* solitary, dry, acropleurogenous, simple, straight, pyriform or clavate, mid to dark brown, smooth or rugulose, with transverse and longitudinal or oblique septa.

Type species: Mystrosporiella litseae Munjal & Kulshrestha.

Mystrosporiella litseae Munjal & Kulshrestha, 1969, *Mycopath. Mycol. appl.*, **39**: 355–357.

(Fig. 208)

Conidiophores up to 500μ long, stipe 6–9μ thick. *Conidia* with 2–7 transverse and 1–4 longitudinal or oblique septa, 22–36 (mostly 24–28)μ long, 10–13μ thick in the broadest part, 2·5–3·5μ wide at the base.

On leaves of *Litsea*; India.

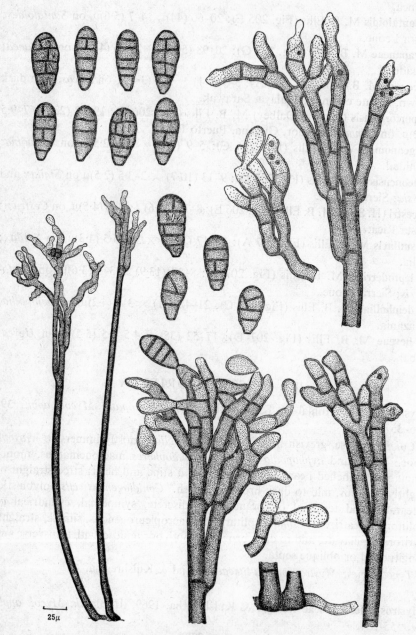

FIG. 208. *Mystrosporiella litseae* (× 650 except where indicated by the scale).

188. CLADOSPORIELLA

Cladosporiella Deighton, 1965, *Mycol. Pap.*, **101**: 35–37.

Colonies hyperparasitic, loosely floccose. *Mycelium* superficial, some hyphae tightly coiled around conidiophores and conidia of host. *Stroma* none. *Setae* and *hyphopodia* absent. *Conidiophores* macronematous, mononematous, straight or flexuous, unbranched, rather pale olivaceous, smooth. *Conidiogenous cells* polyblastic, integrated, terminal, sympodial, more or less cylindrical, cicatrized, scars distinctly thickened, slightly prominent, often situated at the end of short pegs or shoulders around the apex. *Conidia* catenate, in branched chains, acropleurogenous, simple, cylindrical with rounded ends, filiform or obclavate, pale olive, smooth, thin-walled, 0–9-septate.

Type species: Cladosporiella cercosporicola Deighton.

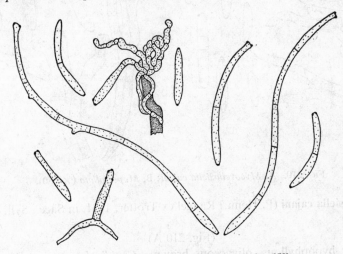

Fig. 209. *Cladosporiella cercosporicola* (× 650).

Cladosporiella cercosporicola Deighton, 1965 (ref. as for genus).
(Fig. 209)

Conidiophores 8–40 × 3–4μ. *Conidia* 20–136 × 2–4·5 (rarely up to 5·2)μ. Parasitizing *Cercospora koepkei* on sugarcane; Sabah.

189. MYCOVELLOSIELLA

Mycovellosiella Rangel, 1917, *Archos Jard. bot., Rio de J.*, **2**: 71.

Colonies effuse, greyish or olivaceous brown, velvety. *Mycelium* partly superficial, partly immersed. *Stroma* none or rudimentary. *Setae* and *hyphopodia* absent. *Conidiophores* macronematous or semi-macronematous, mononematous, usually much branched, flexuous, intertwining, sometimes forming ropes and frequently climbing over leaf hairs, pale to mid pale olivaceous brown, smooth. *Conidiogenous cells* polyblastic, terminal becoming intercalary, sympodial, cylindrical to clavate, cicatrized, scars conspicuous; *ramo-conidia* sometimes present. *Conidia* catenate, chains often branched, acropleurogenous, simple, cylindrical with rounded ends, narrowly ellipsoidal, fusiform or obclavate, subhyaline to mid pale brown or olivaceous brown, smooth, usually 0–3-septate but occasionally with more than 3 septa.

Type species: Mycovellosiella cajani (P. Henn.) Rangel ex Trotter.
Muntañola, M., *Lilloa*, **30**: 165–209, 1960.

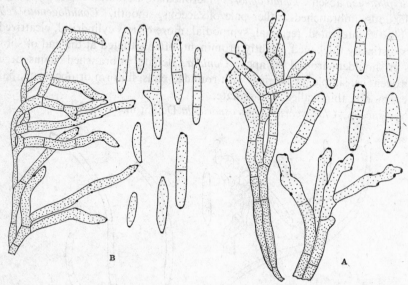

FIG. 210. A, *Mycovellosiella cajani*; B, *M. perfoliata* (× 650).

Mycovellosiella cajani (P. Henn.) Rangel ex Trotter, 1931, in Sacc., Syll. Fung., **25**: 942.

(Fig. 210 A)

Colonies hypophyllous, olivaceous brown. *Conidiophores* much branched, climbing leaf hairs, very variable in length, 1–3μ thick near the base, broadening above to 4–7μ, pale or mid pale olivaceous brown. *Conidia* mostly cylindrical, rounded at the ends, hyaline to mid pale olivaceous brown, 1–3-septate, 20–30 × 4–6μ; scars conspicuous.

On leaves of *Cajanus*, causing round or angular brown spots, often pale in the middle with a dark border; Ethiopia, Haiti, India, Jamaica, Kenya, Malawi, Mauritius, Sierra Leone, Sudan, Taiwan, Tanzania, Trinidad, Uganda, Venezuela, Zambia.

Mycovellosiella perfoliata (Ellis & Everh.) Muntañola, 1960, *Lilloa*, **30**: 201–202.
    Ragnhildiana agerati (Stevens) Stevens & Solheim, 1931, *Mycologia*, **23**: 402.
(Fig. 210 B)

Colonies hypophyllous, irregular, effuse, greyish brown. *Conidiophores* climbing leaf hairs, variable in length, 1–5μ thick, pale to mid pale olivaceous brown. *Conidia* straight or curved, cylindrical rounded at the ends or broadly fusiform, hyaline to pale olivaceous brown, usually 0–1-septate, occasionally 2–3-septate, 10–45 × 3–5μ.

On *Eupatorium* (*Ageratum*); Ceylon, India, Jamaica, Kenya, Malawi, New Caledonia, Papua, Sudan, Taiwan, Tanzania, Trinidad, Uganda.

190. STENELLA

Stenella Sydow, 1930, *Annls mycol.*, **28**: 205–206.
Biharia Thirumalachar & Mishra, 1953, *Sydowia*, 7: 79–80.
Colonies effuse, olive, olivaceous brown or brown, velvety or cottony. *Mycelium* mostly superficial; hyphae usually verruculose. *Stroma* often present, generally small, mostly superficial. *Setae* and *hyphopodia* absent. *Conidiophores* macronematous, mononematous, solitary on hyphae and caespitose on stromata, unbranched or occasionally loosely branched, straight or flexuous, olivaceous

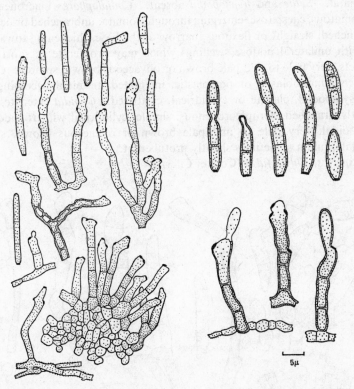

FIG. 211. *Stenella araguata* (× 650 except where indicated by the scale).

or brown, smooth. *Conidiogenous cells* polyblastic, integrated, terminal, sometimes becoming intercalary, sympodial, cylindrical, sometimes geniculate, cicatrized; scars usually conspicuous. *Conidia* mostly in simple or branched acropetal chains, but sometimes solitary, dry, fusiform or obclavate, pale to mid olive, olivaceous brown or brown, smooth, rugulose or verruculose, with 0, 1 or several transverse septa.

Type species: Stenella araguata Syd.

Stenella araguata Syd., 1930, *Annls mycol.*, **28**: 205–206.

(Fig. 211)

Colonies hypophyllous, olive or olivaceous brown. *Conidiophores* often loosely branched, up to 65μ long, 2–4μ thick, sometimes swollen to 5–6μ near the apex, pale to mid olive or olivaceous brown. *Conidia* mostly cylindrical, rarely obclavate, pale to mid olive or olivaceous brown, smooth or verruculose, 0–4 septate, 7–24 × 2–4μ.

On leaves of *Pithecolobium*; Venezuela.

191. FULVIA

Fulvia Ciferri, 1954, *Atti Ist. bot. Univ. Lab. crittogam. Pavia*, Ser. 5, **10**: 245–246.

Colonies effuse, velvety, buff to brown or purplish. *Stroma* present, pale, substomatal. *Setae* and *hyphopodia* absent. *Conidiophores* macronematous, mononematous, caespitose, emerging through stomata, unbranched or occasionally branched, straight or flexuous, narrow at the base, thickening towards the apex, with unilateral nodose swellings which may proliferate as short lateral branchlets, very pale to mid pale brown or olivaceous brown, smooth. *Conidiogenous cells* monoblastic or polyblastic, integrated, terminal becoming intercalary, sympodial, clavate or cylindrical, cicatrized. *Conidia* catenate, chains frequently branched, acropleurogenous, simple, cylindrical with rounded ends or ellipsoidal, very pale to mid pale brown or olivaceous brown, smooth, 0–3-septate, hilum sometimes slightly protuberant.

Type species: Fulvia fulva (Cooke) Ciferri.

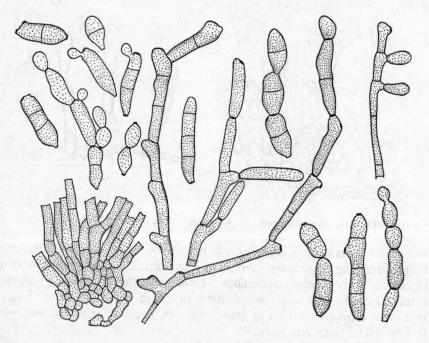

FIG. 212. *Fulvia fulva* (× 650).

Fulvia fulva (Cooke) Ciferri, 1954, *Atti Ist. bot. Univ. Lab. crittogam. Pavia,* Ser. 5, **10**: 245–246.

Cladosporium fulvum Cooke, 1883, *Grevillea*, **12**: 32.

(Fig. 212)

Colonies hypophyllous, effuse, velvety, at first pale buff with whitish margin, later brown and finally often purplish; the upper surface of the leaf above infected areas is at first yellowish, later reddish brown. *Conidiophores* on leaves up to 200μ long but usually 100μ or less, 2–4μ thick near the base broadening to 5–6μ or up to 7–8μ at the nodes. *Conidia* 12–47 × 4–10μ.

On tomato (*Lycopersicon*) causing leaf mould, an important disease especially of plants grown under glass; cosmopolitan [CMI Distribution Map 77].

Barr, Rita & Tomes, M. L., *Am. J. Bot.*, **48**: 512–515, 1961.

Chamberlain, E. E., *N.Z. Jl Agric.*, **45**: 136–142, 1932.

192. PHAEORAMULARIA

Phaeoramularia Muntañola, 1960, *Lilloa*, **30**: 182–183, 209–220.

Colonies effuse, olivaceous or brown, often velvety. *Mycelium* mostly immersed. *Stroma* present but small. *Setae* and *hyphopodia* absent. *Conidiophores* macronematous, mononematous, caespitose, emerging through stomata, unbranched or loosely branched, straight or flexuous, rather pale olivaceous or brown, smooth. *Conidiogenous cells* polyblastic, integrated, terminal becoming intercalary or occasionally discrete, sympodial, cylindrical, cicatrized; *ramoconidia* sometimes present. *Conidia* dry, in branched or unbranched chains,

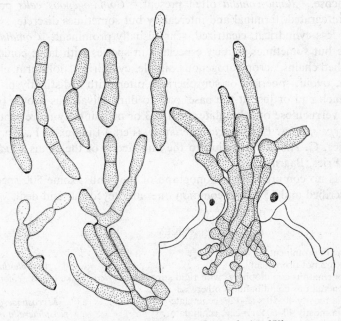

Fig. 213. *Phaeoramularia gomphrenicola* (× 650).

acropleurogenous, simple, cylindrical with rounded ends, ellipsoidal or broadly fusiform, hyaline or olivaceous, smooth, with 0, 1 or several transverse septa.

Type species: Phaeoramularia gomphrenicola (Speg.) Muntañola.

Phaeoramularia gomphrenicola (Speg.) Muntañola, 1960, *Lilloa*, **30**: 209–213.
(Fig. 213)

Conidiophores up to 80μ long but usually shorter, 3·5–7μ thick. *Conidia* cylindrical, mostly rounded at the ends, sometimes constricted in the middle, 0–3-septate, 11–40 × 4·5–8μ.

On living leaves of *Gomphrena*; Argentina.

193. CLADOSPORIUM

Cladosporium Link ex Fries; Link, 1815, *Magazin Ges. naturf. Freunde Berlin*, **7**: 37–38; Fries, 1821, Syst. mycol., **1**: XLVI.

S. J. Hughes in *Can. J. Bot.*, **36**: 750, 1958 cites the following as synonyms: *Sporocladium* Chevallier, 1826; *Myxocladium* Corda, 1837; *Didymotrichum* Bonorden, 1851; *Heterosporium* Klotzsch in Cooke, 1877.

Colonies effuse or occasionally punctiform, often olivaceous but also sometimes grey, buff, brown or dark blackish brown, velvety, floccose or hairy. *Mycelium* immersed and often also superficial. *Stroma* sometimes present. *Setae* and *hyphopodia* absent. *Conidiophores* macronematous or semimacronematous and sometimes also micronematous; *macronematous conidiophores* straight or flexuous, mostly unbranched or with branches restricted to the apical region forming a stipe and head, olivaceous brown or brown, smooth or verrucose. *Ramo-conidia* often present. *Conidiogenous cells* polyblastic, usually integrated, terminal and intercalary but sometimes discrete, sympodial, more or less cylindrical, cicatrized, scars usually prominent. *Conidia* catenate as a rule but sometimes solitary especially in species with large conidia, often in branched chains, acropleurogenous, simple, cylindrical, doliiform, ellipsoidal, fusiform, ovoid, spherical or subspherical, often with a distinctly protuberant scar at each end or just at the base, pale to dark olivaceous brown or brown, smooth, verruculose or echinulate, with 0–3 or occasionally more septa.

Lectotype species: Cladosporium herbarum (Pers.) Link ex S. F. Gray.

de Vries, G. A. Contribution to the knowledge of the genus *Cladosporium* Link ex Fries, Baarn, 1952.

There is no comprehensive monograph of the genus; some 500 species have been described and 15 of the common ones are keyed out and dealt with here.

KEY

Conidial scars prominent 1
Conidial scars not prominent . . . . *Amorphotheca resinae*
  1. Macronematous conidiophores with rigid stipes and *Periconia*-like heads   *chlorocephalum*
     Macronematous conidiophores otherwise 2
  2. Conidia mostly 40–50 × 12–15μ, echinulate . . . *Mycosphaerella dianthi*
     Conidia mostly 40–60 × 17–22μ, echinulate . . . *Mycosphaerella macrospora*
     Conidia smaller, smooth or verruculose 3

3. Conidiophores often with terminal and intercalary swellings (nodose) . . . **4**
   Conidiophores not nodose **8**
4. Conidia mostly 3–6μ thick, conidiophores often 500μ long . . . *oxysporum*
   Conidia mostly 6μ or more thick, conidiophores shorter **5**
5. Conidia mostly 6–8μ thick **6**
   Conidia mostly 7–10μ thick **7**
6. Conidia smooth, often constricted between septa . . . *colocasiae*
   Conidia densely verruculose . . . . *herbarum*
7. Aerial hyphae spirally twisted . . . . *variabile*
   Aerial hyphae not spirally twisted . . . . . *macrocarpum*
8. Conidia mostly spherical or subspherical 3–4·5μ diam. . . . *sphaerospermum*
   Conidia not spherical or subspherical **9**
9. Conidia usually with 3 thick, dark transverse septa . . . *spongiosum*
   Septa when present not noticeably thick and dark **10**
10. Parasitic on *Prunus* . . . . *Venturia carpophila*
    Parasitic on *Musa* . . . . *musae*
    Parasitic on *Cucumis* and other Cucurbitaceae . . . . *cucumerinum*
    Saprophytic—some strains parasitic . . . . . . *cladosporioides*

Cladosporium state of **Amorphotheca resinae** Parbery, 1969, *Aust. J. Bot.*, **17**: 331–357.

Cladosporium resinae (Lindau) de Vries, 1955, *Antonie van Leeuwenhoek*, **21**: 167.

(Fig. 214 A)

Colonies buff, brown or olive green. *Conidiophores* up to 2 mm. long, 2·5–6μ thick, often distinctly warted, the warts irregularly scattered. *Ramo-conidia*, when present, clavate or cylindrical, 8–20μ long, 3–7μ thick, generally smooth with usually 3 protuberant points of attachment for conidia. *Conidia* when borne on ramo-conidia solitary or in chains of rarely more than 3 conidia, broadly ellipsoidal or ovoid, 0-septate, brown or olivaceous brown, smooth, commonly 3–6 × 2–3·5μ. Conidia when borne at the ends of conidiophores and branches without ramo-conidia develop in long chains of up to 20 or sometimes more; they are usually narrowly ellipsoidal, 0-septate, smooth, 3–12 × 2–4μ.

Frequently isolated from aviation kerosene and soil; found also on creosoted wood and in resin, face cream, etc.; Australia, France, Germany, Great Britain, Netherlands, S. Africa, Sweden, U.S.A.

Cladosporium chlorocephalum (Fresen.) Mason & M. B. Ellis, 1953, *Mycol. Pap.*, **56**: 123–126 (with full synonymy).

(Fig. 214 B)

Colonies effuse, olivaceous brown or black on natural substrata, olive green in culture. *Stromata* on stems flat, dark brown to black, 50–300μ long, 15–30μ thick. *Macronematous conidiophores*: stipe dark brown to black, up to 680μ long, 14–24μ thick at base, 8–15μ thick immediately below head; conidial heads spherical or oval, olive green by reflected light, brown by transmitted light, 40–70μ diam.; *primary branches* 9–22 × 6–9μ. *Conidia* olive or pale brown, smooth or verruculose, 0-septate, proximal and intermediate ones ellipsoidal or limoniform 6–14 × 4–9μ, distal or terminal ones spherical 3·5–7μ diam.

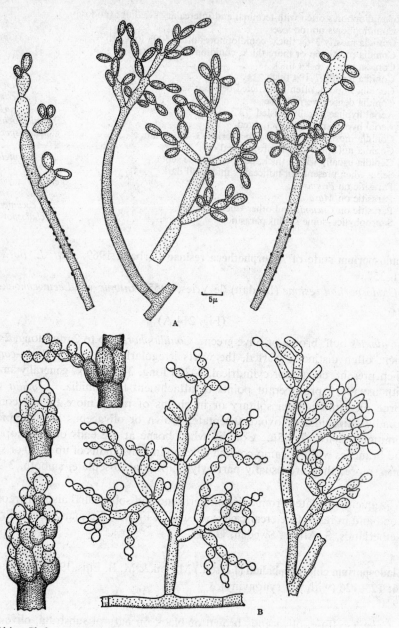

FIG. 214. *Cladosporium* species (1): A, state of *Amorphotheca resinae*; B, *chlorocephalum* (× 650 except where indicated by the scale).

In young cultures 0–2-septate *ramo-conidia* measuring 8–34 × 4–6μ and long, branched chains of ellipsoidal and spherical conidia 4–8 × 3·5–6μ or 3–6μ diam. are formed abundantly on micronematous and short semi-macronematous conidiophores. Macronematous conidiophores in slide cultures are narrower than those formed on the natural substratum.

On dead stems and leaves of peony; Belgium, France, Germany, Great Britain, Italy, U.S.A.

Cladosporium state of **Mycosphaerella dianthi** (Burt.) Jørstad, 1945, *Meld. St. plpatol. Inst.*, **1**: 17.

Cladosporium echinulatum (Berk.) de Vries, 1952, Contribution to the knowledge of the genus *Cladosporium* Link ex Fr.: 49.

Heterosporium echinulatum (Berk.) Cooke, 1877, *Grevillea*, **5**: 123.

(Fig. 215 A)

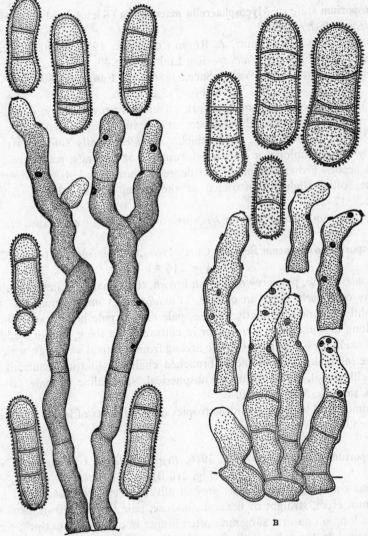

Fig. 215. *Cladosporium* species (2): A, state of *Mycosphaerella dianthi*; B, state of *M. macrospora* (× 650).

Colonies effuse, sometimes orbicular, olivaceous grey, hairy. *Conidiophores* often arising in dense fascicles through stomata, simple or occasionally once branched, flexuous, often geniculate, pale to mid brown, smooth, up to 200μ long, 8–13μ thick. *Conidia* solitary or in short chains, straight or slightly curved, oblong or cylindrical, rounded at the ends, pale or mid pale brown or olivaceous brown, distinctly and densely echinulate, 1–5- (commonly 2–4-) septate, 25–55 × 8–17 (mostly 40–50 × 12–15)μ.

Common and widespread on leaves and sometimes also inflorescences of various species of *Dianthus*.

Cladosporium state of **Mycosphaerella macrospora** (Kleb.) Jørstad, 1945, *Meld. St. plpatol. Inst.*, **1**: 20.

Cladosporium iridis (Fautr. & Roum.) de Vries, 1952, Contribution to the knowledge of the genus *Cladosporium* Link ex Fr.: 49.

Heterosporium gracile (Wallr.) Sacc., 1886, Syll. Fung., **4**: 480.
(Fig. 215 B)

Colonies effuse or punctiform, dark olivaceous brown, hairy. *Conidiophores* often arising in fascicles through stomata, flexuous, often geniculate, brown, smooth, up to 150μ long, 8–20μ thick. *Conidia* usually solitary, straight or slightly curved, oblong or cylindrical, rounded at the ends, pale to mid brown or olivaceous brown, distinctly and densely echinulate, 1–4- (commonly 2–3-) septate, often slightly constricted at the septa, 35–70 × 13–25 (commonly 40–60 × 17–22)μ.

Common and widespread on *Iris* spp., often causing well-defined leaf spots.

Cladosporium oxysporum Berk. & Curt., 1868, *J. Linn. Soc.*, **10** (46): 362.
(Fig. 216 A)

Colonies effuse, pale grey or greyish brown, thinly hairy on natural substrata, cottony or loosely felted in culture. *Conidiophores* macronematous, straight or slightly flexuous, distinctly nodose, pale or mid pale brown, smooth, up to 500μ long or sometimes even longer in culture, 3–5μ thick, terminal and intercalary swellings 6–8μ diam. *Conidia* arising from terminal swellings, which later become intercalary, in simple or branched chains, cylindrical rounded at the ends, ellipsoidal, limoniform or subspherical, subhyaline or pale olivaceous brown, smooth, 5–30 × 3–6μ.

Common and widespread in the tropics on dead parts of leaves and stems of herbaceous and woody plants.

Cladosporium colocasiae Sawada, 1916, *Rep. nat. Hist. Ass. Formosa*, No. 25.
(Fig. 216 B)

Colonies amphigenous, effuse, greyish olive, velvety. *Conidiophores* macronematous, erect, straight or flexuous, nodose, pale to mid brown, smooth, up to 180μ long on natural substrata, often longer in culture, 4–6μ thick, terminal and intercalary vesicular swellings 8–10μ diam. *Conidia* arising from terminal swellings, which later become intercalary, in simple or branched chains, cylindrical or oblong rounded at the ends, or ellipsoidal, often constricted in the middle or between septa, pale to mid brown, smooth, 1–3- (occasionally 5-) septate, 12–32 × 6–9 (mostly 15–20 × 6–8)µ; scars at each end markedly protuberant.

On leaves of *Colocasia* spp., causing brown, orbicular or irregular spots very variable in size; Ethiopia, Guinea, Hong Kong, Mauritius, Nepal, New Caledonia, Nigeria, Pakistan, Papua, Sabah, Sarawak, Taiwan. This species is well

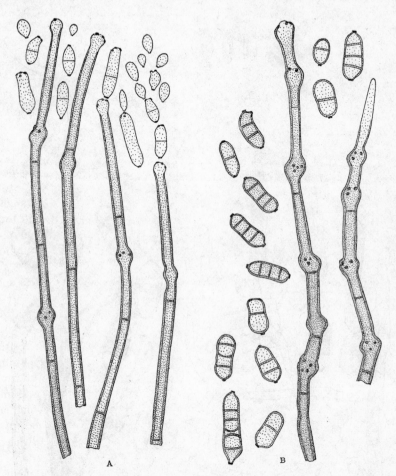

FIG. 216. *Cladosporium* species (3): A, *oxysporum*; B, *colocasiae* (× 650).

described, illustrated and documented by F. Bugnicourt in *Revue Mycol.*, **23**: 233–236, 1958.

Cladosporium herbarum (Pers.) Link ex S. F. Gray, 1821, Nat. Arr. Br. Pl., **1**: 556. For synonymy see S. J. Hughes, *Can. J. Bot.*, **36**: 750–751, 1958.

(Fig. 217 A)

Colonies on natural substrata and in culture, effuse, olive green or olivaceous brown, velvety; reverse on malt agar greenish black. *Stroma* often well developed, especially on herbaceous stems. *Conidiophores* mostly macronematous, straight or flexuous, sometimes geniculate, often nodose, pale to mid olivaceous brown or brown, smooth, up to 250μ long, 3–6μ thick, terminal and intercalary vesicular swellings when present 7–9μ diam. *Conidia* in fairly long, often branched chains, ellipsoidal or oblong rounded at the ends, pale to mid

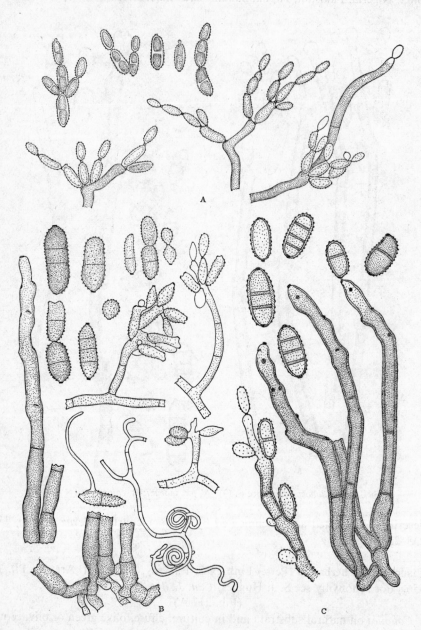

FIG. 217. *Cladosporium* species (4): A, *herbarum*; B, *variabile*; C, *macrocarpum* (× 650).

brown or olivaceous brown, rather thick-walled and distinctly verruculose with low warts, nearly always 0–1-septate although additional septa are formed occasionally, 5–23 × 3–8 (mostly 8–15 × 4–6)μ with scars at one end or both ends small but clearly protuberant.

A very common cosmopolitan species, especially abundant in temperate regions on dead herbaceous and woody plants; frequently isolated from air, soil, foodstuffs, paint, textiles, etc.

Cladosporium variabile (Cooke) de Vries, 1952, Contribution to the knowledge of the genus *Cladosporium* Link ex Fr.: 85–89.
(Fig. 217 B)

Colonies effuse, blackish olive, sometimes with reddish purple pigment, velvety or felted. *Aerial hyphae* tortuous and spirally coiled. *Conidiophores* straight or flexuous, sometimes geniculate and nodose, up to 350μ long but usually less than 150μ, 3–5μ thick in culture, up to 6μ or 8μ on the natural substratum, pale to mid brown, smooth. *Conidia* usually in short chains, oblong rounded at the ends, ellipsoidal or subglobose, pale to mid brown or olivaceous brown, densely verrucose, 0–3-septate, 5–30 × 3–13 (commonly 15–25 × 7–10)μ.

On *Spinacea oleracea* causing white to pale yellow leaf spots each with a dark fruiting area in the middle. Closely related to *C. macrocarpum*; the spirally coiled aerial hyphae and pathogenicity to spinach being held sufficient grounds for keeping it separate from that species.

Cladosporium macrocarpum Preuss, 1848, in Sturm's Deut. Fl., **3**: 27–28.
(Fig. 217 C)

Colonies effuse, olive green, velvety. *Stroma* sometimes well developed. *Conidiophores* mostly macronematous, straight or flexuous, often geniculate and nodose, pale to mid brown or olivaceous brown, smooth or in part verruculose, up to 300μ long, 4–8μ thick, terminal and intercalary swellings when present 9–11μ diam. *Conidia* usually in rather short chains, oblong rounded at the ends or ellipsoidal, 0–3-septate, pale to mid brown or olivaceous brown, thick-walled, densely verrucose, 9–28 × 5–13 (mostly 15–25 × 7–10)μ.

A common cosmopolitan species, especially abundant in temperate regions on dead herbaceous and woody plants; allied to *C. herbarum* but distinguished by its broader, frequently 2- and 3-septate conidia.

Cladosporium sphaerospermum Penz., 1882, *Michelia*, **2**: 473.
(Fig. 218 A)

Colonies olive green to olivaceous brown, pulverulent, often furrowed; reverse on malt agar greenish black. *Conidiophores* macronematous and micronematous, sometimes up to 300μ long but generally much shorter, 3–5μ thick, pale to mid or sometimes dark olivaceous brown, smooth or verruculose. *Ramo-conidia* 0–3-septate, up to 33μ long, 3–5μ thick, smooth or verruculose. *Conidia* mostly globose or subglobose 3–4·5μ diam., mid to dark olivaceous brown, verrucose, the warts more clearly visible in air bubbles or water, much less distinct in lactic acid or lacto-phenol.

A very common cosmopolitan species; it occurs as a secondary invader on many different plants and has been isolated from air, soil, foodstuffs, paint, textiles and occasionally from man and other animals.

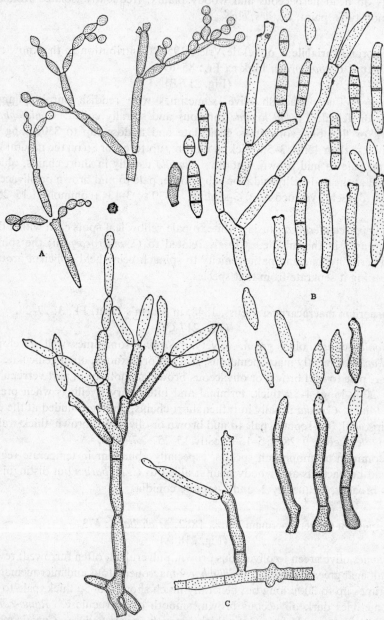

FIG. 218. *Cladosporium* species (5): A, *sphaerospermum*; B, *spongiosum*; C, state of *Venturia carpophila* (× 650).

Cladosporium spongiosum Berk. & Curt., 1868, *J. Linn. Soc.*, **10** (46): 362.

(Fig. 218 B)

Colonies effuse, dark olivaceous brown, closely felted. *Conidiophores* short, usually less than 100μ long, semi-macronematous, simple or branched, 3–6μ thick, smooth, pale golden brown. *Ramo-conidia* up to 60μ long. *Conidia* in simple or branched chains, cylindrical, ellipsoidal or subglobose, rather pale golden brown, smooth or minutely verruculose, 0–5- (commonly 3-) septate, 5–40 × 2·5–7μ, septa thick, dark brown.

On inflorescences of grasses, especially species of *Cenchrus* and *Setaria* in the tropics.

Cladosporium state of **Venturia carpophila** Fisher, 1961, *Trans. Br. mycol. Soc.*, **44**: 337–342.

Cladosporium carpophilum Thüm., 1877, *Öst. bot. Z.*, **27**: 12.

Fusicladium carpophilum (Thüm.) Oudem., 1900, *Verh. K. Akad. Wet.*, 1900: 388.

(Fig. 218 C)

Colonies effuse or punctiform, dark olivaceous brown, velvety. *Conidiophores* macronematous and micronematous, simple or branched towards the apex, straight or slightly flexuous, pale to mid brown or olivaceous brown, up to 100μ long, 4–6μ thick, sometimes swollen to 8–9μ at the base. *Ramo-conidia* when present cylindrical, up to 32μ long, 4–5μ thick, smooth or minutely verruculose. *Conidia* sometimes solitary but often in simple or branched chains of 3–4, especially in culture, mostly cylindrical to fusiform, occasionally obclavate, with protruding scars at one end or both ends, 0-septate, pale olivaceous brown, smooth or minutely verruculose, 9–29 × 4–5 (usually 12–20 × 4–5)μ.

On fruit, leaves and sometimes other parts of apricot, plum, peach and almond, causing a disease commonly known as ' freckle '; cosmopolitan [CMI Distribution Map 198]. Of considerable economic importance especially in parts of Australia and S. Africa.

Cladosporium musae Mason, 1945, apud Martyn in *Mycol. Pap.*, **13**: 2–3.

(Fig. 219 A)

Colonies effuse, greyish brown, brown or dark blackish brown, on leaves thinly hairy, in culture velvety. *Conidiophores* macronematous and micronematous. *Macronematous conidiophores* erect, straight or slightly flexuous, up to 500μ long, 4–6μ thick, sometimes swollen to 9μ at the base, branched at the apex, brown to dark blackish brown, often paler above the middle, smooth; branches up to 50 × 3–4μ. *Conidia* cylindrical, ellipsoidal or fusiform, subhyaline, smooth, 0–1-septate, 6–22 × 3–5μ.

On leaves of *Musa* spp., causing large, elliptical, brown or blackish brown, sometimes zonate spots; British Solomon Islands, Ghana, Guinea, Jamaica, Hong Kong, Sabah, Sierra Leone, Sudan, Uganda, Western Samoa.

Cladosporium cucumerinum Ellis & Arth., 1889, *Bull. agr. Exp. Stn. Indiana*, **19**: 9. For synonymy see de Vries, 1952.

(Fig. 219 B)

Colonies on natural substrata and in culture effuse, rather pale greyish olive, velvety or felted; reverse on malt agar greenish black. *Hyphae* sometimes spirally twisted, immersed hyphae usually with a slime coat. *Conidiophores* macronematous and micronematous, up to 400µ long, 3–5µ thick, sometimes swollen

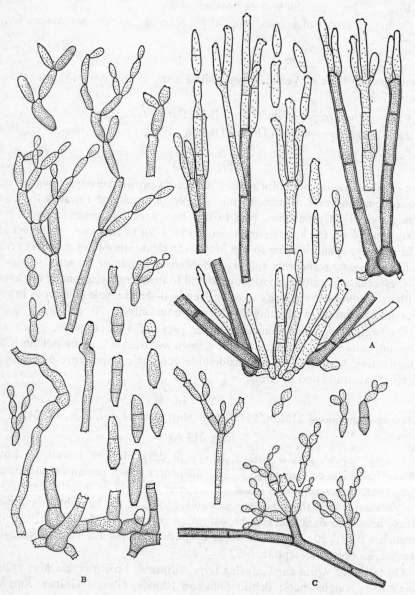

FIG. 219. *Cladosporium* species (6): A, *musae*; B, *cucumerinum*; C, *cladosporioides* (× 650).

to 8μ at the base, pale olivaceous brown, smooth. *Ramo-conidia* 0–2 septate, up to 30μ long, 3–5μ thick, smooth. *Conidia* formed in long, branched chains, mostly 0-septate, occasionally 1-septate, cylindrical rounded at the ends, ellipsoidal, fusiform or subspherical, smooth to minutely verruculose, pale olivaceous brown, 4–25 × 2–6 (mostly 4–9 × 3–5)μ.

On leaves, stems and fruits especially of *Cucumis sativus* causing a disease known as cucumber gummosis or scab; young seedlings are sometimes seriously damaged. It has been recorded on cucumber and other Cucurbitaceae in Africa, Asia, Europe, North and Central America [CMI Distribution Map 310].

Cladosporium cladosporioides (Fresen.) de Vries, 1952, Contribution to the knowledge of the genus *Cladosporium* Link ex Fr.: 57.

(Fig. 219 C)

Colonies effuse, olive green or olivaceous brown, velvety; reverse on malt agar greenish black. *Conidiophores* macronematous and micronematous, sometimes up to 350μ long but generally much shorter, 2–6μ thick, pale to mid olivaceous brown, smooth or verruculose. *Ramo-conidia* 0–1-septate, up to 30μ long, 2–5μ thick, smooth or occasionally minutely verruculose. *Conidia* formed in long branched chains, mostly 0-septate, ellipsoidal or limoniform, 3–11 × 2–5 (mostly 3–7 × 2–4)μ, pale olivaceous brown, most commonly smooth but verruculose in some strains. It is closely related to *C. cucumerinum* and *C. sphaerospermum*.

A very common cosmopolitan species; it occurs as a secondary invader on many different plants and has been isolated from air, soil, textiles, etc.

194. NIGROSPORA

Nigrospora Zimmermann, 1902, *Zentbl. Bakt. ParasitKde*, Abt. 2, **8**: 220.

Basisporium Molliard, 1902, *Bull. Soc. mycol. Fr.*, **18**: 167–170.

Dichotomella Sacc., 1914, *Annls mycol.*, **12**: 312.

Colonies at first white with small, shining black conidia easily visible under a low-power dissecting microscope, later brown or black when sporulation is abundant. *Mycelium* all immersed or partly superficial. *Stroma* none. *Setae* and *hyphopodia* absent. *Conidiophores* micronematous or semi-macronematous, branched, flexuous, colourless to brown, smooth. *Conidiogenous cells* monoblastic, discrete, solitary, determinate, ampulliform or subspherical, colourless. *Conidia* solitary, with a violent discharge mechanism (fully described by Webster, 1952), acrogenous, simple, spherical or broadly ellipsoidal, compressed dorsiventrally, black, shining, smooth, 0-septate.

Type species: Nigrospora panici Zimm.

Mason, E. W., *Trans. Br. mycol. Soc.*, **12**: 152–165, 1927; *Mycol. Pap.*, **3**: 60–61, 1933.

Webster, J., *New Phytol.*, **51**: 229–235, 1952.

KEY

Conidia 10–16 (mostly 12–14)μ diam. . . . . *Khuskia oryzae*
Conidia 14–20 (mostly 16–18)μ diam. . . . . *sphaerica*
Conidia 17–24 (mostly 20–22)μ diam. . . . . *sacchari*
Conidia 25–30μ diam. . . . . *panici*

Nigrospora state of **Khuskia oryzae** Hudson, 1963, *Trans. Br. mycol. Soc.*, **46**: 355–360.

N. oryzae (Berk. & Br.) Petch, 1924, *J. Indian bot. Soc.*, **4**: 24.

(Fig. 220 A)

Conidiophores 3–7μ thick. *Conidiogenous cells* 6–9μ diam. *Conidia* 10–16 (mostly 12–14)μ diam.

Very common in tropical countries and occasionally recorded from temperate regions on many different kinds of plants and especially common on *Oryza*; it has been isolated from air and soil.

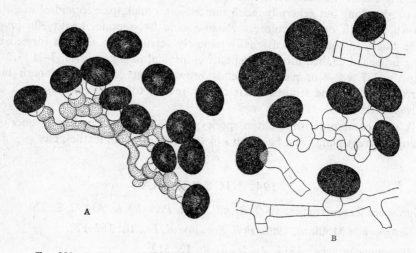

FIG. 220. A, *Nigrospora* state of *Khuskia oryzae*; B, *N. sphaerica* (× 650).

Nigrospora sphaerica (Sacc.) Mason, 1927, *Trans. Br. mycol. Soc.*, **12**: 158.

(Fig. 220 B)

Conidiophores 4–8μ thick. *Conidiogenous cells* 8–11μ diam. *Conidia* 14–20 (mostly 16–18)μ diam.

A very common cosmopolitan species, especially widespread in tropical countries on many different kinds of plants; isolated occasionally from foodstuffs and soil.

Specimens of **Nigrospora sacchari** (Speg.) Mason, mostly on *Saccharum* but also on other plants have been received from Argentina, Fiji, India, Netherlands New Guinea, New Zealand, Nigeria, Pakistan, Tanzania and Zambia. **N. panici** Zimm. is less common; specimens have been seen only from India Java Malaya and Papua.

195. PAATHRAMAYA

Paathramaya Subramanian, 1956, *J. Indian bot. Soc.*, **35**: 68–70.

Colonies effuse, dark brown or blackish brown, with numerous scattered synnemata. *Mycelium* immersed. *Stroma* none seen. *Setae* and *hyphopodia* absent. *Conidiophores* macronematous, synnematous; synnemata dark brown to blackish brown, individual threads unbranched, splaying out to form a brush-like head, straight or flexuous, straw coloured or pale brown, smooth, lower part cylindrical, sterile, upper part clavate, fertile, bearing conidiogenous vesicles at the apex and along the sides. *Conidiogenous vesicles* usually constricted at the base and not cut off by septa, monoblastic, discrete, determinate, at first subspherical, collapsing to become calyciform when conidia are released. *Conidia* solitary, dry, acrogenous (one to each vesicle), simple, ellipsoidal to subspherical, very dark brown, smooth, 0-septate.

Type species: Paathramaya sundara Subram.

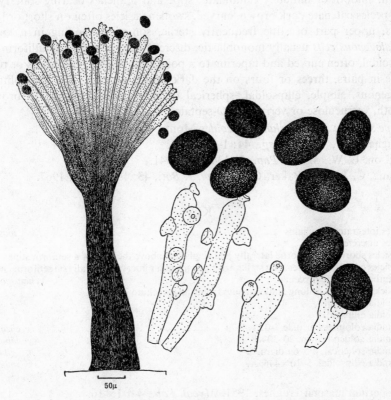

50μ

Fig. 221. *Paathramaya sundara* (× 650 except where indicated by the scale).

Paathramaya sundara Subram., 1956, *J. Indian Bot. Soc.*, **35**: 68–70.
(Fig. 221)

Synnemata up to 750μ long, 100–300μ thick at the base, narrowing to 40–150μ then splaying out again in the head to 250–1000μ; individual threads 5–7μ thick

D.H.—11

near the base, 10–15µ at the apex. *Conidiogenous vesicles* 5–9µ diam. *Conidia* 20–27 × 15–20µ.

On dead wood; India.

196. ZYGOSPORIUM

Zygosporium Montagne, 1842, *Ann. Sci. nat.*, Sér. 2, **17**: 120–121.
  Pimina Grove, 1888, *J. Bot., Lond.*, **26**: 206.
  Urobasidium Giesenhagen, 1892, *Flora, Jena*, **76**: 139–141.
  Urophiala Vuillemin, 1910, *Bull. Séanc. Soc. Sci. Nancy*, Sér. 3, **11**: 158.

Colonies effuse or sometimes compact, often thin, grey, brown, blackish brown or black. *Mycelium* mostly superficial, reticulate. *Stroma* none. Separate *setae* absent but in a number of species the upper part of the conidiophore is sterile and setiform. *Hyphopodia* absent. *Conidiophores* macronematous or sometimes micronematous, mononematous, scattered, unbranched or branched, brown, smooth or minutely echinulate; stipe and branches bearing solitary or in 1 species catenate dark brown, curved, swollen vesicles often on short or long stalks, upper part of stipe frequently sterile, sometimes ending in a knob. *Conidiogenous cells* usually monoblastic, discrete, determinate, ampulliform or ellipsoidal, often curved and tapering to a point, thin-walled, colourless or pale, borne in pairs, threes or fours on the dark brown vesicles. *Conidia* solitary, acrogenous, simple, ellipsoidal, spherical or subspherical, hyaline to brown, smooth, verruculose or verrucose, O-septate.

Type species: Zygosporium oscheoides Mont.

Hughes, S. J., *Mycol. Pap.*, **44**: 1–18, 1951.
Mason, E. W., *Mycol. Pap.*, **5**: 134–144, 1941.
Wang, C. J. K. & Baker, G. E., *Can. J. Bot.*, **45**: 1945–1952, 1967.

KEY

Vesicles integrated, in chains	*masonii*
Vesicles discrete, solitary	1
1. Vesicles short-stalked borne laterally and singly just above the base of a setiform stipe	2
Vesicles on usually rather long stalks arising at irregular intervals laterally on setiform stipes and also on the mycelium	*echinosporum*
Vesicles on short or long stalks arising from the mycelium only . . .	4
2. Conidia spherical	*minus*
Conidia ellipsoidal	3
3. Conidia colourless or pale 7–12 × 4–7µ	*oscheoides*
Conidia golden brown, 20–30 × 8–11µ	*geminatum*
4. Conidia spherical, 4·5–6µ diam.	*gibbum*
Conidia ellipsoidal, 5–10 × 4–6µ	*mycophilum*

Zygosporium masonii Hughes, 1951, *Mycol. Pap.*, **44**: 15–16.

(Fig. 222 A)

Conidiophores up to 100µ long, about 2µ thick, with chains of up to 6 integrated vesicles and a hyaline, sterile apical region which terminates in a small knob. *Vesicles* 7–12µ long, 4–5µ thick in the broadest part. *Conidia* ellipsoidal, hyaline, 6–8 × 3–4µ.

On dead leaves and occasionally other parts of *Alstonia, Artocarpus, Cocos,
Cola, Daniellia, Fagara, Ficus, Gliricidia, Gossypium, Gymnosporia, Musa,
Pancovia, Persea, Polypodium, Saccharum, Solanum* and isolated from soil;
Australia, Ghana, Guinea, Hong Kong, India, Jamaica, Sierra Leone, Tanzania,
Venezuela.

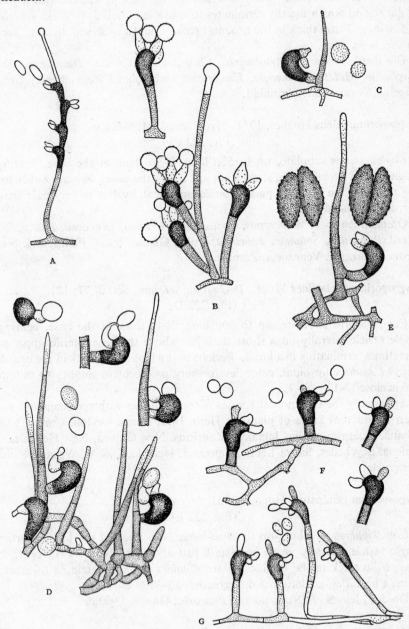

FIG. 222. *Zygosporium* species: A, *masonii*; B, *echinosporum*; C, *minus*; D, *oscheoides*;
E, *geminatum*; F, *gibbum*; G, *mycophila* (× 650).

Zygosporium echinosporum Bunting & Mason, 1941, apud Mason in *Mycol. Pap.*, **5**: 135–137.
(Fig. 222 B)

Conidiophores sometimes up to 300μ long but usually 50–150μ, 2–3μ thick, with 1 or several vesicles borne laterally on rather long stalks, with a sterile apical region which usually terminates in a hyaline knob 5–7μ diam. *Vesicles* 9–15μ long, 7–10μ thick in the broadest part. *Conidia* spherical, hyaline, verruculose, 6–9μ diam.

On dead leaves of *Artocarpus*, *Cocos*, *Cola*, *Cordia*, *Desplatzia*, *Ficus*, *Hyphaene*, *Millettia*, *Pancovia*, *Theobroma*, and isolated from cheese; Ghana, Nigeria, Sierra Leone, Trinidad.

Zygosporium minus Hughes, 1951, *Mycol. Pap.*, **44**: 6–7.
(Fig. 222 C)

Conidiophores subulate, up to 50μ long, 1–3μ thick at the base, bearing a single vesicle laterally on a short stalk just above the base. *Vesicles* 8–12μ long, 4–8μ thick in the broadest part. *Conidia* spherical, hyaline to very pale brown, verruculose, 6–9μ diam.

On dead leaves of *Artocarpus*, *Cocos*, *Cola*, *Crinum*, *Dracaena*, *Elaeis*, *Ficus*, *Musa*, *Polyalthia*, *Solanum*, *Tabebuia;* Cuba, Ghana, India, Philippines, Sierra Leone, Tanzania, Venezuela, Zambia.

Zygosporium oscheoides Mont., 1842, *Ann. Sci. nat.*, Sér. 2, **77**: 121.
(Fig. 222 D)

Conidiophores subulate, up to 80μ long, 3–4μ thick at the base, bearing a single vesicle laterally on a short stalk just above the base; sterile upper part sometimes terminating in a knob. *Vesicles* 9–18μ long, 7–9μ thick in the broadest part. *Conidia* ellipsoidal, colourless or very pale brown, smooth to minutely verruculose, 7–12 × 4–7μ.

Apparently the commonest species of *Zygosporium* with specimens on more than 70 different kinds of plants in Herb. IMI; Brazil, Ceylon, Congo, Cuba, Ghana, Guernsey, India, Jamaica, Mauritius, New Guinea, New Hebrides, St. Helena, Seychelles, Sierra Leone, Tanzania, Uganda, U.S.A., Western Samoa, Venezuela.

Zygosporium geminatum Hughes, 1951, *Mycol. Pap.*, **44**: 5–6.
(Fig. 222 E)

Conidiophores subulate, up to 110μ long, 3–5μ thick at the base, bearing a single vesicle laterally on a short stalk just above the base. *Vesicles* 12–15μ long, 6–8μ thick in the broadest part. *Conidia* usually remaining together in pairs, ellipsoidal, golden brown, verrucose, 20–30 × 8–11μ.

On dead leaves of *Dracaena* and *Pancovia*; Ghana, Uganda.

Zygosporium gibbum (Sacc., Rouss. & Bomm.) Hughes, 1958, *Can. J. Bot.*, **36**: 825.

Z. *parasiticum* (Grove) Bunting & Mason, apud Mason in *Mycol. Pap.*, **5**: 137.

(Fig. 222 F)

Stalked *vesicles* arising directly from the mycelium, 10–15µ long, 7–9µ thick in the broadest part. *Conidia* spherical, hyaline, smooth to minutely verruculose, 4·5–6µ diam.

On dead leaves of *Artocarpus, Borassus, Cola, Culcasia, Dioscorea, Elaeis, Eucalyptus, Euonymus, Ficus, Hydnocarpus, Hyphaene, Landolphia, Laurus, Litsea, Metroxylon, Musa, Nephelium, Ochthocosmus, Passiflora, Polypodium, Pteridium, Smilax;* British Solomon Islands, Costa Rica, Europe, Ghana, Hong Kong, India, Pakistan, Sabah, Sierra Leone.

Zygosporium mycophilum (Vuill.) Sacc., 1911, *Annls mycol.*, **9**: 256.

(Fig. 222 G)

Stalked *vesicles* arising directly from the mycelium, 12–15µ long, 6–8µ thick

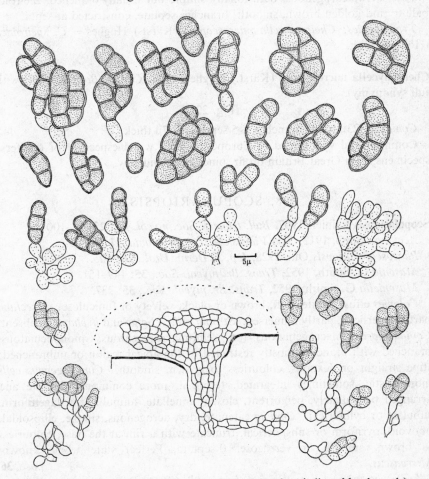

FIG. 223. *Cheiromycella microscopica* (× 650 except where indicated by the scale).

in the broadest part. *Conidia* ellipsoidal, hyaline, smooth or minutely verruculose, 5–10 × 4–6µ.

On dead leaves of *Gaultheria* and found as a contaminant in cultures of *Gonatobotryum fuscum*; Europe, India.

197. CHEIROMYCELLA

Cheiromycella Höhnel, 1910, *Sber. Akad. Wiss. Wien*, **119**: 664.

Sporodochia punctiform, brown. *Mycelium* immersed. *Stroma* erumpent. *Setae* and *hyphopodia* absent. *Conidiophores* macronematous, mononematous, short, branched irregularly and repeatedly, arising from cells of the stroma and composed mainly of swollen conidiogenous cells which are sometimes connected to one another by narrow isthmi, very pale golden brown, smooth. *Conidiogenous cells* polyblastic or monoblastic, integrated or discrete, determinate, clavate, doliiform, spherical or subspherical. *Conidia* aggregated in firm slimy masses, acropleurogenous, occasionally simple but usually branched, cheiroid, pale to mid golden brown, smooth; branches septate, constricted at septa.

Type species: Cheiromycella microscopica (Karst.) Hughes = *C. speiroidea* (Höhn.) Höhn.

Cheiromycella microscopica (Karst.) Hughes, 1958, *Can. J. Bot.*, **36**: 747 (with full synonymy).

(Fig. 223)

Conidia 8–23µ long; branches 1–5 septate, 3–7µ thick.

Common and widespread in Europe on dead wood especially of conifers; specimens from Great Britain on fir, pine, larch and yew.

198. SCOPULARIOPSIS

Scopulariopsis Bainier, 1907, *Bull. trimest. Soc. mycol. Fr.*, **23**: 98–100.
Acaulium Sopp, 1912, *Skr. VidenskSelsk. Christiania*, **11**: 42–53.
Phaeoscopulariopsis Ôta, 1928, *Jap. J. Derm. Urol.*, **28**: 405.
Masonia G. Smith, 1952, *Trans. Br. mycol. Soc.*, **35**: 149–151.
Masoniella G. Smith, 1952, *Trans. Br. mycol. Soc.*, **35**: 237.

Colonies effuse, white, buff, brown or black, velvety or funiculose. *Mycelium* partly superficial, partly immersed. *Stroma* none. *Setae* and *hyphopodia* absent. *Conidiophores* macronematous or semi-macronematous, mononematous, branched with branches mostly restricted to the apical region or unbranched; stipe straight or flexuous, colourless to brown, smooth. *Conidiogenous cells* monoblastic, sometimes integrated, terminal, more commonly discrete and arranged penicillately, percurrent, closely annellate, ampulliform, lageniform, subulate or cylindrical. *Conidia* catenate, dry, acrogenous, simple, ellipsoidal, obovoid, pyriform or subspherical, truncate with a rim at the base, colourless to brown, smooth or verrucose, 0-septate. Perfect state when known, *Microascus*.

Lectotype species: Scopulariopsis brevicaulis (Sacc.) Bainier.

Morton, F. J. & Smith, G., *Mycol. Pap.*, **86**: 1–69 and 85–96, 1963; a monograph of the genus with keys.

Cole, G. T. & Kendrick, Bryce, *Can. J. Bot.*, **47**: 925–929.

Scopulariopsis brevicaulis (Sacc.) Bainier, 1907, *Bull. trimest. Soc. mycol. Fr.*,
23: 98–100. For synonymy, see Morton & Smith.

(Fig. 224 A)

Colonies at first whitish, later buff to nut brown with a narrow white margin.
Annellides sometimes arising singly from hyphae but more frequently in groups
of 2–3 on short stipes or sometimes arranged penicillately in more complex
conidiophores, 10–25μ long, colourless or very pale. *Conidia* brown in mass but
very pale when viewed singly, subspherical or obovoid, truncate at the base,
smooth when young, coarsely verrucose when mature, 5–8 × 5–7μ.

Common and widespread, isolated from air, animals including man, foodstuffs,
paper, sewage, soil, textiles, wood and various plants including *Arachis, Brassica,
Clematis, Curcuma, Dioscorea, Gossypium, Nicotiana, Oryza* and *Triticum*;
Australia, Ceylon, Cuba, Egypt, Europe, India, Israel, Kenya, Pakistan, Sudan,
Tanzania, U.S.A.

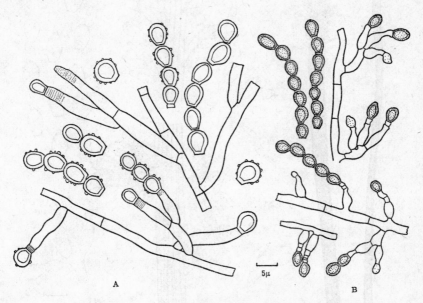

FIG. 224. A, *Scopulariopsis brevicaulis*; B, *S. brumptii*.

Scopulariopsis brumptii Salvanet-Duval, 1935, *Thèse Fac. Pharm. Paris*, **23**: 58.
Masoniella grisea (G. Smith) G. Smith, 1952, *Trans. Br. mycol. Soc.*, **35**: 237.

(Fig. 224 B)

Colonies slow-growing, at first white, then grey, finally dark blackish brown.
Annellides often arising singly from the sides of hyphae or ropes of hyphae, but
sometimes formed in groups of 2–3 on short stipes, ampulliform or lageniform,
pale grey or olivaceous brown, 4–10μ long, swollen part 2·5–3·5μ thick. *Conidia*

black or dark brown in mass, pale when viewed singly, obovoid to subspherical, truncate at the base, smooth or sometimes slightly verruculose, 4–6 × 3·5–4·5μ.

Isolated from air, soil, straw, *Oryza* and *Ricinus*; Brazil, Europe, Hong Kong, India, Pakistan, U.S.A.

199. DORATOMYCES

Doratomyces Corda, 1829, in J. Sturm's Deut. Fl., III, **2** (7): 65–66.
Stysanus Corda, 1837, Icon. Fung., **1**: 21–22.
Colonies effuse, grey, brown, blackish brown or black, velvety, floccose or powdery. *Hyphae* superficial and immersed. *Stroma* none. *Setae* and *hyphopodia*

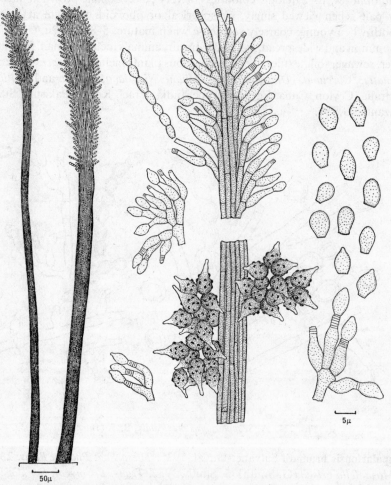

FIG. 225. *Doratomyces stemonitis* (× 650 except where indicated by scales).

absent. *Conidiophores* typically macronematous, synnematous, dark brown to black, threads straight or flexuous, individually pale brown to brown, mostly smooth, branched towards the apex with the branches splaying out to form a head. *Conidiogenous cells* monoblastic, sometimes integrated terminal on branches but mostly discrete, penicillately arranged, percurrent, ampulliform or lageniform. *Conidia* catenate, dry, acrogenous, simple, ellipsoidal, ovoid, obovoid, spherical or subspherical, truncate at the base, 0-septate. *D. stemonitis* also has an *Echinobotryum* state.

Type species: Doratomyces stemonitis (Pers. ex Fr.) Morton & Smith = *D. neesii* Corda.

Morton, F. J. & Smith, G., *Mycol. Pap.*, **86**: 1–96, 1963.

KEY

Conidia smooth 1
Conidia verrucose 3
1. Conidia mostly larger than $5 \times 3 \cdot 5 \mu$ 2
   Conidia $3–5 \times 2–3\mu$ . . . . *microsporus*
2. Heads ellipsoidal or cylindrical, conidia often pointed at apex; usually with an *Echinobotryum* state . . . . *stemonitis*
   Heads spherical or subspherical, conidia mostly rounded at apex; no *Echinobotryum* state . . . . *purpureofuscus*
3. Conidia ovoid, $6–8 \cdot 5 \times 5–6\mu$ . . . . *nanus*
   Conidia spherical or subspherical $8–12\mu$ diam. . . . . . *phillipsii*

Doratomyces microsporus (Sacc.) Morton & Smith, 1963, *Mycol. Pap.*, **86**: 77–80.
(Fig. 226 A)

Colonies grey to almost black. *Synnemata* up to 600μ high, usually with long cylindrical or ellipsoidal heads. *Conidia* usually ovoid, with truncate base and rounded or acutely pointed apex, $3–5 \times 2–3\mu$.

On dead wood, dead leaves of grasses and sedges, herbaceous stems, also isolated from feathers, dung and soil; Europe.

Doratomyces stemonitis (Pers. ex Fr.) Morton & Smith, 1963, *Mycol. Pap.*, **86**: 70–74.
(Fig. 225)

Colonies at first grey, becoming dark blackish brown to black. *Synnemata* up to 1200μ high, with ellipsoidal or cylindrical heads. *Conidia* usually ovoid, with truncate base and usually pointed apex, $6–8 \cdot 5 \times 4–4 \cdot 5\mu$. *Echinobotryum* state usually present and often dominant in young cultures.

Common on dead wood, herbaceous stems, seed oats, plywood, dung, etc.; Europe and N. America.

Doratomyces purpureofuscus (Fr.) Morton & Smith, 1963, *Mycol. Pap.*, **86**: 74–77.
(Fig. 226 B)

Colonies at first greenish grey, becoming dark grey to dark blackish brown. *Synnemata* up to 900μ high, with spherical or subspherical heads, superficially resembling *Periconia*. *Conidia* ovoid to oblong, truncate at the base, usually rounded at the apex, $5–7 \times 3 \cdot 5–4 \cdot 5\mu$. No *Echinobotryum* state.

Common on dead herbaceous stems, especially *Heracleum* and *Brassica*, also on dead wood and isolated from dung, soil, etc.; Europe and N. America.

Doratomyces nanus (Ehrenb. ex Link) Morton & Smith, 1963, *Mycol. Pap.*, **86**: 80–82.

(Fig. 226 C)

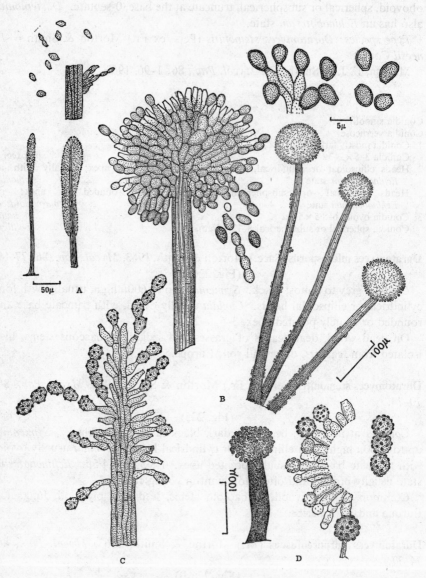

FIG. 226. *Doratomyces* species: A, *microsporus*; B, *purpureofuscus*; C, *nanus*; D, *phillipsii* (× 650 except where indicated by scales).

Colonies grey to brown or blackish brown. *Synnemata* up to 900μ high, with subspherical or ellipsoidal heads. *Conidia* ovoid, base truncate often with a collar, apex rounded or bluntly pointed, distinctly verrucose, 6–8·5 × 5–6μ.

On dead wood and bark, leaves, herbaceous stems, dung, etc.; Europe.

Doratomyces phillipsii (Berk. & Leighton) Morton & Smith, 1963, *Mycol, Pap.*, **86**: 82–83.

(Fig. 226 D)

Synnemata up to 200µ high with subspherical heads. *Conidia* spherical or subspherical, distinctly verrucose, 8–12µ diam.

On moss and soil; England.

200. TRICHURUS

Trichurus Clements & Shear, 1896, University of Nebraska Botanical Survey, **4**: 7.

Colonies effuse, greyish brown, velvety, floccose or powdery. *Mycelium* partly superficial, partly immersed. *Stroma* none. Separate *setae* absent but the conidiophore threads often have sterile setiform apices or bear setiform branches.

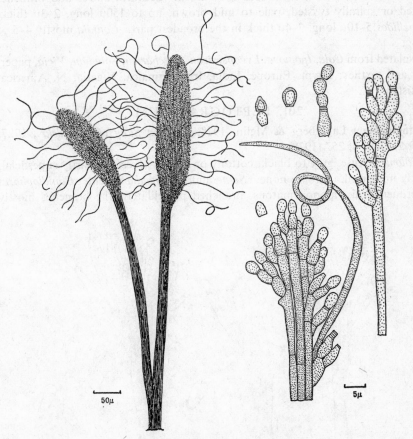

FIG. 227. *Trichurus spiralis.*

Hyphopodia absent. *Conidiophores* typically macronematous, synnematous, dark brown or blackish brown; individual threads straight or flexuous, pale to mid brown, smooth, branched towards the apex, the branches splaying out to form more or less cylindrical heads, often with long sterile, subulate, acutely pointed, straight or coiled setiform apices or with setiform branches. *Conidiogenous cells* monoblastic, sometimes integrated and terminal on branches but mostly discrete, penicillately arranged, percurrent, ampulliform or lageniform. *Conidia* catenate, dry, acrogenous, simple, ellipsoidal, ovoid, obovoid or subspherical, truncate at the base, pale to mid brown, usually smooth, 0-septate.

Type species: Trichurus cylindricus Clements & Shear.

Lodha, B. C., *J. Indian bot. Soc.*, **42**: 135–142, 1963.

Swart, H. J., *Antonie van Leeuwenhoek*, **30**: 257–260, 1964.

Trichurus spiralis Hasselbring, 1900, *Bot. Gaz.*, **29**: 321.
(Fig. 227)

Synnemata up to 3 mm. high, 10–80μ thick, often expanded in the head to 200μ or more; individual threads 2–3μ thick. Setiform apices and branches coiled or spirally twisted, pale to mid brown, up to 150μ long, 2–3μ thick. *Annellides* 5–10μ long, 3–4μ thick in the broadest part. *Conidia* mostly 4–5 × 3–4μ.

Isolated from *Cola, Ipomoea, Lycopersicon, Saccharum, Solanum, Vicia,* paper, soil and textiles; Egypt, Europe, India, Iraq, Jamaica, Nigeria, N. America, Pakistan.

201. LEPTOGRAPHIUM

Leptographium Lagerberg & Melin, 1928, *Svenska SkogsvFör. Tidskr.*, 1927, Häft 2, Och. 4: 257 (1928).

Colonies effuse, grey to black, cottony or hairy. *Mycelium* partly superficial, partly immersed. *Stroma* none. *Setae* and *hyphopodia* absent. *Conidiophores* macronematous, mononematous, branched penicillately, the branches mostly

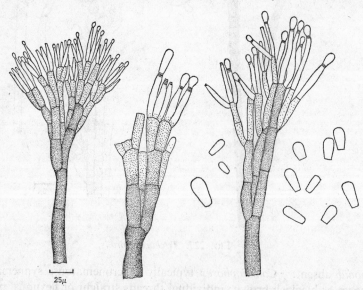

25μ

FIG. 228. *Leptographium lundbergii* (× 650 except where indicated by the scale).

restricted to the apical region; stipe erect or suberect, straight or flexuous, mid to dark brown, smooth. *Conidiogenous cells* monoblastic, mostly discrete, arranged penicillately, percurrent, cylindrical or subulate. *Conidia* aggregated in slimy heads, acrogenous, simple, cuneiform or oblong, rounded at the apex, truncate at the base, hyaline, smooth, 0-septate.

Type species: Leptographium lundbergii Lagerb. & Melin.

Davidson, R. W., *Mycologia,* **34**: 650–662, 1942.

Grossmann, H., *Hedwigia,* **72**: 183–194, 1932.

Hughes, S. J., *Can. J. Bot.,* **31**: 620–621, 1953.

Webb, Shirley, *Proc. R. Soc. Vict.,* **57**: 57–80, 1946.

Leptographium lundbergii Lagerb. & Melin., 1928, *Svenska SkogsvFör. Tidskr.,* 1927, Häft 2, Och. 4: 257 (1928).

(Fig. 228)

Stipes up to 200μ long, cylindrical or subulate, 7–16μ thick. *Annellides* 10–25 × 2–4μ. *Conidia* 8–12 × 3–6μ.

On conifer wood often associated with bark beetles; Europe, N. America.

202. GRAPHIUM

Graphium Corda, 1837, Icon. Fung., **1**: 18.

Colonies effuse, grey, olivaceous brown or black, with erect synnemata clearly visible under a low-power dissecting microscope. *Mycelium* mostly immersed. *Stroma* none. *Setae* and *hyphopodia* absent. *Conidiophores* macronematous, synnematous, each synnema capped by a slimy head; individual threads narrow, straight or flexuous, olivaceous brown, smooth, splaying out and bearing towards the apex a variable number of primary branches which may themselves branch penicillately. *Conidiogenous cells* monoblastic, percurrent, integrated or discrete, subulate or cylindrical. *Conidia* aggregated in slimy heads but those produced from the same annellide often hanging together in quite long chains, acrogenous, simple, straight or curved, cylindrical rounded at the apex, cuneiform or ellipsoidal, usually with a flat base, colourless or pale olivaceous brown, smooth, 0-septate.

Lectotype species: Graphium penicillioides Corda.

KEY

Synnemata up to 200μ high, conidia 4–7×1–2μ   .   .   .   .   .   *penicillioides*
Synnemata 300–1000μ high, conidia 5–11×2–4μ   .   .   .   .   .   . *putredinis*
Synnemata up to 5mm. high, conidia 1·5–3×1–2μ   .   .   .   .   .   . *calicioides*

Graphium penicillioides Corda, 1837, Icon. Fung., **1**: 18.

(Fig. 229 A).

Synnemata black, paler at the apex, up to 200 × 15–30μ; individual threads 1–2μ thick, olivaceous brown. *Annellides* subulate, subhyaline, 50–70 × 1–2μ.

Conidia straight or curved, cuneiform or cylindrical rounded at the apex truncate at the base, 4–7 × 1–2μ.

On wood of *Populus*; Europe, N. America.

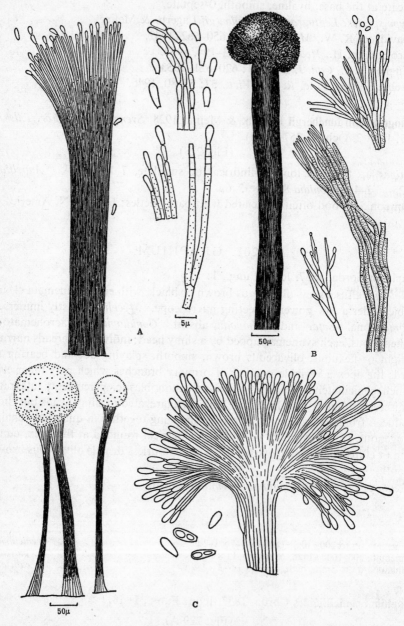

FIG. 229. *Graphium* species: A, *penicillioides*; B, *calicioides*; C, *putredinis* (× 650 except where indicated by scales).

Graphium putredinis (Corda) Hughes, 1958, *Can. J. Bot.*, **36**: 770.

Stysanus putredinis Corda, 1839, Icon. Fung., **3**: 12.

G. cuneiferum (Berk. & Br.) Mason & Ellis, 1953, *Mycol. Pap.*, **56**: 41.

(Fig. 229 C)

Synnemata olivaceous brown or reddish brown, up to 1 mm. long, up to 40μ thick at the base, tapering upwards then splaying out again at the paler apex; individual threads 1–2μ thick. *Annellides* cylindrical or subulate, 10–30 × 1–2μ. *Conidia* ellipsoidal to cuneiform, pale olivaceous brown, smooth, 5–11 × 2–4μ.

On dead herbaceous stems, especially common on *Brassica*, also isolated from soil; Europe.

Graphium calicioides (Fr.) Cooke & Massee, 1887, *Grevillea*, **16**: 11.

(Fig. 229 B)

Synnemata black, up to 5 mm. long, 40–100μ thick; individual threads 1–2μ thick, olivaceous brown. *Annellides* subulate or cylindrical, 7–25 × 1–2μ. *Conidia* ellipsoidal to obovoid, hyaline to olivaceous brown, black in mass, smooth, 1·5–3 × 1–2μ.

Common on rotten wood; Europe.

203. ARTHROBOTRYUM

Arthrobotryum Ces., 1854, *Hedwigia*, **1**: Tab. 4, fig. 1.

Mycelium mostly immersed; hyphae hyaline or brown. *Stroma* none. *Setae* and *hyphopodia* absent. *Conidiophores* macronematous, synnematous, each dark synnema capped by a slimy head; individual threads bearing towards the apex a few primary branches which are themselves sometimes branched. *Conidiogenous cells* integrated and terminal or discrete, monoblastic, percurrent. *Conidia* produced in basipetal succession in slime, ellipsoidal, cylindrical, clavate or fusiform, truncate at base, usually very pale brown, with several transverse septa.

Type species: Arthrobotryum stilboideum Ces.

Hughes, S. J., *Naturalist, Hull*, 1951: 171–173, 1951.

Arthrobotryum stilboideum Ces., 1854, *Hedwigia*, **1**: Tab. 4, fig. 1.

(Fig. 230)

Mycelium mostly immersed, hyphae hyaline or brown, 1·5–3μ thick. *Synnemata* formed singly or in small groups, erect, straight or slightly curved, subulate or cylindrical, black, up to 2·5 mm. high, 15–40μ thick, sometimes up to 80μ at the base, each with a globose or subglobose head up to 150μ in diameter. *Conidiophore* threads usually adhering closely to each other along most of their length, brown or dark brown near the base, paler above, septate, 2–3·5μ thick, bearing towards the apex a few primary branches which are themselves sometimes branched; the end cells of the branches are conidiogenous and percurrent. *Conidia* produced in basipetal succession in slime at the tips of the branches, each as a blown-out end, ellipsoidal or cylindrical, rounded at the apex, truncate at the base, very pale brown, smooth, with 1–3 (usually 3) transverse septa, 10–16 (12)μ long, 3–4 (3·9)μ thick.

On wood, especially oak; Europe.

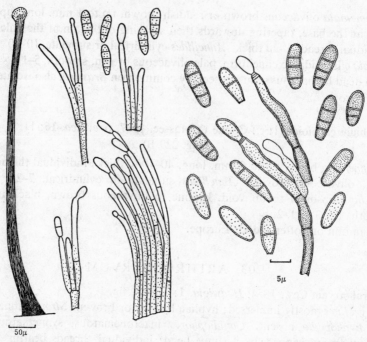

FIG. 230. *Arthrobotryum stilboideum* (× 650 except where indicated by scales).

204. TORULA

Torula Persoon ex Fries; Persoon, 1795, *Annln Bot.* (*Usteri*), 15 Stück: 25–26; Fries, 1821, Syst. mycol., **1**: XLVI.

Hormiscium Kunze ex Fries; Kunze, 1817 in Kunze & Schmidt, Mykol. Hefte, **1**: 12–13; Fries, 1821, Syst. mycol., **1**: XLVI.

Colonies usually effuse but sometimes small and discrete, olive, brown, dark blackish brown or black, often velvety. *Mycelium* superficial and immersed. *Stroma* none. *Setae* and *hyphopodia* absent. *Conidiophores* micronematous or semi-macronematous, unbranched or irregularly branched, straight or flexuous, subhyaline to mid brown, smooth or verruculose. *Conidiogenous cells* polyblastic or sometimes monoblastic, integrated and terminal, or more commonly discrete, determinate, usually spherical, sometimes becoming cupulate, smooth, verruculose or echinulate, distal fertile part thin-walled, sometimes collapsing, proximal sterile part dark brown or reddish brown, thick-walled. *Conidia* dry, in simple or branched chains arising from the surface of the upper half of the very characteristic conidiogenous cells, cylindrical with rounded ends, ellipsoidal or subspherical, brown or olivaceous brown, smooth, verruculose or echinulate, with 0, 1 or several transverse septa, usually strongly constricted at the septa.

The terminal cell of a multiseptate conidium is frequently a conidiogenous cell.

Type species: Torula herbarum (Pers.) Link ex S. F. Gray.

Conidia 3–10- (mostly 4–5-) septate, 5–9 (7)μ thick   .   .   .   . *herbarum*
Conidia 1–3 (mostly 2-) septate, 5–7μ thick   .   .   .   .   *herbarum* f. *quaternella*
Conidia 3–6 (mostly 4-) septate, 8–13μ thick   .   .   .   . *terrestris*
Conidia almost all 3-septate, 7–9μ thick   .   .   .   . *caligans*
Conidia often 0-septate, 4–6μ thick   .   .   .   . *graminis*

Torula herbarum (Pers.) Link ex S. F. Gray, 1821, Nat. Arr. Br. Pl., **1**: 557.
(Fig. 231 A)

Colonies very variable in size, sometimes only a few mm. diam., at others completely encircling stems and extending along them for several centimetres, olive when young, black when old, velvety. *Conidiophores* 2–6μ thick except for the conidiogenous cells which are 7–9μ diam. *Conidia* straight or slightly curved, more or less cylindrical, rounded at the ends, pale olive to brown, verruculose or finely echinulate, 3–10- (mostly 4–5-) septate, constricted at the septa, 20–70 (29) × 5–9 (7)μ.

Very common on dead herbaceous stems, also found occasionally on leaves, wood, old sacking, etc., and isolated from air and soil; cosmopolitan but found most frequently in temperate regions.

Torula herbarum f. **quaternella** Sacc., 1913, *Annls mycol.*, **11**: 556–557.
(Fig. 231 B)

Resembles *T. herbarum* in many ways but the conidia are smoother, almost all 1–3- (mostly 2-) septate and 10–17 × 5–7μ.

This is the common form in the tropics; specimens have been seen from Cuba, India, New Caledonia, Pakistan, Philippines, Tanzania, Venezuela and Zambia.

Torula terrestris Misra, 1967, *Can. J. Bot.*, **45**: 367–369.
(Fig. 231 C)

Colonies effuse, dark olivaceous brown to black, velvety. *Conidiophores* 2–4μ thick, except for the spherical conidiogenous cells which are 4–6μ diam., hyaline to pale yellowish brown, smooth or minutely verruculose. *Conidia* solitary or catenate, ellipsoidal, 3–6-(mostly 4-) septate, often slightly constricted at the septa, verrucose, end cells pale to mid brown, intermediate cells mid to dark brown, 20–45 (mostly 25–35)μ long, 8–13 (mostly 9–11)μ thick in the broadest part.

Isolated from soil and dung; Egypt, India, Netherlands Antilles, Pakistan.

Torula caligans (Batista & Upadhyay) M. B. Ellis comb. nov.
Bahusandhika caligans Batista & Upadhyay, 1965, *Atas Inst. Micol.*, **2**: 321.
(Fig. 231 D)

Colonies effuse, velvety, greyish olive, olivaceous brown or dark blackish brown. *Conidiophores* 1–2μ thick except for the spherical conidiogenous cells which are 3–4μ diam. *Conidia* broadly fusiform to ellipsoidal, almost always 3-septate, constricted at the septa, verruculose or echinulate, end cells small, hyaline or pale, intermediate cells much bigger, mid to dark olivaceous brown, 17–25μ long, 7–9μ thick in the broadest part.

Isolated from *Arachis* and soil; Australia, Brazil, India, Nigeria.

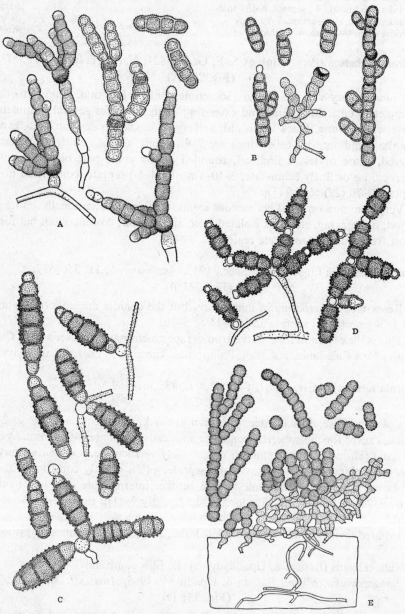

FIG. 231. *Torula* species: A, *herbarum*; B, *herbarum* f. *quaternella*; C, *terrestris*; D, *caligans*; E, *graminis* (× 650).

Torula graminis Desm., 1834, *Annls Sci. nat.*, II, **2**: 72.
(Fig. 231 E)

Colonies round or oval, up to 1·5 × 0·5 mm., dark brown. *Conidiophores* including conidiogenous cells 2–5µ thick. *Conidia* in very long, sometimes branched chains which break up into segments with 0, 1, 2 or many septa, brown, minutely verruculose, cells or 0-septate conidia almost spherical but often slightly broader than long, 4–5 × 4–6µ.

On grasses; Europe.

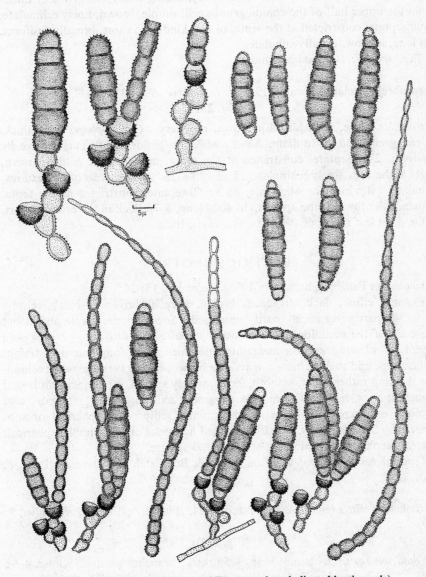

Fig. 232. *Dwayabeeja sundara* (× 650 except where indicated by the scale).

205. DWAYABEEJA

Dwayabeeja Subramanian, 1958, *J. Indian bot. Soc.*, **37**: 53–57.

Mycelium partly superficial, partly immersed. *Stroma* none. *Setae* and *hyphopodia* absent. *Conidiophores* micronematous or semi-macronematous, straight or flexuous, unbranched or irregularly branched, pale brown, smooth, verruculose or finely echinulate. *Conidiogenous cells* polyblastic or monoblastic. discrete, determinate, at first subspherical, later cupulate; lower part dark brown, thick-walled, frequently echinulate; upper part pale, thin-walled, smooth, becoming invaginated. *Conidia* solitary, dry, arising singly or in twos or threes from the upper half of the conidiogenous cell, simple, brown, finely echinulate, multiseptate, constricted at the septa, of two kinds: (1) short, broadly fusiform, (2) long, narrow, usually subulate.

Type species: D. sundara Subram.

Dwayabeeja sundara Subram., 1958, *J. Indian bot. Soc.*, **37**: 53–57.
(Fig. 232)

Colonies effuse, dark blackish brown, powdery. *Conidiophores* 2–6μ thick. *Conidiogenous cells* 6–8μ diam. *Short conidia* straight or slightly curved, broadly fusiform, 2–12-septate, constricted at the septa, mid to dark golden brown, paler at the ends, finely echinulate, 25–65 × 8–12μ. *Long conidia* often flexuous, subulate, cylindrical or obclavate, up to 50-septate, constricted at the septa, brown, paler towards the apex, up to 400μ long, 5–7μ thick in the broadest part.

On dead leaf rachis of *Phoenix canariensis*; India.

206. TRICHOBOTRYS

Trichobotrys Penzig & Saccardo, 1902, *Malpighia*, **15**: 245.

Colonies effuse, dark olivaceous brown, purplish brown or black, velvety. *Mycelium* partly superficial, partly immersed. *Stroma* none. *Setae* absent but the ends of the conidiophores are usually sterile and setiform. *Hyphopodia* absent. *Conidiophores* macronematous, mononematous, long, narrow, straight or flexuous, mid to dark brown or reddish brown, closely verruculose or echinulate, bearing rather short, smooth, fertile, widely spaced, often unciform lateral branches; apex sterile, setiform. *Conidiogenous cells* polyblastic, integrated and terminal or discrete on branches, determinate, ellipsoidal, spherical or subspherical. *Conidia* in dry, usually branched acropetal chains, simple, spherical, brown, verruculose or minutely echinulate, 0-septate.

Type species: Trichobotrys effusa (Berk. & Br.) Petch = *T. pannosa* Penz. & Sacc.

Trichobotrys effusa (Berk. & Br.) Petch, 1924, *Ann. R. bot. Gdns Peradeniya*, **9**: 169.
(Fig. 233)

Conidiophores up to 1 mm. long, 3–5μ thick; branches up to 25μ long, 4–6μ thick. *Conidiogenous cells* 5–6μ diam. or up to 8 × 6μ. *Conidia* 3–5μ diam.

On *Ananas*, *Andropogon*, *Bambusa*, *Hyparrhenia*, palm, *Panicum*, *Phragmites*; Gabon, Ghana, India, Java, New Guinea, Pakistan, Zambia.

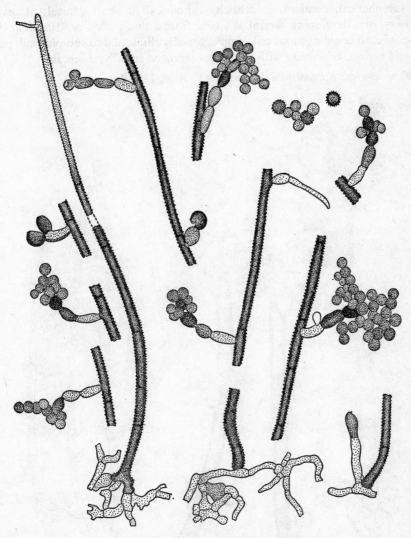

FIG. 233. *Trichobotrys effusa* (× 650).

207. KUMANASAMUHA

Kumanasamuha P. R. Rao & Dev Rao, 1964, *Mycopath. Mycol. appl.*, **22**: 330–334.

Colonies effuse, brown, cottony or felted. *Mycelium* partly superficial, partly immersed. *Stroma* none. *Setae* and *hyphopodia* absent. *Conidiophores* macronematous, mononematous, irregularly branched, mid to dark brown, spinulose or verruculose, each main branch long, straight or flexuous, bearing laterally a number of short, solitary, curved fertile branches but with the upper part usually sterile and rounded at the apex. *Conidiogenous cells* polyblastic, discrete, clustered or arranged penicillately on the lateral branches, determinate, spherical or subspherical, denticulate; denticles cylindrical or short conical. *Conidia* solitary, dry, developing several at a time from a thin-walled protrusion at the apex of each conidiogenous cell, simple, broadly ellipsoidal to subspherical, pale to mid brown, echinulate with the spines arranged spirally, 0-septate.

Type species: Kumanasamuha sundara P. R. & Dev Rao.

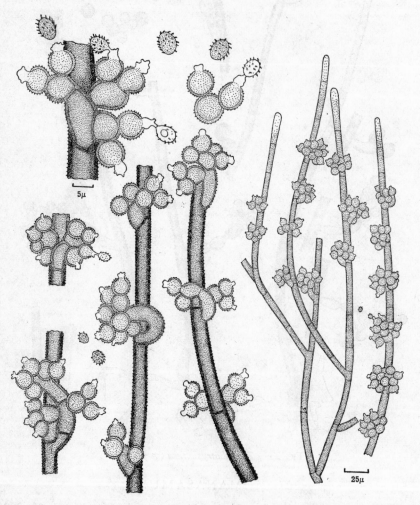

FIG. 234. *Kumanasamuha sundara* (× 650 except where indicated by scales).

Kumanasamuha sundara P. R. & Dev Rao, 1964, *Mycopath. Mycol. appl.*, **22**: 330–334.

(Fig. 234)

Conidiophores up to just over 1 mm. long, branches 7–10µ thick, fertile lateral

branches up to 40μ long. *Conidiogenous cells* 7–9μ diam. *Conidia* 5–7 × 4–5μ.
On rotten wood; French Equatorial Africa, India.

208. SADASIVANIA

Sadasivania Subramanian, 1957, *J. Indian bot. Soc.*, **36**: 64–67.

Colonies effuse, mid to dark blackish brown; conidiophores scattered, each

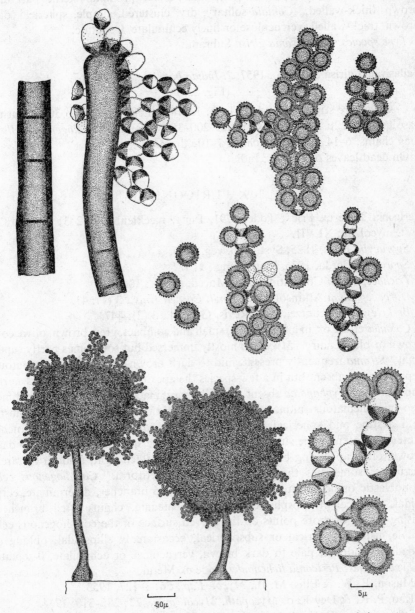

FIG. 235. *Sadasivania girisa* (× 650 except where indicated by scales).

with a large dark head of conidiogenous cells and conidia. *Mycelium* immersed. *Stroma* none seen. *Setae* and *hyphopodia* absent. *Conidiophores* macronematous, mononematous, unbranched, straight or flexuous, swollen at the base, dark brown, smooth. *Conidiogenous cells* arising in simple or branched acropetal chains from the apex and the sides of the upper part of the conidiophore stipe, polyblastic, discrete, determinate, obturbinate or subspherical, smooth or verruculose, distal fertile part pale, thin-walled, proximal sterile part dark brown, thick-walled. *Conidia* solitary, dry, clustered, simple, spherical, dark brown, thick-walled, verruculose or finely echinulate, 0-septate.

Type species: Sadasivania girisa Subram.

Sadasivania girisa Subram., 1957, *J. Indian bot. Soc.*, **36**: 64–67.
(Fig. 235)

Conidiophore stipes up to 450μ long, 13–18μ thick, swollen to 30–40μ at the base and 19–22μ at the apex; heads 200–350μ diam. *Conidiogenous cells* in long chains, 6–14 × 6–9μ. *Conidia* 7–10μ diam.

On dead leaves of grasses; India.

209. PERICONIA

Periconia Tode ex Fries; Tode, 1791, Fung. mecklenb., **2**: 2–3; Fries, 1821, Syst. mycol., **1**: XLVII.
Sporocybe Fries, 1825, Syst. Orb. veg.: 170.
Sporodum Corda, 1836, Icon. Fung., **1**: 18.
Trichocephalum Costantin, 1887, Mucéd. simpl.: 106.
Harpocephalum Atkinson, 1897, *Bull. Cornell Univ.*, **3** (1): 41.
Berkeleyna O. Kuntze, 1898, Revis. Gen. Pl., **3** (2): 447.

Colonies effuse or, in a few species, small and compact, grey, brown, olivaceous brown or black, hairy. *Mycelium* mostly immersed but sometimes partly superficial. *Stroma* frequently present, mid to dark brown, pseudoparenchymatous. Separate *setae* absent but in a few species the apex of the conidiophore is sterile and setiform. *Hyphopodia* absent. *Conidiophores* macronematous and sometimes also micronematous, mononematous. Macronematous conidiophores mostly with a stipe and spherical head, looking like round-headed pins, branches present or absent, stipe straight or flexuous, in one species torsive, pale to dark brown, often appearing black and shining by reflected light, smooth or rarely verrucose; sometimes the apex is sterile and setiform. *Conidiogenous cells* monoblastic or polyblastic, discrete on stipe and branches, determinate, ellipsoidal, spherical or subspherical. *Conidia* catenate, chains often branched, arising at one or more points on the curved surface of the conidiogenous cell, simple, usually spherical or subspherical, occasionally ellipsoidal, oblong or broadly cylindrical, pale to dark brown, verruculose or echinulate, 0-septate.

Type species: Periconia lichenoides Tode ex Mérat.
Mason, E. W. & Ellis, M. B., *Mycol. Pap.*, **56**: 1–127, 1953.
Rao, P. R. & Dev Rao, *Mycopath. Mycol. appl.*, **22**: 285–310, 1964.
Subramanian, C. V., *J. Indian bot. Soc.*, **34**: 339–361, 1955.

KEY

Conidia in a well-defined, fairly compact head at the apex of the stipe . . . 1
Conidia formed along the side of the stipe or if at the apex not in a compact head . 11
  1. Conidiophores without concolorous branches 2
      Conidiophores with concolorous branches 4
  2. Stipe with a short apical cell cut off by a septum; conidiogenous cells formed over the
      apex and in a ring below the septum 3
      Stipe without a short apical cell, conidiogenous cells formed over the swollen apex
                                               cookei
  3. Conidia 10–15µ diam. . . . . *byssoides*
      Conidia 16–22µ diam. . . . . *shyamala*
      Conidia 25–45µ diam. . . . . *manihoticola*
  4. Conidia oblong or broadly cylindrical . . . . . . *sacchari*
      Conidia ellipsoidal . . . . *echinochloae*
      Conidia spherical 5
  5. Conidia more than 14µ diam., distinctly echinulate 6
      Conidia less than 14µ diam., verruculose or with short spines 7
  6. Conidiophores circinate at apex. . . . . *circinata*
      Conidiophores not circinate at apex . . . . *macrospinosa*
  7. Branches short, close together, usually in verticils of 4–7 . . . *atra*
      Branches longer, often widely spaced, irregular 8
  8. Distal branches smooth-walled 9
      Distal branches roughly warted 10
  9. Conidia mostly 4–6µ diam. . . . . *minutissima*
      Conidia 7–11µ diam. . . . . *digitata*
  10. Stipe usually less than 200µ long, conidia with very short spines . . *curta*
      Stipe usually more than 300µ long, conidial spines 1µ long . *Didymosphaeria igniaria*
  11. Apex of conidiophore sterile, setiform 12
      Apex of conidiophore fertile, stipe torsive . . . . . . *funerea*
  12. Conidia formed unilaterally, conidiophores scattered . . . . *lateralis*
      Conidia not unilateral, conidiophores closely packed together . . . 13
  13. Setiform apex usually very long, not curved . . . . . *hispidula*
      Setiform apex short, often curved . . . . *atropurpurea*

Periconia cookei Mason & M. B. Ellis, 1953, *Mycol. Pap.*, **56**: 72–77.
(Fig. 236 A)

Conidiophores up to 800µ long, 14–25µ thick at the base, 17–32µ at the apex, 11–16µ immediately below the head. *Conidia* spherical, brown, verrucose, 13–16µ diam.

Fairly common on blackened areas on dead herbaceous stems; Europe. There is little doubt that either this species or the next one *P. byssoides* is conspecific with the type of the genus *P. lichenoides*. Tode's figure shows the fungus growing on blackened areas of stem and the low-power drawings clearly represent a *Periconia*; his drawings made at higher magnification are somewhat imaginative and unfortunately do not show whether there was a short apical cell or not. No type specimen is known to exist.

Periconia byssoides Pers. ex Mérat, 1821, Nouv. Fl. Environs Paris, Ed. 2, **1**: 18–19.
(Fig. 236 B)

Conidiophores 200–1400µ long or occasionally up to 2 mm., 12–23µ thick at the base, 9–18µ immediately below the head; subhyaline apical cell 12–26 × 11–28µ. Heads 44–120µ diam. *Conidia* spherical, brown, verrucose, 10–15µ diam.

Extremely common on blackened areas on dead herbaceous stems and on leaf spots where it is almost always associated with other fungi, e.g. *Corynespora cassiicola* and *Epicoccum purpurascens*; cosmopolitan.

Periconia shyamala A. K. Roy, 1965, *Indian Phytopath.*, **18**: 332.

Structurally similar to *P. byssoides* but with thicker conidiophores and conidia regularly 16–22μ diam.

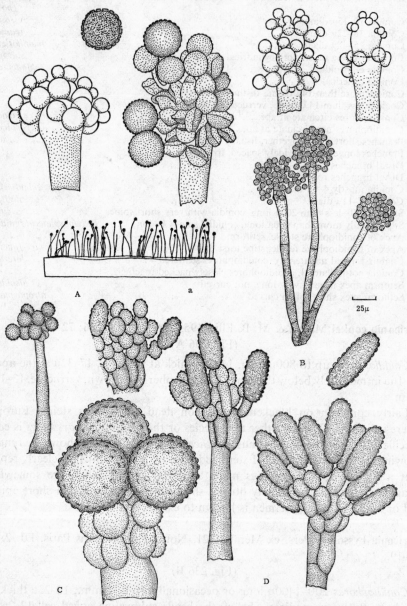

FIG. 236. *Periconia* species (1): A, *cookei*; B, *byssoides*; C, *manihoticola*; D, *sacchari* (a, habit sketch; other figures × 650 except where indicated by the scale).

Quite common causing large, pale tan spots on leaves of *Manihot* and occasionally found on other plants; British Solomon Islands, Ghana, India, New Hebrides, Sarawak, Sierra Leone, Uganda, Zambia.

Periconia manihoticola (Vincens) Viégas, 1955, *Bragantia*, **14**: 63–69.

Haplographium manihoticola Vincens, 1916, *Bull. Soc. Path. vég. Fr.*, **3**: 22–25.
P. heveae Stevenson & Imle, 1945, *Mycologia*, **37**: 571–581.

(Fig. 236 C)

Conidiophores up to 500μ long, 25–40μ thick at the base, 18–26μ immediately below the head; apical cell 18–40 × 15–26μ. *Conidia* spherical, brown, verrucose, 25–45μ diam.

On *Hevea* and *Manihot* causing severe leaf spotting and blight of leaves, petioles and twigs; Brazil, Costa Rica, Mexico.

Periconia sacchari Johnston, 1917, apud Johnston & Stevenson, *J. Dep. Agric. P. Rico*, **1**: 225.

(Fig. 236 D)

Conidiophores up to 450μ long, 10–20μ thick at the base, 6–10μ immediately below the head. Branches up to 40μ long, 5–10μ thick. *Conidia* oblong or cylindrical, brown, verrucose, 15–30 × 8–14μ.

On leaves of *Saccharum*; Brazil, Ghana, Malaya, Puerto Rico, Sierra Leone.

Periconia echinochloae (Batista) M. B. Ellis comb. nov.

Periconiella echinochloae Batista, 1952, *Bolm Secr. Agric. Ind. Com. Est. Pernambuco*, **19**: 174–175.

(Fig. 237 A)

Conidiophores dark brown or dark blackish brown, up to 500μ long, 10–20μ thick at the base, 7–10μ immediately below the head. Branches up to 50 × 6–8μ. *Conidia* ellipsoidal, sometimes slightly curved, golden brown, verruculose, 10–33 × 6–12 (mostly 10–20 × 6–9)μ.

On many grasses including *Andropogon*, *Brachiaria*, *Cenchrus*, *Echinochloa*, *Heteropogon*, *Oryza*, *Panicum*, *Saccharum*, *Sorghum* and *Zea*, also isolated from air and soil; Australia, Brazil, British Solomon Islands Protectorate, Cuba, Ethiopia, India, Jamaica, Kenya, New Guinea, Sierra Leone, Sudan Republic, Uganda, Zambia.

Periconia circinata (Mangin) Sacc., 1906, Syll. Fung., **18**: 569.

(Fig. 237 B)

Conidiophores up to 200μ long, 6–11μ thick at the base, 7–9μ immediately below the head, characteristically circinate at the apex. *Conidia* spherical, dark brown, echinulate, 15–22μ diam.

On *Triticum* and *Sorghum*, the cause of milo disease; Europe, N. America [CMI Distribution Map 282].

Periconia macrospinosa Lefebvre & A. G. Johnson, 1949, apud Lefebvre, Johnson & Sherwin, *Mycologia*, **41**: 416–419.

(Fig. 237 C)

Conidiophores very dark brown, up to 420μ long, 7–12μ thick at the base, 6–10μ immediately below the head. *Conidia* 18–35μ diam., coarsely echinulate; the spines which are 2–7μ long sometimes adhere closely to one another in groups.

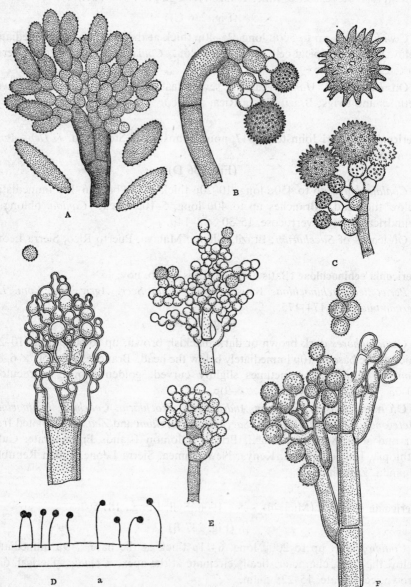

FIG. 237. *Periconia* species (2): A, *echinochloae*; B, *circinata*; C, *macrospinosa*; D, *atra*; E, *minutissima*; F, *digitata* (a, habit sketch; other figures × 650).

Isolated from *Chenopodium*, *Prunus*, *Trifolium*, *Triticum* and soil; Australia, Canada, Europe, Hong Kong, India, Iraq, U.S.A.

Periconia atra Corda, 1837, Icon. Fung., **1**: 19.
(Fig. 237 D)
This species is characterised by its rather large heads of tightly compacted conidia, and by the rings of short, fat branches which arise close together beneath the apex of the conidiophore. *Conidiophores* up to 2·5 mm. long, 15–40μ thick at the base, 9–16μ immediately below the head. *Conidia* spherical, straw-coloured when young, brown when mature, verruculose, 5–9μ diam.
On dead leaves of grasses, rushes and sedges; Europe.

Periconia minutissima Corda, 1837, Icon. Fung., **1**: 19.
(Fig. 237 E)
Conidiophores up to 550μ long, 8–14μ thick at the base, 5–10μ immediately below the head; branches at first adpressed, later spreading. *Conidia* spherical, straw-coloured to pale brown, verruculose, mostly 4–6 (occasionally 7)μ diam.
Common and widely distributed on dead stems, sticks, leaves and other plant parts, usually close to or on the ground; Cuba, Europe, Ghana, Kenya, Lebanon, New Zealand, Pakistan, Sierra Leone, Sudan, Tanzania, Zambia.

Periconia digitata (Cooke) Sacc., 1886, Syll. Fung., **4**: 274.
P. paludosa Mason & M. B. Ellis, 1953, *Mycol. Pap.*, **56**: 94–98.
(Fig. 237 F)
Conidiophores up to 660μ long, 9–15μ thick at the base, 6–9μ immediately below the head. Branches seen clearly in mature heads where the conidia are relatively loosely compacted. *Conidia* spherical, brown, verruculose to shortly echinulate, 7–11μ diam.
On dead leaves and culms of *Andropogon*, *Borassus*, *Carex*, *Chloris*, *Cladium*, *Eriophorum*, *Juncus*, *Musa*, *Oryza*, *Panicum*, *Phragmites*, *Sorghum* and *Zea*; Europe, India, Israel, Java, Kenya, Malawi, Pakistan, Sabah, Sierra Leone, U.S.A.

Periconia curta (Berk.) Mason & M. B. Ellis, 1953, *Mycol. Pap.*, **56**: 98–104.
(Fig. 238 A)
This species is characterised by the very large heads of conidia and the complex system of rough-walled branches inside the heads. *Conidiophores* 100–330μ long, 11–15μ thick immediately above the swollen base, 9–14μ just below the head. Heads of conidia up to 450μ diam. *Conidia* spherical, brown, shortly echinulate, 6–8μ diam.
On *Carex*, *Cladium* and *Juncus*; Europe.

Periconia state of **Didymosphaeria igniaria** Booth, 1968, *Trans. Br. mycol. Soc.*, **51**: 803–805.
P. igniaria Mason & M. B. Ellis, 1953, *Mycol. Pap.*, **56**: 104–108.
(Fig. 238 B)

Distinguished from *P. curta* by its longer, narrower conidiophores, shorter branches, much more strongly echinulate conidia and, in culture, by the production of rose madder or wine-coloured pigments. *Conidiophores* up to 550μ long, 9–13μ thick at the base, 6–10μ immediately below the head. *Conidia* spherical, brown, 7–10μ diam. with spines approximately 1μ long.

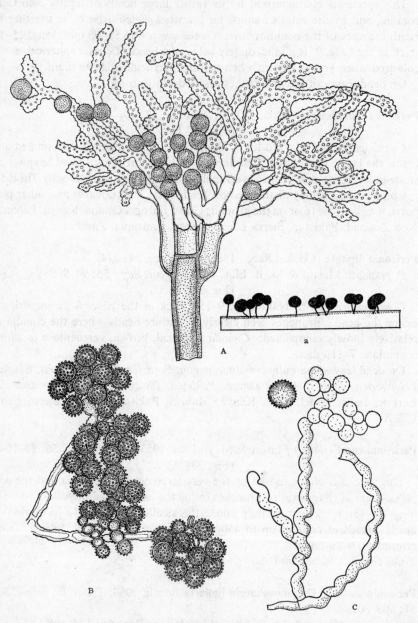

FIG. 238. *Periconia* species (3): A, *curta*; B, state of *Didymosphaeria igniaria*; C, *funerea* (a, habit sketch; other figures × 650).

On *Ammophila, Artemisia, Borassus, Cajanus, Carex, Citrus, Dactylis, Lyco-persicon, Phalaris, Phragmites, Saccharum, Spartina* and *Zea*, often following burning, and isolated from air, soil and copper sulphate solution; Australia, Europe, Ghana, India, Jamaica, U.S.A., Zambia.

Periconia funerea (Ces.) Mason & M. B. Ellis, 1953, *Mycol. Pap.*, **56**: 117–120.
(Fig. 238 C)

Characterised by its spirally twisted conidiophores which are up to 260μ long, 3–4·5μ thick at the base, 4–6μ at the apex. *Conidia* spherical, dark brown, shortly echinulate, 8–12μ diam.

On *Carex* and *Juncus*; Europe.

Periconia lateralis Ellis & Everh., 1886, *J. Mycol.*, **2**: 104.
Harpocephalum dematioides Atkinson, 1897, *Bull. Cornell Univ.*, **3** (1): 41.
P. obliqua Subram., *J. Indian bot. Soc.*, **34**: 348.
(Fig. 239 A)

Characterised by the formation of conidia below the sterile, setiform apex and on one side only of the conidiophore. *Conidiogenous cells* are borne directly on the stipe and on unilateral branches which sometimes have sterile, setiform apices. *Conidiophores* often curved, subulate, brown, up to 360μ long, 10–16μ thick at the base, 7–11μ just above the basal swelling, tapering to 1·5–6μ at the apex. *Conidia* spherical, rather pale brown, verruculose or shortly echinulate, 8–15 (10·4)μ diam.

On *Bambusa, Chloris, Cinna, Coriandrum, Cynodon, Digitaria, Heeria, Musa, Oryza, Panicum, Pappophorum*, and isolated from soil; Ghana, Guinea, Hong Kong, India, Kenya, Mauritius, Pakistan, Sierra Leone, Sudan, Uganda, U.S.A., Venezuela, Zambia.

Periconia hispidula (Pers. ex Pers.) Mason & M. B. Ellis, 1953, *Mycol. Pap.*, **56**: 112–117.
(Fig. 239 B)

Conidiophores close together forming almost black, bristly mats; they are subulate, up to 900μ long, 8–12μ thick at the base, 5–8μ immediately above the basal swelling, tapering to 2·5–4μ at the apex, and the sterile setiform part is usually longer than the fertile lower part. *Conidia* spherical, brown, thick-walled, shortly echinulate, 10–16μ diam.

On *Arundo, Carex, Deschampsia, Festuca, Glyceria, Phalaris, Phragmites, Sesleria* and *Typha*; Europe.

Periconia atropurpurea (Berk. & Curt.) Litvinov, 1967, Opredelitel' mikro-skopieuskikh Pouvennȳkh Gribov, Moniliales, Aspergillaceae: 147.
(Fig. 239 C)

Colonies pulvinate, dark chocolate to black, often staining the substratum reddish purple and producing a similar pigment in culture. *Conidiophores* up to 260μ long, 3–4μ thick at the base, 4–8μ at the apex, upper sterile part much shorter than the fertile lower part and often curved or bent over like the handle of an umbrella. *Conidia* spherical, dark reddish brown, verruculose to shortly echinulate, 8–10µ diam.

On *Andropogon, Arundinaria, Areca, Borassus, Citrus, Cynodon, Dahlia, Eragrostis, Lantana, Nitraria, Panicum, Punica, Saccharum, Sporobolus* and

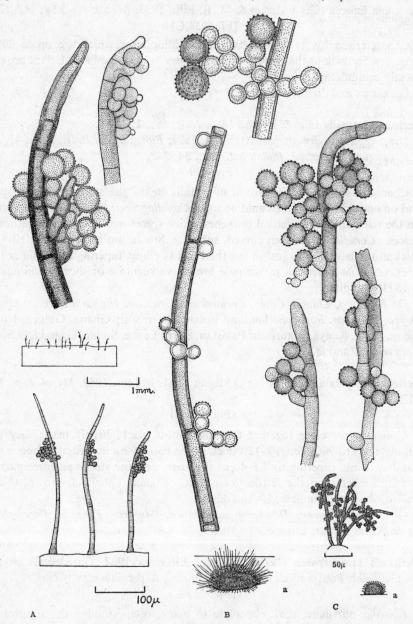

FIG. 239. *Periconia* species (4): A, *lateralis*; B, *hispidula*; C, *atropurpurea* (a, habit sketches, other figures × 650 except where indicated by scales).

Vigna, also isolated from soil; Congo, Ghana, India, Israel, New Caledonia, Nigeria, Pakistan, Philippines, Sierra Leone, S. Africa, U.S.A. and Zambia.

210. HAPLOBASIDION

Haplobasidion Eriksson, 1889, *Bot. Zbl.*, **38**: 786–787.

Colonies compact or effuse, brown or olivaceous brown, velvety or hairy. *Mycelium* immersed. *Stroma* none. *Setae* and *hyphopodia* absent. *Conidiophores* macronematous, mononematous, unbranched, straight or flexuous, subhyaline to brown, smooth. Each conidiophore terminates in a vesicle or ampulla the lower part of which is dark brown; the upper part, which frequently becomes flattened or invaginated, is much paler. Conidiophores often proliferate percurrently forming further ampullae at higher levels. *Conidiogenous cells* polyblastic, usually discrete, determinate, spherical, subspherical, ellipsoidal or clavate, covering the upper surface of the ampulla but sometimes integrated, terminal, the ampulla itself then being the conidiogenous cell. When conidia and conidiogenous cells become detached from the ampulla they sometimes leave annular scars on its surface. *Conidia* catenate, chains simple or branched, dry, borne on the rounded upper part of the conidiogenous cell, simple, spherical or subspherical, subhyaline to brown, verruculose, 0-septate.

Type species: Haplobasidion thalictri Erikss.

Ellis, M. B., *Mycol. Pap.*, **67**: 1–6, 1957.

KEY

Conidiogenous cells ellipsoidal, clavate or subspherical	*thalictri*
Conidiogenous cells when present spherical or subspherical	1
1. Conidia 7–13µ diam.	*lelebae*
Conidia 4–6µ diam.	*musae*

Haplobasidion thalictri Erikss., 1889, *Bot. Zbl.*, **38**: 786–787.

Oedemium thalictri Jaap, 1905, *Annls mycol.*, **3**: 401.

H. pavoninum Höhnel, 1905, *Annls mycol.*, **3**: 407.

(Fig. 240 A)

Colonies amphigenous, olivaceous brown, effuse. *Conidiophores* 25–60 × 4–9µ, ampullae 10–18µ diam. *Conidiogenous cells* ellipsoidal, clavate or subspherical, 7–25 × 4–10µ or 5–10µ diam. *Conidia* subhyaline to pale golden brown, sparsely verruculose, 5–10µ diam.

On leaves of *Thalictrum* and *Aquilegia* causing well-defined buff spots up to 2 cm. across, often with a broad purple border; Europe, India.

Haplobasidion lelebae [Sawada ex] M. B. Ellis, 1957, *Mycol. Pap.*, **67**: 4.

(Fig. 240 B)

Colonies hypophyllous, ellipsoidal, 0·5–1·5 mm. long, brown. *Conidiophores* 90–150 × 5–10µ, ampullae 11–15µ diam. *Conidiogenous cells* 7–10µ diam. *Conidia* brown, verruculose, 7–13µ diam.

On leaves of *Bambusa*; Formosa.

Haplosbasidion musae M. B. Ellis, 1957, *Mycol. Pap.*, **67**: 5.
(Fig. 240 C)

Colonies hypophyllous, ellipsoidal or round, olivaceous brown, velvety. *Conidiophores* 50–110 × 4–6μ, ampullae 9–12μ diam. *Conidiogenous cells* spherical, 4–8μ diam. *Conidia* brown, verruculose, 4–6μ diam.

On living leaves of *Musa* causing diamond-shaped, white, grey or pale brown spots each with a dark purple border, the spots often very pale on the upper, darker on the lower surface; Fiji, Malaya, W. Samoa.

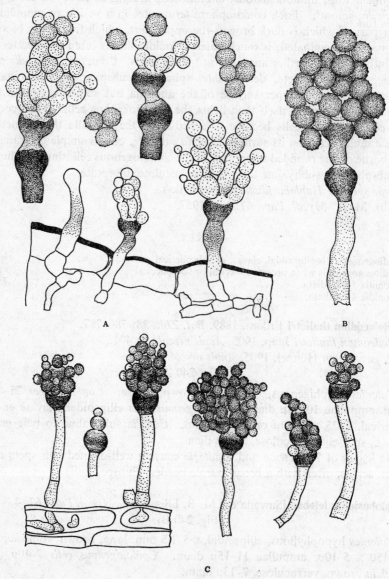

FIG. 240. *Haplobasidion* species: A, *thalictri*; B, *lelebae*; C, *musae* (× 650).

211. LACELLINOPSIS

Lacellinopsis Subramanian, 1953, *Proc. Indian Acad. Sci.*, Sect. B, **37**: 100–105.

Colonies small and discrete or effuse, grey, olive, brown or black. *Mycelium* mostly immersed. *Stroma* small, brown; sometimes flat plates of cells are formed. *Setae* unbranched or occasionally branched, straight or flexuous, subulate, mid to dark brown, paler near the apex, smooth. *Hyphopodia* absent. *Conidiophores* macronematous, mononematous, often caespitose, unbranched, straight or flexuous, in one species torsive, pale to mid brown, smooth with terminal and often also intercalary ampullae which may bear conidiogenous cells or themselves function as conidiogenous cells; the lower part of the ampulla is dark and non-fertile, the upper part pale and fertile. *Conidiogenous cells* polyblastic, integrated, terminal and intercalary or discrete and spread over the upper surface of the ampulla, spherical, determinate or when integrated often percurrent and

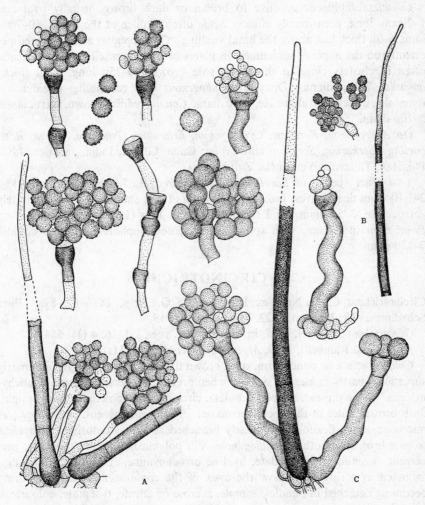

Fig. 241. *Lacellinopsis* species: A, *sacchari*; B, *levispora*; C, *spiralis* (× 650).

sometimes becoming calyciform. *Conidia* dry, catenate, simple, spherical or subspherical, brown, smooth or verruculose, 0-septate.

Type species: Lacellinopsis sacchari Subram.

Ellis, M. B., *Mycol. Pap.*, **67**: 6–10, 1957.

KEY

Conidiophores torsive . . . . *spiralis*
Conidiophores straight or flexuous 1
1. Conidia 5–10μ diam. . . . . *sacchari*
   Conidia 3–5μ diam. . . . . *levispora*

Lacellinopsis sacchari Subram., 1953, *Proc. Indian Acad. Sci.*, Sect. B, **37**: 100–105.

(Fig. 241 A)

Colonies amphigenous, olive to brown or dark brown, usually oval and 1–2 mm. long, occasionally effuse. *Setae* often swollen at the base, 150–700μ long, 6–9μ thick just above the basal swelling. *Conidiophores* arising sometimes directly on the hyphae, sometimes on plates of cells or small brown stromata, often in clusters close to the setae, pale brown, 20–75μ long, 3–5μ thick; *ampullae* 7–10μ diam. Discrete *conidiogenous cells* resembling conidia but remaining attached to ampullae, 4–8μ diam. *Conidia* reddish brown, verruculose, 5–10μ diam.

On *Andropogon, Borassus, Cymbopogon, Erianthus, Gynerium, Oryza, Rottboellia, Saccharum, Sorghum* and *Typha*; Cuba, Ghana, India, Jamaica, Java, Pakistan, Tanzania, Venezuela, Zambia.

Two other species of *Lacellinopsis* are figured: **L. levispora** Subram. (Fig. 241 B) from dead leaves, India, with conidia at first smooth but later minutely verruculose, 3–5μ diam.; and **L. spiralis** M. B. Ellis (Fig. 241 C) collected on *Pennisetum* in Ghana, with spirally twisted conidiophores, and the conidia 7–12μ diam.

212. CIRCINOTRICHUM

Circinotrichum C. G. Nees ex Persoon; C. G. Nees, 1816–17, Syst. Pilze Schwämme: 18; Persoon, 1822, Mycol. eur., **1**: 19.

Gyrotrichum Sprengel, 1827, in Linné's Syst. veg., Ed. 16, **4** (1): 554.

Dephilippia Rambelli, 1959, *Mycopath. Mycol. appl.*, **11**: 136–138.

Colonies effuse or punctiform, dark brown to black, velvety. *Mycelium* partly superficial, partly immersed. *Stroma* when present, immersed, pseudoparenchymatous. *Setae* present, simple, subulate, circinate and sometimes also straight, dark brown, paler at the apex, verrucose. *Hyphopodia* absent. *Conidiophores* micronematous, flexuous, irregularly branched and anastomosing, subhyaline to pale brown, smooth. *Conidiogenous cells* polyblastic, discrete, solitary, percurrent, lageniform or subulate, hyaline or subhyaline. *Conidia* solitary, dry, arranged in a ring just below the apex of the conidiogenous cell and often becoming detached in bundles, simple, acerose or falcate, 0-septate, colourless, smooth.

Type species: Circinotrichum maculiforme C. G. Nees ex Pers.
Pirozynski, K. A., *Mycol. Pap.*, **84**: 3–8, 1962.

Circinotrichum maculiforme C. G. Nees ex Persoon, 1822, Mycol. eur., **1**: 19.
(Fig. 242)

Setae circinate, 90–180μ high, 3–4μ thick just above the swollen base, tapering
to 1·5–2μ. *Conidiogenous cells* mostly 8–12μ long, 3–4·5μ thick at base, 1·5–2μ
at the apex. *Conidia* acerose, straight or slightly curved, 9–17 (mostly 12–15) ×
1–1·5μ.

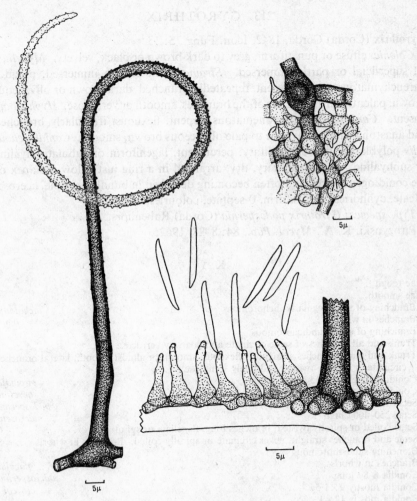

FIG. 242. *Circinotrichum maculiforme* (del. K. Pirozynski).

On twigs, Czechoslovakia; on fallen leaves of *Platanus*, France, Switzerland,
U.S.A.
Pirozynski (1962) described and illustrated 2 other species:

(1) **Circinotrichum olivaceum** (Speg.) Pirozynski on fallen leaves of avocado, coconut, *Drimys*, *Pinus longifolia* and dead stems of *Bignonia magnifica* from Africa, Asia and S. America with circinate setae 35–75µ high, 2·5–3·5µ thick near the base tapering to 1µ, *conidiogenous cells* 5–8 × 2–4µ and cylindrical to fusiform, straight or slightly curved conidia 8·5–13 (mostly 10–12) × 1·3–1·6µ.

(2) **Circinotrichum falcatisporum** Pirozynski on fallen leaves of *Trichilia heudelotii* from Ghana with circinate setae 75–125µ high, 4·5–5µ thick near the base, tapering to 1–1·5µ and straight setae 250–350µ long, *conidiogenous cells* 6–12 × 3–5µ, *conidia* falcate, 17·5–21 × 1·5–1·8µ.

213. GYROTHRIX

Gyrothrix (Corda) Corda, 1842, Icon. Fung., **5**: 13.

Colonies effuse or punctiform, grey to dark brown or black, velvety. *Mycelium* all superficial or partly immersed. *Stroma* when present immersed, pseudo-parenchymatous. *Setae* present, repeatedly branched, dark brown or olivaceous brown, paler towards the ends of the branches, smooth or verrucose. *Hyphopodia* absent. *Conidiophores* micronematous, repent, flexuous, irregularly branched and anastomosing, subhyaline to pale olivaceous brown, smooth. *Conidiogenous cells* polyblastic, discrete, solitary, percurrent, lageniform or subulate, hyaline or subhyaline. *Conidia* solitary, dry, arranged in a ring just below the apex of the conidiogenous cell and often becoming detached in bundles, simple, acerose, falcate, cylindrical or fusiform, 0-septate, colourless, smooth.

Type species: Gyrothrix podosperma (Corda) Rabenhorst.

Pirozynski, K. A., *Mycol. Pap.*, **84**: 8–28, 1962.

KEY

Setae rough	1
Setae smooth	5
1. Branching of setae regularly dichotomous	*dichotoma*
Branches in whorls	*inops*
Branching of setae subdichotomous	2
2. Trunk and all branches of setae circinate and coarsely verrucose . . .	3
Trunk and main branches straight or flexuous, smooth or almost smooth, lateral branches circinate or spirally twisted and finely spinulose	4
3. Conidia straight, 13–15µ long	*circinata*
Conidia falcate, 7–10µ long	*thevetiae*
4. Setae 120–260µ high	*podosperma*
Setae 250–400µ high	*macroseta*
5. Setae coiled or spirally twisted, branches long, flagellate or spiral . . .	6
Setae and branches straight, never circinate or spirally coiled. Branches in whorls .	8
6. Branching subdichotomous	7
Branches in whorls	*flagella*
7. Conidia 6–8µ long	*microsperma*
Conidia mostly 9 × 1·5µ	*citricola*
Conidia mostly 12 × 1	*grisea*
8. Setae stout, 70–120µ high, branches erect	*hughesii*
Setae slender, 150–300µ high, branches horizontal . . .	*verticillata*

Gyrothrix podosperma (Corda) Rabenhorst, 1844, Krypt.-Fl., **1**: 72.

(Fig. 243 A)

Colonies dark reddish brown, up to 5 mm. diam. *Setae* 4–6 times subdichotomously branched, 120–260μ high, trunk 3·5–4·5μ thick just above the basal swelling. *Conidiogenous cells* 6–14μ long, 3–4μ thick at the base, tapering to 1μ. *Conidia* straight or slightly curved, 8–16 × 1·5–2μ.

On dead leaves and stems of *Carex*, *Eugenia*, *Mangifera*, *Persea*, *Platanus* and *Sambucus*; America, Asia, Europe.

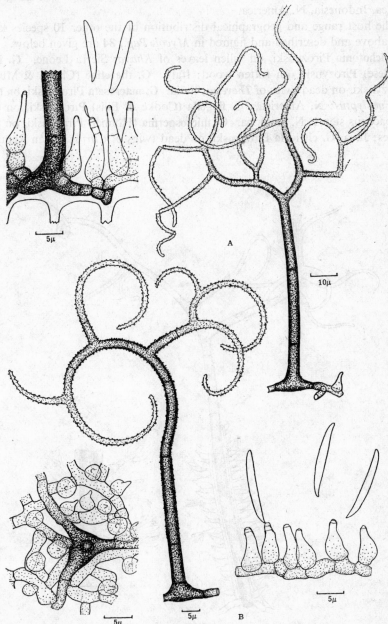

FIG. 243. A, *Gyrothrix podosperma*; B, *G. circinata* (del. K Pirozynski).

Gyrothrix circinata (Berk. & Curt.) Hughes, 1958, *Can. J. Bot.*, **36**: 771.
(Fig. 243 B)

Colonies dark brown to black, 1–5 mm. diam. *Setae* 80–140μ high, trunk 4μ thick just above the basal swelling. *Conidiogenous cells* 8μ long, 3–4μ thick at the base. *Conidia* straight or slightly curved, 12–15 × 1·5–1·8μ.

On fallen leaves of *Artocarpus*, *Carapa*, *Cocos*, *Crataegus* and *Magnolia*; Africa, Indonesia, N. America.

The host range and geographical distribution of the other 10 species keyed out above and described and figured in *Mycol. Pap.*, **84** are given below.

G. dichotoma Pirozynski, on fallen leaves of *Khaya*; Sierra Leone. **G. inops** (Berlese) Pirozynski, on rotten wood; Italy. **G. thevetiae** (Chona & Munjal) Pirozynski, on dead twigs of *Thevetia*; India. **G. macroseta** Pirozynski, on twigs of *Eucalyptus*; N. America. **G. flagella** (Cooke & Ellis) Pirozynski, on dead herbaceous stems; N. America. **G. microsperma** (Höhnel) Pirozynski, on dead leaves; Java. **G. citricola** Pirozynski, on dead twigs of *Citrus*; Sudan Republic.

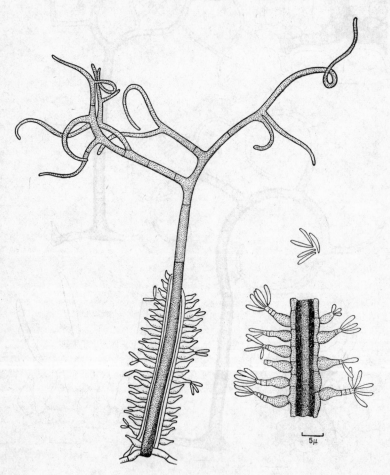

Fɪɢ. 244. *Ceratocladium microspermum* (× 650 except where indicated by the scale).

G. grisea Pirozynski, on leaves; India. **G. hughesii** Pirozynski, on *Ananas*, *Axonopus*, *Borassus*, *Musa* and *Prosopis*; Ghana, Pakistan, Sierra Leone, Sudan Republic. **G. verticillata** Pirozynski, on dead stems of *Urtica* and leaves of *Khaya*; England, Sierra Leone.

214. CERATOCLADIUM

Ceratocladium Corda, 1839, Pracht-Fl.: 41–42.

Colonies effuse, velvety, pale olive brown to dark brown. *Mycelium* superficial and immersed. *Stroma* none. *Setae* present, erect, repeatedly branched at the apex; stipe dark brown, smooth, septate; branches flexuous, tapering, pale at the tips. *Hyphopodia* absent. *Conidiophores* micronematous, narrow, branched, pale brown, smooth, encasing the lower part of setae. *Conidiogenous cells* discrete, numerous, lateral, solitary, polyblastic, percurrent, ampulliform or lageniform. *Conidia* solitary, formed in a ring around the apex of the conidiogenous cell, simple, cylindrical with rounded ends or fusiform, straight or curved, colourless, 0-septate, smooth.

Type species: C. microspermum Corda.
Hughes, S. J., *Mycol. Pap.*, **47**: 5–8, 1951.

Ceratocladium microspermum Corda, 1839, Pracht-Fl.: 41–42.
(Fig. 244)

Hyphae very pale brown, 1–3μ thick. *Setae* up to 300μ long, stipe 4–5μ thick, branches tapering to 1μ. *Conidiogenous cells* 6–11μ long, 3–4μ thick in the broadest part, tapering to 1μ. *Conidia* 4–7 × 0·5–1μ.

On wood and inside bark, especially of beech (*Fagus*); Czechoslovakia, Germany, Great Britain.

215. PSEUDOBOTRYTIS

Pseudobotrytis Krzemieniewska & Badura, 1954, *Acta Soc. Bot. Pol.*, **23**: 761–762.

Umbellula Morris, 1955, *Mycologia*, **47**: 603.

Colonies effuse, velvety, brown or greyish brown in the centre, often with a broad yellow margin. *Mycelium* immersed; rows of dark brown, spherical or ellipsoidal *chlamydospores* are sometimes formed in the cultures. *Stroma* none. *Setae* and *hyphopodia* absent. *Conidiophores* macronematous, mononematous, stipe unbranched, straight or flexuous, swollen at the base and at the apex, pale to dark brown, smooth. *Conidiogenous cells* polyblastic, discrete, umbellately arranged over the swollen apex of the stipe, determinate, clavate, denticulate; denticles cylindrical. *Conidia* solitary, dry, formed on denticles which cover the swollen tip of each conidiogenous cell, simple, oblong rounded at the ends or ellipsoidal, pale brown, smooth, 1-septate; hilum slightly protuberant.

Type species: Pseudobotrytis terrestris (Timonin) Subram. = *P. fusca* Krzemieniewska & Badura.

Pseudobotrytis terrestris (Timonin) Subram., 1956, *Proc. Indian Acad. Sci.*, Sect. B, **43**: 277.

    Spicularia terrestris Timonin, 1940, *Can. J. Res.*, **18** (C): 314–315.

    Umbellula terrestris (Timonin) Morris, 1955, *Mycologia*, **47**: 603–605.

(Fig. 245)

Chlamydospores 3–6μ diam. or 6–7 × 3–4μ. *Conidiophores* up to 500μ long, 4–5μ thick just above the basal swelling. *Conidiogenous cells* in umbels of 6–12, 14–23μ long, 2–3μ thick, swollen to 4–6μ at the apex. *Conidia* 7–9 × 3–3·5μ.

    On dead stems of *Brassica*, wood of *Fagus* and isolated from *Saccharum* and soil; Congo, Europe, Jamaica, New Zealand, N. America.

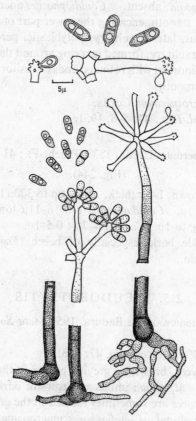

FIG. 245. *Pseudobotrytis terrestris* (× 650 except where indicated by the scale).

216. WIESNERIOMYCES

Wiesneriomyces Koorders, 1907, *Bot. Untersuch.*: 246–247.

    Colonies effuse, consisting of scattered, pulvinate sporodochia, each with a brown stromatic base supporting hyaline conidiophores and a hemispherical golden yellow, slimy conidial mass encircled by curved, dark brown setae. *Mycelium* immersed. *Stroma* erumpent or superficial, brown. *Setae* simple, long, inwardly curved, subulate, swollen at the base, acutely pointed at the apex, septate, brown, smooth. *Hyphopodia* absent. *Conidiophores* macronematous, arising close to one another and forming sporodochia, narrow, branched, at the apex, straight or flexuous, hyaline, smooth. *Conidiogenous cells* formed usually in threes at the end of short branches, polyblastic, discrete, determinate, clavate or cylindrical, usually with 2 or 3 terminal protuberances or denticles on

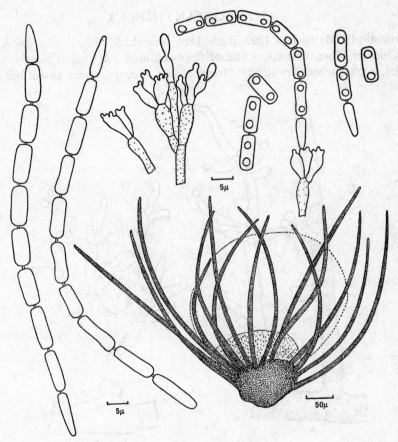

Fig. 246. *Wiesneriomyces javanicus.*

which the chains of conidia are borne. *Conidia* formed in acropetal chains but becoming aggregated in slimy golden yellow masses, individually colourless, smooth, 0-septate; the conidium at each end of a chain tapered, intermediate ones more or less cylindrical. The conidia remain attached to one another for a long time by narrow isthmi or connectives and secede only with difficulty.

Type species: Wiesneriomyces javanicus Koorders.

Maniotis, J. & Strain, J. W., *Mycologia*, **60**: 203–208, 1968.

Wiesneriomyces javanicus Koorders, 1907, Bot. Untersuch.: 246–247.

(Fig. 246)

Setae up to 600μ long, 9–15μ thick at the base. Conidiophores up to 50 ×
2–3μ. Conidiogenous cells 8–12 × 3–4μ. Conidia in chains of up to 15, mostly
10–12 × 3–4·5μ.

On rotten leaves and twigs; Java. This species is probably widely distributed
in the tropics but some of the collections seen have much smaller conidia and
may be different.

217. COSTANTINELLA

Costantinella Matruchot, 1892, Rech. Dév. Mucéd.: 97.

Colonies effuse, cottony, white or fawn-coloured. Mycelium mostly superficial, hyphae often very thick. Stroma absent. Separate setae absent but ends

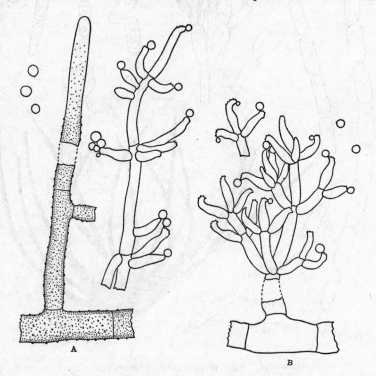

FIG. 247. A, *Costantinella terrestris*; B, *C. micheneri* (× 650).

of conidiophores in one species are long, sterile and setiform. Hyphopodia
absent. Conidiophores macronematous or semi-macronematous, mononematous,
in one species with sterile setiform apex, straight or flexuous, verticillately or
irregularly branched, colourless or yellowish brown, smooth or verrucose.
Conidiogenous cells polyblastic, discrete, arranged in verticils, sympodial, lageniform or subcylindrical, denticulate; denticles small, tapered, restricted to a small
region at the apex which is recurved and resembles a crest or cock's comb.
Conidia solitary at ends of denticles, dry, acropleurogenous, simple, spherical
or subspherical, colourless, smooth, 0-septate.

Type species: Costantinella terrestris (Link ex S. F. Gray) Hughes = *C. cristata* Matr.

Nannfeldt, J. A. & Eriksson, J., *Svensk bot. Tidskr.*, **46**: 109–128, 1952; a full and well illustrated account in English.

Costantinella terrestris (Link ex S. F. Gray) Hughes, 1958, *Can. J. Bot.*, **36**: 758, with full synonymy.

Stachylidium terrestre Link ex S. F. Gray; Link, 1809, *Magazin Ges. naturf. Freunde Berl.*, **3**: 15; S. F. Gray, 1821, Nat. Arr. Br. Pl., **1**: 553.

Costantinella cristata Matr., 1892, Rech. Dév. Mucéd.: 97.

C. tilletii (Desm.) Mason & Hughes, 1952, apud Nannf. & Erikss. in *Svensk bot Tidskr.*, **46**: 117.

(Fig. 247 A)

Colonies at first white becoming fawn-coloured or greyish brown. *Conidiophores* up to 1 mm. long, yellowish brown, verrucose and 9–18μ thick near base, hyaline, smooth and 4–15μ thick above; often with a sterile setiform apex. *Conidiogenous cells* 10–25 × 3–7μ, colourless. *Conidia* 3·5–4·5μ diam.

Common on decaying wood and other plant debris, less frequent on living or withering plants and bare soil; Europe and N. America.

Costantinella micheneri (Berk. & Curt.) Hughes, 1953, *Can. J. Bot.*, **31**: 605.

C. athrix Nannf. & Erikss., 1952, *Svensk bot. Tidskr.*, **46**: 122.

(Fig. 247 B)

Colonies white, loose, cottony. *Conidiophores* up to 250μ long, 8–10μ thick at base, tapering to 3·5–4μ, colourless, smooth; no setiform apex. *Conidiogenous cells* 8–17 × 3–5μ. *Conidia* 3·5–4·5μ diam.

On wood debris, rarely on bare soil; Europe, N. America. Not dematiaceous but included for comparison with *C. terrestris*.

218. HAPLOGRAPHIUM

Haplographium Berkeley & Broome, 1859, *Ann. Mag. nat. Hist.*, III, **3**: 360.

Colonies effuse, dark brown to black, with numerous raised, spherical, hyaline or cinnamon, glistening slimy heads clearly visible under a binocular dissecting microscope. *Mycelium* immersed. *Stroma* none. *Setae* and *hyphopodia* absent. *Conidiophores* macronematous, mononematous, erect, branched at the apex forming stipe and head; stipe straight or flexuous, often swollen at the base, smooth, mid to very dark brown except at the apex which is subhyaline, often inflated and bears several primary branches which themselves bear secondary branches arranged penicillately. *Conidiogenous cells* polyblastic, discrete, borne at the ends of secondary branches, determinate, cylindrical. *Conidia* aggregated in slimy heads, acropleurogenous, simple, cylindrical, rounded at the ends or ellipsoidal, hyaline, smooth, 0-septate.

Type species: Haplographium state of *Hyaloscypha dematiicola* (Berk. & Br.) Nannf. = *H. delicatum* Berk. & Br.

Hughes, S. J., *Can. J. Bot.*, **31**: 588–589, 1953.

Mason, E. W., *Mycol. Pap.*, **3**: 61–63, 1933.

Haplographium state of **Hyaloscypha dematiicola** (Berk. & Br.) Nannf., 1936, *Trans. Br. mycol. Soc.*, **20**: 205. For synonymy see Hughes, *Can. J. Bot.*, **36**: 771, 1958.

(Fig. 248)

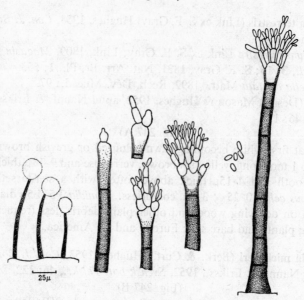

FIG. 248. *Haplographium* state of *Hyaloscypha dematiicola* (×650 except where indicated by the scale).

Conidiophores up to 250μ long, 4–7μ thick, often swollen at base and apex to 8–10μ. *Conidia* 2·5–4 × 1–2μ.

Very common on dead wood and bark; Europe.

219. VERTICICLADIELLA

Verticicladiella Hughes, 1953, *Can. J. Bot.*, **31**: 653.

Colonies effuse, brown, dark blackish brown or black. *Mycelium* partly superficial, partly immersed. *Stroma* none. *Setae* and *hyphopodia* absent. *Conidiophores* macronematous, mononematous, branched towards the apex forming a stipe and head; stipe straight or flexuous, mid to dark brown, smooth; branches arranged penicillately, usually in several series, paler than the stipe. *Conidiogenous cells* polyblastic, discrete, arranged penicillately on branches, sympodial, cylindrical or subulate. *Conidia* formed solitarily but becoming aggregated in slimy heads, acropleurogenous, simple, often curved, allantoid, clavate, cuneiform, ellipsoidal, fusiform or pyriform, individually colourless or very pale, sometimes becoming brown in mass with age, smooth, 0-septate. Perfect states where known mostly *Ceratocystis*.

Type species: Verticicladiella abietina (Peck) Hughes.

Kendrick, W. B., *Can. J. Bot.*, **40**: 771–797, 1962; an excellent illustrated account of 7 species.

Verticicladiella abietina (Peck) Hughes, 1953, *Can. J. Bot.*, **31**: 653.

(Fig. 249)

Colonies dark blackish brown. *Conidiophore* stipes up to 1300μ long, 5–12μ thick at the base, 4–6·5μ just below the head; branches usually in 2–5 series, primary branches pale brown 10–20 × 3–7μ. *Conidiogenous cells* usually subulate, hyaline, 8–25 × 1–1·5μ. *Conidia* usually curved, often clavate, 2·5–5 × 1–1·7μ, forming cream-coloured slimy heads which become brown with age.

On *Picea* and other trees; N. America.

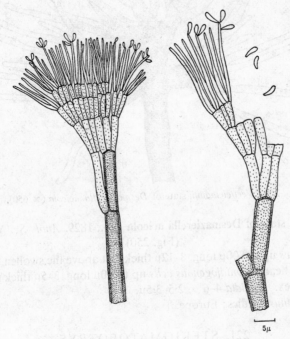

5μ

FIG. 249. *Verticicladiella abietina* (× 650 except where indicated by the scale).

220. VERTICICLADIUM

Verticicladium Preuss, 1851, *Linnaea*, **24**: 127.

Colonies effuse, greyish, hairy. *Mycelium* immersed. *Stroma* none. *Setae* and *hyphopodia* absent. *Conidiophores* macronematous, mononematous, solitary or rarely caespitose, emerging through stomata, branched towards the apex forming a stipe and head; stipe straight, often swollen at the base, mid to dark brown, smooth; primary branches in verticils and usually at right-angles to the main axis, secondary and tertiary branches also arranged verticillately. *Conidiogenous cells* polyblastic, sometimes integrated and terminal but mostly discrete on branches, often in threes, sympodial, subulate. *Conidia* solitary, dry, acropleurogenous, simple, ellipsoidal, hyaline or pale brown, minutely verruculose, 0-septate.

Type species: Verticicladium state of *Desmazierella acicola* Lib. = *V. trifidum* Preuss.

Hughes, S. J., *Mycol. Pap.*, **43**, 6–13, 1951.

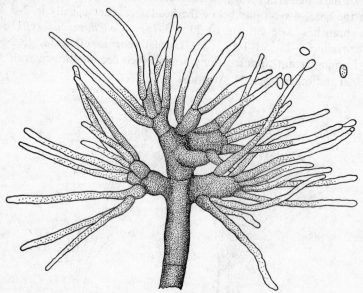

FIG. 250. *Verticicladium* state of *Desmazierella acicola* (× 650).

Verticicladium state of **Desmazierella acicola** Lib., 1829, *Annls Sc. Nat.*, **17**: 82. (Fig. 250)

Conidiophores up to 800μ long, 8–12μ thick just above the swollen base, 8–10μ just below the head. *Conidiogenous cells* up to 60μ long, 3–5μ thick at the base, 2–3μ at the apex. *Conidia* 4–6 × 2·5–3·5μ.

On rotten *Pinus* needles; Europe.

221. STERIGMATOBOTRYS

Sterigmatobotrys Oudemans, 1886, *Ned. kruidk. Archf*, Ser. 2, **4**: 548–549.
Phragmostachys Costantin, 1888, Mucéd. simpl.: 97.
Atractina Höhnel, 1904, *Hedwigia*, **43**: 298–299.

Colonies effuse, dark blackish brown, thinly hairy. *Mycelium* immersed. *Stroma* a small brown plate at the base of the conidiophore. *Setae* and *hyphopodia* absent. *Conidiophores* macronematous, mononematous, solitary, scattered, branched penicillately at the apex forming a stipe and head; stipe straight or flexuous, dark blackish brown, smooth; branches at an acute angle to the stipe, closely adpressed, hyaline or pale brown. *Conidiogenous cells* polyblastic, integrated and terminal on branches or discrete, sympodial, cylindrical to subulate. *Conidia* formed solitarily but becoming aggregated in slimy heads, acropleurogenous, simple, straight or slightly curved, broadly fusiform or subclavate, smooth, 2-septate, at first hyaline the middle cell later becoming brown and the ends cells pale brown.

Lectotype species: Sterigmatobotrys macrocarpa (Corda) Hughes = *S. elata* (Sacc.) Oud.

Sterigmatobotrys macrocarpa (Corda) Hughes, 1958, *Can. J. Bot.*, **36**: 814, with full synonymy.

(Fig. 251)

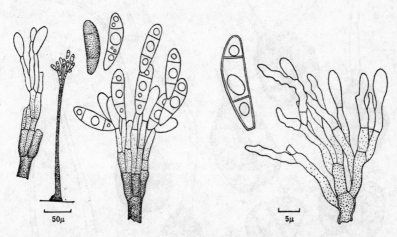

FIG. 251. *Sterigmatobotrys macrocarpa* (× 650 except where indicated by scales).

Conidiophores 170–500μ long, 4–7μ thick immediately below the head, 7·5–12μ at the base. *Conidia* 16–25 × 6–8μ.

On dead wood of *Abies, Picea, Taxus*, etc.; Europe, N. America.

222. PIRICAUDA

Piricauda Bubák, 1914, *Annls mycol.*, **12**: 217–218.

Colonies effuse, thin, arachnoid, scarcely visible to the naked eye. *Mycelium* superficial. *Stroma* none. *Setae* and *hyphopodia* absent. *Conidiophores* semi-macronematous, mononematous, branched, forming arched, sometimes anasto-mosing loops slightly thicker than the ordinary hyphae. *Conidiogenous cells* monotretic, integrated, terminal and intercalary, determinate, cylindrical or clavate, unciform, cicatrized, scars prominent. *Conidia* solitary, dry, pleuro-genous, broadly ellipsoidal, ovoid, spherical, subspherical or obpyriform, often becoming rostrate, golden brown, smooth or verruculose, muriform.

Type species: Piricauda paraguayensis (Speg.) R. T. Moore = *P. uleana* (Sacc. & Syd.) Bubák.

Hughes, S. J., *Can. J. Bot.*, **38**: 921–924, 1960.

Piricauda paraguayensis (Speg.) R. T. Moore, 1959, *Mycologia*, **50**: 691 [as ' *paraguayense* '].

(Fig. 252)

Conidiophore loops 5–7µ thick. *Conidia* with body dark golden brown, 25–42 × 20–32µ and beak hyaline or pale brown, subulate, up to 120µ long, 4–6µ thick at the base tapering to 1–1·5µ at the apex.

On leaves of *Bignonia*, *Citharexylum* and *Duranta*; Brazil, Cuba, Paraguay.

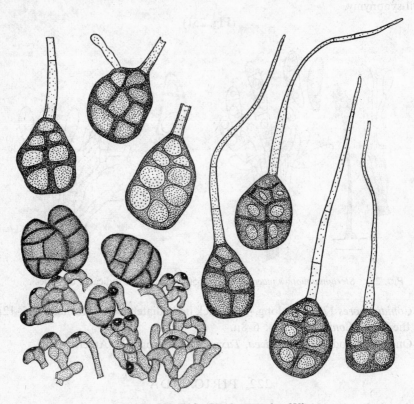

FIG. 252. *Piricauda paraguayensis* (× 650).

223. ACAROCYBELLA

Acarocybella M. B. Ellis, 1960, *Mycol. Pap.*, **76**: 5.

Colonies effuse, dark blackish brown. *Mycelium* superficial. *Stroma* none. *Setae* and *hyphopodia* absent. *Conidiophores* macronematous, pale brown or brown, with ascending and descending branches. The conidiophore commences development as an ascending or erect hypha which grows for some distance and forms a single conidium through a pore at its apex; slightly below and to one side of the apex a lateral branch develops which almost immediately branches dichotomously to form a descending branch and an ascending branch. The descending branch grows down to mingle with the vegetative mycelium, the ascending branch swells and forms a conidium at its apex; as before a lateral branch develops which branches dichotomously to form a descending and an ascending branch and the same process is often repeated several times. *Conidiogenous cells* integrated, monotretic, terminal and intercalary. *Conidia* solitary, dry, straight or curved, obclavate, rostrate, brown, smooth, with numerous transverse septa.

Type species: Acarocybella jasminicola (Hansf.) M. B. Ellis.

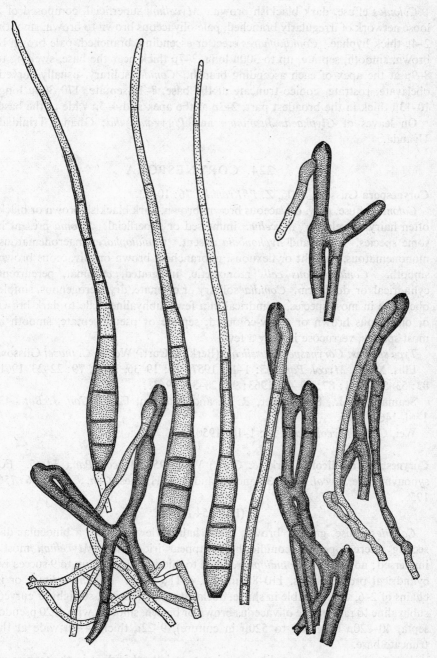

Fig. 253. *Acarocybella jasminicola* (× 650).

Acarocybella jasminicola (Hansf.) M. B. Ellis, 1960, *Mycol. Pap.*, **76**: 5.

Cercospora (?) *jasminicola* Hansford, 1944, *Proc. Linn. Soc. Lond.*, **1943–44**: 121.

(Fig. 253)

Colonies effuse, dark blackish brown. *Mycelium* superficial, composed of a loose network of irregularly branched, pale olivaceous brown to brown, smooth, 2–4µ thick hyphae. *Conidiophores* erect or ascending, branched, pale brown to brown, smooth, septate, up to 300µ long, 3–7µ thick near the base, swelling to 8–9µ at the apex of each ascending branch. *Conidia* solitary, usually curved, obclavate, rostrate, conico-truncate at the base, 7–13-septate, 170–300µ long, 10–13µ thick in the broadest part, 2–3µ at the apex and 4–5µ wide at the base.

On leaves of *Glyphaea*, *Jasminum* and *Phryganocydia*; Ghana, Trinidad, Uganda.

224. CORYNESPORA

Corynespora Güssow, 1906, *Z. PflKrankh.*, **76**: 10–13.

Colonies effuse, grey, olivaceous brown, brown, dark blackish brown or black, often hairy or velvety. *Mycelium* immersed or superficial. *Stroma* present in some species. *Setae* and *hyphopodia* absent. *Conidiophores* macronematous, mononematous, straight or flexuous, unbranched, brown or olivaceous brown, smooth. *Conidiogenous cells* monotretic, integrated, terminal, percurrent, cylindrical or doliiform. *Conidia* solitary or catenate, dry, acrogenous, simple, obclavate in most species, cylindrical in a few, subhyaline, pale to dark brown or olivaceous brown or straw-coloured, septate or pseudoseptate, smooth in most species, verrucose in only a few.

Type species: Corynespora cassiicola (Berk. & Curt.) Wei = *C. mazei* Güssow.

Ellis, M. B., *Mycol. Pap.*, **65**: 1–15, 1957; **76**: 19–36, 1960; **79**: 22–23, 1961; **82**: 53–55, 1961; **87**: 39–42, 1963; **93**: 28–31, 1963.

Seaman, W. L., Shoemaker, R. A. and Peterson, E. A., *Can. J. Bot.*, **43**: 1461–1469, 1965.

Wei, C. T., *Mycol. Pap.*, **34**: 1–10: 1950.

Corynespora cassiicola (Berk. & Curt.) Wei, 1950, *Mycol. Pap.*, **34**: 5. For synonymy, see *Mycol. Pap.*. **34** and **65** and S. J. Hughes, *Can. J. Bot.*, **36**: 756, 1958.

(Fig. 254 A)

Colonies effuse, grey or brown, thinly hairy; viewed under a binocular dissecting microscope the conidiophores appear iridescent. *Mycelium* mostly immersed; no stroma. *Conidiophores* pale to mid brown, with up to 9 successive cylindrical proliferations, 110–850µ long, 4–11µ thick. *Conidia* solitary or in chains of 2–6, very variable in shape, obclavate to cylindrical, straight or curved, subhyaline to rather pale olivaceous brown or brown, smooth, with 4–20 pseudosepta, 40–220µ long (up to 520µ in culture), 9–22µ thick, 4–8µ wide at the truncate base.

A very common cosmopolitan species, especially abundant in the tropics on a wide range of host plants; it often causes distinct leaf spots and is recognised as a pathogen on cowpea (*Vigna sinensis*), cucumber, melon, sesame, soy bean (*Glycine max*), etc. Other plants of economic importance upon which *C. cassiicola* is frequently found include cassava (*Manihot*), papaw (*Carica papaya*), rubber (*Hevea brasiliensis*) and tomato.

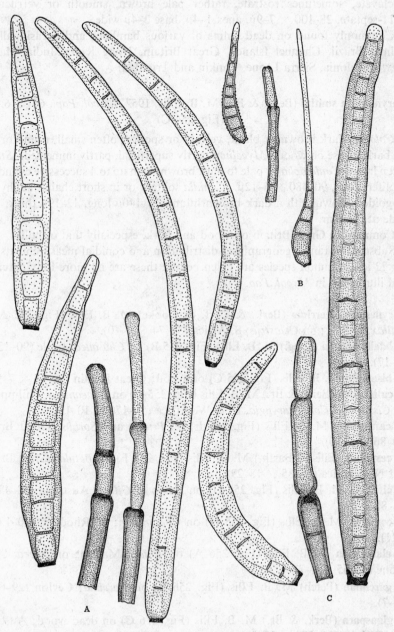

FIG. 254. *Corynespora* species (1): A, *cassiicola*; B, *foveolata*; C, *smithii* (× 650).

Corynespora foveolata (Pat.) Hughes, 1958, *Can. J. Bot.*, **36**: 757.
(Fig. 254 B)

Colonies effuse, dark chocolate brown. *Mycelium* partly superficial, partly immersed; no stroma. *Conidiophores* pale to mid brown, with up to 7 successive cylindrical proliferations, 25–420 × 4–6μ. *Conidia* solitary, straight or curved, obclavate, sometimes rostrate, rather pale brown, smooth or verruculose, 4–11-septate, 28–100 × 7–9μ, apex 1–4μ, base 3–4μ wide.

Commonly found on dead culms of various bamboos and occasionally on palms; Brazil, Channel Islands, Great Britain, Hong Kong, India, Malaya, New Caledonia, Sierra Leone, Tonkin and Trinidad.

Corynespora smithii (Berk. & Br.) M. B. Ellis, 1957, *Mycol. Pap.*, **65**: 3–6.
(Fig. 254 C)

Colonies dark brown or black, velvety or spongy, often small, round or oval on bark, effuse on wood. *Mycelium* partly superficial, partly immersed. *Stroma* often large. *Conidiophores* pale to dark brown, with up to 4 successive cylindrical proliferations, 90–480 × 6–12μ. *Conidia* solitary or in short chains, subhyaline to golden brown with a dark brown hilum, 70–140μ long, 12–19μ thick, 6–9μ wide at the scar.

Common in Great Britain on wood and bark, especially that of holly.

Substratum range, geographical distribution and conidial measurements in μ for 22 less common species are given below; these are all more fully described and illustrated in *Mycol. Pap.*

Corynespora aterrima (Berk. & Curt. ex Cooke) M. B. Ellis (Fig. 255 A) on *Smilax*, U.S.A.; on *Celastrus*, S. Africa (33–74 × 8–10).
C. bdellomorpha (Speg.) M. B. Ellis (Fig. 255 B) on *Chusquea*; Chile (90–138 × 12–17).
C. biseptata M. B. Ellis (Fig. 255 C) on wood; Great Britain (20–35 × 7–9).
C. calicioidea (Berk. & Br.) M. B. Ellis (Fig. 255 D) on *Allaeanthus*, Philippines; on *Cissus* and *Conopharyngia*, E. & W. Africa (50–170 × 10–15).
C. cambrensis M. B. Ellis (Fig. 255 E) on *Prunus* and *Sorbus*; Great Britain (20–86 × 5–10).
C. cespitosa (Ellis & Barth.) M. B. Ellis (Fig. 255 F) on *Betula*; Great Britain and N. America (55–85 × 18–29).
C. citricola M. B. Ellis (Fig. 255 G) on leaves of *Citrus*; Australia (48–480 × 4·5–8).
C. combreti M. B. Ellis (Fig. 255 H) on *Combretum*; N. Rhodesia (40–122 × 8–11).
C. elaeidicola M. B. Ellis (Fig. 256 A) on *Elaeis*, Malaya; on *Areca*, Sierra Leone (43–65 × 4–7).
C. garciniae (Petch) M. B. Ellis (Fig. 256 B) on *Garcinia*; Ceylon (29–60 × 4·5–7).
C. gigaspora (Berk. & Br.) M. B. Ellis (Fig. 256 C) on dead wood; Australia and Ceylon (100–270 × 19–28).

C. hansfordii M. B. Ellis (Fig. 256 D) on wood, Uganda; on *Nauclea*, Sierra Leone (70–100 × 9–13).

C. homaliicola Deighton & M. B. Ellis (Fig. 256 E) on *Homalium*; Sierra Leone (160–220 × 11–15).

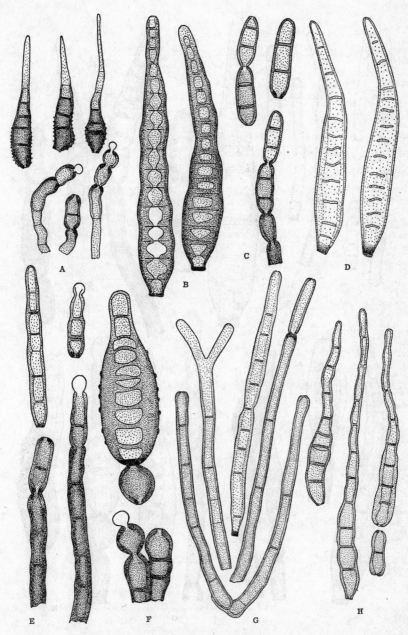

FIG. 255. *Corynespora* species (2): A, *aterrima*; B, *bdellomorpha*; C, *biseptata*; D, *calicioidea*; E, *cambrensis*; F, *cespitosa*; G, *citricola*; H, *combreti* (× 650).

C. lanneicola Deighton & M. B. Ellis (Fig. 256 F) on *Lannea*; Sierra Leone (40–58 × 10–15).

C. leptoderridicola Deighton & M. B. Ellis (Fig. 256 G) on *Leptoderris*; Sierra Leone (70–120 × 14–17).

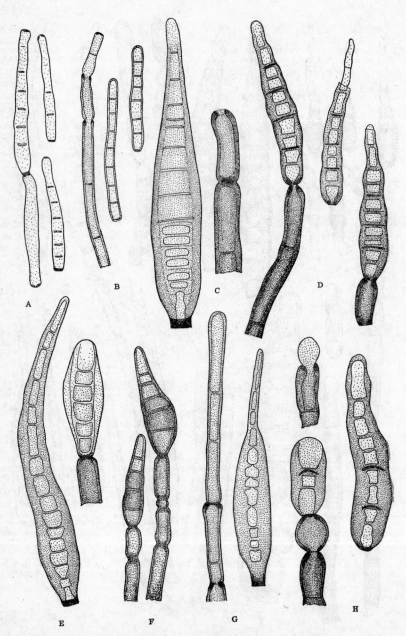

FIG. 256. *Corynespora* species (3): A, *elaeidicola*; B, *garciniae*; C, *gigaspora*; D, *hansfordii*; E, *homaliicola*; F, *lanneicola*; G, *leptoderridicola*; H, *olivacea* (× 650).

C. olivacea (Wallr.) M. B. Ellis (Fig. 256 H) on *Tilia*; Europe and N. America (50–105 × 12–19).

C. polyphragmia (Syd.) M. B. Ellis (Fig. 257 A) on *Camellia*; Japan (110–280 × 14–17).

C. pruni (Berk. & Curt.) M. B. Ellis (Fig. 257 B) on *Acer*, *Alnus*, *Magnolia* and *Prumus*, U.S.A.; on *Fagus*, Great Britain (50–130 × 10–16).

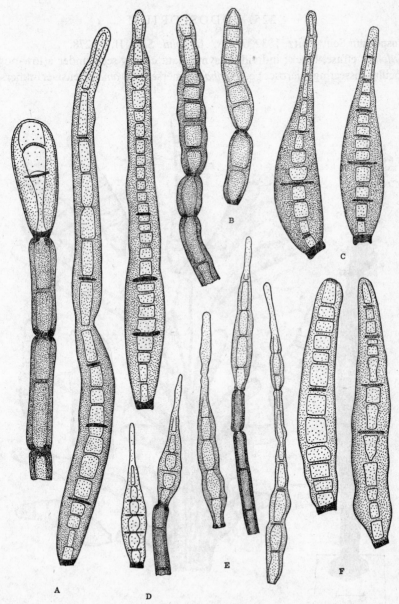

FIG. 257. *Corynespora* species (4): A, *polyphragmia*; B, *pruni*; C, *siwalika*; D, *trichiliae*; E, *vismiae*; F, *yerbae* (× 650).

C. siwalika (Subram.) M. B. Ellis (Fig. 257 C) on *Helicteres*; India (88–140 × 15–20).

C. trichiliae M. B. Ellis (Fig. 257 D) on *Trichilia*; Sierra Leone (53–74 × 9–11).

C. vismiae M. B. Ellis (Fig. 257 E) on leaves of *Vismia*; Sierra Leone (55–107 × 6–9).

C. yerbae (Speg.) M. B. Ellis (Fig. 257 F) on *Ilex*; Argentina (72–110 × 16–18).

225. PODOSPORIUM

Podosporium Schweinitz, 1832, *Trans. Am. phil. Soc.*, II, **4**: 278.

Colonies effuse, black; individual synnemata clearly seen under a low-power binocular dissecting microscope. *Hyphae* immersed. *Stroma* extensive, immersed,

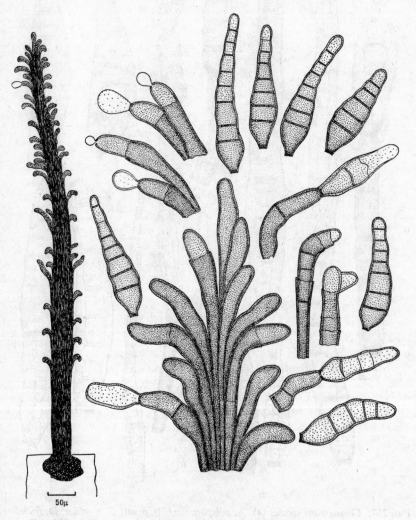

FIG. 258. *Podosporium rigidum* (× 650 except where indicated by the scale).

dark brown, pseudoparenchymatous. *Setae* and *hyphopodia* absent. *Conidiophores* macronematous, synnematous, with individual threads unbranched, brown, smooth, narrow along most of their length but markedly swollen and often sharply bent near the apex, splaying out laterally and at the top of the synnema. *Conidiogenous cells* monotretic, integrated, terminal, percurrent, rarely sympodial, clavate or cuneiform. *Conidia* solitary, dry, acrogenous, simple, obclavate, brown, smooth, multiseptate.

Lectotype species: Podosporium rigidum Schw.

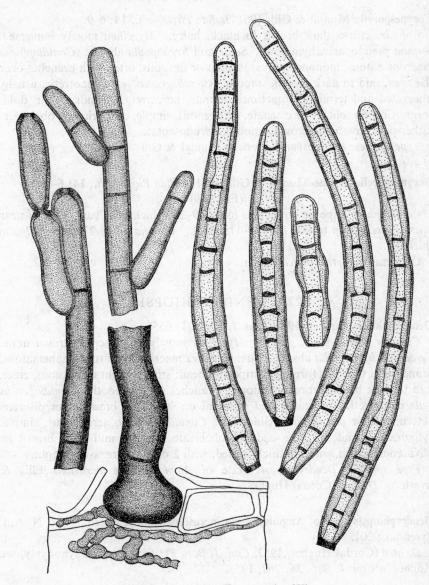

Fig. 259. *Corynesporella urticae* (× 650).

Podosporium rigidum Schw., 1832, *Trans. Am. phil. Soc.*, II, **4**: 278.
(Fig. 258)

Synnemata subulate or cylindrical, black, up to 2 mm. high, 20–80μ thick; individual threads 3–4μ thick, often bent and swollen to 8–10μ near the apex. *Conidia* 40–70μ long, 10–14μ thick in the broadest part, 3–4μ at the base.

On dead stems and branches of *Ampelopsis* and *Rhus*; U.S.A.

226. CORYNESPORELLA

Corynesporella Munjal & Gill, 1961, *Indian Phytopath.*, **14**: 6–9.
Colonies effuse, dark brown to black, hairy. *Mycelium* mostly immersed. *Stroma* pseudoparenchymatous. *Setae* and *hyphopodia* absent. *Conidiophores* macronematous, mononematous, straight or flexuous, often with branches near the apex, mid to dark brown, smooth. *Conidiogenous cells* monotretic, usually integrated and terminal, sometimes discrete, percurrent, cylindrical or dolii-form. *Conidia* solitary or catenate, acrogenous, simple, cylindrical to obclavate, rather pale olivaceous brown, smooth, pseudoseptate.

Type species: Corynesporella urticae Munjal & Gill.

Corynesporella urticae Munjal & Gill, 1961, *Indian Phytopath.*, **14**: 6–9.
(Fig. 259)

Conidiophore stipe up to 1 mm. long, 19–25μ thick near base, 11–14μ near apex; branches up to 90μ long, 9–11μ thick. *Conidia* up to 270μ long, 7–20μ thick.

On stems of *Urtica dioica*; India.

227. DENDRYPHIOPSIS

Dendryphiopsis Hughes, 1953, *Can. J. Bot.*, **31**: 655.
Colonies effuse, black, hairy. *Mycelium* mostly immersed. *Stroma* none. *Setae* and *hyphopodia* absent. *Conidiophores* macronematous, mononematous, branched at the apex forming a stipe and head; stipe straight or flexuous, erect, mid to dark blackish brown, smooth, branches paler, smooth. *Conidiogenous cells* monotretic, integrated and terminal on stipe and branches or discrete, determinate or percurrent, cylindrical. *Conidia* solitary, acrogenous, simple, cylindrical rounded at the ends or obclavate, pale to mid dark brown or olivaceous brown, smooth, thick-walled, with 2 or more transverse septa.

Type species: Dendryphiopsis state of *Amphisphaeria incrustans* Ellis & Everh. = *D. atra* (Corda) Hughes.

Dendryphiopsis state of **Amphisphaeria incrustans** Ellis & Everh., 1892, N. Am. Pyrenom.: 201.
D. atra (Corda) Hughes, 1953, *Can. J. Bot.*, **31**: 655. For full synonymy, see Hughes in *Can. J. Bot.*, **36**: 762, 1958.

(Fig. 260)

Stipes up to 500μ long, 8–11μ thick; branches 6–9μ thick. *Conidia* pale to mid-dark smoky or olivaceous brown, 40–80 × 12–25μ.

Not uncommon on wood; Europe, Kenya, New Zealand, N. America; a number of collections have been made in Great Britain on beech (*Fagus*).

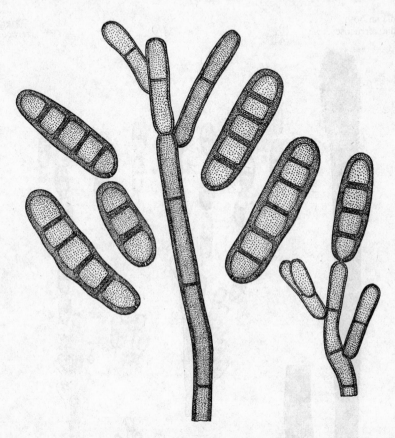

FIG. 260. *Dendryphiopsis* state of *Amphisphaeria incrustans* (× 650).

228. DIDYMOBOTRYUM

Didymobotryum Saccardo, 1886, Syll. Fung., **4**: 626–627.

Colonies effuse, olivaceous brown to black, velvety or hairy; large synnemata very distinct under the low-power binocular dissecting microscope. *Mycelium* immersed. *Stroma* usually present, immersed. *Setae* and *hyphopodia* absent. *Conidiophores* macronematous, synnematous, straight or flexuous; threads branched, with the branches restricted to the apical region forming a stipe and head, smooth or verrucose; compound stipes often black, separate threads brown or olivaceous brown. *Conidiogenous cells* monotretic, integrated and terminal or discrete, determinate, clavate or cylindrical. *Conidia* catenate, dry, acrogenous, simple, clavate, cylindrical rounded at the ends or ellipsoidal, pale to mid greyish olive, olivaceous brown or dark brown, 1-septate, in some species with a thick dark brown to black band at the septum, smooth or verrucose.

Lectotype species: Didymobotryum rigidum (Berk. & Br.) Sacc.

KEY

Conidia smooth . . . . *rigidum*
Conidia verrucose . . . . *verrucosum*

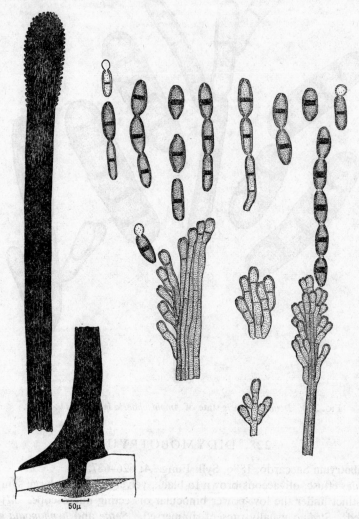

FIG. 261. *Didymobotryum rigidum* (× 650 except where indicated by the scale).

Didymobotryum rigidum (Berk. & Br.) Sacc., 1886, Syll. Fung., **4**: 627.
Periconia rigida Berk. & Br., 1873, *J. Linn. Soc.*, **14**: 99.
D. atrum Pat., 1891, *J. Bot.*, Paris, **5**: 320.

(Fig. 261)

Colonies dark blackish brown to black. *Synnemata* black, up to 2 mm. high, 30–100μ thick in the middle, expanded slightly at the apex and up to 200μ at the base; separate threads, 2–3μ thick. *Conidiogenous cells* usually clavate, swollen to 5μ. *Conidia* broadly ellipsoidal or cylindrical rounded at the ends, smooth, mid to dark brown with a very dark band at the septum, 11–17 × 4–8μ.

On bamboo and dead wood; Ceylon, Tonkin.

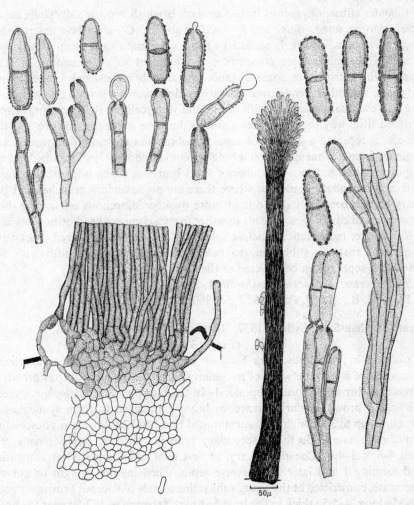

FIG. 262. *Didymobotryum verrucosum* (× 650 except where indicated by the scale).

Didymobotryum verrucosum Hino & Katumoto, 1959, *Bull. Fac. Agric. Yamaguti Univ.*, **10**: 1181.

(Fig. 262)

Colonies dark olivaceous brown. *Synnemata* up to 1 mm. high, 25–45μ thick; separate threads pale to mid olivaceous brown, 2–3μ thick. *Conidiogenous cells* usually clavate, swollen to 4–5µ. *Conidia* clavate, cylindrical rounded at the ends or ellipsoidal, pale to mid greyish olive, 15–26 × 5–7µ.

On dead wood of *Sasa*; Japan.

229. ACAROCYBE

Acarocybe H. Sydow, 1937, *Annls mycol.*, **35**: 285, emend. M. B. Ellis, 1960, *Mycol. Pap.*, **76**: 2.

Colonies effuse, olivaceous brown to dark blackish brown. *Mycelium* superficial. *Stroma* none. *Setae* and *hyphopodia* absent. *Conidiophores* macronematous, olivaceous brown, branched to form synnemata each surmounted by a head. The conidiophore commences development as an ascending or erect hypha which grows for some distance and then branches, one branch grows upwards, the other bends over and grows downwards towards the surface of the leaf, eventually mingling with the vegetative mycelium; the ascending branch grows a little way then branches again in the same manner as before and this process is repeated a number of times; the branches are closely adpressed and together form a synnema. After a time the conidiophore stops forming descending hyphae and puts out a number of short branches which often branch again and each secondary branch or where there are no secondary branches each primary branch bears at its tip one or more oval conidiogenous cells. The short branches and conidiogenous cells together form a compact head. *Conidiogenous cells* discrete, monotretic. *Conidia* solitary, dry, straight or curved, obclavate, sometimes rostrate, subhyaline to pale olivaceous brown, smooth with 1–6 transverse septa, often constricted at the septa.

Type species: Acarocybe hansfordii H. Sydow.

Ellis, M. B., *Mycol. Pap.*, **76**: 2–5, 1960; **82**: 50–51, 1961.

Acarocybe hansfordii Sydow, 1937, *Annls mycol.*, **35**: 285.
(Fig. 263 A)

Colonies hypophyllous, dark blackish brown, effuse. *Mycelium* superficial, composed of a loose network of irregularly branched, pale olivaceous brown to olivaceous brown, smooth, septate, 1–4µ thick hyphae. *Conidiophores* erect, olivaceous brown or dark olivaceous brown, branched to form synnema-like structures up to 300µ high each surmounted by a head 20–45µ diam.; descending branches septate, 3–5µ thick; secondary branches shorter; *conidiogenous cells* oval 4–6 × 3–4µ. *Conidia* solitary, at first oval and 0-septate but elongating and forming 1 and later 2 transverse septa, when mature straight or curved, obclavate, constricted at the septa, subhyaline to pale olivaceous brown, smooth, 17–35µ long, 4–5·5µ thick in the broadest part, tapering to 1–2·5µ near the apex.

On living leaves of *Bridelia micrantha*; Uganda.

A. deightonii M. B. Ellis, 1960, *Mycol. Pap.*, **76**: 4.
(Fig. 263 B)

A much smaller species than *A. hansfordii*, found on *Lychnodiscus dananensis* in Sierra Leone. *Conidia* 10–20µ long, 3–4µ thick in the broadest part, tapering to 1–2µ.

A. formosa (Batista & Bezerra) M. B. Ellis, 1961, *Mycol. Pap.*, **82**: 50.
(Fig. 263 C)

Larger than *A. hansfordii* with synnemata up to 700μ high; *conidia* 4–6-septate, 55–93 (68)μ long, 6–9 (8·2)μ thick in the broadest part, tapering to 1–2μ; originally described as *Arthrobotryum formosum* on an unknown plant; Brazil.

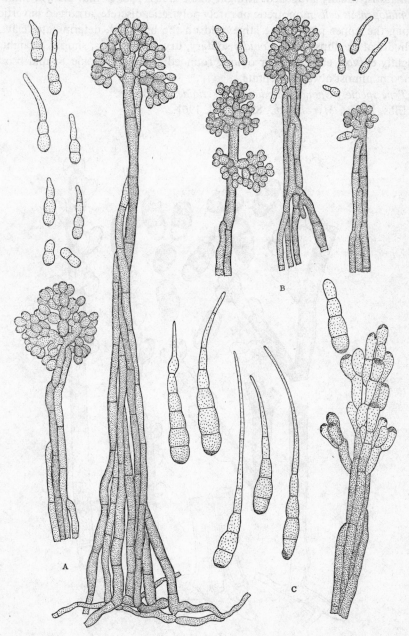

FIG. 263. *Acarocybe* species: A, *hansfordii*; B, *deightonii*; C, *formosa* (× 650).

230. SPONDYLOCLADIELLA

Spondylocladiella Linder, 1934, *Mycologia*, **26**: 437.

Colonies effuse, mid to dark brown. *Mycelium* mostly immersed. *Stroma* none. *Setae* and *hyphopodia* absent. *Conidiophores* macronematous, mononematous, usually branched, straight or flexuous, pale to mid brown, smooth. *Conidiogenous cells* monotretic or rarely polytretic, discrete, arranged in verticils along the upper part of the stipe, and on the branches, determinate, clavate, cylindrical or ellipsoidal. *Conidia* solitary, dry, acrogenous, simple, straight or slightly curved, ellipsoidal or oblong rounded at the ends, pale to mid brown, smooth, almost always 2-septate.

Type species: Spondylocladiella botrytioides Linder.

Ellis, M. B., *Mycol. Pap.*, **82**: 43–44, 1961.

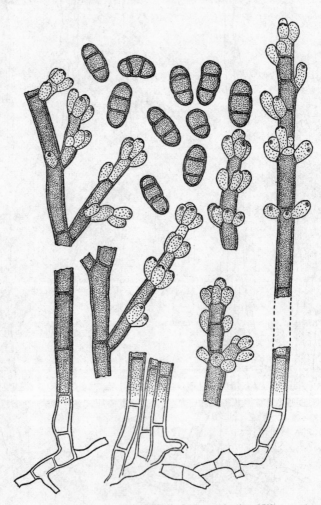

FIG. 264. *Spondylocladiella botrytioides* (× 650).

Spondylocladiella botrytioides Linder, 1934, *Mycologia*, **26**: 437.
(Fig. 264)

Conidiophores pale to mid brown except near the base where they are hyaline or subhyaline, up to 400μ long, 6–8μ thick. *Conidiogenous cells* pale brown, smooth, 7–14 × 4–6μ. *Conidia* sometimes constricted at the septa, 13–20 (18) × 8–10 (8·7)μ.

On resupinate hymenomycetes (*Corticium*); Europe, N. America.

231. DICHOTOMOPHTHORA

Dichotomophthora Mehrlich & Fitzpatrick ex M. B. Ellis.

Colonies effuse, greyish brown, to dark olivaceous brown, hairy or velvety. *Mycelium* immersed and sometimes also superficial. *Stroma* none. *Setae* and

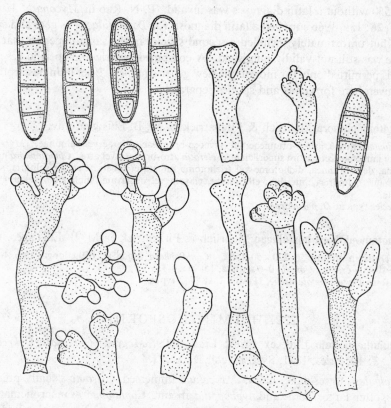

FIG. 265. *Dichotomophthora portulacae* (× 650).

hyphopodia absent. Dark brown or black *sclerotia* often formed in culture. *Conidiophores* macronematous, mononematous, sometimes swollen and repeatedly dichotomously or trichotomously branched or lobed at the apex, forming a stipe and head; stipe hyaline to brown, smooth, branches usually short. *Conidiogenous cells* polytretic, integrated and terminal or discrete, determinate, lobed, cicatrized. *Conidia* solitary, dry, borne one at the tip of each lobe of the

conidiogenous cell, simple, usually cylindrical rounded at the ends, sometimes
ellipsoidal, clavate or obclavate, subhyaline to brown, 0–6-septate.

Type species: Dichotomophthora portulacae Mehrlich & Fitzpatrick ex M. B.
Ellis.

Dichotomophthora portulacae Mehrlich & Fitzpatrick ex M. B. Ellis.
(Fig. 265)

Sclerotia in culture 50–220 × 50–150μ. *Conidiophores* up to 650μ long, 4–10μ
thick at the base, 10–20μ at the apex. *Conidia* 15–90 × 6–15μ.

Parasitic, causing a leaf-spot disease of purslane (*Portulaca oleracea*); Hawaii,
Jamaica, Sudan, Venezuela.

The generic name *Dichotomophthora* published in 1935 in *Mycologia*, **27**:
543–550 without a latin diagnosis was invalid. P. N. Rao in *Mycopath. Mycol.
appl.*, **28**: 139, 1966 supplied a latin diagnosis for *D. portulacae* as gen. and spec.
nov. but unfortunately included a second species in the same paper so that the
name was still not validly published. A combined generic and specific description is permitted only for monotypic new generic names. Brief latin descriptions
are given here for genus and species separately.

Dichotomophthora Mehrlich & Fitzpatrick ex M. B. Ellis gen. nov.

Coloniae effusae, griseo-brunneae vel olivaceo-brunneae, pilosae vel velutinae. *Mycelium*
partim immersum et partim superficiale. *Sclerotia* atro-brunnea vel atra. *Conidiophora* primo
hyalina, dein brunnea, dichotome vel trichotome furcata et lobata. *Conidia* solitaria, sicca,
plerumque cylindrica, interdum ellipsoidea, clavata vel obclavata, subhyalina vel brunnea,
septata.
Species typica: *D. portulacae*.

Dichotomophthora portulacae Mehrlich & Fitzpatrick ex M. B. Ellis spec. nov.

Sclerotia in cultura 50–220×50–150μ. *Conidiophora* usque ad 650μ longa, basi 4–10μ,
apice 10–20μ crassa. *Conidia* 0–6-septata, 15–90×6–15μ.
In foliis *Portulacae oleraceae*, Hawaii, IMI 8742 typus.

232. HELMINTHOSPORIUM

Helminthosporium Link ex Fries; Link, 1809, *Magazin Ges. naturf. Freunde
Berl.*, **3**: 10; Fries, 1821, Syst. mycol., **1**: XLVI.

Colonies effuse, dark, hairy. *Mycelium* immersed. *Stroma* usually present,
dark, often large. *Setae* and *hyphopodia* absent. *Conidiophores* macronematous,
mononematous, unbranched, often caespitose, straight or flexuous, cylindrical
or subulate, mid to very dark brown, smooth or occasionally verruculose, with
small pores at the apex and laterally beneath the septa. *Conidiogenous cells*
polytretic, integrated, terminal and intercalary, determinate, cylindrical. *Conidia*
solitary, catenate in one species, acropleurogenous, developing laterally often
in verticils through very small pores beneath the septa whilst the tip of the
conidiophore is actively growing, growth of the conidiophore ceasing with the
formation of terminal conidia, simple, usually obclavate, sometimes rostrate,

subhyaline to brown, smooth, pseudoseptate, frequently with a prominent, dark brown or black scar at the base.

Type species: *Helminthosporium velutinum* Link ex Ficinus & Schubert.
Ellis, M. B., *Mycol. Pap.*, **82**: 2–21, 1961.

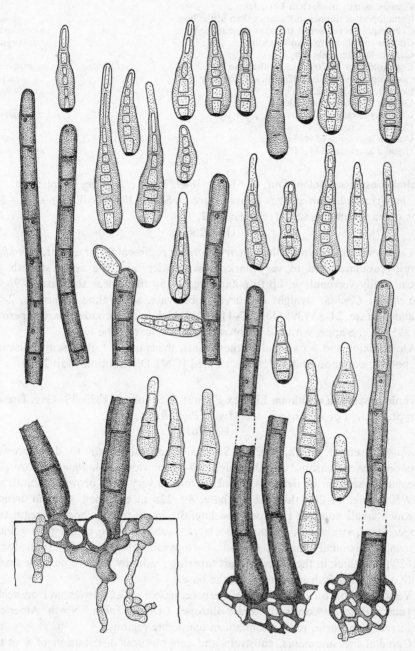

FIG. 266. *Helminthosporium solani* (× 650).

KEY

Conidia obclavate, not in chains 1
Conidia not obclavate, often in short chains . . . . . . *hypselodelphyos*
1. Conidia usually less than 10μ thick 2
   Conidia usually more than 10μ but less than 14μ thick 3
   Conidia usually more than 14μ thick 5
2. Conidiophores thinner at the apex than at the base . . . . *solani*
   Coniodiophores thicker at the apex than at the base . . . . *chlorophorae*
3. Conidia mostly more than 90μ long . . . . *dalbergiae*
   Conidia less than 60μ long 4
4. Conidiophores 14–20μ thick at the base . . . . . *mauritianum*
   Conidiophores less than 12μ thick at the base . . . . . *acaciae*
5. Conidia more than 23μ thick . . . . *ahmadii*
   Conidia less than 23μ thick 6
6. Conidia rostrate . . . . *bauhiniae*
   Conidia not rostrate 7
7. Conidia averaging 68 × 15μ . . . . *velutinum*
   Conidia averaging 114 × 17μ . . . . *microsorum*

Helminthosporium solani Dur. & Mont., 1849, Flore d'Algérie, Crypt.: 356.
   Spondylocladium atrovirens (Harz) Harz ex Sacc., 1886, Syll. Fung., **4**: 428.
For other synonyms, see *Mycol. Pap.*, **82**.
(Fig. 266)

   Colonies dark brown to black, hairy. *Stroma* rudimentary or absent. *Conidio-phores* subulate, mid to very dark brown, paler near the apex, smooth or occasionally verruculose, up to 600μ long, 9–15μ thick near the base, 6–9μ at the apex. *Conidia* straight or curved, obclavate, subhyaline to brown, 2–8-pseudoseptate, 24–85 (39)μ long, 7–11 (9·4)μ thick in the broadest part, tapering to 2–5μ at the apex, with a dark brown to black scar at the base.

   On *Solanum* and occasionally other plants, the cause of ' silver scurf ' disease of potato; common and widely distributed [CMI Distribution Map 233].

Helminthosporium velutinum Link ex Ficinus & Schubert, 1823, Fl. Geg. Dresd. Krypt.: 283. For synonymy, see *Mycol. Pap.*, **82**.
(Fig. 267)

   Colonies effuse, black, hairy. *Stroma* erumpent, brown to dark brown, pseudoparenchymatous, 20–400μ high, 40–700μ wide. *Conidiophores* usually caespitose, straight or flexuous, subulate, mid to very dark brown, smooth, up to 950μ long, 14–26μ thick at the base, 8·5–12μ at the apex. *Conidia* arising through small pores at the apex and laterally in up to 12 whorls beneath the upper septa, straight or flexuous, obclavate, subhyaline to rather pale golden brown, occasionally darker, smooth, 6–16-pseudoseptate, 40–118 (68)μ long, 11–20 (15)μ thick in the broadest part tapering gradually to 5–7μ near the apex, with a large blackish brown scar at the base.

   Very common on dead stems of herbaceous plants and twigs and branches of many different kinds of trees; Europe, Ceylon, India, North America, Pakistan, Venezuela, most abundant in temperate regions.

   Conidial measurements, substrata and geographical distribution of 8 other species of *Helminthosporium* are given below.

H. hypselodelphyos M. B. Ellis (Fig. 268 A): 15–28 (25) × 6·5–8 (7·1)μ, on *Hypselodelphys*; Sierra Leone.

H. chlorophorae M. B. Ellis (Fig. 268 B): 50–102 (73) × 8–11 (9·5)μ, on *Chlorophora*; Sierra Leone.

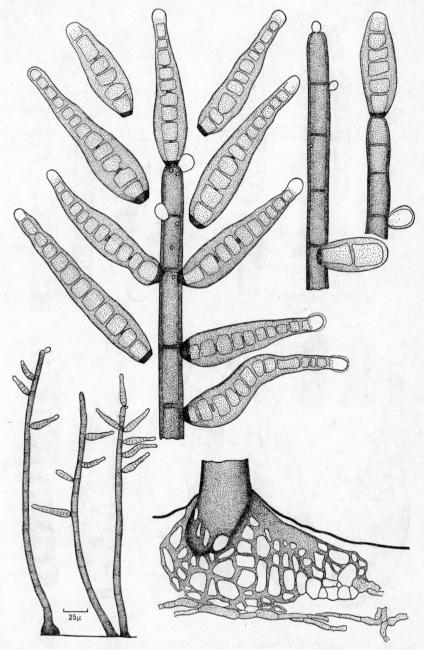

FIG. 267. *Helminthosporium velutinum* (× 650 except where indicated by the scale).

H. dalbergiae M. B. Ellis (Fig. 268 C): 58–125 (93) × 12–14 (13·2)μ, on *Dalbergia*; Pakistan.

H. mauritianum Cooke (Fig. 268 D): 27–55 (39) × 8–13 (11·1)μ, on *Alnus*, *Leucaena*, *Manihot*, *Phyllanthus*, etc.; Brazil, Cuba, Jamaica, Mauritius, Sabah, U.S.A., Zambia.

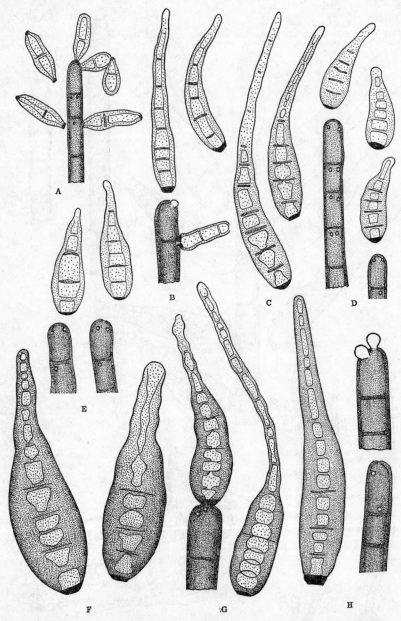

FIG. 268. *Helminthosporium* species: A, *hypselodelphyos*; B, *chlorophorae*; C, *dalbergiae*; D, *mauritianum*; E, *acaciae*; F, *ahmadii*; G, *bauhiniae*; H, *microsorum* (× 650).

H. acaciae M. B. Ellis (Fig. 268 E): 31–49 (44) × 10–14 (12)μ, on *Acacia*; Sierra Leone.

H. ahmadii M. B. Ellis (Fig. 268 F): 95–150 (110) × 25–30 (28)μ, on *Quercus*; Pakistan.

H. bauhiniae M. B. Ellis (Fig. 268 G): 55–145 (86) × 16–18 (17·2)μ on *Bauhinia* and *Macrolobium*; Sierra Leone.

H. microsorum D. Sacc. (Fig. 268 H): 60–160 (114) × 12–22 (17)μ, on *Quercus ilex*; Crete, Great Britain, Italy.

233. SPADICOIDES

Spadicoides Hughes, 1958, *Can. J. Bot.*, **36**: 805–806.

Colonies effuse, dark olivaceous brown, blackish brown or black, hairy or velvety. *Stroma* none. *Setae* and *hyphopodia* absent. *Mycelium* partly superficial, partly immersed. *Conidiophores* macronematous, mononematous, generally unbranched, straight or flexuous, pale to very dark brown or olivaceous brown, smooth. *Conidiogenous cells* polytretic, integrated, terminal and intercalary, determinate, cylindrical. *Conidia* solitary, dry, acropleurogenous, developing through minute channels in the thick wall of the conidiogenous cell, simple, ellipsoidal, oblong rounded at the ends or obovoid, mid pale to dark brown or reddish brown, smooth, 0–3-septate, sometimes with thick, black or dark brown bands at the septa.

Type species: Spadicoides bina (Corda) Hughes.

Ellis, M. B., *Mycol. Pap.*, **93**: 6–14, 1963.

KEY

Conidia 0-septate	*atra*
Conidia 1-septate	*bina*
Conidia 2-septate	*obovata*
Conidia 3-septate	1
1. Conidia ellipsoidal, rounded at the base	*xylogena*
Conidia obovoid or clavate, truncate at the base	*grovei*

Spadicoides bina (Corda) Hughes, 1958, as ' *binum* ', *Can. J. Bot.*, **36**: 806.
(Fig. 269 A)

Colonies dark blackish brown. *Conidiophores* erect or ascending, mid to very dark brown, often paler towards the apex, 20–550 × 2·5–4·5μ. *Conidia* developing singly through many small channels in the thick outer wall of the conidiophore, usually in the upper half, straight, usually oblong and rounded at the ends, occasionally ellipsoidal, mid pale to dark reddish brown, almost always 1-septate, with a wide, dark blackish brown or black band at the septum, 7–12 (9·3) × 3–5 (4)μ.

On dead wood and bark of various trees including *Betula*, *Celtis*, *Fagus*, *Pinus*, *Prunus*, *Quercus* and *Ulmus;* Europe, N. America.

Conidial measurements, host range and geographical distribution of 4 other species of *Spadicoides* are given below; they are fully described in *Mycol. Pap.*, **93**.

S. atra (Corda) Hughes (Fig. 269 B): 4–6·5 (5·2) × 3–4 (3·4)μ, on dead wood and bark of *Larix*, *Picea*, *Populus*, *Pseudotsuga* and *Quercus*; Europe, N. America.

S. obovata (Cooke & Ellis) Hughes (Fig. 269 C): 12–15 (13·4) × 6–8·5 (7·5)μ, on dead wood of *Magnolia*; U.S.A.

S. xylogena (A. L. Smith) Hughes (Fig. 269 D): 16–34 (21) × 7–10·5 (8·6)μ, on resupinate hymenomycetes and dead wood; Europe.

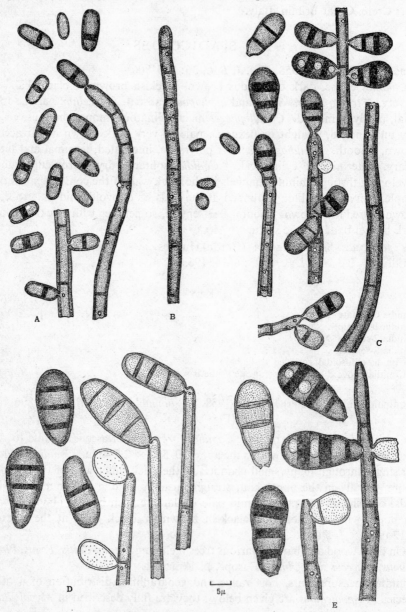

FIG 269. *Spadicoides* species: A, *bina*; B, *atra*; C, *obovata*; D, *xylogena*; E, *grovei*.

S. grovei M. B. Ellis (Fig. 269 E): 16–26 (22) × 8–13 (10·5)μ, on dead wood of *Fagus* and other trees; Great Britain.

234. DIPLOCOCCIUM

Diplococcium Grove, 1885, *J. Bot.*, *Lond.*, **23**: 167.

Colonies effuse, dark, cottony or velvety. *Mycelium* partly superficial, partly

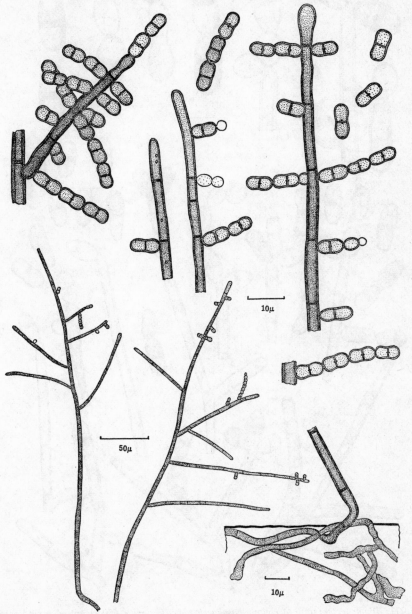

FIG. 270. *Diplococcium spicatum.*

immersed. *Stroma* none. *Setae* and *hyphopodia* absent. *Conidiophores* macronematous, mononematous, straight or flexuous, brown, smooth, branched, branches often long. *Conidiogenous cells* polytretic, integrated, terminal and

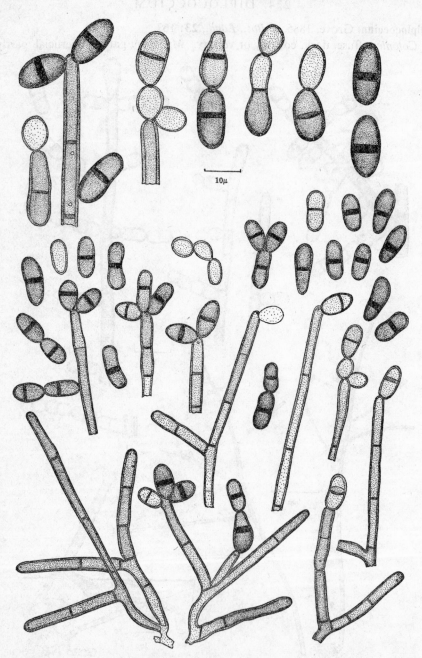

FIG. 271. *Diplococcium* state of *Helminthosphaeria clavariarum* (× 650 except where indicated by the scale).

intercalary, determinate, cylindrical. *Conidia* catenate, acropleurogenous, simple, short, clavate, ellipsoidal or oblong rounded at the ends, pale or dark brown, smooth, 0–2-septate but almost all 1-septate, in one species with dark bands often formed at the septa.

Type species: D. spicatum Grove.
Ellis, M. B., *Mycol. Pap.*, **93**: 2–6, 1963.

KEY

Conidia 6–9 × 3–4µ . . . . *spicatum*	
Conidia 13–23 × 6–8·5µ . . . . . . *Helminthosphaeria clavariarum*	

Diplococcium spicatum Grove, 1885, *J. Bot., Lond.*, **23**: 167.
(Fig. 270)

Colonies dark brown. *Hyphae* brown, 1–4µ thick. *Conidiophores* erect or ascending, 200–400µ or occasionally up to 900µ long, 2·5–4µ thick, sometimes expanded to 5µ at the apex, with branches up to 200µ long forming a wide angle with the main stipe; secondary branches sometimes formed. Minute pores scattered along the upper part of the stipe and along the branches indicate where conidia have been borne. *Conidia* oblong rounded at the ends, very pale brown, 1-septate, usually constricted at the septum, 6–9 × 3–4µ.

On dead, often rotting wood and bark of various trees including *Betula*, *Fagus*, *Pinus*, *Quercus* and *Sorbus*; only specimens from Great Britain seen.

Diplococcium state of **Helminthosphaeria clavariarum** (Tul.) Fuckel [as ' *clavariae* '], 1870, Symb. mycol.: 166; for synonymy, see *Mycol. Pap.*, **93**.
(Fig. 271)

Colonies dark olivaceous brown, charcoal or blackish brown, velvety. *Conidiophores* 50–120µ long, 3–5µ thick, branches up to 50µ long. *Conidia* usually ellipsoidal but sometimes clavate or dumb-bell-shaped, pale to dark brown often with a dark blackish brown band at the septum, 13–23 × 6–8·5µ.

On *Clavaria*; Europe.

235. BLASTOPHORELLA

Blastophorella Boedijn, 1937, *Blumea* (Suppl. 1): 140–141.

Colonies effuse, dark grey or greyish brown, hairy. *Mycelium* partly superficial, partly immersed. *Stroma* none. *Setae* and *hyphopodia* absent. *Conidiophores* macronematous, mononematous, unbranched, straight or flexuous, clavate, smooth, pale brown to brown. *Conidiogenous cells* integrated, terminal, polytretic, determinate, clavate. *Conidia* solitary, acropleurogenous, colourless, 1-septate, smooth, cylindrical or oblong, rounded at the ends.

Type species: Blastophorella smithii Boedijn.

Blastophorella smithii Boedijn, 1937, *Blumea* (Suppl. 1): 140–141.
(Fig. 272)

Hyphae pale brown to brown, 5–7μ thick. *Conidiophores* up to 2 mm. long, 11–12μ thick below, swelling to 19–31μ near the apex. *Conidia* mostly 15–22 (17·6) × 4–5μ but occasionally up to 8μ thick.

On fallen male flowers of *Arenga pinnata*; Java.

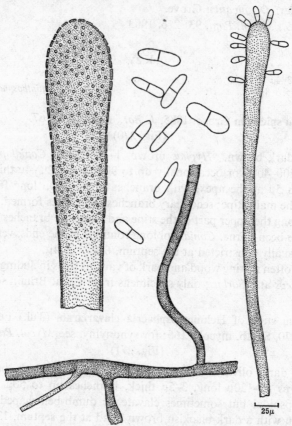

FIG. 272. *Blastophorella smithii* (× 650 except where indicated by the scale).

236. DENDROGRAPHIUM

Dendrographium Massee, 1892, *Grevillea*, **21**: 5.

Colonies effuse, dark blackish brown to black, tufted, with large, black, erect synnemata. *Mycelium* immersed. *Stroma* very well developed, erumpent, pseudoparenchymatous, dark brown. *Setae* and *hyphopodia* absent. *Conidiophores* macronematous, synnematous; individual threads closely adpressed along most of their length but splaying out at the apex, unbranched, straight or flexuous, mid to dark brown, smooth, with numerous small channels or pores at successive apices and also in verticils beneath septa. *Conidiogenous cells* polytretic, integrated, terminal becoming intercalary, sympodial, cylindrical. *Conidia* solitary, dry, acropleurogenous, simple, obclavate, pale olivaceous brown, smooth, pseudoseptate.

Type species: *Dendrographium atrum* Massee.

Dendrographium atrum Massee, 1892, *Grevillea*, **21**: 5.

(Fig. 273)

Synnemata up to 1·5 mm. high, 40–80μ thick, broadening sometimes up to 200μ at the base; individual threads 6–7μ thick near the base, 7–10μ at the apex.

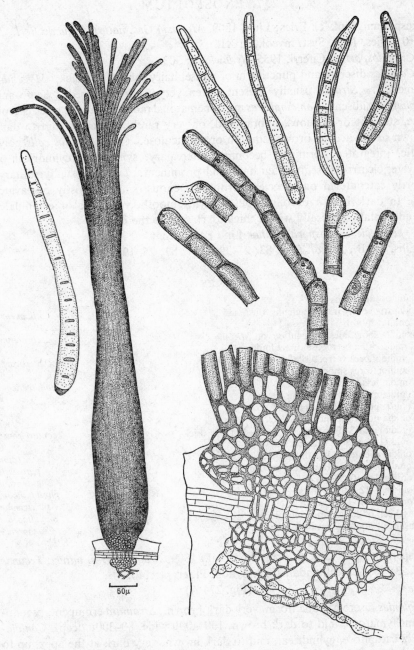

50μ

FIG. 273. *Dendrographium atrum* (× 650 except where indicated by the scale).

Conidia 6–15-pseudoseptate, 40–130μ long, 7–9μ thick in the broadest part, 4–5μ at the apex.

On dead twigs; Argentina, Brazil.

237. EXOSPORIUM

Exosporium Link ex Fries; Link, 1809, *Magazin Ges. naturf. Freunde Berl.*, **3**: 9–10; Fries, 1821, Syst. mycol., **1**: XL.

Cuspidosporium Ciferri, 1955, *Sydowia*, **9**: 303.

Colonies discrete and punctiform or effuse, hairy, brown to black. *Mycelium* immersed. *Stroma* usually present, often very well developed. *Setae* and *hyphopodia* absent. *Conidiophores* macronematous, mononematous, often caespitose, straight or flexuous, unbranched, or very rarely branched, mid to dark brown or olivaceous brown, smooth or verruculose. *Conidiogenous cells* polytretic, integrated, terminal, becoming intercalary, sympodial, cylindrical or clavate, cicatrized, scars often dark and prominent. *Conidia* usually solitary, shortly catenate in one species, acropleurogenous, simple, mostly obclavate, pale to dark brown or olivaceous brown, smooth, verrucose or echinulate, pseudoseptate, generally with a thick, dark scar at the base.

Type species: Exosporium tiliae Link ex Schlecht.

Ellis, M. B., *Mycol. Pap.*, **82**: 21–39, 1961; **87**: 25, 1963.

KEY

Conidia not in chains	1
Conidia frequently in short chains, on *Elaeis*	*stilbaceum*
1. Conidia smooth	2
Conidia sometimes verruculose, on *Miconia*	*insuetum*
Conidia often verrucose or echinulate	8
2. Conidiophores verruculose, especially at the apex	*pterocarpi*
Conidiophores smooth	3
3. Conidia less than 30μ long	*leptoderridicola*
Conidia more than 30μ long	4
4. Conidia not less than 14μ thick	5
Conidia not more than 14μ thick	7
5. Conidial scar 3–5μ wide, conidia 14–17 (15)μ thick	*cantareirense*
Conidial scar more than 6μ wide	6
6. Conidia averaging 16μ thick, on *Tilia*	*tiliae*
Conidia averaging 18μ thick, on *Croton*	*nattrassii*
7. Conidia 7·5–10 (9)μ thick	*extensum*
Conidia 9–12 (10·3)μ thick	*phyllantheum*
Conidia 11–14 (12·5)μ thick	*mexicanum*
8. Basal cell of conidium subhyaline	*ampullaceum*
Basal cell of conidium dark brown	*coonoorense*

Exosporium tiliae Link ex Schlecht.; Link, 1809, *Magazin Ges. naturf. Freunde Berl.*, **3**: 10; Schlechtendal, 1824, Synop. Pl. crypt.:140.

(Fig. 274 A)

Colonies discrete, punctiform, very dark brown. *Stromata* erumpent, pseudoparenchymatous, mid to dark brown, 100–400μ wide, 80–300μ deep. *Conidiophores* caespitose, cylindrical, mid to dark brown, very dark at the apex, up to 150μ long, 8–12μ thick, occasionally swollen at the base or apex up to 15μ.

Often apparently one conidium only is formed on a conidiophore but sometimes 2 or more conidia are formed. The outer wall of the conidiophore is thickened at the apex and the first conidium develops through a channel in the middle of this thickening. After the first conidium has fallen the conidiophore grows out

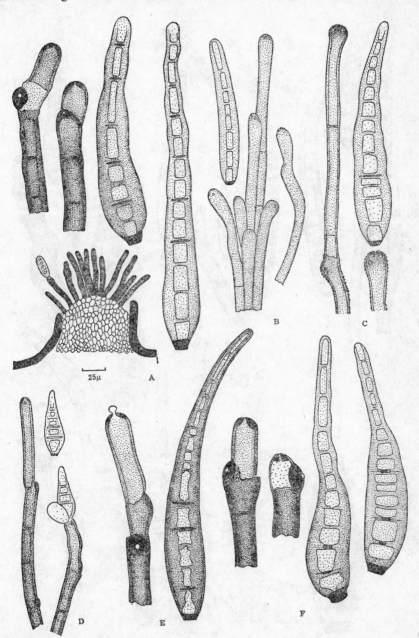

FIG. 274. *Exosporium* species (1): A, *tiliae*; B, *stilbaceum*; C, *pterocarpi*; D, *leptoderridicola*; E, *cantareirense*; F, *nattrassii* (× 650 except where indicated by the scale).

laterally, splitting the side wall then growing on obliquely for some distance before forming another conidium at the newly constituted apex. *Conidia* straight or slightly curved, obclavate, pale to mid golden brown, smooth, with a thick, black, truncate scar at the base, 7–18-pseudoseptate, 70–195 (113)µ long, 12–21

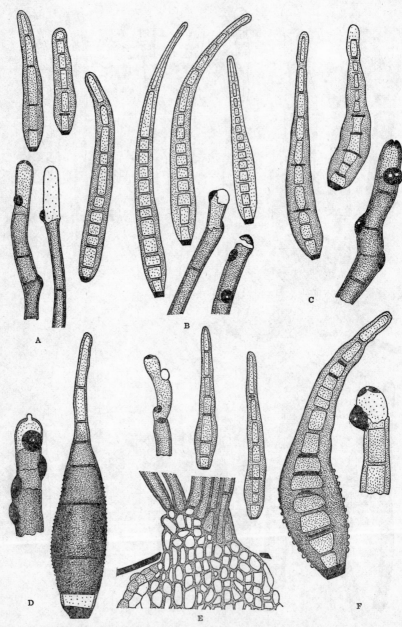

Fig. 275. *Exosporium* species (2): A, *extensum*; B, *phyllantheum*; C, *mexicanum*; D, *ampullaceum*; E, *insuetum*; F, *coonoorense* (× 650).

(16·4)μ thick in the broadest part, tapering to 6–9μ near the apex, hilum 6–7μ wide.

On branches and twigs of *Tilia*; Europe.

Conidial measurements, substrata and distribution of 11 other species of *Exosporium* are given below.

E. stilbaceum (Moreau) M. B. Ellis (Fig. 274 B): 35–104 (63) × 7–12 (9)μ, on leaves of *Elaeis*; Congo, Ghana, Sierra Leone, Sudan, Zambia.

E. pterocarpi M. B. Ellis (Fig. 274 C): 60–115 (84) × 11–16 (13·6)μ, on leaves of *Pterocarpus*; Malaya.

E. leptoderridicola M. B. Ellis (Fig. 274 D): 19–27 (24) × 6·5–8 (7·3)μ, on dead twigs of *Leptoderris* and *Uvaria*; Sierra Leone.

E. cantareirense (P. Henn.) M. B. Ellis (Fig. 274 E): 80–220 (140) × 14–17 (15)μ, on leaves and dead branches of *Afzelia*, *Paludia* and *Vismia*; Brazil, Ghana, Philippines, Sierra Leone, Uganda.

E. nattrassii M. B. Ellis (Fig. 274 F): 70–127 (103) × 17–20 (18)μ, on dead branches of *Croton*; Kenya.

E. extensum (Petch) M. B. Ellis (Fig. 275 A): 45–86 (63·5) × 7·5–10 (9)μ, on dead branches of *Erythrina*; Ceylon.

E. phyllantheum (Sacc.) M. B. Ellis (Fig. 275 B): 65–120 (96) × 9–12 (10·3)μ, on dead twigs of *Phyllanthus*, *Ochthocosmus*, etc.; Ghana, Sierra Leone.

E. mexicanum (Ellis & Everh.) M. B. Ellis (Fig. 275 C): 48–97 (75) × 11–14 (12·5)μ; on dead stems and branches of *Erythrina*, *Mascagnia*, *Smilax* and *Uvaria*; Ceylon, India, Mexico, New Guinea, Philippines, Sierra Leone, U.S.A.

E. ampullaceum (Petch) M. B. Ellis (Fig. 275 D): 93–220 (124) × 16–24 (22)μ, on dead twigs and branches of *Funtumia*, *Rauwolfia*, *Theobroma*, etc.; Ceylon, Ghana, Sierra Leone.

E. insuetum (Petrak) M. B. Ellis (Fig. 275 E): 45–130 (69) × 8–13 (10·4)μ, on leaves of *Miconia*; Ecuador.

E. coonoorense Subram. (Fig. 275 F): 95–200 (148) × 21–28 (23·7)μ, on twigs of *Eugenia*; India.

238. DRECHSLERA

Drechslera Ito, 1930, *Proc. imp. Acad. Japan*, **6**: 355.

Bipolaris Shoemaker, 1959, *Can. J. Bot.*, **37**: 882.

Colonies effuse, grey, brown or blackish brown, often hairy, sometimes velvety. *Mycelium* mostly immersed. *Stroma* present in some species. *Sclerotia* or *protothecia* often formed in culture. *Setae* and *hyphopodia* absent. *Conidiophores* macronematous, mononematous, sometimes caespitose, straight or flexuous, often geniculate, unbranched or in a few species loosely branched, brown, smooth in most species. *Conidiogenous cells* polytretic, integrated, terminal, frequently becoming intercalary, sympodial, cylindrical, cicatrized. *Conidia* solitary, in certain species also sometimes catenate or forming secondary conidiophores which bear conidia, acropleurogenous, simple, straight or curved, clavate, cylindrical rounded at the ends, ellipsoidal, fusiform or obclavate, straw-coloured or pale to dark brown or olivaceous brown, sometimes with cells unequally coloured, the end cells then being paler than intermediate ones, mostly smooth, rarely verruculose, pseudoseptate.

Lectotype species: Drechslera state of *Pyrenophora tritici-repentis* (Died.) Drechsl. = *D. tritici-vulgaris* (Nisikado) Ito ex Hughes.

Shoemaker, R. A., *Can. J. Bot.*, **40**: 809–836, 1962.

KEY

Conidiophores formed on large, erect, black stromata on natural substrata	
	Pyrenophora semeniperda
Conidiophores not formed on erect stromata on natural substrata	1
1. Conidia each with a protuberant hilum	2
Conidia without a protuberant hilum	8
2. Conidia mostly with thick dark septa cutting off pale cells at one end or both ends	3
Conidia without thick dark septa	5
3. Conidia frequently more than 25µ thick	*Trichometasphaeria holmii*
Conidia always less than 25µ thick	4
4. Conidia becoming rostrate	*rostrata*
Conidia not becoming rostrate	*halodes*
5. Conidia broadly fusiform with a long, pedicel-like extension at the base	
	Trichometasphaeria pedicellata
Conidia without pedicel-like extension	6
6. Conidia mostly more than 20µ thick	*Trichometasphaeria turcica*
Conidia mostly less than 20µ thick	7
7. Conidia fusiform tapering gradually to the base	*monoceras*
Conidia tapering more abruptly to the base	*frumentacei*
8. Conidia mostly less than 40µ long	9
Conidia mostly more than 40µ long	13
9. Conidia typically curved	*papendorfii*
Conidia straight	10
10. Conidia mostly less than 10µ thick	11
Conidia mostly more than 10µ thick	12
11. Conidia with 3 pseudosepta	*australiensis*
Conidia with 5 pseudosepta	*hawaiiensis*
12. Conidia always with 3 pseudosepta	*Cochliobolus spicifer*
Conidia with 2–3 pseudosepta	*biseptata*
Conidia with 2–7 (mostly 3–4) pseudosepta	*dematioidea*
13. Conidiophores often branched	14
Conidiophores very rarely branched	16
14. Conidia often curved	*miyakei*
Conidia straight or rarely slightly curved	15
15. Conidia obclavate or sometimes ovoid	*Cochliobolus nodulosus*
Conidia ellipsoidal tapered slightly to the rounded ends	*ravenelii*
16. Conidia mostly straight	17
Conidia mostly curved	32
17. Conidia with end cells often much paler than the others	18
Conidia with end cells not paler	19
18. Conidia less than 20µ thick	*Cochliobolus bicolor*
Conidia frequently more than 20µ thick	*iridis*
19. Conidia mostly cylindrical or subcylindrical	20
Conidia obclavate, fusiform or ellipsoidal, usually tapered gradually towards the apex	27
20. Basal cell of conidium conical or shaped like a snake's head	*Pyrenophora tritici-repentis*
Base of conidium rounded	21
21. Conidia mostly 11–13µ thick	*erythrospila*
Conidia more than 13µ thick	22

22. Conidia usually more than 200μ long . . . . *gigantea*
    Conidia mostly 150–190μ long . . . . . *Pyrenophora bromi*
    Conidia less than 150μ long 23
23. Conidia mostly more than 18μ thick 24
    Conidia mostly less than 18μ thick 25
24. Short secondary conidiophores and conidia formed . . *Pyrenophora graminea*
    No secondary conidiophores and conidia . . . . . *Pyrenophora teres*
25. Conidiophores frequently geniculate, scars often widely spaced,
    not very conspicuous . . . . *Pyrenophora avenae*
    Conidiophores not or only occasionally geniculate, scars close together,
    numerous, conspicuous 26
26. Conidiophores often up to 1 mm. long, subulate, very dark brown except
    at the apex . . . . *avenacea*
    Conidiophores up to 400μ long, chestnut brown . . . . . *siccans*
27. Conidia 9–12μ thick in the broadest part . . . . . . *cactivora*
    Conidia more than 12μ thick 28
28. Conidia mostly less than 17μ thick 29
    Conidia mostly more than 17μ thick 30
29. Conidia broadest at 1st septum and tapering uniformly from there
    to the apex . . . . *Pyrenophora dictyoides*
    Conidia broadest at second cell from the base, tapering to a narrow apex . *phlei*
30. Conidia 40–60μ long . . . . *coicis*
    Conidia 60–100μ long 31
31. Basal cell of conidium usually longer than wide . . . . . *fugax*
    Basal cell of conidium often isodiametric . . . . . . *poae*
32. Conidia with hilum minute, often slightly protruding and papillate . *Cochliobolus*
        miyabeanus
    Conidia with hilum not protruding or papillate 33
33. Secondary conidiophores and conidia often formed . . . . *sorghicola*
    No secondary conidiophores or conidia 34
34. Conidia mostly less than 17μ thick 35
    Conidia mostly more than 17μ thick 39
35. Conidia averaging 50μ long . . . . . *Cochliobolus cynodontis*
    Conidia mostly more than 50μ long 36
36. Conidia mostly 12–14μ thick 37
    Conidia 14–17μ thick 38
37. On *Euphorbia* . . . . *euphorbiae*
    On *Saccharum* . . . . *sacchari*
    On *Setaria* . . . . *Cochliobolus setariae*
38. Conidia dark, end cells sometimes paler . . . . *Cochliobolus carbonus*
    Conidia pale . . . . *Cochliobolus victoriae*
39. Conidia not more than 20μ thick . . . *Cochliobolus heterostrophus*
    Conidia often more than 20μ thick 40
40. Conidia mostly 60–100 × 18–23μ . . . . . *Cochliobolus sativus*
    Conidia frequently more than 100μ long 41
41. On *Cocos* . . . . *incurvata*
    On *Hevea* . . . . *heveae*
    On *Musa* . . . . *musae-sapientum*

Common substrata for *Drechslera* spp.

Gramineae and sometimes other substrata: *Cochliobolus bicolor, C. sativus, C. spicifer, D. australiensis, D. biseptata, D. halodes, D. hawaiiensis, D. papendorfii, D. rostrata, Pyrenophora semeniperda.*

Agropyron: P. tritici-repentis.

Agrostis: D. erythrospila, D. fugax.

Anthoxanthum: D. dematioidea.

Avena: C. victoriae, D. avenacea, P. avenae.

Bromus: P. bromi.

Cactaceae: *D. cactivora.*
Cocos: *D. incurvata.*
Coix: *D. coicis.*
Cynodon: *C. cynodontis, D. gigantea.*
Dactyloctenium: *Trichometasphaeria holmii.*
Echinochloa: *D. monoceras.*
Eleusine: *C. nodulosus.*
Eragrostis: *D. miyakei.*
Euphorbia: *D. euphorbiae.*
Festuca: *P. dictyoides.*
Hevea: *D. heveae.*
Hordeum: *C. sativus, P. graminea, P. teres.*
Iris: *D. iridis.*
Lolium: *D. siccans.*
Musa: *D. musae-sapientum.*
Oryza: *C. miyabeanus.*
Panicum: *D. frumentacei.*
Phleum: *D. phlei.*
Poa: *D. poae.*
Saccharum: *D. sacchari.*
Setaria: *C. setariae.*
Sorghum: *C. heterostrophus, D. sorghicola, T. turcica.*
Sporobolus: *D. ravenelii.*
Triticum: *C. sativus, P. tritici-repentis, T. pedicellata.*
Zea: *C. carbonus, C. heterostrophus, T. pedicellata, T. turcica.*

Drechslera state of **Pyrenophora semeniperda** (Brittlebank & Adam) Shoemaker, 1966, *Can. J. Bot.,* **44**: 1451–1456.
   D. verticillata (O'Gara) Shoemaker, 1966, *Can. J. Bot.,* **44**: 1451.
   Helminthosporium cyclops Drechsler, 1923, *J. Agric. Res.,* **24**: 729–731.

(Fig. 276)

Stromata erect, simple or branched, cylindrical, dark blackish brown to black, up to 2 × 1 mm., formed on seeds. *Conidiophores* arising from surface of stromata, straight or flexuous, sometimes geniculate, mid to dark brown, up to 170µ long, 6–9µ thick, sometimes swollen to 11–13µ at the apex. *Conidia* straight or slightly curved, occasionally cylindrical but usually obclavate, 7–12-pseudoseptate, sometimes with dark septa cutting off 1 or 2 cells at the base, golden brown except for colourless or pale areas immediately above the dark scar and at the apex, 70–160 (mostly 90–120)µ long, 13–17 (mostly 15)µ thick in the broadest part; hilum 5–8µ wide.

On wheat, oats and many other grasses; Australia, N. America, S. Africa (see Wallace, H. A. H., *Can. J. Bot.,* **37**: 509–515, 1959).

Drechslera state of **Trichometasphaeria holmii** Luttrell, 1963, *Phytopathology,* **53**: 281–285.

D. holmii (Luttrell) Subram. & Jain, 1966, *Curr. Sci.*, **35**: 354.

Helminthosporium holmii Luttrell, 1963, *Phytopathology*, **53**: 285.

(Fig. 277)

Conidiophores solitary or in small groups, straight or flexuous, pale to dark

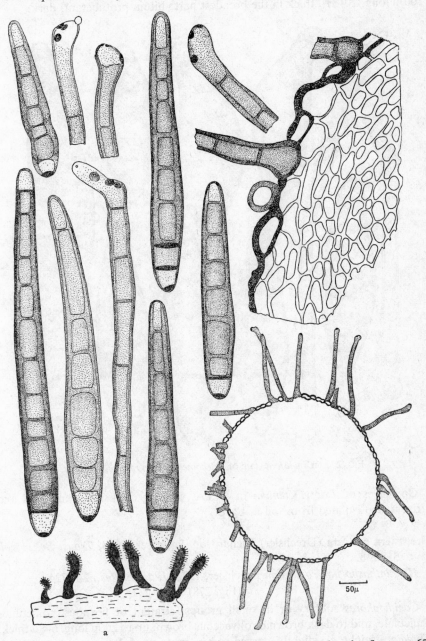

FIG. 276. *Drechslera* state of *Pyrenophora semeniperda* (a, habit sketch; other figures × 650 except where indicated by the scale).

brown, up to 280μ long, 6–10μ thick, sometimes with a basal swelling up to 18μ diam. *Conidia* straight or curved, obclavate and rostrate or broadly ellipsoidal, with 6–11 pseudosepta, end cells often rather pale and cut off by dark, thick septa, intermediate cells mid-dark golden brown, smooth, 60–130 (mostly 75–100)μ long, 20–31μ thick in the broadest part; hilum protuberant, dark.

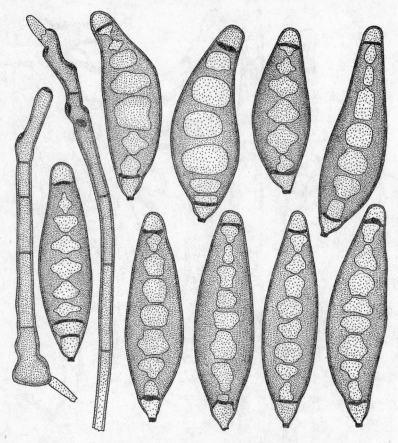

FIG. 277. *Drechslera* state of *Trichometasphaeria holmii* (× 650).

On leaves of *Dactyloctenium* in U.S.A. causing reddish-brown spots and stripes; also isolated from soil in Egypt.

Drechslera rostrata (Drechsler) Richardson & Fraser, 1968, *Trans. Br. mycol. Soc.*, **51**: 148.

Helminthosporium rostratum Drechsler, 1923, *J. agric. Res.*, **24**: 724.
(Fig. 278)

Conidiophores solitary or in small groups, straight or flexuous, sometimes geniculate, mid to dark brown or olivaceous brown, up to 200μ long, 6–8μ thick. *Conidia* straight or slightly curved, when mature obclavate, rostrate, 6–16-pseudoseptate, end cells often hyaline or very pale and cut off by thick, dark septa, intermediate cells golden brown, 40–180μ long, 14–22μ thick in the broadest part; hilum distinctly protuberant.

Common on grasses and many other substrata, isolated from soil; cosmopolitan.

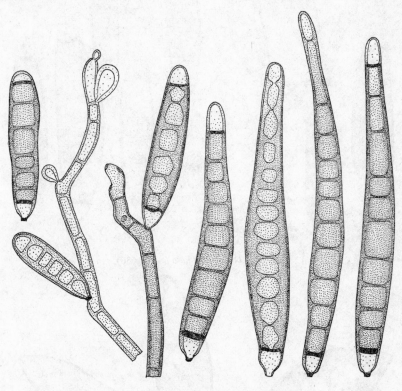

FIG. 278. *Drechslera rostrata* (× 650).

Drechslera halodes (Drechsler) Subram. & Jain, 1966, *Curr. Sci.*, **35**: 354.
   Helminthosporium halodes Drechsler, 1923, *J. agric. Res.*, **24**: 709.

(Fig. 279)

Conidiophores arising singly or in pairs, straight or flexuous, upper part often geniculate, brown, up to 150μ long, 5–8μ thick. *Conidia* straight or slightly curved, cylindrical to ellipsoidal with up to 12 but commonly 6–8 pseudosepta, end cells hyaline or very pale and cut off by thick, dark septa, intermediate cells golden brown, 30–100 (in culture usually 60–90)μ long, 11–20μ thick in the broadest part; hilum distinctly protuberant.

Frequently isolated from grasses, many other plants, soil and textiles; Ceylon, Egypt, Ethiopia, India, Jamaica, Kenya, Malawi, Nigeria, Pakistan, Rhodesia, Seychelles, Tanzania.

Drechslera state of **Trichometasphaeria pedicellata** Nelson, 1965, *Mycologia*, **57**: 665.

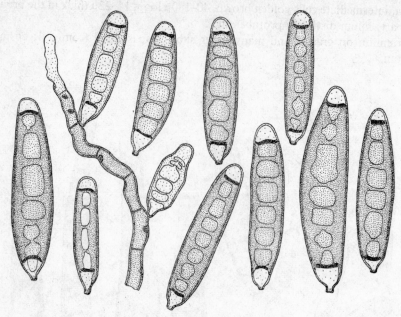

FIG. 279. *Drechslera halodes* (× 650).

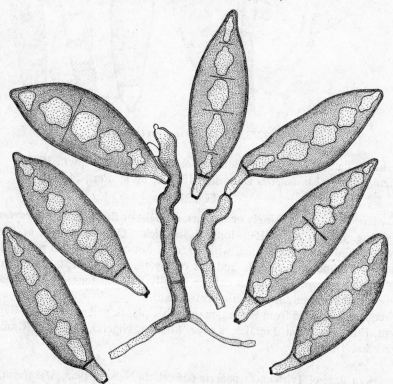

FIG. 280. *Drechslera* state of *Trichometasphaeria pedicellata* (× 650).

D. pedicellata (Henry) Subram. & Jain, 1966, *Curr. Sci.*, **35**: 354.

Helminthosporium pedicellatum Henry, 1924, *Univ. Minn. Agric. Exp. Stn tech. Bull.*, **22**: 42–43.

(Fig. 280)

Conidiophores mostly solitary, flexuous, brown, up to 200μ long, 5–9μ thick. *Conidia* straight, broadly fusiform, with a narrow, pedicel-like extension at the base, golden brown, smooth 4–9-pseudoseptate, 40–90 (mostly 70–85)μ long, 16–29 (mostly 19–26)μ thick in the broadest part; hilum protuberant.

Isolated from wheat roots in U.S.A., rice in India and maize in S. Africa.

Drechslera state of **Trichometasphaeria turcica** Luttrell, 1958, *Phytopathology*, **48**: 281–287.

Helminthosporium turcicum Pass., 1876, *Boll. Comizio Agr.*, *Parma*, No. 10.

D. turcica (Pass.) Subram & Jain, 1966, *Curr. Sci.*, **35**: 355.

H. inconspicuum Cooke & Ellis, 1878, *Grevillea*, **6**: 88.

(Fig. 281)

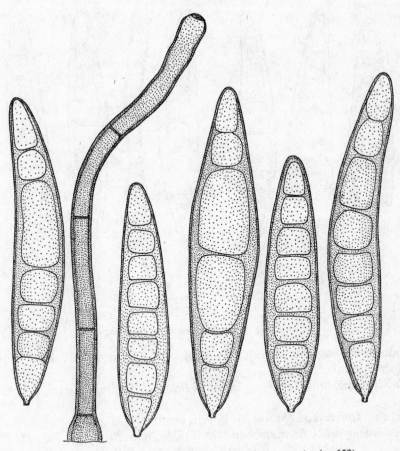

FIG. 281. *Drechslera* state of *Trichometasphaeria turcica* (× 650).

Conidiophores emerging singly or in groups of 2–6 through stomata, straight or flexuous, brown, up to 300µ long, 7–11 (mostly 8–9)µ thick. *Conidia* straight or slightly curved, ellipsoidal to obclavate, pale to mid straw-coloured, smooth, 4–9-pseudoseptate, 50–144 (115)µ long, 18–33 (commonly 20–24)µ thick in the broadest part; hilum distinctly protuberant.

Parasitic on leaves of *Sorghum* and *Zea* and occasionally other grasses, causing elongated, straw-coloured spots often 10 × 4 cm. or larger, sometimes with a narrow brown margin; cosmopolitan [CMI Distribution Map 257].

Drechslera monoceras (Drechsler) Subram & Jain, 1966, *Curr. Sci.*, **35**: 354.
   Helminthosporium monoceras Drechsler, 1923, *J. agric. Res.*, **24**: 706.
(Fig. 282)

Conidiophores solitary or in groups of 2–3, straight or flexuous, sometimes geniculate, mid to dark brown or olivaceous brown, up to 300µ long, 6–9µ thick. *Conidia* straight or very slightly curved, fusiform, tapering gradually towards the base, pale to dark straw-coloured, smooth, 4–10- (commonly 5–7-)

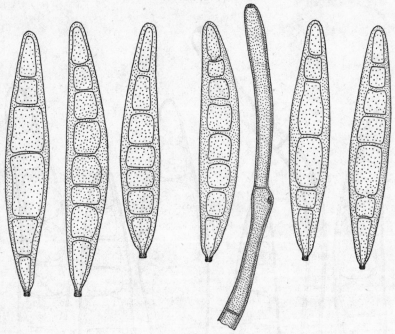

FIG. 282. *Drechslera monoceras* (× 650).

pseudoseptate, 60–150 (mostly 100–120) × 16–25 (mostly 17–20)µ; hilum distinctly protuberant.

Causing dark brown spots on leaves of *Echinochloa*; Cuba, Israel, U.S.A.

Drechslera frumentacei (Mitra) M. B. Ellis comb. nov.
   Helminthosporium frumentaceum Mitra, 1931, *Trans. Br. mycol. Soc.*, **15**: 288.
(Fig. 283)

Conidiophores arising singly or in small groups, straight or flexuous, mid to dark brown, up to 300μ long, 6–9μ thick. *Conidia* straight or curved, broadest near the middle, tapering towards the ends but often rather abruptly at the base, pale to mid-dark straw-coloured, smooth, 5–11-pseudoseptate, 50–160

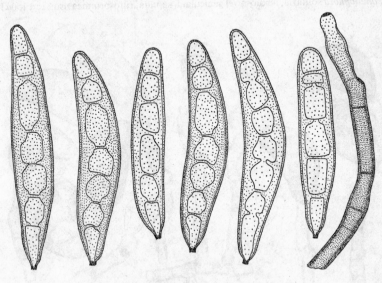

Fig. 283. *Drechslera frumentacei* (× 650).

(103)μ long, 15–20 (18)μ thick in the broadest part; hilum distinctly protuberant. On leaves and leaf sheaths of *Panicum frumentaceum*; India.

Drechslera papendorfii (van der Aa) M. B. Ellis comb. nov.

Curvularia papendorfii van der Aa, 1967, *Persoonia*, **5**: 45–46.

C. siddiquii S. I. Ahmad & Quraishi, 1960, *Pakist. J. scient. Res.*, **3**: 177; name earlier but without a latin diagnosis and therefore not validly published.

(Fig. 284)

Colonies effuse, grey or black. *Stromata* when present small. *Conidiophores* solitary, straight or flexuous, often geniculate, pale to mid brown, up to 200μ long, 4–9μ thick, with large, dark scars. *Conidia* typically curved, navicular or obpyriform, usually broadest at the second cell from the base, mid-dark to dark olivaceous brown, paler at the extremities, smooth, 3-pseudoseptate, 30–50 (mostly 35–45)μ long, 17–30 (mostly 20–24)μ thick in the broadest part.

Isolated from the air, coriander seed, cotton roots, pomegranate, sorghum, tobacco, leaf litter and soil; Australia, Egypt, Europe, India, Malawi, Pakistan, Rhodesia, S. Africa, Sudan.

Drechslera australiensis (Bugnicourt) Subram. & Jain ex M. B. Ellis; Subram. & Jain, 1966, *Curr. Sci.*, **35**: 354 [as ' *australiense* '].

Helminthosporium australiense Bugnicourt, 1955, *Revue gén. Bot.*, **62**: 242 [name published invalidly without a latin diagnosis].
(Fig. 285)

Coloniae effusae, griseae vel fuscae, velutinae. *Stromata* erecta, recta, cylindrica, atra. *Mycelium* ex hyphis pallide brunneis vel atro-brunneis, laevibus, septatis, 2–4µ crassis compositum. *Conidiophora* solitaria, flexuosa vel geniculata, septata, rufo-brunnea usque ad 150µ longa,

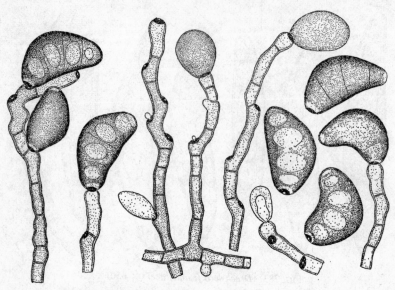

FIG. 284. *Drechslera papendorfii* (× 650).

3–7µ crassa. *Conidia* recta, ellipsoidea vel oblonga, in extremis rotundata, pallide brunnea vel rufo-brunnea, plerumque 3-pseudoseptata aliquando 4–5-pseudoseptata, 13–40×6–11µ, plerumque 18–33×8–10µ. Ex seminibus *Oryzae sativae*, Australia, F. Bugnicourt, Herb. IMI 53994 typus.

Colonies effuse, grey to dark blackish brown, velvety. *Stromata* erect, straight, cylindrical, black, formed in culture on rice grains. *Hyphae* pale to dark brown,

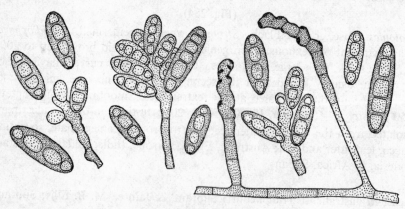

FIG. 285. *Drechslera australiensis* (× 650).

smooth, septate, 2–4μ thick. *Conidiophores* solitary flexuous or geniculate, septate, reddish brown, sometimes up to 150μ long but usually shorter, 3–7μ thick. *Conidia* straight, ellipsoidal or oblong, rounded at the ends, pale brown to mid reddish brown, at least 80% 3-pseudoseptate, rarely with 4 or 5 pseudosepta, 13–40 (mostly 18–33) × 6–11 (mostly 8–10)μ.

Isolated from rice grains, tomato, *Chloris* and *Pennisetum*; Australia, Kenya, India.

Drechslera hawaiiensis (Bugnicourt) Subram. & Jain ex M. B. Ellis; Subram. & Jain, 1966, *Curr. Sci.*, **35**: 354 [as ' *hawaiiense* '].

Helminthosporium hawaiiense Bugnicourt, 1955, *Revue gén. Bot.*, **62**: 238 [name published invalidly without a latin diagnosis].

(Fig. 286)

Coloniae effusae, griseae, fuscae vel atrae. *Stromata* erecta, recta, cylindrica, atra. *Mycelium* ex hyphis pallide brunneis vel brunneis, laevibus, septatis, 1–3μ crassis compositum. *Conidiophora* solitaria, flexuosa vel geniculata, septata, pallide brunnea vel brunnea, usque ad 120μ longa, 2–7μ crassa. *Conidia* recta, ellipsoidea, oblonga vel cylindrica, in extremis rotundata, pallide brunnea vel brunnea, 2–7- (plerumque 5-) pseudosepta, 12–37 (24·5)×5–11 (8·2)μ. Ex seminibus *Oryzae sativae*, Hawaii, F. Bugnicourt, Herb. IMI 53993 typus.

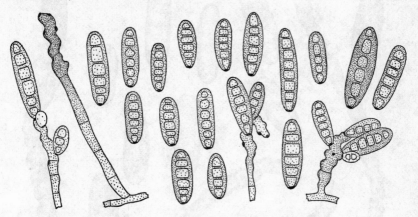

Fig. 286. *Drechslera hawaiiensis* (× 650).

Colonies effuse, grey, dark blackish brown or black. *Stromata* sometimes formed in culture, erect, straight, cylindrical, black. *Hyphae* pale to mid brown, smooth, septate, 1–3μ thick. *Conidiophores* solitary, flexuous or geniculate, septate, pale to mid brown, up to 120μ long but usually much shorter, 2–7μ thick. *Conidia* straight, ellipsoidal, oblong or cylindrical, rounded at the ends, pale to mid brown, 2–7- (mostly 5-) pseudoseptate, 12–37 (24·5) × 5–11 (8·2)μ.

First described from rice grains in Hawaii. It has been isolated from many different plants and from soil, textiles and other substrata; Ceylon, Cuba, Egypt, India, Jamaica, Kenya, Nepal, New Caledonia, Pakistan, Rhodesia, Tanzania.

Drechslera state of **Cochliobolus spicifer** Nelson, 1964, *Mycologia*, **56**: 198.

Brachycladium spiciferum Bainier, 1908, *Bull. trimest. Soc. mycol. Fr.*, **24**: 81–82.

Curvularia spicifera (Bainier) Boedijn, 1933, *Bull. Jard. bot. Buitenz.*, III, **13** (1): 127.

Helminthosporium spiciferum (Bainier) Nicot, 1953, *Öst. Bot. Z.*, **100**: 482.
(Fig. 287)

Conidiophores solitary or in small groups, flexuous, repeatedly geniculate with numerous well-defined scars, often torsive, mid to dark brown, up to 300μ, long or sometimes longer, 4–9μ thick. *Conidia* straight, oblong or cylindrical rounded at the ends, when mature golden brown except for a small area just

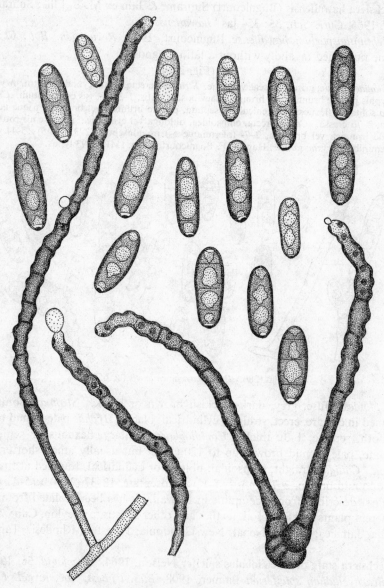

FIG. 287. *Drechslera* state of *Cochliobolus spicifer* (× 650).

above the dark scar which remains hyaline or very pale, smooth, constantly 3-pseudoseptate, 20–40 × 9–14 (mostly 30–36 × 11–13)μ; hilum 2–3μ wide.

A very common species, more than 150 isolates from 77 different plants and from air and soil have been examined; cosmopolitan but most abundant in tropical and subtropical countries.

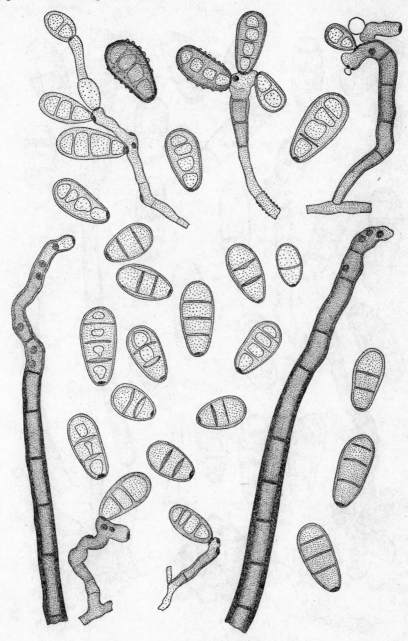

FIG. 288. *Drechslera biseptata* (× 650).

Drechslera biseptata (Sacc. & Roum.) Richardson & Fraser, 1968, *Trans. Br. mycol. Soc.*, **51**: 148.

Helminthosporium biseptatum Sacc. & Roum., 1881, *Revue mycol.*, **3** (11) 56.

H. biforme Mason & Hughes, 1948, apud Chesters in *Trans. Br. mycol. Soc.*, **30**: 114–117.

(Fig. 288)

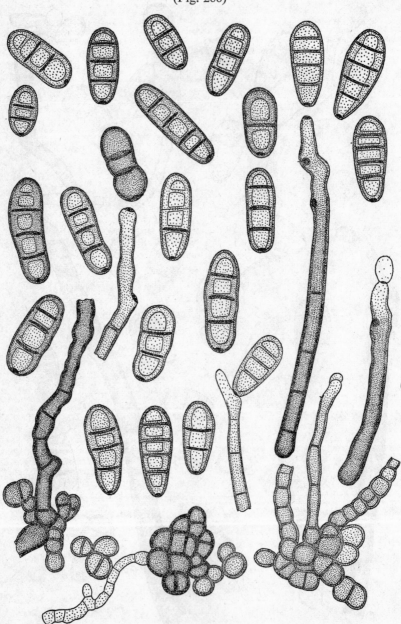

FIG. 289. *Drechslera dematioidea* (× 650).

Conidiophores arising either solitarily from hyphae or in large fascicles from very dark brown pulvinate stromata. Conidiophores are of 2 kinds, flexuous or geniculate, pale to mid brown, thin-walled ones up to about 80μ long and 3–8μ thick, and subulate, straight or flexuous, dark brown, thick-walled ones up to 800μ long, 10–14μ thick at the base tapering to 6–8μ at the rather paler apex. *Conidia* straight, typically obovoid or broadly clavate, occasionally ellipsoidal, pale to mid brown, smooth or verrucose, with almost always 2–3 pseudosepta, 20–42μ long, 11–19μ thick in the broadest part, mostly 23–33 × 14–17μ.

On or isolated from *Apium*, *Dactylis*, *Polytrichum*, *Pteridium* and *Triticum*, also isolated from soil; Australia, Europe, N. America.

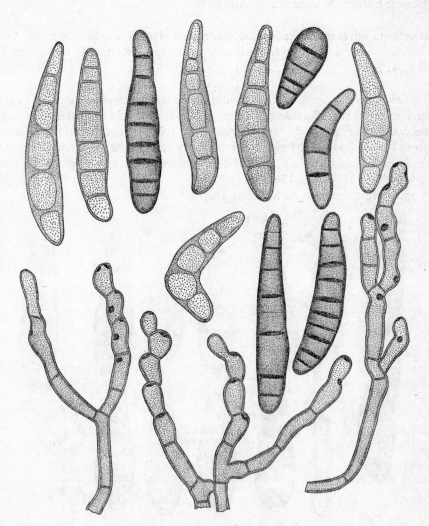

Fig. 290. *Drechslera miyakei* (× 650).

Drechslera dematioidea (Bubák & Wróblewski) Subram. & Jain, 1966, *Curr. Sci.*, **35**: 354.

Helminthosporium dematioideum Bubák & Wróblewski, 1916, apud Bubák in *Hedwigia*, **57**: 337.

(Fig. 289)

Conidiophores arising singly or in small groups often from mid to dark brown cells which form rather loose stromata, straight or flexuous, sometimes geniculate, brown, up to 350 × 5–9 (usually 60–150 × 5–6)μ. *Conidia* straight, cylindrical to clavate, rounded at the ends, golden brown to dark brown, smooth, thick-walled, with 2–7 (usually 3–4) pseudosepta, 20–70 (36) × 10–16 (14·3)μ; hilum 2·5–3·5μ wide.

On dead leaves and inflorescences of *Anthoxanthum* and occasionally on other grasses; Europe, N. America, S. Africa.

Drechslera miyakei (Nisikado) Subram. & Jain, 1966, *Curr. Sci.*, **35**: 354.

Helminthosporium miyakei Nisikado, 1928, *Spec. Rept. Ohara Inst. agric. Res.*, **4**: 145.

(Fig. 290)

Conidiophores arising in small groups on leaves but close together on superficial mycelium in inflorescences forming black spongy colonies, often branched, flexuous, geniculate, with well-defined scars, mid-pale to dark brown or olivaceous brown, up to 300μ long, 4–11μ thick. *Conidia* straight or curved, obclavate, fusiform or clavate, pale to mid-dark golden brown, 4–12-pseudoseptate, mostly 50–90 (72) × 12–15 (14)μ.

In inflorescences and on leaves of *Eragrostis*; Ethiopia, Japan.

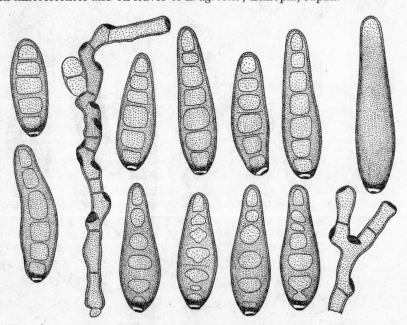

FIG. 291. *Drechslera* state of *Cochliobolus nodulosus* (× 650).

Drechslera state of **Cochliobolus nodulosus** Luttrell, 1957, *Phytopathology*, **47**: 547.

D. nodulosa (Berk. & Curt.) Subram. & Jain, 1966, *Curr. Sci.*, **35**: 354.

Helminthosporium nodulosum Berk. & Curt. apud Sacc., 1886, in Syll. Fung., **4**: 421.

H. leucostylum Drechsler, 1923, *J. agric. Res.*, **24**: 711.

(Fig. 291)

Conidiophores sometimes solitary but frequently in quite large groups, often branched, flexuous, markedly geniculate with large, mid to dark brown scars, very pale straw-coloured at first becoming darker later, up to 150μ long in inflorescences but usually much shorter on leaves, 5–9μ thick. *Conidia* straight, ovoid to obclavate, pale to mid-dark golden brown, 5–7-pseudoseptate, 43–78 (most commonly 50–65)μ long, 16–19μ thick in the broadest part; hilum 3–4μ wide, usually with a hyaline zone just above it and then a dark zone.

On *Eleusine* causing spotting and striping of leaves and a sooty growth in the inflorescences, also found once on *Pennisetum* and isolated from soil; Brazil, Ethiopia, India, Kenya, Mauritius, Nigeria, Papua, Sierra Leone, S. Africa, Tanzania, Uganda, Zambia, U.S.A.

Drechslera ravenelii (Curt.) Subram. & Jain, 1966, *Curr. Sci.*, **35**: 354.

Helminthosporium ravenelii Curt., 1848, *Am. J. Sci.*, Ser. 2, **6**: 352.

H. tonkinense Karst. & Roum., 1890, *Revue mycol.*, **12**: 78.

(Fig. 292)

Conidiophores closely packed together, flexuous, nodose, geniculate with distinct dark scars, often branched, very pale to mid golden brown, up to 600μ

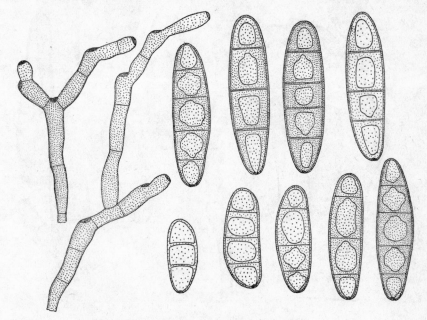

FIG. 292. *Drechslera ravenelii* (× 650).

long, 5–10µ thick. *Conidia* straight or rarely slightly curved, more or less oblong, rounded at the ends or ellipsoidal but frequently tapering slightly towards the base, pale to mid golden brown, smooth, 1–6- (usually 3–4-) pseudoseptate, 30–80 (most commonly 50–70)µ long, 13–19 (mostly 16–18)µ thick in the broadest part; hilum 2–3µ wide.

Forming spongy, olivaceous brown to black colonies in inflorescences of *Sporobolus* and occasionally other grasses; Australia, Burma, Ceylon, Costa Rica, Ghana, Hong Kong, India, Jamaica, Malaya, Mauritius, New Zealand, Philippines, Rhodesia, Sabah, Sarawak, S. Africa, Tonkin, Trinidad, Uganda, Uruquay, U.S.A.

Drechslera state of **Cochliobolus bicolor** Paul & Parbery, 1966, *Trans. Br. mycol. Soc.*, **49**: 386.

    Helminthosporium bicolor Mitra, 1931, *Trans. Br. mycol. Soc.*, **15**: 286.

    Drechslera bicolor (Mitra) Subram. & Jain, 1966, *Curr. Sci.*, **35**: 354.

(Fig. 293)

Conidiophores emerging singly or in small groups, straight or flexuous, some-

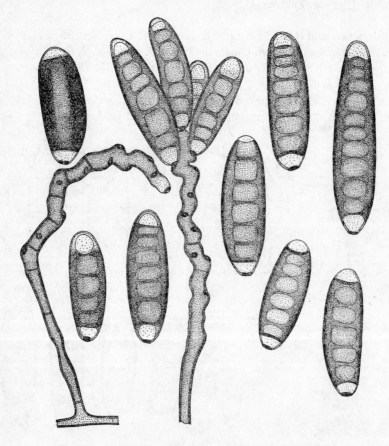

FIG. 293. *Drechslera* state of *Cochliobolus bicolor* (× 650).

times swollen at the base, the upper part often repeatedly geniculate with large, dark scars, golden brown, up to 400μ long, 5–10μ thick. *Conidia* straight or rarely slightly curved, cylindrical or rather broader in the middle tapering towards the ends, rarely obclavate, rounded at the apex, often truncate at the base, with 3–14 pseudosepta, 20–135 × 12–20μ, mostly 40–80 × 14–18μ with 5–9 pseudosepta, central cells of mature conidia often dark brown or smoky brown and sometimes quite opaque but the cell at each end remains hyaline or very pale and is frequently cut off by a very dark septum; hilum flat, dark, 3–5μ wide.

On or isolated from *Coriandrum*, *Cordyline*, *Eichhornia*, *Musa*, *Pennisetum*, *Sorghum* and *Triticum*; Australia, E. & W. Africa, India.

Drechslera iridis (Oud.) M. B. Ellis comb. nov.
Clasterosporium iridis Oud., 1898, *Hedwigia*, **37**: 318.
Bipolaris iridis (Oud.) Dickinson, 1966, *Trans. Br. mycol. Soc.*, **49**: 578.
Mystrosporium adustum Massee, 1899, *Gdnrs' Chron.*, 3, **25**: 412.
(Fig. 294)

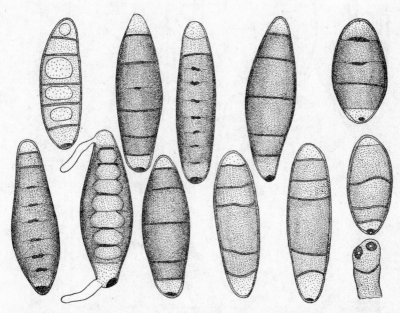

FIG. 294. *Drechslera iridis* (× 650).

Conidiophores in groups, flexuous, mid to dark brown, up to 150μ long, 10–14μ thick. *Conidia* straight, ovoid to broadly fusiform or obclavate mid to dark golden brown, end cells often very pale, 4–9-pseudoseptate, 45–90 × 16–29μ; hilum 3–6μ wide.

On leaves and especially leaf bases of *Iris*, the cause of 'ink disease', the oval or elongated colonies brown at first but becoming black on senescent leaves; Europe.

Drechslera state of **Pyrenophora tritici-repentis** (Died.) Drechsler, 1923, *J. agric. Res.*, **24**: 667.

Helminthosporium tritici-repentis Diedicke, 1902, *Zentbl. Bakt. ParasitKde*, Abt. 2, **9**: 329.

Drechslera tritici-repentis (Died.) Shoemaker, 1959, *Can. J. Bot.*, **37**: 880.

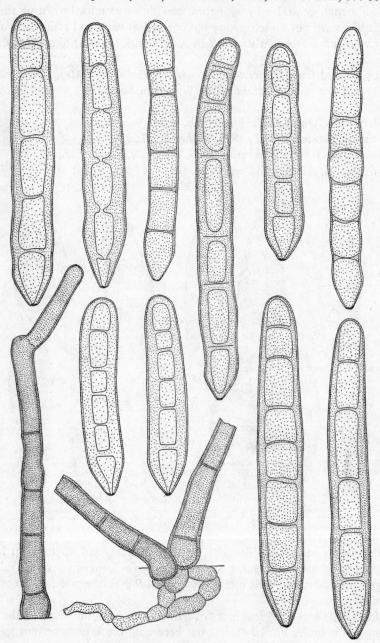

Fig. 295. *Drechslera* state of *Pyrenophora tritici-repentis* (× 650).

Helminthosporium tritici-vulgaris Nisikado, 1928, *Ann. phytopath. Soc. Japan*, **2**: 96.

Drechslera tritici-vulgaris (Nisikado) Ito, 1930, *Proc. imp. Acad. Japan*, **6**: 355.

(Fig. 295)

Conidiophores arising singly or in groups of 2–3, emerging through stomata or between epidermal cells, erect, simple, straight or flexuous, sometimes geniculate, cylindrical or slightly tapered, often swollen at the base, mid-pale to mid brown, smooth, usually up to 250µ long but occasionally as much as 400µ, 6–12µ thick, basal swelling up to 15µ wide, with usually one or a few rather inconspicuous conidial scars. *Conidia* solitary, straight or slightly curved, cylindrical, rounded at the apex, the basal segment distinctly and characteristically conical or the shape of a snake's head, typically subhyaline or rather pale straw-coloured, smooth, thin-walled, with 1–9 (usually 5–7) pseudosepta, old conidia often constricted at the pseudosepta, 80–250 (117)µ long, 14–20 (17·7)µ thick in the broadest part, 2–4 (3)µ wide at the base.

Common and widespread on *Agropyron repens* and wheat, occasionally on barley and rye and recorded on many other grasses including species of *Agrostis*, *Arrhenatherum*, *Beckmannia*, *Bromus*, *Calamagrostis*, *Elymus*, *Leersia*, *Phalaris* and *Spartina*; Australia, Canada, China, Cyprus, Denmark, England, Germany, India, Israel, Italy, Japan, Kenya, Poland, Sweden, Tanzania, Thailand, U.S.A., U.S.S.R. Sometimes causes severe leaf wilt and spotting especially on durum wheat. Leaves of *Agropyron repens* attacked gradually lose their colour and wither from the tips backwards; they become at first pale yellow, later grey. On wheat fusiform, oval or lanceolate spots, 0·5–2 cm. long, 2–4 mm. broad are formed; these are at first yellow but later turn brown or greyish brown often with a yellow halo. The leaves die prematurely from the tips backwards. In culture on p.d.a. colonies cottony white to grey; brown pigment in agar; protothecia sometimes formed.

Drechslera erythrospila (Drechsler) Shoemaker, 1959, *Can. J. Bot.*, **37**: 880.

Helminthosporium erythrospilum Drechsler, 1935, *Phytopathology*, **25**: 344–361.

(Fig. 296)

Conidiophores emerging usually singly or in pairs, cylindrical, sometimes repeatedly geniculate near the apex, mid to dark brown, up to 340µ long, usually 6–8µ thick, swollen at the base to 10–15µ. *Conidia* typically straight, occasionally slightly curved, cylindrical or almost cylindrical rounded at the ends, yellowish to mid-pale olivaceous brown, with 2–10, most frequently 4–6 pseudosepta, occasionally up to 100 × 16µ but most commonly 40–70 × 11–13µ; the basal cell is often slightly longer than the others and may be cut off by a dark transverse septum.

On *Agrostis* and occasionally on *Triticum* and other grasses, sometimes causing straw-coloured spots up to 2·5 mm. long each surrounded by a reddish border or rusty brown spots with pale centres; Australia, Europe, N. America. Numerous very small (0·1–0·3 mm. diam.) dark brown or black protothecia formed in culture and dark brown and sometimes pink pigments diffuse into the agar.

Drechslera gigantea (Heald & Wolf) Ito, 1930, *Proc. imp. Acad. Japan*, **6**: 355.
    Helminthosporium giganteum Heald & Wolf, 1911, *Mycologia*, **3**: 21.
    Conidiophores brown, up to 400µ long, 9–12µ thick, sometimes swollen to 15µ near the apex which often has a saucer-like depression at the scar. *Conidia*

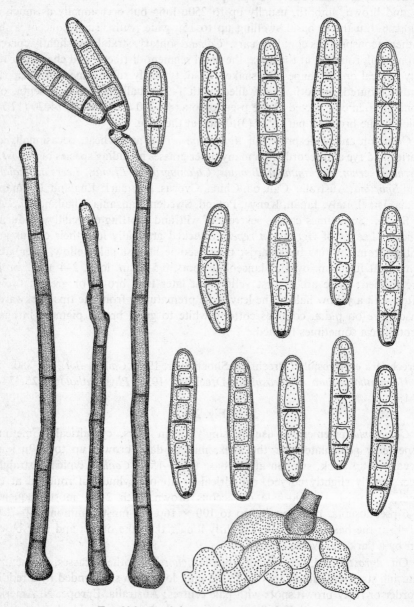

FIG. 296. *Drechslera erythrospila* (× 650).

straight, cylindrical, very pale, thin-walled, with 3–6 (commonly 5) pseudo-septa, usually 200–320 × 15–25µ but occasionally up to 390 × 30µ; hilum 1–2µ wide.

Causes a zonate eyespot disease of grasses in N. & S. America; it is especially common on *Cynodon dactylon*. A full account with host range of this species was given by C. Drechsler in *J. agric. Res.*, **37**: 473–492, 1928. D. S. Meredith in *Ann. appl. Biol.*, **51**: 29–41, 1963 recorded *D. gigantea* as the cause of eyespot disease of banana in Jamaica and described how the conidia are dispersed.

Drechslera state of **Pyrenophora bromi** (Died.) Drechsler, 1923, *J. agric. Res.*, **24**: 672.

D. bromi (Died.) Shoemaker, 1959, *Can. J. Bot.*, **37**: 881.

Helminthosporium bromi (Died.) Died., 1903, *Zentbl. Bakt. ParasitKde*, Abt. 2, **11**: 56.

(Fig. 297)

Conidiophores emerging singly or in pairs, straight or flexuous, yellowish brown, paler near the apex, usually 100–150µ long but occasionally up to 300µ, 7–11µ thick, swollen at the base to 14–20µ. *Conidia* straight, cylindrical, rounded

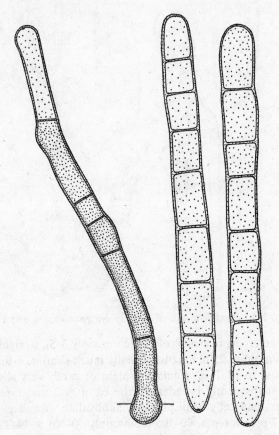

Fig. 297. *Drechslera state of Pyrenophora bromi* (× 650).

and sometimes very slightly narrower at the ends, with 2–10 (commonly 4–8) pseudosepta, subhyaline to very pale golden brown, thin-walled, hilum inconspicuous, 100–250 × 14–26μ, mostly 150–190 × 16–22μ.

On *Bromus*; Australia, Europe, N. America. Leaf spots at first small, dark brown with a yellow halo, later elongating and coalescing; finally large areas turn yellow and the leaves wither.

Drechslera state of **Pyrenophora graminea** Ito & Kuribayashi, 1930, apud Ito in *Proc. imp. Acad. Japan*, **6**: 353.

    D. graminea (Rabenh. ex Schlecht.) Shoemaker, 1959, *Can. J. Bot.*, **37**: 881.

    Helminthosporium gramineum Rabenh. ex Schlecht., 1857, *Bot. Ztg.*, **15**: 94.

(Fig. 298)

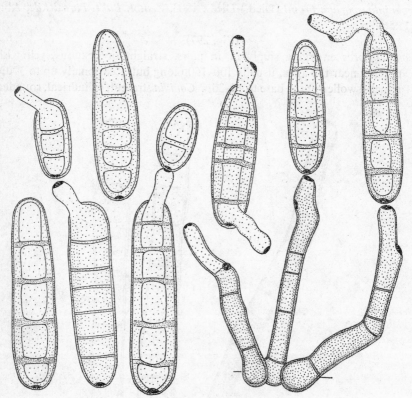

FIG. 298. *Drechslera* state of *Pyrenophora graminea* (× 650).

Conidiophores emerging in groups of 2–6 (commonly 3–5), straight or flexuous, pale to mid brown, up to 250μ long but usually much shorter, 6–9μ thick, often swollen at the base to 12–16μ. *Conidia* straight or rarely very slightly curved, subcylindrical but frequently broadest in the basal part and tapering slightly towards the apex, end cells hemispherical, subhyaline to mid golden brown, smooth, with 1–7 pseudosepta, 40–105 (commonly 50–80) × 14–22 (commonly 18–20)μ, cells usually shorter than they are broad; hilum 3–6μ wide. Short

secondary conidiophores formed from apical and often also basal cells and bear
ing *secondary conidia* are regularly produced.

Causes leaf stripe disease of barley (*Hordeum*); cosmopolitan.

Drechslera state of **Pyrenophora teres** Drechsler, 1923, *J. agric. Res.*, **24**: 656.

    D. teres (Sacc.) Shoemaker, 1959, *Can. J. Bot.*, **37**: 881.

    Helminthosporium teres Sacc., 1882, *Michelia*, **2**: 558.
(Fig. 299)

Conidiophores solitary or in groups of 2–3, straight or flexuous, sometimes
geniculate, often swollen at the base, pale to mid brown or olivaceous brown,
up to 200µ long, 7–11µ thick. *Conidia* straight, cylindrical, rounded at the ends,
subhyaline to straw-coloured, smooth, 1–10-(commonly 4–6-) pseudoseptate,

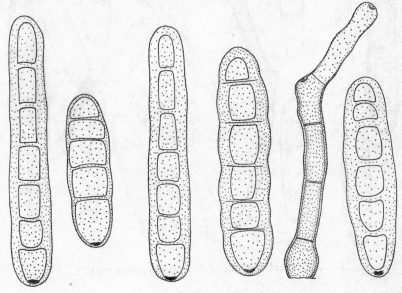

Fig. 299. *Drechslera* state of *Pyrenophora teres* (× 650).

frequently with constriction, 70–160 (mostly 90–120) × 16–23 (mostly 19–21)µ;
hilum 3–7µ wide. There are no secondary conidiophores or conidia.

Causes net-blotch lesions on leaves of barley and other species of *Hordeum*;
Africa, Asia, Europe, N. & S. America [CMI Distribution Map 364].

Drechslera state of **Pyrenophora avenae** Ito & Kuribayashi, 1930, apud Ito in
Proc. imp. Acad. Japan, **6**: 354.

    D. avenae (Eidam) Scharif, 1963, Studies on graminicolous species of *Helmin-
thosporium*, Teheran: 72.

    Helminthosporium avenae Eidam, 1891, *Der Landwirt*, **27**: 509.
(Fig. 300)

Conidiophores emerging singly or in groups of 2–4, more or less cylindrical,
straight or flexuous, often geniculate, sometimes swollen at the base, brown,

smooth, up to 350μ long, 8–11μ thick. *Conidia* solitary or occasionally catenate, straight, cylindrical, sometimes slightly tapered, rarely obclavate, pale to mid yellowish or olivaceous brown, smooth, 30–170 × 11–22μ with 1–9 pseudo-septa, on the host usually 50–110 × 15–19μ with 2–6 pseudosepta, in culture usually 30–60 × 12–15μ with 2–5 pseudosepta; hilum 4–6μ wide.

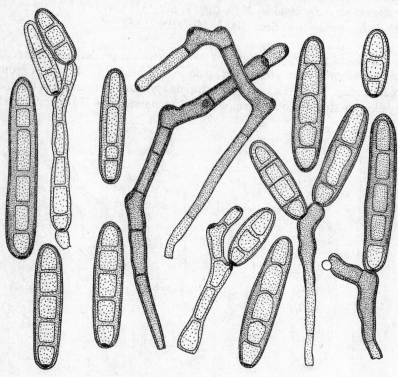

FIG. 300. *Drechslera* state of *Pyrenophora avenae* (× 650).

On oat (*Avena*) and occasionally on other grasses, cosmopolitan, causing eye-spots which are at first small (1–3 × 1–2 mm.) with a white centre surrounded by a reddish brown halo with a red border; later these spots coalesce and elon-gate to form short longitudinal stripes. On p.d.a. and other media characteristic white, splayed out 'coremia' appear on the surface of both plate and tube cultures.

Drechslera avenacea (Curtis ex Cooke) Shoemaker, 1959, *Can. J. Bot.*, **37**: 880.
 Helminthosporium avenaceum Curtis ex Cooke, 1889, *Grevillea*, **17**: 67.
(Fig. 301)

Conidiophores emerging singly or in pairs, erect, straight, usually subulate, rather closely septate, dark brown to dark blackish brown except at the apex where they are pale and bear a number of scars close together, up to 1 mm. long, 14–16μ thick immediately above the swollen base, 10–13μ thick, near the apex, basal swelling 17–24μ diam. *Conidia* solitary or occasionally in very short chains, straight, typically almost cylindrical and rounded at the ends but also sometimes clavate or obclavate, straw-coloured to dark brown, smooth, with 3–9 (usually 4–6) pseudosepta, 25–120 × 12–21 (mostly 50–95 × 13–17·5)μ; there is a large (5–8μ wide) dark hilum at the base.

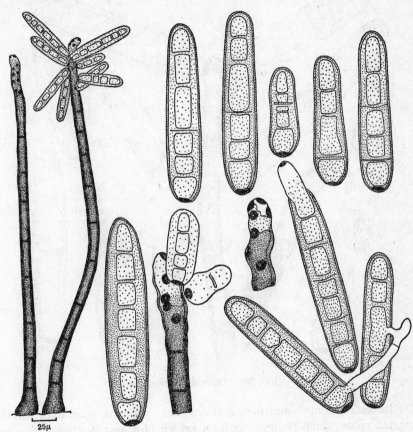

25μ

FIG. 301. *Drechslera avenacea* (× 650 except where indicated by the scale).

On straw of oat and occasionally wheat; Europe, N. America. A few proto-thecia are formed in culture. In old cultures small, scattered white tufts or 'coremia' are sometimes formed but these are not nearly so large or distinct as they are in *Pyrenophora avenae*.

Drechslera siccans (Drechsler) Shoemaker, 1959, *Can. J. Bot.*, **37**: 881.
Helminthosporium siccans Drechsler, 1923, *J. agric. Res.*, **24**: 682.
(Fig. 302)

Conidiophores solitary or occasionally in small groups, more or less straight, sometimes geniculate near the apex, chestnut brown, up to 400μ long, 7–11μ thick, swollen at the base to 19–25μ; scars large and often close together. *Conidia* usually straight, almost cylindrical or very slightly tapered towards the base or the apex, pale to mid-pale straw-coloured or olivaceous brown, 3–11–(mostly 4–6-) pseudoseptate, 30–170 (mostly 60–100) × 14–22 (mostly 16–18)μ; hilum 5–8μ wide.

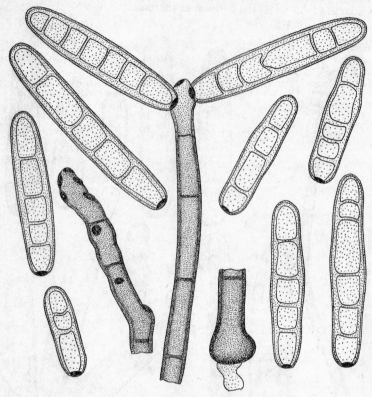

FIG. 302. *Drechslera siccans* (× 650).

On leaves of *Lolium* and occasionally other grasses causing dark brown, elongated spots which coalesce to form brown blotches; Australia, Europe, N. America.

Drechslera cactivora (Petrak) M. B. Ellis comb. nov.
Helminthosporium cactivorum Petrak, 1931, Gartenbauwissenschaft, **5**: 226–228.

(Fig. 303)

Conidiophores caespitose, straight or flexuous, often swollen at the base, usually swollen and often irregularly lobed at the apex, pale to mid golden brown, up to 250μ long, 4–6μ thick except at the apex and base where they are swollen to 8–10μ. *Conidia* straight, ellipsoidal, fusiform or obclavate, pale to mid golden brown, with 2–4 pseudosepta, 30–65μ long, 9–12μ thick in the broadest part.

On Cactaceae, Europe. No spots are formed; colonies black, velvety or hairy.

Drechslera state of **Pyrenophora dictyoides** Paul & Parberry, 1968, *Trans. Br. mycol. Soc.*, **51**: 707–710.

D. *dictyoides* (Drechsler) Shoemaker, 1959, *Can. J. Bot.*, **37**: 881.

Helminthosporium dictyoides Drechsler, 1923, *J. agric. Res.*, **24**: 679.

(Fig. 304)

Conidiophores emerging through stomata, singly or in groups of 2–6, straight or flexuous, sometimes geniculate, mid to dark brown or olivaceous brown,

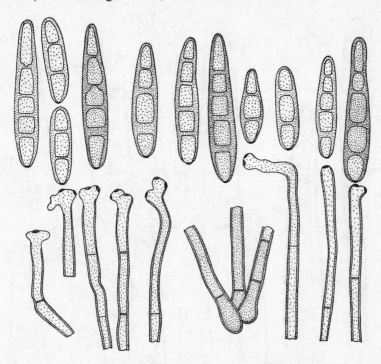

FIG. 303. *Drechslera cactivora* (× 650).

up to 250μ long, 6–10μ thick, sometimes with a swollen base. *Conidia* solitary on natural substrata but often becoming catenate when leaves are placed in a damp chamber and also in culture, straight, with maximum thickness usually at the first septum or just above this, tapering uniformly to the apex, pale to mid-pale straw-coloured, smooth with 0–14 (usually 4–7) pseudosepta, 20–250 (mostly 50–90)μ long, 14–20 (mostly 15–17)μ thick in the broadest part, 6–9μ at the apex; hilum 3–5μ wide. Conidia occasionally forked.

Common on meadow fescue (*Festuca pratensis*) causing net-blotch lesions and occasionally found on *Lolium* and other grasses; Australia, Europe, N. America.

Drechslera phlei (Graham) Shoemaker, 1959, *Can. J. Bot.*, **37**: 881.

Helminthosporium dictyoides Drechsler var. *phlei* Graham, 1955, *Phytopathology*, **45**: 228.

(Fig. 305)

Conidiophores solitary or occasionally in groups of 2–4, straight or flexuous, pale to mid brown or yellowish brown, sometimes up to 600μ long but usually 60–150μ, 6–8μ thick, often with a basal swelling 10–16μ diam. *Conidia* solitary or catenate, straight or slightly flexuous, obclavate, broadest at the second cell

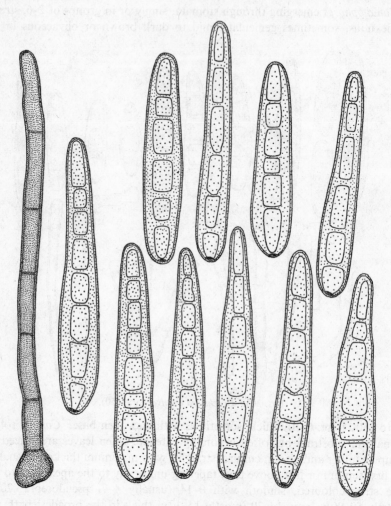

FIG. 304. *Drechslera* state of *Pyrenophora dictyoides* (× 650).

from the base and tapering to a narrow apex, pale to mid straw-coloured, smooth, usually 4–6-pseudoseptate, sometimes up to 240μ long but most frequently 50–90μ, 12–17 (commonly 14–16)μ thick in the broadest part; hilum 2–4μ wide.

On *Phleum pratense* causing irregular, pale brown necrotic streaks and blotches with chlorotic borders, occasionally also on other grasses; Europe, N. America.

Drechslera coicis (Nisikado) Subram. & Jain, 1966, *Curr. Sci.*, **35**: 354.

Helminthosporium coicis Nisikado, 1928, *Spec. Rept. Ohara Inst. agric. Res.*, **4**: 136.

Curvularia coicis Castellani, 1956, *Nuovo G. bot. ital*. N.S., **62**: 555.

(Fig. 306)

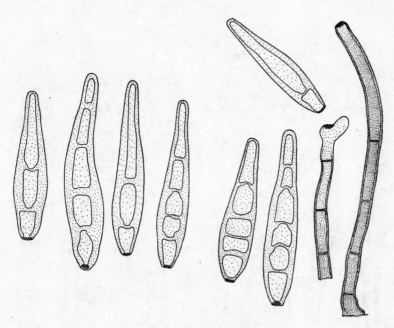

FIG. 305. *Drechslera phlei* (× 650).

Conidiophores emerging in small fascicles through stomata, straight or flexuous, brown, up to 200μ long, 6–11μ thick. *Conidia* straight or occasionally slightly curved, broadly ellipsoidal to obclavate, pale to mid golden brown, smooth, with usually 4–5 pseudosepta, 40–64 (56)μ long, 17–23 (20)μ thick in the broadest part; hilum 3–5μ wide.

On leaves of *Coix*; Brazil, Japan.

Drechslera fugax (Wallr.) Shoemaker, 1958, apud Hughes in *Can. J. Bot.*, **36**: 765.

Helminthosporium fugax Wallr., 1833, Fl. crypt. Ger., **2**: 164.

H. stenacrum Drechsler, 1923, *J. agric. Res.*, **24**: 683.

(Fig. 307)

Stroma often present. *Conidiophores* emerging singly or in pairs, cylindrical, dark brown or olivaceous brown, up to 250μ long, 7–12μ thick, swollen at the base to 11–16μ. *Conidia* straight, sometimes almost cylindrical but usually thickest at or just below the middle and tapering moderately towards the ends, subhyaline to golden brown with 4–8 (occasionally up to 10) pseudosepta,

50–170 × 14–24 (mostly 60–90 × 18–20)μ; the basal cell is usually longer than wide and may be slightly paler than the other cells.

On withered leaves of *Agrostis* and occasionally other grasses; Europe, N. America. Numerous large (2 mm. diam.) dark brown to black protothecia formed in culture; a yellowish brown pigment diffuses into the agar.

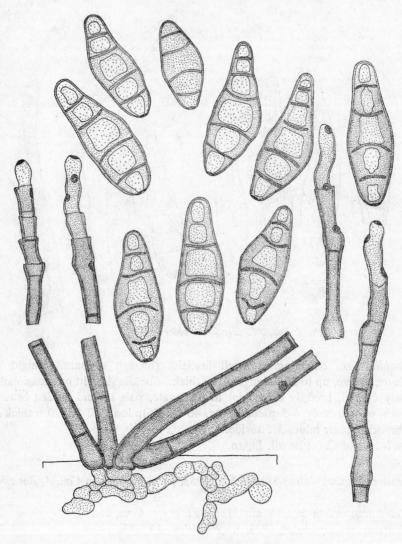

FIG. 306. *Drechslera coicis* (× 650).

Drechslera poae (Baudys) Shoemaker, 1962, *Can. J. Bot.*, **40**: 827.
　　Helminthosporium poae Baudys, 1916, *Lotos*, **64**: 87.
　　H. vagans Drechsler, 1923, *J. agric. Res.*, **24**: 688.

(Fig. 308)

Conidiophores solitary or in small groups, straight or flexuous, often geniculate, mid to dark brown, up to 250μ long, 8–12μ thick, basal cell sometimes swollen to 14–16μ. *Conidia* straight, subcylindrical, broadly fusiform or obclavate, yellowish brown when mature, smooth, 1–12 (mostly 5–8-) pseudoseptate, 30–160 × 13–28 (mostly 60–100 × 17–23)μ; hilum 3·5–5μ wide.

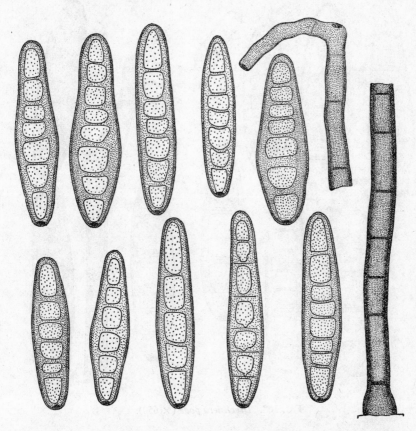

Fig. 307. *Drechslera fugax* (× 650).

On *Poa* and occasionally other grasses, causing dark brown or reddish brown elongated spots the centres of which become bleached to form eye–spots; Australia, Europe, N. America. In culture on p.d.a. colonies slow-growing, dark green.

Drechslera state of **Cochliobolus miyabeanus** (Ito & Kuribayashi) Drechsler ex Dastur, 1942, *Indian J. agric. Sci.*, **12**: 733.

  D. oryzae (Breda de Haan) Subram. & Jain, 1966, *Curr. Sci.*, **35**: 354.

  Helminthosporium oryzae Breda de Haan, 1900, *Bull. Inst. bot. Buitenz.*, **6**: 11–13.

(Fig. 309)

Conidiophores solitary or in small groups, straight or flexuous, sometimes geniculate, pale to mid brown or olivaceous brown, up to 600μ long, 4–8μ thick. *Conidia* usually curved, navicular, fusiform or obclavate, occasionally almost cylindrical, pale to mid golden brown, smooth, 6–14-pseudoseptate, 63–153

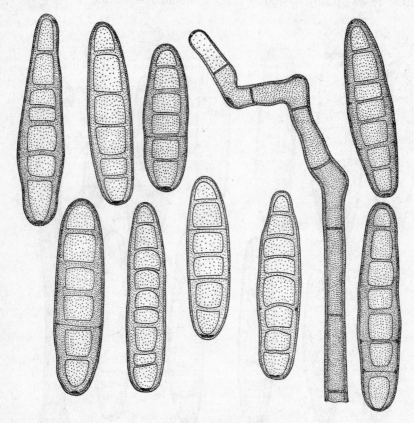

FIG. 308. *Drechslera poae* (× 650).

(109) × 14–22 (17)μ; hilum minute, often protruding slightly, papillate. Conidia formed on glumes and in culture on wheat straw t.w.a. under near u.v. light are as a rule larger and darker than those formed on leaf spots.

On leaves, leaf sheaths and glumes of rice (*Oryza*), the cause of brown spot disease; cosmopolitan [CMI Distribution Map 92]. Leaf spots, when small brown or purplish brown, later pale in the centre with a mid to dark brown margin.

Drechslera sorghicola (Lefebvre & Sherwin) Richardson & Fraser, 1968, *Trans. Br. mycol. Soc.*, **51**: 148.

Helminthosporium sorghicola Lefebvre & Sherwin, 1948, *Mycologia*, **40**: 708–716.

(Fig. 310)

Conidiophores usually solitary, sometimes in small groups, straight or flexuous, mid to dark brown or olivaceous brown, up to 700μ long, 6–9μ thick, sometimes swollen at the base. *Conidia* slightly curved, fusiform, pale to mid golden brown, smooth, 3–8- (mostly 6-) pseudoseptate, 30–100 (mostly 50–90) × 12–19 (mostly 14–17)μ; hilum 3–4μ wide. Primary conidia whilst still attached to the conidiophore frequently bear long *secondary conidiophores* with *secondary conidia.*

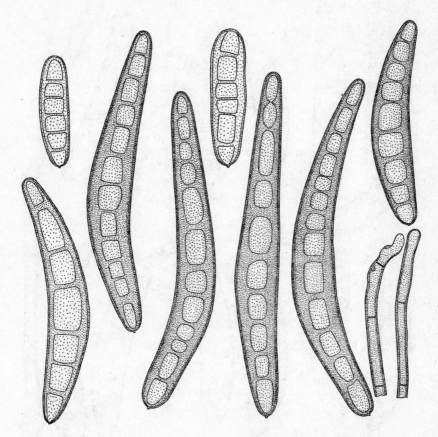

FIG. 309. *Drechslera* state of *Cochliobolus miyabeanus* (× 650).

On *Sorghum* causing at first small reddish purple or tan spots, later zonate target spots; Australia, Ethiopia, India, Malaya, Pakistan, Nigeria, Rhodesia, Sudan, U.S.A.

Drechslera state of **Cochliobolus cynodontis** Nelson, 1964, *Mycologia*, **56**: 67.

   D. cynodontis (Marignoni) Subram. & Jain, 1966, *Curr. Sci.*, **35**: 354.

   Helminthosporium cynodontis Marignoni, 1909, Micromiceti di Schio: 27.

(Fig. 311)

Conidiophores arising singly or in small groups, flexuous, pale to mid brown, up to 170μ long, 5–7μ thick. *Conidia* mostly slightly curved, sometimes almost cylindrical but usually broadest in the middle tapering towards the rounded ends, pale to mid golden brown, smooth, thin-walled, with 3–9 (usually 7–8) pseudosepta, 30–75 (50) × 10–16 (13)μ; hilum 2·5–3μ wide.

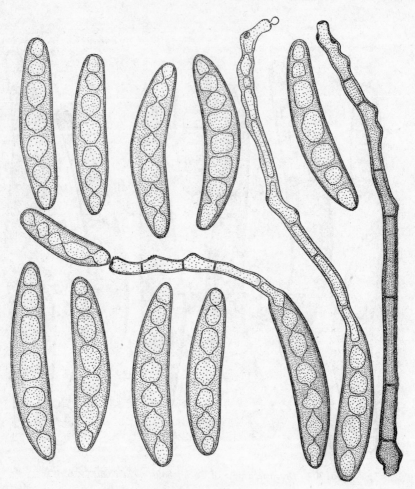

FIG. 310. *Drechslera sorghicola* (× 650).

Common on *Cynodon* and recorded on many other grasses, also isolated from apple, pine and *Ipomoea*; cosmopolitan.

Drechslera euphorbiae (Hansford) M. B. Ellis comb. nov.

Helminthosporium euphorbiae Hansford, 1943, *Proc. Linn. Soc. Lond.*, 1942–43: 49.

(Fig. 312)

Conidiophores arising singly or in small groups, straight or flexuous, often geniculate, brown, up to 200μ long, 5–8μ thick. *Conidia* mostly slightly curved, usually thickest near the middle and tapering towards the rounded ends, pale to mid golden brown, 5–11-pseudoseptate, 50–130 × 11–17μ; hilum 2·5–4μ wide.

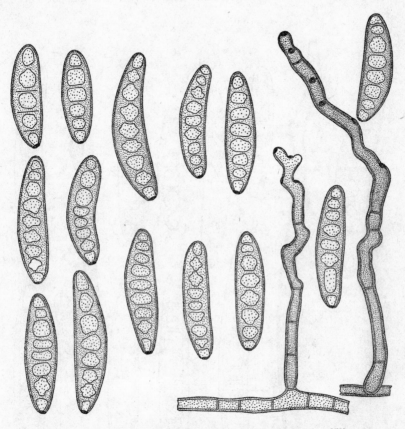

Fig. 311. *Drechslera* state of *Cochliobolus cynodontis* (× 650).

On leaves of *Euphorbia* causing orbicular, pale to mid brown spots each surrounded by a dark border; E. & W. Africa, India.

Drechslera sacchari (Butler) Subram. & Jain, 1966, *Curr. Sci.*, **35**: 354.

Helminthosporium sacchari Butler, 1913, apud Butler & Khan, *Mem. Dep. Agric. India*, bot. ser., **6** (6): 204–208.

(Fig. 313)

Conidiophores arising singly or in small fascicles, often from groups of dark cells which form loose stromata, straight or flexuous, mid to dark brown or olivaceous brown, paler towards the apex, up to 200μ long, 5–8μ thick (up to

700 × 10μ in culture). *Conidia* slightly curved, occasionally straight, cylindrical or narrowly ellipsoidal, mid pale to mid golden brown, with 5–9 (mostly 8) pseudosepta, 35–96 (65) × 9–17 (13·8)μ; hilum 2–3μ wide.

On leaves of sugar-cane (*Saccharum*); cosmopolitan [CMI Distribution Map 349]. Infected leaves at first show small red spots which spread longitudinally and may run together to form long streaks. The centre of the spot soon becomes

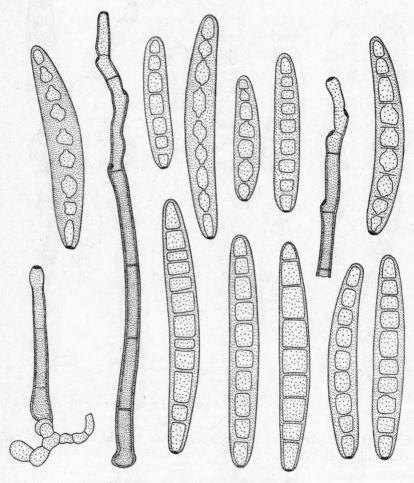

FIG. 312. *Drechslera euphorbiae* (× 650).

straw-coloured, the margin remains red for some time but eventually changes to dark brown. When infection is severe the leaf tissues outside the spots are affected, the leaf tips and long strips down the margin withering. Conidia in culture tend to be more curved than they are on natural substrata; groups of dark cells are formed.

Drechslera state of **Cochliobolus setariae** (Ito & Kuribayashi) Drechsler ex
Dastur, 1942, *Indian J. agric. Sci.*, **12**: 733.

D. *setariae* (Sawada) Subram. & Jain, 1966, *Curr. Sci.*, **35**: 354.
(Fig. 314)

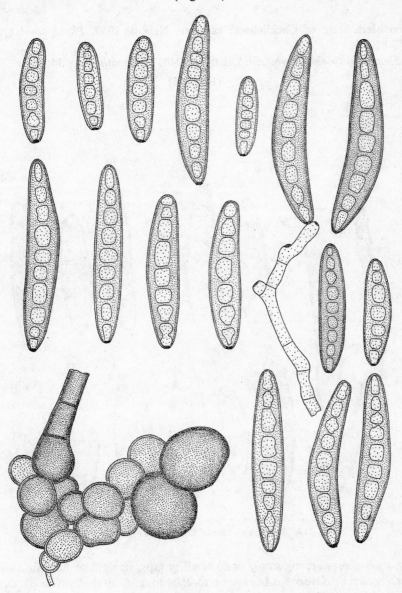

FIG. 313. *Drechslera sacchari* (× 650).

Conidiophores solitary or in small groups, straight or flexuous, sometimes
geniculate, pale to mid brown or olivaceous brown, up to 200μ long, 5–9μ
thick, sometimes swollen at the base to 11μ. *Conidia* slightly curved or sometimes

straight, fusiform or navicular, pale to mid golden brown, smooth, 5–10-pseudoseptate, 45–100 (mostly 50–70) × 10–15 (mostly 12–14)µ.

Most records are on *Setaria italica* from Europe, N. America and Taiwan but it has been isolated occasionally from other plants and from soil.

Drechslera state of **Cochliobolus carbonus** Nelson, 1959, *Phytopathology*, **49**: 807–810.

Helminthosporium carbonum Ullstrup, 1944, *Phytopathology*, **34**: 219.

(Fig. 315)

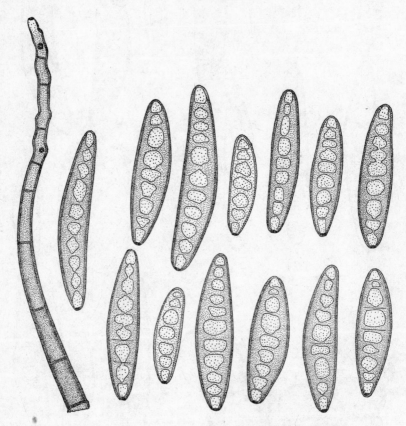

Fig. 314. *Drechslera* state of *Cochliobolus setariae* (× 650).

Conidiophores arising singly or in small groups, straight or flexuous, mid to dark brown or olivaceous brown, up to 250µ long, 5–8µ thick. *Conidia* curved or sometimes straight, occasionally almost cylindrical but usually broader in the middle and tapering towards the rounded ends, 30–100 × 12–18 (mostly 60–80 × 14–16)µ, with 6–12 (usually 7–8) pseudosepta, often finally becoming dark or very dark brown or olivaceous brown, the end cells sometimes remaining paler than the middle ones; hilum not very conspicuous.

On *Zea mays*, affecting all aerial parts; diseased ears are covered with very dark brown or black mycelium which gives them a characteristic charred appearance. Similar mycelium is formed in culture, especially when the fungus is grown on wheat straw and t.w.a.; the dark, thick-walled hyphae sometimes break up into segments. Specimens seen from N. America, E., S. and W. Africa and India, mostly on *Zea* but occasionally on *Sorghum* and other grasses and once on *Elaeis* [CMI Distribution Map 380].

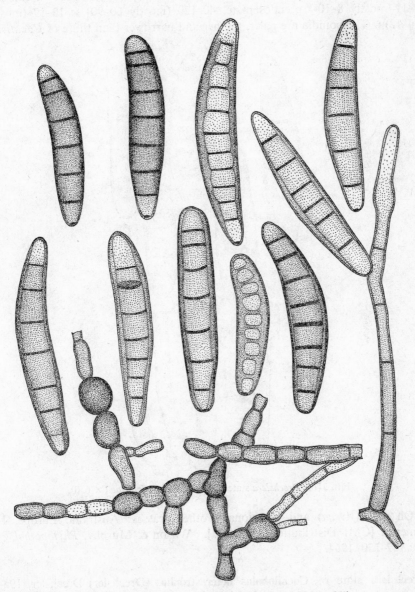

FIG. 315. *Drechslera* state of *Cochliobolus carbonus* (× 650).

Drechslera state of **Cochliobolus victoriae** Nelson, 1960, *Phytopathology*, **50**: 774.

　　Drechslera victoriae (Meehan & Murphy) Subram. & Jain, 1966, *Curr. Sci.*, **35**: 354.

　　Helminthosporium victoriae Meehan & Murphy, 1946, *Science*, **104**: 413–414.
　　　　　　　　　　　　　(Fig. 316)

　　Conidiophores solitary or in groups, straight or flexuous, sometimes geniculate, pale to mid brown, up to 250μ long, 6–10μ thick. *Conidia* slightly curved, broadly fusiform or obclavate-fusiform, pale or mid-pale golden brown, smooth, 4–11-(mostly 8–10-) pseudoseptate, 40–120 (mostly 60–90) × 12–19 (mostly 14–18)μ. The conidia are paler, thinner and narrower than those of *C. sativus*.

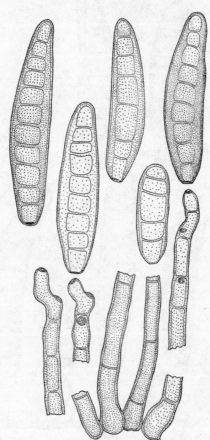

FIG. 316. *Drechslera* state of *Cochliobolus victoriae* (× 650).

　　On oats (*Avena*) and occasionally other grasses; Australia, Europe, N. America [CMI Distribution Map 267]. Wilson & Murphy, *Phytopathology*, **54**: 147–150, 1964.

Drechslera state of **Cochliobolus heterostrophus** (Drechsler) Drechsler, 1934, *Phytopathology*, **24**: 973.

D. maydis (Nisikado) Subram. & Jain, 1966, *Curr. Sci.*, **35**: 354.

Helminthosporium maydis Nisikado, 1926, *Sci. Res. Alumni Assoc. Morioka agric. Col. Japan*, **3**: 46.

(Fig. 317)

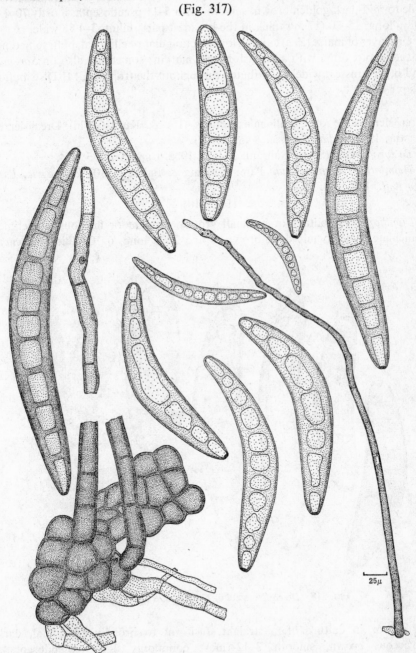

FIG. 317. *Drechslera* state of *Cochliobolus heterostrophus* (× 650 except where indicated by the scale).

Conidiophores arising in groups often from flat, dark brown to black stromata, straight or flexuous, sometimes geniculate, mid to dark brown, pale near the apex, smooth, up to 700μ long, 5–10μ thick. *Conidia* distinctly curved, fusiform, pale to mid-dark golden brown, smooth, with 5–11 pseudosepta, mostly 70–160 (98)μ long, 15–20 (17·3)μ thick in the broadest part; hilum 3–4·5μ wide.

On leaves of maize (*Zea*) causing long, rectangular or elliptical, buff, sometimes zonate spots, often with a reddish brown margin; sometimes also on *Sorghum* and other grasses; widely distributed throughout the tropics [CMI Distribution Map 346].

Drechslera state of **Cochliobolus sativus** (Ito & Kuribayashi) Drechsler ex Dastur, 1942, *Indian J. agric. Sci.*, **12**: 733.

D. sorokiniana (Sacc.) Subram. & Jain, 1966, *Curr. Sci.*, **35**: 354.

Helminthosporium sativum Pammel, King & Bakke, 1910, *Iowa agric. Exp. Stn Bull.*, **116**: 180.

(Fig. 318)

Conidiophores solitary or in small groups, straight or flexuous, sometimes geniculate, pale to mid-dark brown, up to 220μ long, 6–10μ thick. *Conidia*

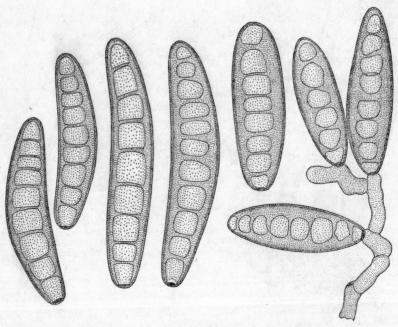

FIG. 318. *Drechslera* state of *Cochliobolus sativus* (× 650).

curved, or in culture often straight, fusiform to broadly ellipsoidal, dark olivaceous brown, smooth, 3–12–(most commonly 6–10-) pseudoseptate, 40–120 (most commonly 60–100)μ long, 17–28 (most commonly 18–23)μ thick in the broadest part.

Very common on grasses including the cereals barley, oats, rye and wheat, causing dark brown elongated leaf spots, seedling blight, foot rot, etc.; cosmopolitan [CMI Distribution Map 322]. Luttrell, E. S., *Am. J. Bot.*, **42**: 57–68, 1965.

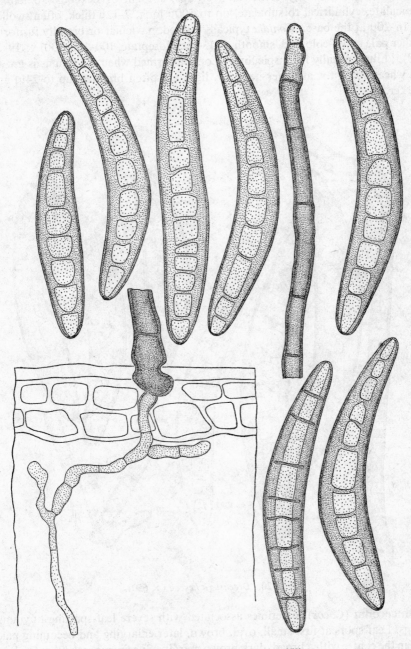

FIG. 319. *Drechslera incurvata* (× 650).

D.H.—15

Drechslera incurvata (Ch. Bernard) M. B. Ellis comb. nov.

Helminthosporium incurvatum Ch. Bernard, 1906, *Bull. Dép. Agric. Indes néerl.*, No. **2**: 31.

(Fig. 319)

Conidiophores solitary or in small groups, straight or flexuous, sometimes geniculate, cylindrical to subulate, up to 500μ long, 7–12μ thick, often swollen to 16–20μ at the base. *Conidia* typically curved, navicular or broadly fusiform, rather pale straw-coloured, smooth, 8–13-pseudoseptate, 100–150 (130) × 19–22 (20)μ; hilum usually inconspicuous. Conidia formed when the fungus is grown on wheat straw t.w.a. under near u.v. light are often broader (up to 24μ) and darker.

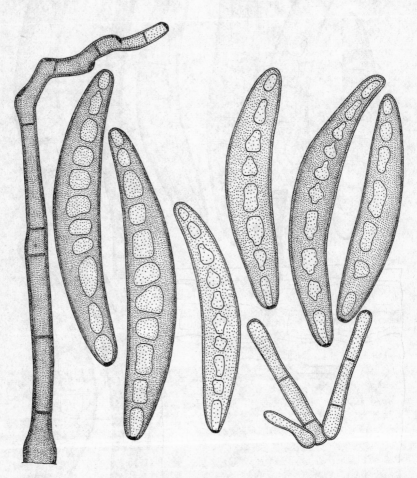

FIG. 320.   *Drechslera heveae* (× 650).

On coconut (*Cocos*) sometimes associated with severe leaf-spotting of young palms. Leaf spots at first small, oval, brown, later enlarging and becoming pale buff in the centre with a broad, dark brown margin. Specimens seen from Ceylon,

Fiji, French Polynesia, Guadalcanal, Malaya, New Guinea, New Hebrides, Papua, Philippines, Sabah, Seychelles.

Drechslera heveae (Petch) M. B. Ellis comb. nov.

Helminthosporium heveae Petch, 1906, *Ann. R. bot. Gdns Peradeniya*, **3** (1): 8–9.

(Fig. 320)

Conidiophores solitary or in small groups, straight or flexuous, sometimes geniculate, pale to mid brown, up to 200µ long, 6–8µ thick; up to 10µ thick and often much darker in culture. *Conidia* usually curved, navicular or fusiform, pale to mid golden or reddish brown, smooth, 6–11-pseudoseptate, mostly 90–130 × 15–21µ; hilum 3–5µ wide. Conidia formed when the fungus is grown on wheat straw t.w.a. under near u.v. light are often up to 140µ long, 22–28µ thick and dark reddish brown.

On leaves of *Hevea brasiliensis*, causing minute (1–5 mm.) orbicular purple spots which later become white each with a brown border; fairly widespread in the tropics [CMI Distribution Map 270].

Drechslera musae-sapientum (Hansford) M. B. Ellis comb. nov.

Helminthosporium musae-sapientum Hansford, 1943, *Proc. Linn. Soc. Lond.*, *1942–43*: 49–50.

(Fig. 321)

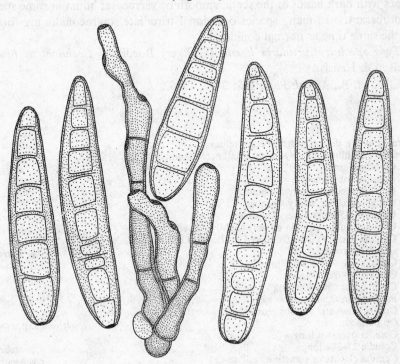

FIG. 321. *Drechslera musae-sapientum* (× 650).

Conidiophores solitary or in small groups, flexuous, sometimes geniculate, pale to mid brown or olivaceous brown, up to 250μ long, 7–12μ thick. *Conidia* mostly slightly curved, subcylindrical, obclavate or broadly fusiform, pale to mid golden brown, smooth, 6–12-pseudoseptate, 70–130 × 17–28μ; hilum 3–4μ wide.

On leaves of *Musa*, causing oval or irregular very pale brown spots each surrounded by a very dark brown border; Ethiopia, Sudan, Uganda.

239. CURVULARIA

Curvularia Boedijn, 1933, *Bull. Jard. bot. Buitenz.*, III, **13** (1): 120–134.

Malustela Batista & Lima, 1960, *Publçoes Inst. Micol. Recife*, **263**: 5–10.

Colonies effuse, brown, grey or black, hairy, cottony or velvety. *Mycelium* immersed in natural substrata. *Stromata* often large, erect, black, cylindrical, sometimes branched, formed by many species in culture, especially on firm substrata such as rice grains. *Conidiophores* macronematous, mononematous, straight or flexuous, often geniculate, sometimes nodose, brown, usually smooth. *Conidiogenous cells* polytretic, integrated, terminal, sometimes later becoming intercalary, sympodial, cylindrical or occasionally swollen, cicatrized. *Conidia* solitary, acropleurogenous, simple, often curved, clavate, ellipsoidal, broadly fusiform, obovoid or pyriform with 3 or more transverse septa, pale or dark brown, often with some cells, usually the end ones, paler than the others, sometimes with dark bands at the septa, smooth or verrucose; hilum in some species protuberant. In many species occasional triradiate stauroconidia are formed at the same time as normal conidia.

Type species: Curvularia lunata (Wakker) Boedijn = *Cochliobolus lunatus* Nelson & Haasis.

Ellis, M. B., *Mycol. Pap.*, **106**: 2–43, 1966.

KEY

Conidia with a distinctly protuberant hilum 1
Conidia with hilum scarcely or not at all protuberant 9
1. Conidia sigmoid . . . . *deightonii*
   Conidia not sigmoid 2
2. Conidia predominantly 3-septate 3
   Conidia predominantly 4-septate 7
   Conidia 3–5-septate, very variable . . . . . . *stapeliae*
3. Conidia usually straight or only slightly curved 4
   Conidia usually distinctly curved at the third cell from the base . . 5
4. Conidia clavate . . . . *borreriae*
   Conidia cylindrical or ellipsoidal . . . . . . *harveyi*
5. Conidia 45–66μ long . . . . *andropogonis*
   Conidia less than 40μ long 6
6. Parasitic on *Ophioglossum vulgatum* . . . . . . *crepinii*
   On *Trifolium* and various grasses . . . . . . *trifolii*
   On *Gladiolus* . . . . *trifolii* f.sp. *gladioli*
7. Conidia over 40μ long 8
   Conidia 27–38μ long . . . . *protuberata*
8. Conidia clavate or sometimes ellipsoidal . . . . *cymbopogonis*
   Conidia obclavate to ellipsoidal . . . . . *comoriensis*

 9. Conidia becoming rough-walled 10
    Conidia remaining smooth-walled 12
10. Conidia 4-septate . . . . *verruciformis*
    Conidia 3-septate 11
11. Conidia straight . . . . *tuberculata*
    Conidia usually curved, asymmetrical . . . . . . *verruculosa*
12. Conidia predominantly 3-septate 13
    Conidia predominantly 4-septate 22
    Conidia 3–4-septate, thick-walled . . . . *prasadii*
13. Middle septum usually truly median 14
    Middle septum not median 16
14. Conidia symmetrical . . . . *eragrostidis*
    Conidia asymmetrical 15
15. Conidia 19–26×10–14μ . . . . *brachyspora*
    Conidia 27–40×13–20μ . . . . *intermedia*
16. All conidial cells usually pale or very pale brown 17
    Some conidial cells always mid or dark brown 18
17. Conidia usually straight or only slightly curved . . . . . *pallescens*
    Conidia strongly geniculate or uncinate . . . . . *leonensis*
18. Conidia straight or almost straight 19
    Conidia curved 20
19. Conidia symmetrical, obclavate to ellipsoidal . . . . . *oryzae*
    Conidia symmetrical, clavate . . . . *clavata*
    Conidia often asymmetrical . . . . *ovoidea*
20. Conidia 29–42×13–20μ . . . . *penniseti*
    Conidia 18–32×8–16μ 21
21. Stromata very rarely formed in culture, colonies on p.d.a. not markedly zonate *lunata*
    Stromata regularly and abundantly formed in culture, colonies on p.d.a. usually zonate
    lunata var. *aeria*
22. Conidia predominantly uncinate . . . . *uncinata*
    Conidia not uncinate 23
23. Conidia averaging 24μ long . . . . *senegalensis*
    Conidia averaging over 45μ long, over 20μ thick . . . . *robusta*
    Conidia averaging over 30μ long, under 17μ thick 24
24. Conidia straight or slightly curved, only tapering a little at each end . *inaequalis*
    Conidia curved or straight, tapering gradually towards each end . . 25
25. Conidia often distinctly geniculate, stromata not formed in culture . *geniculata*
    Conidia often curved but seldom geniculate; stromata regularly and abundantly formed in
    culture 26
26. Stromata simple, conidia averaging 32×10μ . . . . . *affinis*
    Stromata branched, conidia averaging 30×12·2μ . . . . *fallax*

With the exception of *Curvularia ovoidea* (Hiroe & Watan.) Muntañola
[*Revta agron. NE Argent*, **2**: 322, 1957] and *C. robusta* Kilpatrick & Luttrell
[*Mycologia*, **59**: 888–892, 1967] all the species keyed out above are fully described
and illustrated in *Mycol. Pap.*, **106**. Their substratum range, geographical distribution and conidial measurements in μ are given below; S indicates that stromata
have been recorded in culture.

Curvularia affinis Boedijn (Fig. 322 A) on *Durio, Lagenaria, Manihot, Oryza* and
soil; Congo, Java, Malaya, Pakistan, Venezuela; S (27–39 × 8–13).
C. andropogonis (Zimm.) Boedijn (Fig. 322 B) on *Cymbopogon*; India, Indonesia,
Malaysia; S (45–66 × 18–28).
C. borreriae (Viégas) M. B. Ellis (Fig. 322 C) on *Borreria*; Brazil (20–32 × 8–15).

C. brachyspora Boedijn (Fig. 322 D) on *Agave*, *Olea*, *Saccharum*, *Triticum* and isolated from air; Australia, Puerto Rico, Tanzania, Venezuela; S (20–26 × 10–14).

C. clavata Jain (Fig. 322 E) isolated from *Sorghum* and *Tripogon*; Australia, India (17–29 × 7–13).

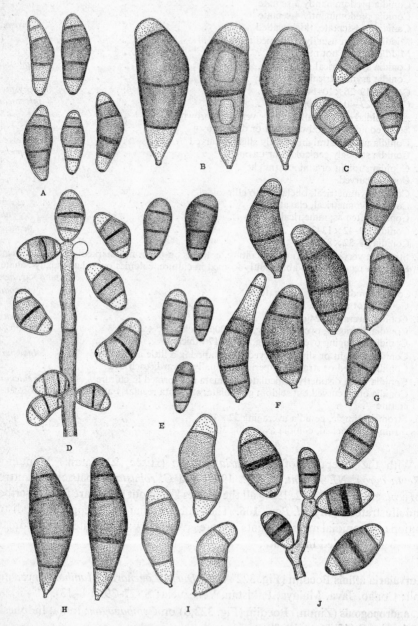

FIG. 322. *Curvularia* species (1): A, *affinis*; B, *andropogonis*; C, *borreriae*; D, *brachyspora*; E, *clavata*; F, *comoriensis*; G, *crepinii*; H, *cymbopogonis*; I, *deightonii*; J, *eragrostidis* (× 650).

C. comoriensis Bouriquet & Jauffret (Fig. 322 F) isolated from *Cymbogon*; Comoro Islands, Congo; S (30–55 × 13–20).

C. crepinii (Westend.) Boedijn (Fig. 322 G) on *Ophioglossum*; Europe (18–32 × 7–16).

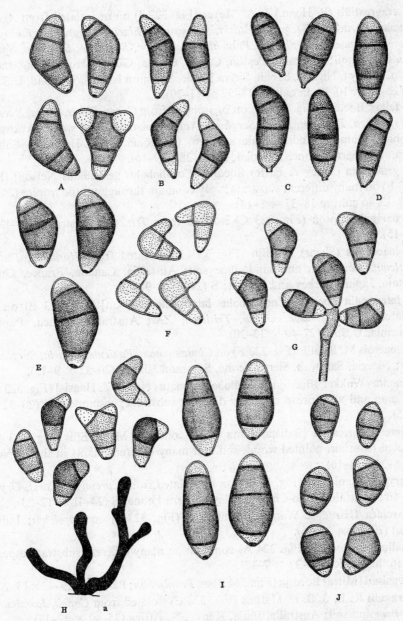

FIG. 323. *Curvularia* species (2): A, *fallax*; B, *geniculata*; C, *harveyi*; D, *inaequalis*; E, *intermedia*; F, *leonensis*; G, *lunata*; H, *lunata* var. *aeria*; I, *oryzae*; J, *ovoidea* (a, habit sketch of stroma; other figures × 650).

C. cymbopogonis (C. W. Dodge) Groves & Skolko (Fig. 322 H) on Andropogoneae; Jamaica, Sudan, W. Africa and isolated from *Pinus;* Tanzania; S (35–60 × 14–20, in culture 28–40 × 12–18).

C. deightonii M. B. Ellis (Fig. 322 I) on *Andropogon*; Sierra Leone (34–47 × 11–19).

C. eragrostidis (P. Henn.) J. A. Meyer (Fig. 322 J) on or isolated from *Agave, Ananas, Arachis, Cocos, Digitaria, Dioscorea, Elaeis, Eragrostis, Furcraea, Hevea, Ipomoea, Oldenlandia, Polygala, Saccharum, Sesamum, Sorghum, Sporobolus, Zea*, soil; Australia, Ceylon, Congo, Ghana, Guinea, Hong Kong, India, Java, Malaya, Nigeria, Sabah, Sierra Leone, Solomon Islands, Trinidad, U.S.A.; S (22–33 × 10–18, in culture 18–37 × 11–20).

C. fallax Boedijn (Fig. 323 A) on or isolated from *Cocos, Elaeis, Musa, Nyctanthes, Oryza, Panicum, Piper, Sorghum, Xanthosoma*, air, soil, wood; Australia, Congo, Gambia, India, Malaya, New Caledonia, New Hebrides, Sabah, Sarawak, Sierra Leone, Trinidad; S (24–38 × 9–16).

C. geniculata (Tracy & Earle) Boedijn [**Cochliobolus geniculatus** Nelson] (Fig. 323 B) on many different substrata; very common throughout the tropics (26–48 ×8–13, in culture 18–37 × 8–14).

C. harveyi Shipton (Fig. 323 C) isolated from *Triticum*; Australia (25–43 × 10–15).

C. inaequalis (Shear) Boedijn (Fig. 323 D) isolated from *Hordeum, Pisum, Triticum, Vaccinium* and sand dune soil; Australia, Canada, France, Great Britain, Japan, Turkey and U.S.A.; S (24–45 × 9–16).

C. intermedia Boedijn [**Cochliobolus intermedius** Nelson] (Fig. 323 E) on or isolated from *Cynodon, Oryza, Triticum, Zea*; Australia, Guinea, Papua, Tanzania, U.S.A. (27–40 × 13–20).

C. leonensis M. B. Ellis (Fig. 323 F) on *Cinnamomum, Panicum, Setaria, Sorghum, Zea*; Nigeria, Sarawak, Sierra Leone, Solomon Islands (20–32 × 9–13).

C. lunata (Wakker)Boedijn [**Cochliobolus lunatus** Nelson & Haasis] (Fig. 323 G) common and widespread on many different substrata; cosmopolitan (20–32 × 9–15).

C. lunata var. **aeria** (Batista, Lima & Vasconcelos) M. B. Ellis (Fig. 323 H) isolated from air, painted wood, soil and many different plants in the tropics; S (18–32 × 8–16).

C. oryzae Bugnicourt (Fig. 323 I) on or isolated from *Curculigo, Elaeis, Oryza* and air; Australia, Indo-China, Malaya, Sierra Leone; S (24–41 × 12–23).

C. ovoidea (Hiroe & Watan.) Muntañola (Fig. 323 J) on *Capsicum*; India, Japan (20–25 × 13–16).

C. pallescens Boedijn (Fig. 324 A) common on many different substrata especially in the tropics (17–32 × 7–12).

C. penniseti (Mitra) Boedijn (Fig. 324 B) on *Pennisetum*; Pakistan (29–42 × 13–20).

C. prasadii R. L. & B. L. Mathur (Fig. 324 C) isolated from *Coffea, Jasminum, Trifolium* and soil; Australia, India, Kenya, N. Africa (25–40 × 12–17).

C. protuberata Nelson & Hodges (Fig. 324 D) isolated from *Deschampsia* and *Phleum*; Canada, Scotland; S (27–35 × 10–14).

C. robusta Kilpatrick & Luttrell (Fig. 324 E) on *Dichanthium*; U.S.A. (25–76 × 15–27).

C. senegalensis (Speg.) Subram. (Fig. 325 A) isolated from *Urena*, *Zea*, paintwork and soil; India, Kenya, Nigeria (19–30 × 10–14).

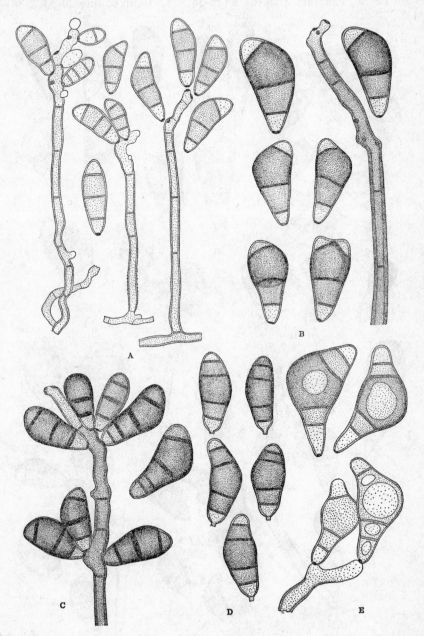

FIG. 324. *Curvularia* species (3): A, *pallescens*; B, *penniseti*; C, *prasadii*; D, *protuberata*; E, *robusta* (× 650).

C. stapeliae (du Plessis) Hughes & du Plessis (Fig. 325 B) on *Dioscorea, Huernia, Stapelia, Trichocaulon*; Nigeria, S. Africa (28–46 × 10–16).

C. trifolii (Kauffm.) Boedijn (Fig. 325 C) most commonly found on *Trifolium* but occasionally also on grasses; Australia, Ghana, Kenya, Papua, Portugal, Sierra Leone, Tanzania, U.S.A.; S (28–38 × 12–16, in culture 20–34 × 8–14).

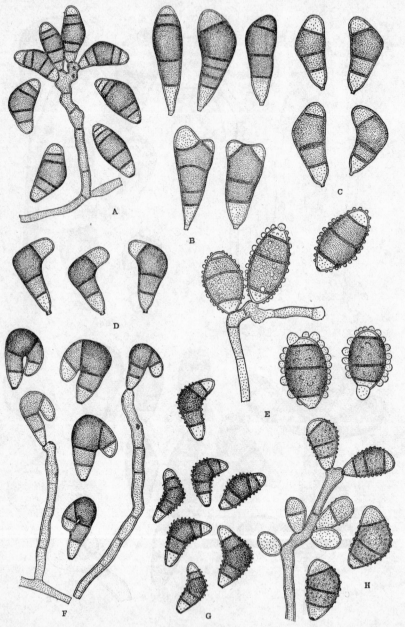

FIG. 325. *Curvularia* species (4): A, *senegalensis*; B, *stapeliae*; C, *trifolii*; D, *trifolii* f. sp. *gladioli*; E, *tuberculata*; F, *uncinata*; G, *verruciformis*; H, *verruculosa* (× 650).

C. trifolii f. sp. **gladioli** Parmelee & Luttrell (Fig. 325 D) on *Gladiolus*; Canada, Malaya, Rhodesia, Sierra Leone.

C. tuberculata Jain (Fig. 325 E) isolated from *Pinus*, air, fabrics and soil; India, Indo-China, Iraq, Tanzania (23–52 × 13–20).

C. uncinata Bugnicourt (Fig. 325 F) isolated from *Oryza* and air; Australia, Indo-China; S (24–35 × 6–15).

C. verruciformis Agarwal & Sahni (Fig. 325 G) isolated from *Triticum* and from a masked plover; Australia, India; S (16–26 × 8–12).

C. verruculosa Tandon & Bilgrami ex M. B. Ellis (Fig. 325 H) on many different plants and other substrata in the tropics; S (25–32 × 12–14, in culture 20–35 × 12–17).

240. BRACHYDESMIELLA

Brachydesmiella Arnaud ex Hughes, 1961, *Can. J. Bot.*, **39**: 1095–1097.

Colonies effuse, black, shining. *Mycelium* partly superficial, partly immersed. *Stroma* none. *Setae* and *hyphopodia* absent. *Conidiophores* macronematous, mononematous, flexuous, unbranched or rarely branched, smooth. *Conidiogenous cells* integrated, terminal, polytretic, sympodial, cylindrical, cicatrized.

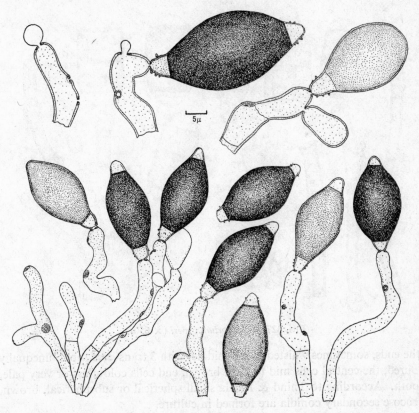

Fɪɢ. 326. *Brachydesmiella biseptata* (× 650 except where indicated by the scale).

Conidia solitary, acropleurogenous, simple, limoniform, nearly always 2-septate, unequally coloured, the central cell large and smooth, brown to almost black, the end cells small, very pale, usually verrucose or echinulate.

Type species: Brachydesmiella biseptata Arnaud ex Hughes.

Brachydesmiella biseptata Arnaud ex Hughes, 1961, *Can. J. Bot.*, **39**: 1095.

(Fig. 326)

Conidiophores up to 70μ long, 3–4μ thick at the base, 5–9μ near the apex, very pale brown. *Conidia* 42–48 (45) × 19–22 (20·8)μ.

On wood of ash (*Fraxinus*), beech (*Fagus*), etc.; Canada, France, Great Britain.

241. DUOSPORIUM

Duosporium Thind & Rawla, 1961, *Am. J. Bot.*, **48**: 859–862.

Mycelium immersed. *Stroma* present, substomatal. *Setae* and *hyphopodia* absent. *Conidiophores* macronematous, mononematous, caespitose, straight or flexuous, unbranched, pale to mid brown, smooth. *Conidiogenous cells* polytretic, integrated, terminal becoming intercalary, sympodial, cylindrical, cicatrized. *Conidia* solitary, dry, acropleurogenous, simple, straight, oblong rounded

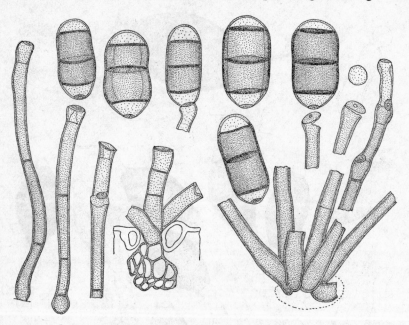

FIG. 327. *Duosporium cyperi* (× 650).

at the ends, sometimes waisted in the middle, with 3 transverse septa, unequally coloured, the central cells mid to dark brown, end cells colourless or very pale, smooth. According to Thind & Rawla small spherical or subspherical, brown, verrucose secondary conidia are formed in culture.

Type species: Duosporium cyperi Thind & Rawla.

Duosporium cyperi Thind & Rawla, 1961, *Am. J. Bot.*, **48**: 859–862.

(Fig. 327)

Conidiophores up to 300µ long, 5–9µ thick; scars large. The proliferating conidiophore tip pushes the cap scar either to one side or sometimes right off. *Conidia* 35–45 × 16–22µ. *Secondary conidia* 6–11µ diam.

On leaves of *Cyperus iria*, causing narrow, reddish-brown stripes; India, Venezuela.

242. ACROCONIDIELLA

Acroconidiella Lindquist & Alippi, 1964, *Darwiniana*, **13**: 612.

Colonies effuse, greyish brown, cottony. *Mycelium* immersed. *Stroma* none. *Setae* and *hyphopodia* absent. *Conidiophores* macronematous, mononematous,

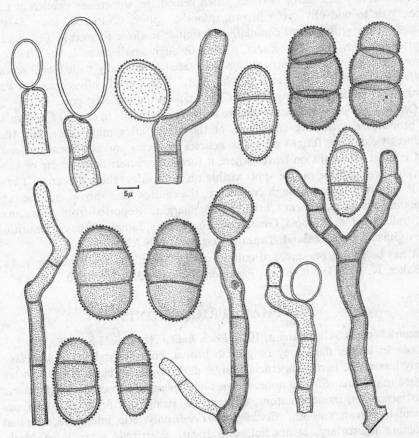

FIG. 328. *Acroconidiella tropaeoli* (× 650 except where indicated by the scale).

simple or occasionally branched, flexuous, often geniculate, sometimes swollen at the apex, pale to mid olivaceous brown; arising singly or in small groups, developing between epidermal cells and through stomata. *Conidiogenous cells* integrated, terminal, polytretic, sympodial, cicatrized. *Conidia* solitary, ellipso-idal, septate and very strongly constricted at the septa, thin-walled, olivaceous brown, echinulate.

Type species: Acroconidiella tropaeoli (Bond) Lindquist & Alippi.

Acroconidiella tropaeoli (Bond) Lindquist & Alippi, 1964, *Darwiniana*, **13**: 613.
    Heterosporium tropaeoli Bond, 1947, *Ceylon J. Sci.*, Sect. A, **12**: 185.

(Fig. 328)

Colonies predominantly hypophyllous, only occasionally also epiphyllous. *Mycelium* immersed; hyphae branched, septate, hyaline, smooth, 3–7μ thick. *Conidiophores* arising singly or in small groups terminally and laterally on the hyphae, developing between epidermal cells and through stomata, erect, simple or occasionally branched, flexuous, often geniculate, sometimes swollen at the apex, pale to mid olivaceous brown, smooth, septate, up to 180μ long, 5–10μ thick, usually with several conidial scars similar to those formed in *Drechslera* and *Curvularia*. *Conidia* solitary, arising through small pores in the conidio-phore wall, ellipsoidal, nearly always 2-septate (occasionally with 1 or 3 septa) and very strongly constricted at the septa, thin-walled, olivaceous brown, echinulate, 30–50 (41)μ long, 15–27 (21)μ thick in the broadest part.

On *Tropaeolum majus*, causes severe losses in seed fields in coastal California. It produces a yellowing and death of the leaves after mid season and this reduces yield. The fungus occurs sometimes on stems and is present on seeds but is most abundant on leaves where it forms characteristic irregular or sub-circular brownish or purple spots visible on both sides; these are up to 1 cm. diam. or often larger through confluence, the centres later shrivel and the sur-rounding tissues may form a broad yellow margin. Reported from Argentina, Australia, Ceylon, Ethiopia, Guatamala, Haiti, India, Jamaica, Kenya, Mauritius New Guinea, New Zealand, Tanzania, Uganda, U.S.A.

It can be grown on standard culture media; sporulates quite well on p.d.a. Baker, K. F. & Davis, L. H., *Phytopathology*, **40**: 553–566, 1950.

243. PHAEOTRICHOCONIS

Phaeotrichoconis Subramanian, 1956, *Proc. Indian Acad. Sci.*, Sect. B., **44**: 2.
    Colonies effuse, dark grey or greyish brown, cottony or hairy. *Mycelium* partly immersed, partly superficial; large, dark brown to black *sclerotia* often formed in culture. *Stroma* none. *Setae* and *hyphopodia* absent. *Conidiophores* macronematous, mononematous, unbranched, straight or flexuous, sometimes geniculate, brown, smooth. *Conidiogenous cells* polytretic, integrated, terminal becoming intercalary, sympodial, cylindrical, cicatrized; scars large, dark. *Conidia* solitary, dry, acropleurogenous, obclavate, rostrate, transversely sep-tate, body of conidium golden brown, fairly thick-walled, smooth, with a large dark brown scar at the base; beak long, narrow, hyaline or very pale brown, thin-walled.

Type species: Phaeotrichoconis crotalariae (Salam & Rao) Subram.

Phaeotrichoconis crotalariae (Salam & Rao) Subram., 1956, *Proc. Indian Acad. Sci.*, Sect. B., **44**: 2.

Trichoconis crotalariae Salam & Rao, 1954, *J. Indian bot. Soc.*, **33**: 191.

(Fig. 329)

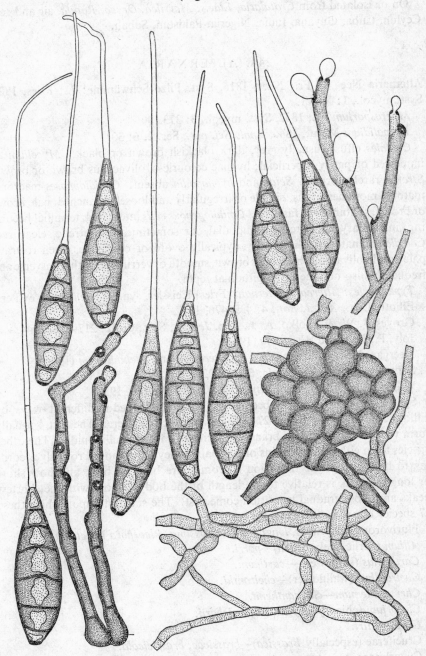

FIG. 329. *Phaeotrichoconis crotalariae* (× 650).

Conidiophores pale to mid brown, up to 150µ long, 4–8µ thick, sometimes swollen at the base to 10–17µ. *Conidia* with body 3–8-(mostly 5–6-)septate, 50–85µ long, 15–22µ thick in the broadest part; scar 3–7µ wide and beak 30–140 × 1–2·5µ.

On or isolated from *Crotalaria, Elaeis, Marsilea, Oryza, Rotala*, air and soil; Ceylon, Cuba, Guyana, India, Nigeria, Pakistan, Sabah.

244. ALTERNARIA

Alternaria Nees ex Fr.; Nees, 1816, Syst. Pilze Schwämme: 72; Fries, 1821, Syst. mycol., **1**: XLVI.

Macrosporium Fr., 1832, Syst. mycol., **3**: 373.

Rhopalidium Mont., 1846, *Annls Sci. nat.*, Ser. 2, **6**: 30.

Colonies effuse, usually grey, dark blackish brown or black. *Mycelium* all immersed or partly superficial; hyphae colourless, olivaceous brown or brown. *Stroma* rarely formed. *Setae* and *hyphopodia* absent. *Conidiophores* macronematous, mononematous, simple or irregularly and loosely branched, pale brown or brown, solitary or in fascicles. *Conidiogenous cells* integrated, terminal becoming intercalary, polytretic, sympodial, or sometimes monotretic, cicatrized. *Conidia* catenate or solitary, dry, typically ovoid or obclavate, often rostrate, pale or mid olivaceous brown, or brown, smooth or verrucose, with transverse and frequently also oblique or longitudinal septa.

Type species: Alternaria alternata (Fries) Keissler, syn. *A. tenuis* Nees ex Pers.

Elliott, J. A., *Am. J. Bot.*, **4**: 439–476, 1917.

Groves, J. W. & Skolko, A. J., *Can. J. Res.*, Sect. C, **22**: 217–234, 1944.

Joly, P., Le genre *Alternaria* (1964).

Neergaard, P., Danish species of *Alternaria* and *Stemphylium* (1945).

Simmons, E. G., *Mycologia*, **59**: 67–92, 1967.

Wiltshire, S. P., *Trans. Br. mycol. Soc.*, **18**: 135–160, 1933.

Species of *Alternaria* have been keyed out and grouped in different ways by Elliott, Joly and Neergaard; 27 of them are described here. The first 4 usually form very long, sometimes branched chains of rather small conidia. The other species form either short chains or no chains; they are arranged roughly according to how abrupt the transition is from spore body to beak and how short or long the beak is relative to the length of the body. Those with the shortest beaks and most gradual transition come first. The substrata upon which these 27 species occur most frequently are listed.

Plurivorous—*alternata, longissima, tenuissima*, Pleospora *infectoria*.

Allium (garlic, leek, onion)—*porri*.

Carthamus (safflower)— *carthami*.

Cheiranthus (wallflower)—*cheiranthi*.

Chrysanthemum—*chrysanthemi*.

Cichorium (chicory and endive)—*cichorii*.

Citrus—*citri*.

Cruciferae (especially *Brassica*)—*brassicae, brassicicola*.

Cucurbitaceae—*cucucumerina*.

Datura (thorn-apple)—*crassa.*
Daucus (carrot)—*dauci, radicina.*
Dianthus (carnation)—*dianthi, dianthicola.*
Gossypium (cotton)—*macrospora.*
Nicotiana (tobacco)—*longipes.*
Oryza (rice)—*padwickii.*
Passiflora (passion fruit)—*passiflorae.*
Raphanus (radish)—*raphani.*
Ricinus (castor-bean)—*ricini.*
Sesamum—sesami.
Solanum (potato, tomato, eggplant)—*solani.*
Sonchus—sonchi.

Alternaria alternata (Fr.) Keissler, 1912, *Beih. Bot. Zbl.*, **29**: 434.
Torula alternata Fr., 1832, Syst. mycol., **3**: 500.
A. tenuis C. G. Nees, 1816/17, Syst. Pilze Schwämme: 72.
[The reasons why the epithet *alternata* should be used instead of the more commonly accepted one *tenuis* are clearly stated by E. G. Simmons in *Mycologia*, **59**: 73, 1967.]

(Fig. 330)

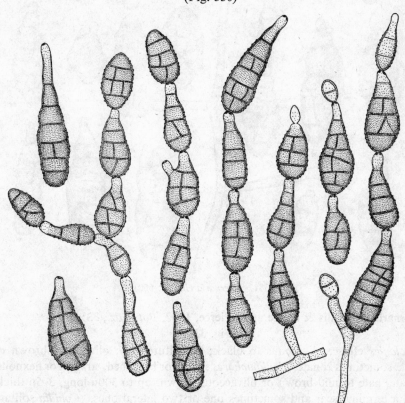

FIG. 330. *Alternaria alternata* (× 650).

Colonies usually black or olivaceous black, sometimes grey. *Conidiophores* arising singly or in small groups, simple or branched, straight or flexuous, sometimes geniculate, pale to mid olivaceous or golden brown, smooth, up to 50μ long, 3–6μ thick with 1 or several conidial scars. *Conidia* formed in long, often branched chains, obclavate, obpyriform, ovoid or ellipsoidal, often with a short conical or cylindrical beak, sometimes up to but not more than one third the length of the conidium, pale to mid golden brown, smooth or verruculose, with up to 8 transverse and usually several longitudinal or oblique septa, overall length 20–63 (37)μ, 9–18 (13)μ thick in the broadest part; beak pale, 2–5μ thick.

An extremely common saprophyte found on many kinds of plants and other substrata including foodstuffs, soil and textiles; cosmopolitan.

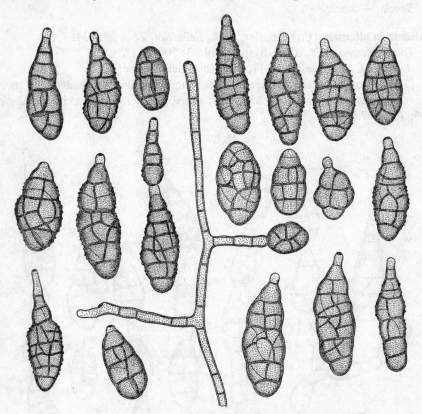

FIG. 331. *Alternaria citri* (× 650).

Alternaria citri Ellis & Pierce apud Pierce, 1902, *Bot. Gaz.*, **33**: 234.
(Fig. 331)

Colonies effuse, olivaceous to black; in culture grey, olivaceous brown or black, sometimes zonate. *Conidiophores* simple or branched, straight or flexuous, septate, pale to mid brown or olivaceous brown, up to 300μ long, 3–5μ thick, with a terminal scar and sometimes one or two lateral ones. *Conidia* solitary or in simple or branched chains of 2–7, straight or slightly curved, variously shaped but commonly obclavate or oval, often rostrate, pale to mid or sometimes dark brown or olivaceous brown, smooth to verruculose with up to 8 transverse and numerous longitudinal or oblique septa, constricted at the septa, 8–60 (42)µ long including beak when present, 6–24 (17)µ thick in the broadest part; beaks mostly 8µ or less long, 2·5–4µ thick, colourless or rather pale brown. Isolates of *A. citri* often become sterile after subculturing.

Responsible for various types of injury to fruits and leaves of *Citrus* spp. including black rot of oranges, stem end rot of lemons and a leaf spot of rough lemon. Black rot of oranges begins at the blossom end of the fruit and causes premature ripening. Underneath the small brown spot on the rind at the blossom end the tissues are discoloured brown, later greenish to black. In advanced cases the spot at the blossom end expands and turns dark green or black, by which time the interior of the fruit is black and the cells beginning to break down. *Alternaria* rot of lemons starts at the stem end and the peel is the last part to be affected. Recorded from Argentina, Australia, Bulgaria, Burma, China, Cuba, Cyprus, Egypt, England, France, Greece, India, Iran, Israel, Italy, Jamaica, Japan, Kenya, Libya, Malawi, Malta, Morocco, Mozambique, Nepal, New Zealand, Pakistan, Paraguay, Portugal, Puerto Rico, Rhodesia, South Africa, Spain, Sudan, Tanzania, Uganda, Uruguay, U.S.A., U.S.S.R., Zambia.

Bartholomew, E. T., *California agric. Exp. Stn Bull.*, **408**, 1926.
Bliss, D. E. & Fawcett, H. S., *Mycologia*, **36**: 469–502, 1944.
Burger, O. F. & Gomme, W., *Florida agric. Exp. Stn Press. Bull.*, **343**, 1922.
Doidge, E. M., *S. Afr. Dep. Agric. sci. Bull.*, **69**, 1929.
Ruehle, G. D., *Phytopathology*, **27**: 863–865, 1937.

Alternaria brassicicola (Schw.) Wiltshire, 1947, *Mycol. Pap.*, **20**: 8.
Helminthosporium brassicicola Schweinitz, 1832 [as ' *brassicola* '], *Trans. Am. Phil. Soc.*, N.S., **4**: 279.
Macrosporium cheiranthi Fr. var. *circinans* Berk. & Curt., 1875, *Grevillea*, **3**: 105.
Alternaria circinans (Berk. & Curt.) Bolle, 1924, *Meded. phytopath. Lab. Willie Commelin Scholten*, **7**: 26.
Alternaria oleracea Milbraith, 1922, *Bot. Gaz.*, **74**: 320.
Full synonymy given by S. P. Wiltshire in *Mycol. Pap.*, **20**, 1947.

(Fig. 332)

Colonies amphigenous, effuse, dark olivaceous brown to dark blackish brown, velvety. *Mycelium* immersed; hyphae branched, septate, hyaline at first, later brown or olivaceous brown, inter- and intracellular, smooth, 1·5–7·5µ thick. *Conidiophores* arising singly or in groups of 2–12 or more, emerging through stomata, usually simple, erect or ascending, straight or curved, occasionally geniculate, more or less cylindrical but often slightly swollen at the base, septate, pale to mid olivaceous brown, smooth, up to 70µ long, 5–8µ thick. *Conidia* mostly in chains of up to 20 or more, sometimes branched, acropleurogenous, arising through small pores in the conidiophore wall, straight, nearly cylindrical, usually tapering slightly towards the apex or obclavate, the basal cell rounded, the beak usually almost non-existent, the apical cell being more or less rectangular or resembling a truncated cone, occasionally better developed but then always short and thick, with 1–11, mostly less than 6 transverse septa and usually few but up to 6 longitudinal septa, often slightly constricted at the septa, pale to dark olivaceous brown, smooth or becoming slightly warted with age, 18–130μ long, 8–20μ thick in the broadest part, with the beak 1/6 the length of the conidium and 6–8μ thick.

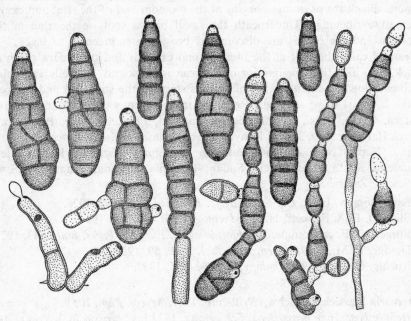

FIG. 332. *Alternaria brassicicola* (× 650).

On leaves of Cruciferae forming dark brown to almost black, circular, zonate spots, 1–10 mm. diam. More common and causing more severe disease than *A. brassicae* in seed crops. White cabbage and cauliflower subject to particularly severe attacks, Savoy cabbage relatively free. May cause considerable damage to cabbage and cauliflower in transit. Widespread and common in N.W. Europe; found also in Australia, Burma, Canada, Ceylon, Cyprus, Ethiopia, Ghana, Great Britain, Guinea, Hong Kong, India, Jamaica, Japan, Libya, Malawi, Malaya, Mauritius, Nepal, Netherlands New Guinea, New Zealand, Nigeria, Rhodesia, Romania, Sabah, Sierra Leone, South Africa, Sudan, Tanzania, Turkey, Uganda, U.S.A. and Zambia.

Morton, F. J., *N.Z. Jl Bot.*, **2**: 19–33, 1964.

Alternaria state of **Pleospora infectoria** Fuckel, 1870, Symb. mycol.: 132.
(Fig. 333)

Colonies grey to black. *Conidiophores* arising singly or in small groups, simple or branched, straight or flexuous, sometimes geniculate, pale to mid golden brown, smooth, up to 80μ long, 3–6μ thick, with 1 or several conidial scars. *Conidia* formed in long, often branched chains, conical, obpyriform or obclavate, tapering gradually to a beak which is up to half the total length of the conidium and often swollen at its tip, pale to mid golden brown, distinctly verrucose and rather thick-walled, with up to 8 transverse septa usually close together and several longitudinal or oblique septa, overall length 20–70 (50)μ, 9–18(14)μ thick in the broadest part; beak 3–5μ wide.

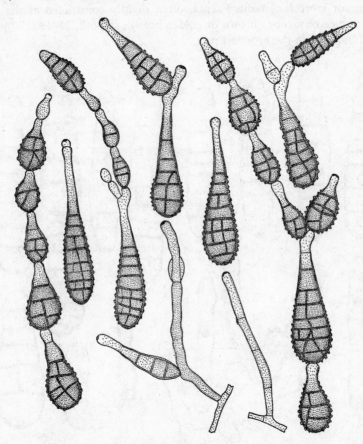

FIG. 333. *Alternaria* state of *Pleospora infectoria* (× 650).

Common and recorded on many different plants, especially cereals such as barley, oats, rye and wheat; Australia, Cyprus, Czechoslovakia, Egypt, England, India, Italy, Kenya, New Zealand, Pakistan, Switzerland, Turkey, U.S.S.R., Wales.

Alternaria chrysanthemi Simmons & Crosier, 1965, *Mycologia*, **57**: 142.
(Fig. 334)

Colonies mostly epiphyllous, effuse, hairy, grey to dark blackish brown. *Mycelium* immersed; hyphae branched, septate, hyaline to rather pale olivaceous

brown, 4–8µ diam. *Conidiophores* arising singly or in fascicles of up to 6 terminally and laterally on the hyphae, simple (occasionally branched in culture),
erect or ascending, straight or flexuous, cylindrical, septate, pale or mid olivaceous or golden brown, smooth, up to 100µ long, 6–11µ thick, usually bearing
1–3 conidial scars. *Conidia* solitary, acropleurogenous, arising through small
pores in the conidiophore wall, straight or very slightly curved, often cylindrical,
sometimes obclavate, with up to 9 (12 in culture) transverse septa and sometimes one or more longitudinal septa, often slightly constricted at the septa,
hyaline to pale olivaceous brown or golden brown, smooth, 25–130 (70)µ long,
10–26 (17)µ thick in the broadest part.

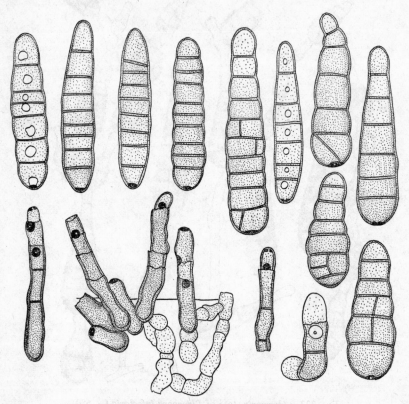

FIG. 334. *Alternaria chrysanthemi* (× 650).

On stems, leaves, flowers and seeds of *Chrysanthemum maximum*; also one
record from India on *C. indicum*. On the leaves spots are round, at first pale
grey and up to 1 cm. diam., later grey or brownish black, often with a whitish
spot in the centre surrounded by pale and dark concentric rings. Recorded
from Austria, India, Netherlands.

Alternaria radicina Meier, Drechsler & Eddy, 1922, *Phytopathology*, **12**: 157–166.
   Thyrospora radicina (Meier, Drechsler & Eddy) Neergaard, 1938, *Bot. Tidsskr.*,
44: 361.

Stemphylium radicinum (Meier, Drechsler & Eddy) Neergaard, 1939, *Aarsberetn. Ohlsens Enkes plantepat. Lab.*, **4**: 3.

Pseudostemphylium radicinum (Meier, Drechsler & Eddy) Subram., 1961, *Curr. Sci.*, **30**: 423.

(Fig. 335)

Conidiophores arising usually singly, simple or occasionally branched, straight or flexuous, cylindrical, septate, pale to mid brown or olivaceous brown, smooth, up to 200µ long, 3–9µ thick, with 1 or several conidial scars. *Conidia* solitary or in chains of 2 or rarely 3, straight, very variable in shape, often ellipsoidal,

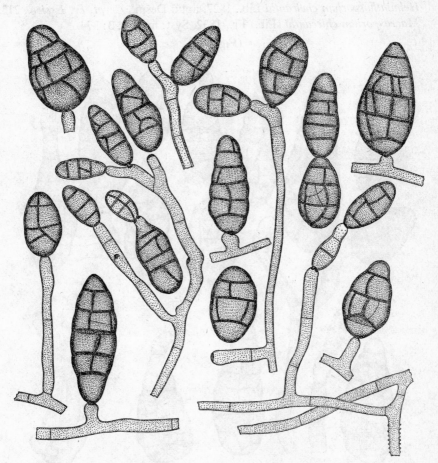

FIG. 335. *Alternaria radicina* (× 650).

obclavate or obpyriform, mid to dark brown or olivaceous brown, smooth, usually with 3–7 transverse and 1 to several longitudinal or oblique septa, sometimes constricted at the septa, 27–57 (38)µ long, 9–27 (19)µ thick in the broadest part.

On carrot, celery, dill, parsnip and occasionally on other plants. On carrot it causes black rot, common in storage, a progressive softening and blackening of the tissue of the root, infection frequently starting at the crown and extending downwards, but sometimes originating at other spots. Seed and soil borne. Recorded from Argentina (?), Australia, Bulgaria, Canada, Czechoslovakia, Denmark, England, Finland, France, Germany, Hungary, Italy, Japan, Netherlands, Norway, Romania, Sweden, U.S.A., U.S.S.R.

Simmons, E. G., *Mycologia*, **59**: 90, 1967.

Alternaria cheiranthi (Lib.) Bolle, 1924, as ' (Fr.) Bolle ', *Meded. phytopath. Lab. Willie Commelin Scholten*, **7**: 55.

Helminthosporium cheiranthi Lib., 1827, apud Desm., *Crypt. Fr. Exsicc.*, 213.

Macrosporium chieranthi (Lib.) Fr., 1832, Syst. mycol., **3**: 374.

(Fig. 336)

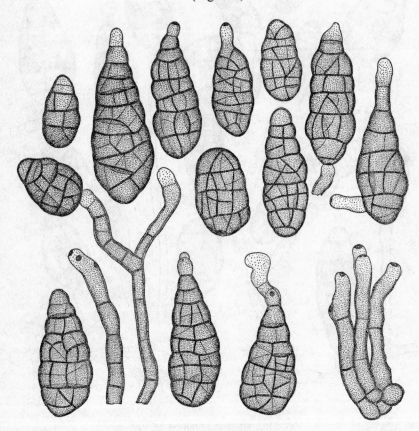

FIG. 336. *Alternaria cheiranthi* (× 650).

Conidiophores arising singly or in groups, mostly simple but sometimes branched, straight or flexuous, septate, rather pale olive, often hyaline at the apex, smooth, up to 130µ long, 5–8µ thick with a single terminal scar at first but later with up to 4 scars which may be borne close together without marked geniculation. *Conidia* mostly solitary, rarely in chains of 2, 3 or more, variously shaped, often pyriform, ovoid or elongate ovoid at first, later becoming irregular, mostly tapering to the apex which may be drawn out into a beak, generally rounded at the base, with numerous transverse, longitudinal and oblique septa, light olive to golden brown, translucent, with the interior walls, which are dark, often plainly visible, smooth or with the wall pitted from the inside, 20–100μ long, 13–32μ thick in the broadest part.

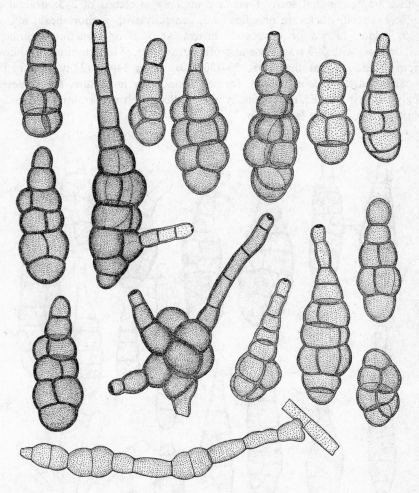

FIG. 337. *Alternaria raphani* (× 650).

Common on wallflower and occasionally recorded on other plants, especially Cruciferae. On the leaves it forms yellowish spots 4–13 by 3–8 mm., frequently occupying half the width of the leaf; conidia more abundant on the upper surface of the leaf where they cover the spots with a black powdery growth except for a marginal zone about 1 mm. wide; also found on wilting stems and fruits. Recorded from Belgium, Denmark, France, Germany, Great Britain, Holland, Ireland, Italy.

Alternaria raphani Groves & Skolko, 1944, *Can. J. Res.*, Sect. C, **22**: 227.

A. matthiolae Neergaard, 1945, Danish species of Alternaria and Stemphylium: 177.

(Fig. 337)

Conidiophores simple or occasionally branched, septate, olivaceous brown, up to 150µ long, 3–7µ thick, sometimes swollen slightly at the tip and usually with a single conidial scar. *Conidia* commonly in chains of 2–3, straight or slightly curved, obclavate or ellipsoidal, generally with a short beak, mid to dark golden brown or olivaceous brown, smooth or sometimes minutely verruculose, with 3–7 transverse and often a number of longitudinal or oblique septa; constricted at the septa, 50–130 (70)µ long, 14–30 (22)µ thick in the broadest part. *Chlamydospores* formed abundantly in culture, sometimes in chains, at first 1-celled, round, finally many-celled and irregular, brown; conidiophores often develop from them.

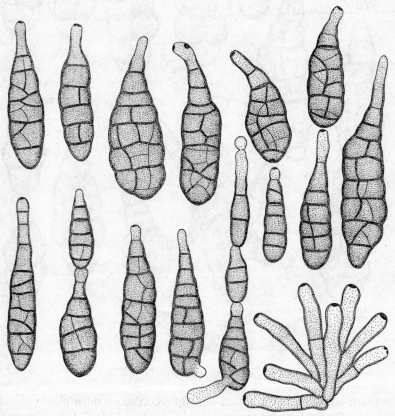

Fig. 338. Alternaria dianthi (× 650).

Produces circular, black spots, up to 4 mm. diam. on seed pods of radish, 'black pod blotch'. Also occurs on *Matthiola incana* and other Cruciferae. Recorded from Canada, Denmark, Egypt, Greece, India, Japan, Netherlands, U.S.A.

Alternaria dianthi Stevens & Hall, 1909, *Bot. Gaz.*, **47**: 409–413.
(Fig. 338)

Colonies amphigenous. *Conidiophores* arising sometimes singly but more commonly in fascicles emerging through stomata, simple, straight or flexuous, more or less cylindrical, septate, pale to mid brown or olivaceous brown, up to 120μ long but usually shorter, 5–8μ thick, with one or a few conidial scars. *Conidia* usually in chains of 2–4, straight or slightly curved, conical to obclavate, rostrate, brown or olivaceous brown, smooth, often rather dark when old, overall 30–120 (64)μ long, 10–25 (16)μ thick in the broadest part, with up to 9 transverse and usually several longitudinal or oblique septa, often constricted at the septa; beak often slightly swollen at the tip, almost the same colour as the body of the spore, 2–70μ long, 3–6μ thick.

Causes carnation blight; common on various species of *Dianthus* and occasionally on other members of the Caryophyllaceae. On the leaves very small purple spots are seen at first and these soon develop a broad yellowish green border. The spots expand, the centre becomes light brown or grey and adjacent spots tend to coalesce. Healthy tissue between the spots often turns yellow. Stem lesions at first on one side of the stem and more or less superficial often eventually extend into the pith, girdle the stem and may kill the whole plant. Recorded from Austria, Brazil, Bulgaria, Canada, Cyprus, Denmark, England, France,

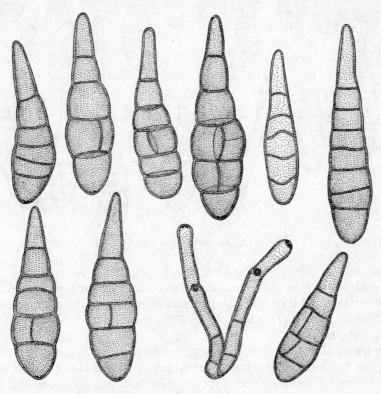

Fig. 339. *Alternaria sonchi* (× 650).

Hawaii, Italy, Jamaica, New Zealand, Romania, South Africa, Sweden, Turkey, U.S.A., U.S.S.R., Yugoslavia, Zambia.

Alternaria sonchi J. J. Davis, 1916, apud J. A. Elliott, *Bot. Gaz.*, **62**: 416.
(Fig. 339)

Colonies amphigenous. *Conidiophores* arising singly or in groups, straight or flexuous, sometimes geniculate, septate, pale to mid pale olivaceous brown or brown, up to 80μ long, 5–9μ thick, with 1 or several conidial scars. *Conidia* solitary or sometimes in very short chains, straight or curved, obclavate or

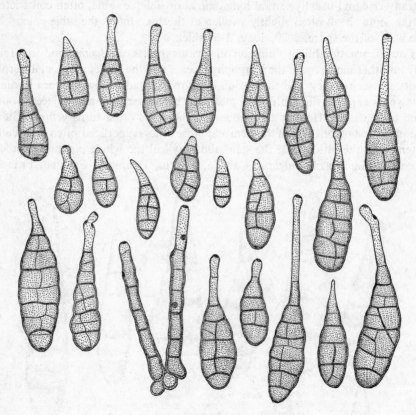

FIG. 340. *Alternaria tenuissima* (× 650).

conical with a rather short, fat beak, broadly rounded at the base, mid pale to mid golden brown or olivaceous brown, smooth or minutely verruculose, with 4–8 transverse and 0 or a few longitudinal or oblique septa, often constricted at the septa, 60–130 (77)μ long, 15–26 (20)μ thick in the broadest part; beak 4–10μ wide, tapering.

On leaves of *Sonchus* spp. including *S. oleraceus* and *S. asper*, and *Lactuca* spp., causing more or less round spots, greyish brown above with a narrow purple margin. Recorded from Cyprus, Ethiopia, Great Britain, Kenya, Libya, Mauritius, Sudan, Uganda, U.S.A., Zambia.

Alternaria tenuissima (Kunze ex Pers.) Wiltshire, 1933, [as ' (Fr.) Wiltshire '], *Trans. Br. mycol. Soc.*, **18**: 157.

Helminthosporium tenuissimum Kunze in C. G. & T. F. L. Nees, 1818, *Nova Acta Acad. Caesar. Leop. Carol.*, **9**: 242; Persoon, 1822, Mycol. eur., **1**: 18.

Macrosporium tenuissimum Fr., 1832, Syst. mycol., **3**: 374.

(Fig. 340)

Conidiophores solitary or in groups, simple or branched, straight or flexuous, more or less cylindrical, septate, pale or mid pale brown, smooth, with 1 or several conidial scars, up to 115µ long, 4–6µ thick. *Conidia* solitary or in short chains, straight or curved, obclavate or with the body of the conidium ellipsoidal

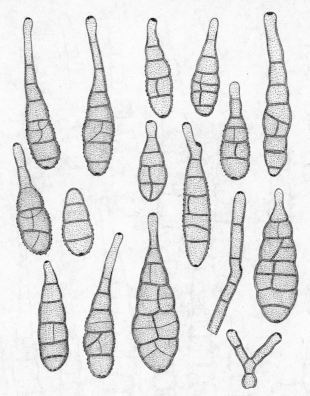

Fig. 341. *Alternaria longipes* (× 650).

tapering gradually to the beak which is up to half the length of the conidium, usually shorter, sometimes tapered to a point but more frequently swollen at the apex where there may be several scars, pale to mid clear golden brown, usually smooth, sometimes minutely verruculose, generally with 4–7 transverse and several longitudinal or oblique septa, slightly or not constricted at the septa, overall length 22–95 (54)µ, 8–19 (13·8)µ thick in the broadest part; beak 2–4µ thick, swollen apex 4–5µ wide.

Extremely common and recorded on a very wide range of plants, usually as a secondary invader rather than a primary parasite; cosmopolitan.

Alternaria longipes (Ellis & Everh.) Mason, 1928, *Mycol. Pap.*, **2**: 19.
  Macrosporium longipes Ellis & Everh., 1892, *J. Mycol.*, **7**: 134.
(Fig. 341)

Colonies amphigenous. *Conidiophores* arising singly or in groups, erect or ascending, simple or loosely branched, straight or flexuous, cylindrical, septate, rather pale olivaceous brown, up to 80μ long, 3–5μ thick, with 1 or several conidial scars. *Conidia* sometimes solitary but usually in chains, obclavate, rostrate, pale to mid pale brown, smooth or verruculose, overall length 35–110 (69)μ, body of conidium 11–21 (14)μ thick in the broadest part, tapering

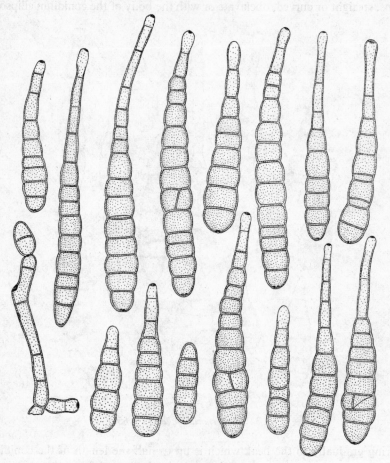

FIG. 342. *Alternaria dianthicola* (× 650).

gradually into the pale brown beak which is usually 1/3 to 1/2 the total length, 2–5μ thick and often slightly swollen at the tip; there are 3–7, usually 5–6 transverse septa and 1 to several longitudinal or oblique septa.

On tobacco, causing brown spot. The spots, which appear first on the lower leaves are orbicular, brown, frequently zonate. The entire leaf eventually becomes brown and the spots then appear a shade paler than the surrounding areas. The upper leaves and, in severe cases, the stalks afterwards become infected. The fungus is said to overwinter as mycelium in the stalks in some countries. Reported from Bolivia, China, Colombia, Congo, Germany, Hungary, India, Italy, Jamaica, Japan, Java, Malawi, Mauritius, Morocco, Mozambique, Nepal, Netherlands, New Guinea, Pakistan, Panama, Poland, Rhodesia, Romania, Sabah, South Africa, Sudan, Tanzania, Uganda, U.S.A., Venezuela, Yugoslavia, Zambia.

Tisdale, W. B. & Wadkins, R. F., *Phytopathology*, **21**: 641–660, 1931.

Alternaria dianthicola Neergaard, 1945, Danish species of Alternaria and Stemphylium: 190.

(Fig. 342)

Conidiophores arising singly or in groups, erect or ascending, usually simple, occasionally branched, straight or flexuous, cylindrical, septate, pale olivaceous brown, up to 150µ long, 4–6µ thick. *Conidia* usually in chains of 4–5, straight or curved, obclavate or almost cylindrical, rostrate, pale olivaceous brown, smooth, with up to 14 transverse and occasionally 1 or 2 longitudinal or oblique septa, constricted at the septa, overall length 55–130 (93)µ, 10–16 (13)µ thick in the broadest part, beak the same colour as the body of the conidium, 3–6µ thick, sometimes inflated at the tip.

On *Dianthus* species including carnation; common cause of flower bud rot. Induces whitish to pale brownish yellow, dark-bordered, usually oval spots 5–10 mm. long, sometimes confluent and forming areas several cm. long on the stems, leaves and buds and producing immediately on exposure of the petals dense olive patches of conidia which render flowers unsaleable. Flower buds attacked at an early stage turn brown and shrivel without opening, the interior being completely destroyed by the pathogen. In those infected later the calyx often bursts at the site of invasion and the petals are extruded through the fissure but usually the flower develops unilaterally and the petals below the point of entry of the fungus also tend to be brown and discoloured. Reported from Austria, Chile, Denmark, France, Germany, Italy, Jamaica, Malawi, Malaya, Netherlands, New Zealand, U.S.A.

Alternaria longissima Deighton & MacGarvie, 1968, *Mycol. Pap.*, **113**: 10.

(Fig. 343)

Mycelium partly superficial, partly immersed. *Conidiophores* erect or ascending, simple or occasionally branched, straight or slightly flexuous, sometimes geniculate, somewhat swollen at the apex, septate, pale to mid pale brown, smooth below, verruculose at and sometimes below the apex, up to 150µ long, 3–5µ thick with 1 to several conidial scars. *Conidia* solitary or catenulate, extremely variable in shape and size, pale straw coloured to brown. Many are very long (up to 500µ), *Cercospora*-like, obclavate or with a basal subcylindric portion of a few to several cells and a very long, narrow septate beak; they have 5–40 transverse septa, are 4–17µ thick in the broadest part and about 2·5µ thick at the apex. Shorter conidia, variable in shape and often with a few longitudinal or oblique septa are also formed. Conidia are thin-walled, smooth except around the base where they are often verruculose. Dark brown, multicellular, muriform *chlamydospores* 16–42 × 16–34µ sometimes occur, both on natural substrata and in culture.

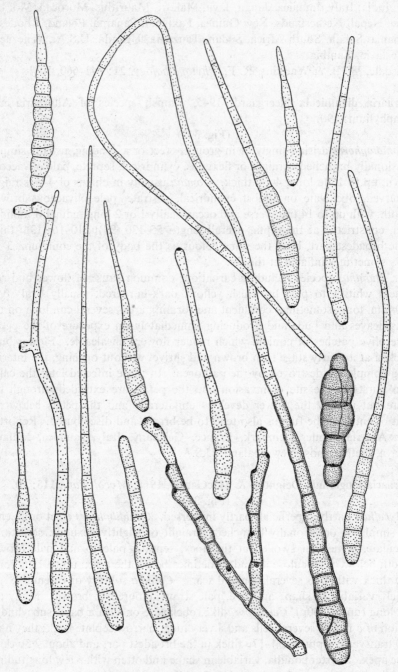

FIG. 343. *Alternaria longissima* (× 650).

On fallen pollen grains of maize, on husks and grain of rice and *Sorghum* and on leaves of a wide range of host plants, often mixed with other fungi. Recorded from Cuba, Egypt, Guinea, India, Laos, Malawi, Nepal, New

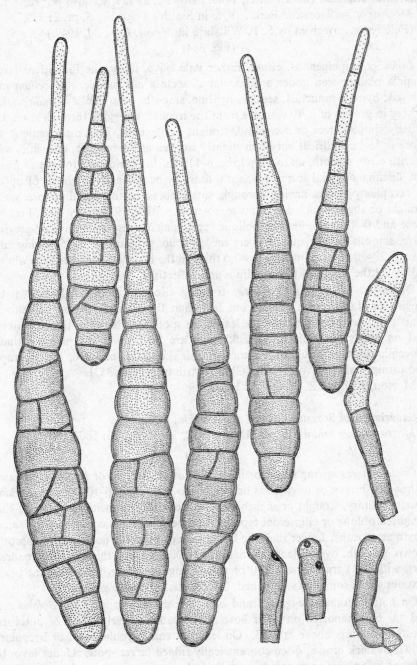

FIG. 344. *Alternaria brassicae* (× 650).

D.H.—16

Georgia, Nigeria, Sierra Leone, Sudan, Tanzania, U.S.A., Zambia. A full and very well illustrated account of this fungus is given in *Mycol. Pap.*, **113** (15 pp.).

Alternaria brassicae (Berk.) Sacc., 1880, *Michelia*, **2**: 129, see also p. 172.
    Macrosporium brassicae Berk., 1836, in Smith's Engl. Fl., **5**, pt. 2: 339.
    (Full synonymy given by S. P. Wiltshire in *Mycol. Pap.*, **20**, 1947.)

(Fig. 344)

Colonies amphigenous, effuse, rather pale olive, hairy, the individual large conidia plainly seen under a binocular dissecting microscope. *Mycelium* immersed; hyphae branched, septate, hyaline, smooth, 4–8μ thick. *Conidiophores* arising in groups of 2–10 or more from the hyphae, emerging through stomata, usually simple, erect or ascending, straight or flexuous, frequently geniculate, more or less cylindrical but often slightly swollen at the base, septate, mid-pale greyish olive, smooth, up to 170μ long, 6–11μ thick, bearing one to several small but distinct conidial scars. *Conidia* solitary or occasionally in chains of up to 4, acropleurogenous arising through small pores in the conidiophore wall, straight or slightly curved, obclavate, rostrate, with 6–19 (usually 11–15) transverse and 0–8 longitudinal or oblique septa, pale or very pale olive or greyish olive, smooth or, infrequently, very inconspicuously warted, 75–350μ long and usually 20–30μ (sometimes up to 40μ) thick in the broadest part, the beak about 1/3 to 1/2 the length of the conidium and 5–9μ thick.

On leaves of various Cruciferae forming circular, zonate, light brown to greyish or dark brown spots from less than 0·5–12 mm. diam., sometimes coalescing; on the mid ribs of the leaves the spots are oblong or linear, sunken and on heads of cauliflower black spots are formed. Host plants include broccoli, cabbage, cauliflower, horse-radish, kohlrabi, mustard, radish, rape and turnip. Almost world-wide [CMI Distribution Map 353].
    Morton, F. J., *N. Z. Jl Bot.*, **2**: 19–33, 1964.

Alternaria solani Sorauer, 1896, *Z. PflKrankh.*, **6**: 6.
    Macrosporium solani Ellis & Martin, 1882, *Am. Nat.*, **16**: 1003.

(Fig. 345)

Conidiophores arising singly or in small groups, straight or flexuous, septate, rather pale brown or olivaceous brown, up to 110μ long, 6–10μ thick. *Conidia* usually solitary, straight or slightly flexuous, obclavate or with the body of the conidium oblong or ellipsoidal tapering to a beak which is commonly the same length as or rather longer than the body, pale or mid pale golden or olivaceous brown, smooth, overall length usually 150–300μ, 15–19μ thick in the broadest part, with 9–11 transverse and 0 or a few longitudinal or oblique septa; beak flexuous, pale, sometimes branched, 2·5–5μ thick tapering gradually.

On potato, tomato, eggplant and other plants belonging to the *Solanaceae* and has been reported on other hosts. The cause of early blight of potatoes affecting all parts above ground. On leaves it causes round, oval or irregular, brown or dark brown, often concentrically ridged target spots. Under favourable conditions leaf spots enlarge rapidly and may eventually involve as much as half a leaf. Thin-leaved early varieties are said to be more susceptible than thick-leaved later ones. McWhorter says that *A. solani* is one of the commonest causes of seedling blight or damping off of tomatoes, causing dark lesions on rootlets. It also often girdles the larger seedlings just below the collar. Stems, fruit and

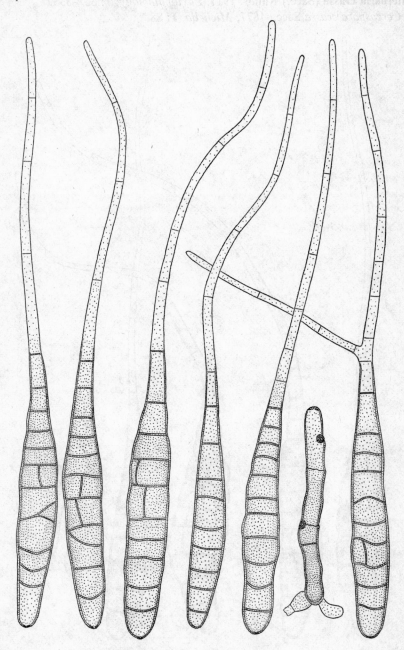

FIG. 345. *Alternaria solani* (× 650).

leaves are affected. Recorded from most parts of the world [CMI Distribution Map 89].

McWhorter, F. P., *Va Truck Exp. Stn Bull.*, **59**: 547–566, 1927.

Alternaria crassa (Sacc.) Rands, 1917, *Phytopathology*, **7**: 327–338.
  Cercospora crassa Sacc., 1877, *Michelia*, **1**: 88.

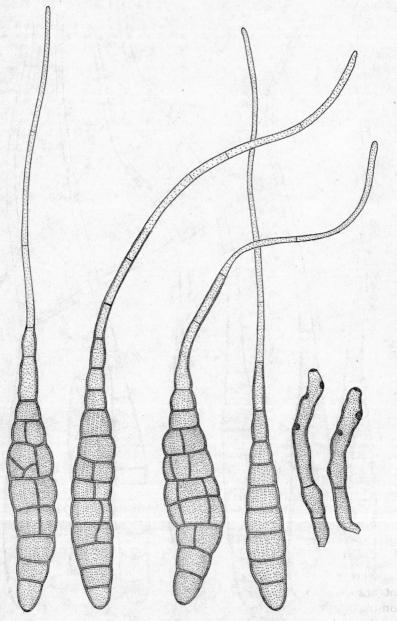

FIG. 346. *Alternaria crassa* (× 650).

Cercospora daturae Peck, 1882, *Rep. N.Y. St. Mus. nat. Hist.*, **35**: 140.

Macrosporium daturae Fautrey, 1894, *Revue mycol.*, **16**: 76.

Alternaria daturae (Fautrey) Bubák & Ranojevič, 1909, Fungi Imperfecti Exsicc., 694.

(Fig. 346)

Conidiophores arising singly or in small groups, erect or ascending, straight or flexuous, sometimes geniculate, septate, pale or mid pale brown, up to 90μ long, 7–10μ thick, with 1 or several scars. *Conidia* usually solitary, occasionally in very short chains, obclavate, rostrate, beak generally greatly exceeding the length of the body of the spore, pale brown, smooth, overall 120–440 (250)μ long, 15–40 (22)μ thick in the broadest part, body up to 140μ long, with 7–10

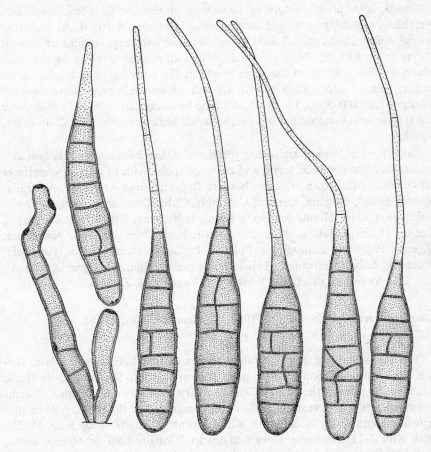

FIG. 347. *Alternaria porri* (× 650).

transverse and usually several longitudinal septa; beak pale brown, septate, not branched, 4–8μ thick at the base, tapering to 2–2·5μ. Cultures on p.d.a. non-chromogenic in contrast to *A. solani* which usually stains the agar deep pink or yellow.

On *Datura stramonium* (Jimson weed, thorn-apple) and other species of *Datura* causing irregular straw coloured, zonate spots which appear first on the lower, more shaded leaves and spread gradually upwards; dark sunken lesions may be formed on the seed pods. Heavily infected leaves are often shed. Recorded from Cuba, Cyprus, Ethiopia, Germany, Ghana, India, Italy, Kenya, Nepal, Nigeria, Pakistan, Rhodesia, Romania, Spain, Sudan, Switzerland, Tanzania, Uganda, U.S.A., Zambia.

Alternaria porri (Ellis) Cif., 1930, *J. Dep. Agric. P. Rico*, **14**: 30.
  Macrosporium porri Ellis, 1879, *Grevillea*, **8**: 12.

(Fig. 347)

Conidiophores arising singly or in groups, straight or flexuous, sometimes geniculate, septate, pale to mid brown, up to 120μ long, 5–10μ thick, with 1 or several well-defined conidial scars. *Conidia* usually solitary, straight or curved, obclavate or with the body of the conidium ellipsoidal tapering to the beak which is commonly about the same length as the body but may be shorter or longer, pale to mid golden brown, smooth or minutely verruculose, overall length usually 100–300μ, 15–20μ thick in the broadest part, with 8–12 transverse and 0 to several longitudinal or oblique septa, beak flexuous, pale, 2–4μ thick, tapering.

On garlic, leek, onion and other species of *Allium* causing ' purple blotch '. Leaf spots often elliptical, large, and coloured some shade of purple, sometimes with a broad pale brown or yellow border. Reported from Argentina, Australia, Austria, Brazil, Bulgaria, Canada, Colombia, Cuba, Denmark, Egypt, Ethiopia, Germany, Ghana, Honduras, Hong Kong, India, Iraq, Israel, Jamaica, Japan, Kenya, Malawi, Malaya, Mauritius, Netherlands, New Zealand, Nicaragua, Nigeria, Pakistan, Philippines, Poland, Portugal, Puerto Rico, Rhodesia, Romania, Sabah, Salvador, Taiwan, Tanzania, Thailand, Uganda, U.S.A., U.S.S.R., Venezuela, Vietnam, West Irian, Yugoslavia, Zambia.

Alternaria carthami Chowdhury, 1944, *J. Indian bot. Soc.*, **23**: 65.

(Fig. 348)

Conidiophores erect, simple, straight or flexuous, sometimes geniculate, septate, brown or olivaceous brown, paler near apex, up to 90μ long, 5–8μ thick, sometimes swollen at the base. *Conidia* solitary or in very short chains, straight or curved, obclavate, rostrate, sometimes constricted at the septa, pale or mid brown, smooth; body of conidium without beak usually 60–110μ long, 15–26μ thick with 7–11 transverse septa and up to 7 longitudinal or oblique septa; beak 25–160μ long, 4–6μ thick at the base, tapering to 2–3μ, with up to 5 transverse septa.

On leaves of safflower causing brown, zonate spots about 1 cm. diam. which frequently coalesce into large, irregular lesions. Infected flower buds often shrivel without opening. Reported from Ethiopia, India, Israel, Kenya, Tanzania, U.S.S.R. and Zambia.

Alternaria sesami (Kawamura) Mohanty & Behera, 1958, *Curr. sci.*, **27**: 493.
Macrosporium sesami Kawamura, 1931, *Fungi, Tokyo*, **1** (2): 26.
(Fig. 349)

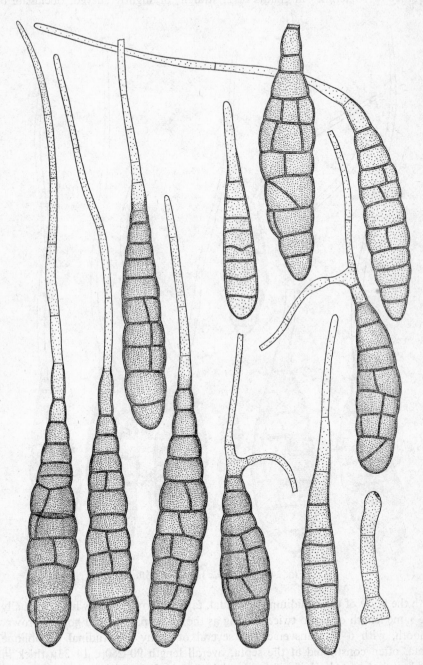

Fig. 348. *Alternaria carthami* (× 650).

Conidiophores arising singly or in small groups, usually simple, straight or flexuous, almost cylindrical, septate, rather pale brown or yellowish brown, smooth, with 1 or several conidial scars, up to 100μ long, 5–9μ thick. *Conidia* solitary or sometimes in chains of 2, straight or slightly curved, obclavate or

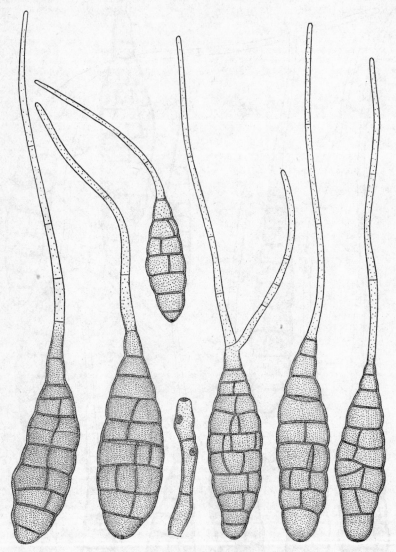

FIG. 349. *Alternaria sesami* (× 650).

with the body of the conidium ellipsoidal, tapering to the beak which is usually the same length or up to twice as long as the body, pale to mid golden brown, smooth, with 6–11 transverse and several or many longitudinal or oblique septa, often constricted at the septa, overall length 90–260μ, 14–33μ thick in the broadest part; beak simple or branched, pale, 2–4μ thick.

On sesame (*Sesamum indicum*), attacking seedlings, stems of young plants, leaves and pods. On leaves causing round or irregular, often zonate spots up to 8 mm. diam. and frequently coalescing. Affected leaves fall prematurely. Recorded from Afghanistan, Argentina, Brazil, China, Greece, India, Iraq,

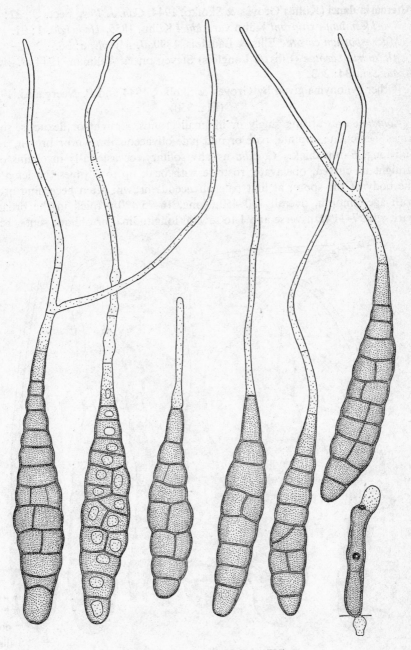

FIG. 350. *Alternaria dauci* (× 650).

Iran, Israel, Japan, Kenya, Mozambique, Nigeria, Pakistan, Somalia, Sudan, Turkey, U.S.A., U.S.S.R., Venezuela.

Berry, S. Z., *Phytopathology*, **50**: 298–304, 1960.

Leppik, E. E. & Sowell, G., *Pl. Prot. Bull.F.A.O.*, **12**: 13–16, 1964.

Alternaria dauci (Kühn) Groves & Skolko, 1944, *Can. J. Res.*, Sect. C., **22**: 222.

Sporidesmium exitiosum Kühn var. *dauci* Kühn, 1855, *Hedwigia*, **1**: 91.

Macrosporium carotae Ellis & Langlois, 1890, *J. Mycol.*, **6**: 36.

Alternaria carotae (Ellis & Langlois) Stevenson & Wellman, 1944, *J. Wash. Acad. Sci.*, **34**: 263.

[Other synonyms given by Groves & Skolko, 1944 and P. Neergaard, 1945.]
(Fig. 350)

Conidiophores arising singly or in small groups, straight or flexuous, sometimes geniculate, septate, pale or mid pale olivaceous brown or brown, up to 80µ long, 6–10µ thick. *Conidia* usually solitary, occasionally in chains of 2, straight or curved, obclavate, rostrate with beak up to 3 times the length of the body of the spore, at first pale olivaceous brown, often becoming brown with age, smooth, overall 100–450µ long, 16–25 (20)µ thick in the broadest part, with 7–11 transverse and 1 to several longitudinal or oblique septa, beaks

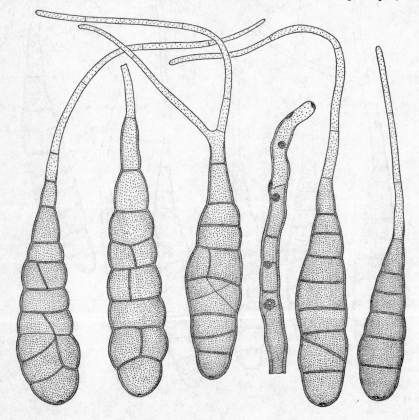

FIG. 351. *Alternaria passiflorae* (× 650).

often once branched, flexuous, hyaline or pale, 5–7μ thick at the base tapering
to 1–3μ.

Causes leaf blight of carrot and has been recorded on other Umbelliferae.
Affected leaves and petioles turn yellow and then brown or black and when
infection is severe the whole top may be killed. Reduction of leaf surface
prevents full development of the root. Seed-borne. Recorded from Australia,
Austria, Bermuda, Brazil, Canada, Ceylon, Congo, Denmark, England, France,
Germany, Ghana, Honduras, India, Israel, Jamaica, Japan, Kenya, Malawi,
Malaya, Nicaragua, Peru, Philippines, Puerto Rico, Rhodesia, Salvador, Scot-
land, South Africa, Tanzania, Tobago, Trinidad, U.S.A., U.S.S.R., Venezuela,
Zambia.

Alternaria passiflorae Simmonds, 1938, *Proc. R. Soc. Qd*, **49**: 151.
(Fig. 351)

Colonies on leaves amphigenous but often most abundant on the lower surface. *Conidiophores* arising singly or in groups, simple or rarely branched,
straight or flexuous, almost cylindrical or tapering towards the apex, septate,
pale to mid pale brown, smooth, with several conidial scars, up to 120μ long,
6–10μ thick. *Conidia* solitary on host, often in chains of up to 5 in culture,
straight or slightly curved, obclavate or with the body of the conidium ellipsoidal
tapering to the beak which is usually about the same length as, or longer than,
the body, pale to mid brown, smooth or occasionally minutely verruculose, with
5–12 transverse and a few longitudinal or oblique septa, constricted at the septa,
overall length 100–250μ, 14–29μ thick in the broadest part; beak simple or
branched, branching conspicuous in culture, sometimes flexuous near base,
pale, 1·5–4μ thick.

On all aerial parts of vines of *Passiflora* spp. including passion fruit (*P. edulis*),
causing brown spot disease. Leaf spots at first small brown dots, later expanding
to 1–2 cm., round or angular and then pale brown or grey in the centre with a
dark margin, sometimes zonate. Recorded from Australia, Kenya, New
Zealand, Tanzania and Zambia.

Simmonds, J. H., *Bull. Div. Ent. Pl. Path. Qd*, **6** (N.S.): 1–15, 1930.

Alternaria cichorii Nattrass, 1937, A first list of Cyprus fungi: 29.
(Fig. 352)

Colonies amphigenous. *Conidiophores* arising singly or in small groups, erect,
simple or occasionally branched, straight or flexuous, cylindrical, septate, rather
pale olive or olivaceous brown, almost hyaline at the apex, smooth, with a
single terminal conidial scar at first, later with several scars, up to 100μ long,
6–11μ thick. *Conidia* mostly solitary, straight or slightly curved, fusiform to
obclavate, rostrate, sometimes slightly constricted at the septa, pale or mid pale
olivaceous brown or brown, smooth; body of conidium without beak 60–120
(usually 60–90)μ long, 15–20 (usually 17–19)μ thick in the broadest part, with
8–11 transverse septa and 0–5 longitudinal or oblique septa; beak simple or
branched, subhyaline, up to 280μ long, 1–2μ thick along most of its length,
broadening to 3–6μ at its base.

On leaves of chicory and endive causing small (1–2 mm. diam.) round, oval or angular, whitish, pale grey or pale brown spots each with a broad, brown or reddish-violet margin, sometimes confluent. Reported from Austria, Cyprus, Denmark, Egypt, Greece, Italy, New Guinea, Pakistan, U.S.A. and Yugoslavia. Seed-borne. Sporulates readily in culture on p.d.a. under near u.v. light.

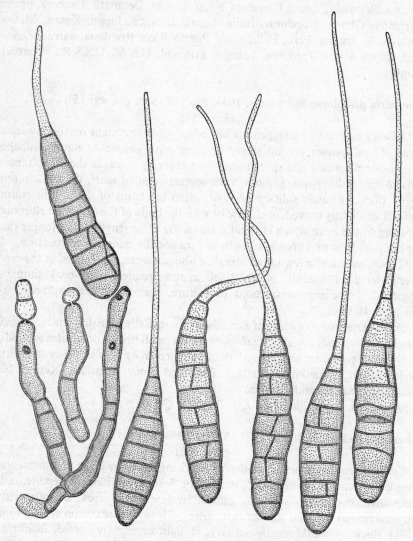

FIG. 352. *Alternaria cichorii* (× 650).

Alternaria cucumerina (Ellis & Everh.) Elliott, 1917, *Am. J. Bot.*, **4**: 472.
Macrosporium cucumerinum Ellis & Everh., 1895, *Proc. Acad. nat. Sci. Philad.*, 1895: 440.

(Fig. 353)

Colonies amphigenous. *Conidiophores* arising singly or in small groups, erect, straight or flexuous, sometimes geniculate, cylindrical, septate, pale to mid brown, up to 110μ long, 6–10μ thick, usually with several well developed conidial scars. *Conidia* solitary or occasionally in chains of 2, obclavate, rostrate, the beak longer, often much longer than the body of the spore, pale to mid golden brown, smooth to verruculose, overall 130–220 (180)μ long, 15–24 (20)μ thick in the broadest part; body with 6–9 transverse and several, sometimes many longitudinal and oblique septa; beak pale brown, septate, not branched, 4–5μ thick at the base rapidly narrowing to 1–2·5μ.

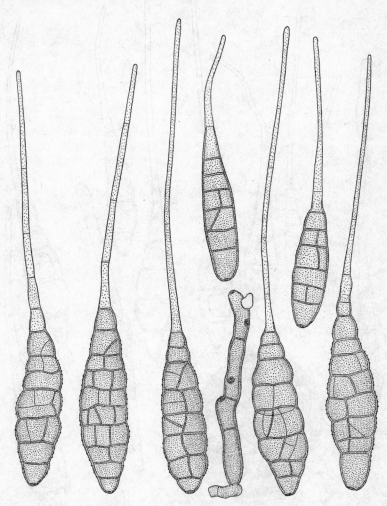

FIG. 353. *Alternaria cucumerina* (× 650).

Causing leaf blight of cucurbits, often of economic importance in U.S.A. Spots at first small, circular, water-soaked, whitish or tan, later expanding and often zonate, with a clear brown margin on the upper side of the leaf. It also occurs on the fruits and is said to overwinter in soil. Host plants include bur gherkin, cantaloupe, cucumber, melon, squash and watermelon. Recorded from Arabia, Australia, Canada, Chile, Cuba, Cyprus, France, Kenya, Libya, New Zealand, Nigeria, Rhodesia, Sierra Leone, S. Africa, Sudan, Trinidad, U.S.A., Venezuela, Zambia.

Brisley, H. R., *Phytopathology*, **13**: 199–204, 1923.

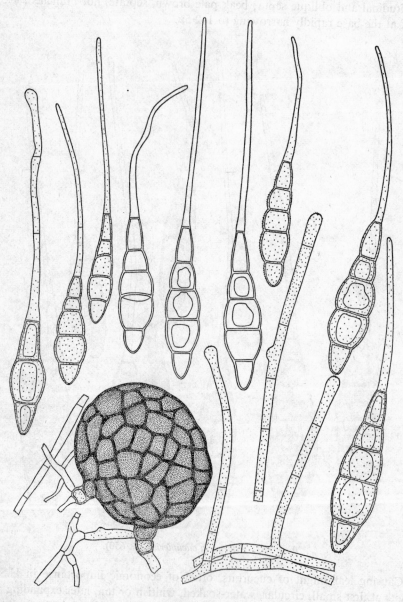

Fig. 354. *Alternaria padwickii* (× 650).

Alternaria padwickii (Ganguly) M. B. Ellis comb. nov.
  Trichoconis padwickii Ganguly, 1947, *J. Indian bot. Soc.*, **26**: 238.
(Fig. 354)

Colonies effuse, thin. *Mycelium* partly superficial, partly immersed. *Sclerotia* spherical or subspherical, black, with reticulate walls, usually 50–200μ diam. *Conidiophores* up to 180 × 3–4μ, often swollen to 5–6μ at the apex, smooth except at the apex, which is often minutely echinulate. *Conidia* straight or curved, fusiform to obclavate and rostrate, the beak at least half and often more than half the length of the conidium, at first hyaline, later straw-coloured to golden brown, smooth or minutely echinulate, often echinulate only near the scar, 95–170 (130)μ long, 11–20 (15·7)μ thick in the broadest part, 1·5–5 (2·7)μ thick in the centre of the beak, with 3–5, most commonly 4, transverse septa in the body of the conidium and often 1 or more septa in the beak, sometimes constricted at the septa.

On leaves of *Oryza* causing round or oval spots 3–9 mm. across, grey in the centre with a dark brown margin, also on grains; Egypt, India, Malaya, Nigeria, Pakistan, Sabah. Colonies on p.d.a. greyish, often deep pink or purple in reverse, sporulating freely under near u.v. light; sclerotia small, sometimes abundant in old cultures. In the type species of *Trichoconis* (*T. caudata*) the conidiogenous cells are polyblastic and colourless conidia are formed at the end of long denticles.

Alternaria macrospora Zimm., 1904, *Ber. Land- u. Forstw. Dt. Ostafr.*, **2**: 24.
  Sporidesmium longipedicellatum Reichert, 1921, *Bot. Jb.*, **56**: 723.
  Alternaria longipedicellata Snowden, 1927, *Uganda Rep. 1926*: 31.
(Fig. 355)

Colonies when on leaves amphigenous. *Conidiophores* arising singly or in groups, erect, simple, straight or flexuous, almost cylindrical or tapering slightly towards the apex, septate, pale to mid brown, smooth, with 1 or several conidial scars, up to 80μ long, 4–9μ thick. *Conidia* solitary or sometimes in chains of 2, straight or curved, obclavate or with the body of the conidium ellipsoidal tapering rather abruptly to a very narrow beak which is equal in length to or up to twice as long as the body, mid to mid-dark reddish brown, usually minutely verruculose, with 4–9 (normally 6–8) transverse and several longitudinal or oblique septa, often slightly constricted at the septa, overall length 90–180 (134)μ, 15–22 (17·7)μ thick in the broadest part; beak simple, pale 1–1·5μ thick along most of its length.

On cotton and is said to occur sometimes on other plants. On leaves causes small circular, brown spots each with a narrow purple border; these spots may expand and later have grey, dry, cracked centres. On buds, flowers and bolls small, round lesions are formed on glandular areas alternating with the points of insertion of the bracts on the receptacle. The disease causes some shedding of flowers and young bolls and on older bolls may produce a mummified condition (Jones, 1928). The fungus sometimes causes severe leaf-shedding (*R.A.M.*, **28**: 65). Recorded on cotton from Australia, Brazil, Central African Republic, China, Congo, Ethiopia, France, Ghana, Guadeloupe, India, Italy,

Japan, Malawi, Morocco, Nigeria, Rhodesia, Romania, Senegal, South Africa, Sudan, Tanzania, Trinidad, Uganda, Venezuela, Zambia.

Jones, G. H., *Ann. Bot.*, **42** (No. 168): 935–947, 1928.

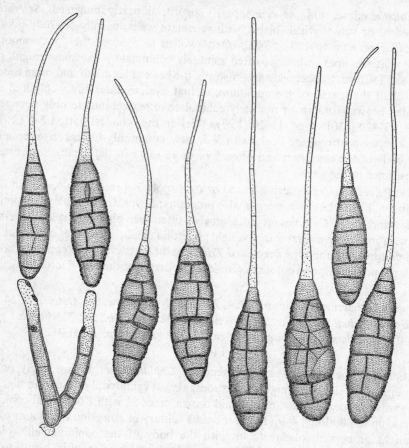

FIG. 355. *Alternaria macrospora* (× 650).

Alternaria ricini (Yoshii) Hansford, 1943, *Proc. Linn. Soc. Lond.* **1942–43**: 52.
Macrosporium ricini Yoshii, 1929, *Bult. sci. Fak. terk. Kjušu Univ.*, 3: 327.
(Fig. 356)

Colonies when on leaves amphigenous. *Conidiophores* arising singly or in groups, erect, simple, straight or flexuous, almost cylindrical or rather thicker towards the base, septate, rather pale brown, smooth, with 1 or a few conidial scars, up to 80μ long, 5–9μ thick. *Conidia* solitary or occasionally in chains of 2, straight or curved, obclavate or with the body of the conidium ellipsoidal, tapering rather abruptly to a very narrow beak which is equal in length to or up to twice as long as the body, pale to mid golden brown or reddish brown, smooth, with 5–10 transverse and several longitudinal or oblique septa, sometimes constricted at the septa, overall length 70–170 (140)μ, 13–27 (19)μ thick in the broadest part; beak simple, pale, 1–1·5μ thick along most of its length.

On castor-bean (*Ricinus communis*) sometimes causing serious damage to seedlings, leaves and inflorescences. Leaf spots are irregular in outline, variable in size but often quite large, brown, zonate, surrounded by a yellow halo. There are two types of symptoms on capsules, according to Stevenson, one involving sudden wilt, purple, or dark brown discoloration, pedicel collapse, few seeds,

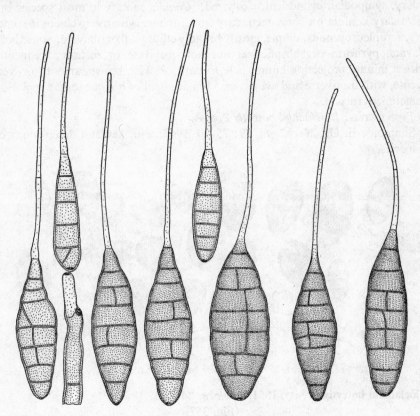

FIG. 356. *Alternaria ricini* (× 650).

failure of normal dehiscence, and the other characterized by a unilateral sunken area which gradually enlarges to cover the whole pod. All racemes and flower primordia may be killed. Premature defoliation common. Seed-borne. Recorded from Bulgaria, Ethiopia, India, Ivory Coast, Japan, Kenya, Malawi, New Guinea, Pakistan, Rhodesia, Romania, Senegal, Sudan, Tanzania, Thailand, Uganda, U.S.A., Vietnam, Zambia.

Stevenson, E. C., *Phytopathology*, **35**: 249–256, 1945; **37**: 184–188, 1947.

245. ULOCLADIUM

Ulocladium Preuss, 1851, Linnaea, **24**: 111 and Sturm's Deut. Fl. III, **6** (30): 83–84.

Pseudostemphylium (Wiltshire) Subramanian, 1961, *Curr. Sci.*, **30**: 423–424.

Colonies effuse, brown, olivaceous brown, dark blackish brown or black. *Mycelium* partly superficial, partly immersed. *Stroma* none. *Setae* and *hyphopodia* absent. *Conidiophores* macronematous, mononematous, unbranched or branched, straight or flexuous, often geniculate, pale to mid brown, smooth or verruculose. *Conidiogenous cells* polytretic, integrated, terminal becoming intercalary, sympodial, cylindrical, cicatrized. *Conidia* solitary in most species but secondary conidia on short secondary conidiophores give rise to chains in some, dry, acropleurogenous, simple, mostly broadly ellipsoidal or obovoid, sometimes clavate, pyriform or subspherical but not obclavate or rostrate, frequently with a minute projecting hilum, pale to dark blackish brown, smooth or verrucose, with transverse and usually also longitudinal or oblique septa; septation sometimes cruciate.

Type species: Ulocladium botrytis Preuss.

Simmons, E. G., *Mycologia*, **59**: 75–90, 1967; with key and descriptions of 8 species.

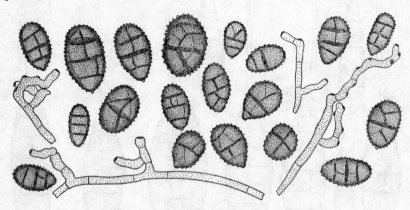

FIG. 357. *Ulocladium botrytis* (× 650).

Ulocladium botrytis Preuss, 1851, *Linnaea*, **24**: 111.

(Fig. 357)

Colonies dark blackish brown to black, velvety. *Conidiophores* often dichotomously branched near the apex, geniculate, pale to mid golden brown, generally smooth, up to 100μ long, 3–5μ thick. *Conidia* solitary, usually broadly ellipsoidal or obovoid, frequently with a minute projecting hilum, golden brown, closely verruculose or verrucose, with 1–3 (most commonly 3) transverse and 1 or more longitudinal septa, 13–30 (most commonly 14–24) × 6–19 (most commonly 9–15)μ; septation rarely cruciate. In a closely allied species **Ulocladium atrum** Preuss the conidia are mostly cruciately septate.

On dead herbaceous plants, rotten wood, paper, textiles, etc. and isolated from soil; Egypt, Europe, India, Kuwait, N. America, Pakistan.

246. DENDRYPHIELLA

Dendryphiella Bubák & Ranojevič, 1914, *Annls mycol.*, **12**: 417–418.

Colonies effuse, rust-coloured, brown or black, hairy or velvety. *Mycelium* mostly immersed. *Stroma* none. *Setae* and *hyphopodia* absent. *Conidiophores* macronematous, mononematous, unbranched or irregularly branched towards the apex, with terminal and intercalary conidiogenous nodose swellings, mid to dark brown or reddish brown, smooth or verruculose. *Conidiogenous cells*

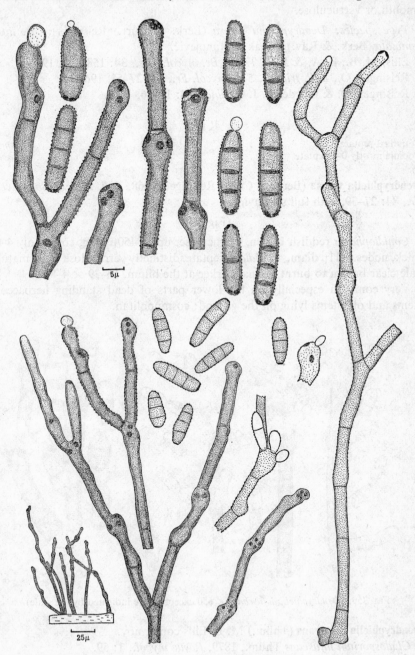

FIG. 358. *Dendryphiella vinosa* (× 650 except where indicated by scales).

polytretic, integrated, terminal becoming intercalary, sympodial, subspherical or clavate, cicatrized; scars usually close together. *Conidia* catenate or solitary, acropleurogenous, simple, cylindrical or oblong rounded at the ends or ellipsoidal, pale to mid brown, burnt sienna or olivaceous brown, 0–3-septate, smooth or verruculose.

Type species: Dendryphiella vinosa (Berk. & Curt.) Reisinger = *D. interseminata* (Berk. & Rav.) Bubák & Ranojevič.

Ellis, M. B., E. A. & J. P., *Trans. Br. mycol. Soc.*, **34**: 158–161, 1951.

Reisinger, O., *Bull. trimest. Soc. mycol. Fr.*, **84**: 27–51, 1968.

Reisinger, O. & Guedenet, J. C., *ibid.*, **84**: 19–26, 1968.

K E Y

Conidia 3-septate . . . . *vinosa*
Conidia mostly 0–2-septate . . . . *infuscans*

Dendryphiella vinosa (Berk. & Curt.) Reisinger, 1968, *Bull. trimest. Soc. mycol. Fr.*, **84**: 27–39, with full synonymy.

(Fig. 358)

Conidiophores reddish brown, verruculose, up to 450µ long, commonly 4–6µ thick, nodes 6–11µ diam. *Conidia* 3–septate, distinctly verruculose when mature, pale clear brown to burnt sienna, darker at the hilum, 16–39 × 4–8µ.

Very common especially on the lower parts of dead standing herbaceous stems and old stems lying on the ground; cosmopolitan.

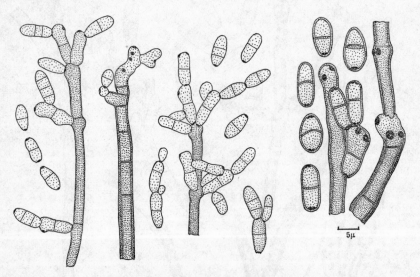

FIG. 359. *Dendryphiella infuscans* (× 650 except where indicated by the scale).

Dendryphiella infuscans (Thüm.) M. B. Ellis comb. nov.
Cladosporium infuscans Thüm., 1879, *Revue mycol.*, **1**: 59.
(Fig. 359)

Conidiophores smooth, up to 500μ long, 4–6μ thick, nodes 7–9μ diam. *Conidia* 0–2-septate, smooth to minutely verruculose, pale brown, darker at the hilum, 9–16 × 4–7μ.

On herbaceous stems; Europe, N. America.

247. DENDRYPHION

Dendryphion Wallroth, 1833, Fl. crypt. Germ., **2**: 300.

Entomyclium Wallroth, 1833, Fl. crypt. Germ., **2**: 189.

Brachycladium Corda, 1838, Icon. Fung., **2**: 14.

Colonies effuse, dark grey, olive, reddish brown or black, hairy or velvety. *Mycelium* immersed. *Stroma* when present immersed or partly superficial. *Setae* and *hyphopodia* absent. *Conidiophores* macronematous, usually branched at the apex forming a stipe and head; stipe straight or flexuous, usually stout, erect, brown to black, smooth or with the upper part verruculose; branches usually paler, smooth or verruculose. *Conidiogenous cells* monotretic or polytretic, usually integrated, terminal and intercalary on branches, occasionally discrete, sympodial, clavate, cylindrical or doliiform, cicatrized; scars usually large and dark. *Conidia* catenate or solitary, dry, acropleurogenous, simple or branched, cylindrical with rounded ends or obclavate, sometimes cheiroid, pale to mid brown or olivaceous brown, multiseptate, smooth or verrucose.

Type species: Dendryphion comosum Wallr.

Reisinger, O., *Bull. trimest. Soc. mycol. Fr.*, **84**: 39–51, 1968.

Shoemaker, R. A., *Can. J. Bot.*, **46**: 1144–1147, 1968.

KEY

Conidia 45–90 × 10–12·5μ, 5–11-septate .   .   .   .   *nanum*
Conidia 10–50 × 5–8μ, 1–7-septate .   .   .   .   *comosum*
Conidia mostly 17–28 × 5–9μ and 3-septate, on Papaveraceae   .   *Pleospora papaveracea*

Dendryphion nanum (C. G. Nees ex S. F. Gray) Hughes, 1958, *Can. J. Bot.*, **36**: 761, with full synonymy.

Helminthosporium nanum C. G. Nees ex S. F. Gray; Nees, 1816, Syst. Pilze Schwämme: 67; S. F. Gray, 1821, Nat. Arr. Br. Pl., **1**: 566.

D. laxum Berk. & Br., 1851, *Ann. Mag. nat. Hist.*, II, **7**: 176.

(Fig. 360)

Colonies black, velvety, variable in size, sometimes encircling stems and extending along them for up to 10 cm. *Hyphae* colourless to pale brown, 1–3μ thick swollen to about 8μ at the point of origin of the conidiophores. *Stromata* partly superficial, partly immersed, dark brown, up to 120μ wide, 8–30μ high. *Conidiophores* arising singly or in groups of 2–6, branched at the apex; stipe cylindrical, straight or flexuous, smooth, dark brown or reddish brown by transmitted light, black and shining by reflected light, 5–14-septate, 80–300μ long, 10–12μ thick at the base, 7–9μ at the apex; primary branches 2, 3 or 4 bearing secondary branches which are themselves sometimes branched. Branches irregular in shape, spreading, the proximal part consisting of a few cylindrical cells in line with one another, the rest of the branch being made up of a number of doliiform cells arranged in a zig-zag manner, each cell ending in a large scar with the next cell above formed obliquely to one side of the scar, pale to mid brown, smooth or verruculose, up to 100µ long. *Conidia* in chains or

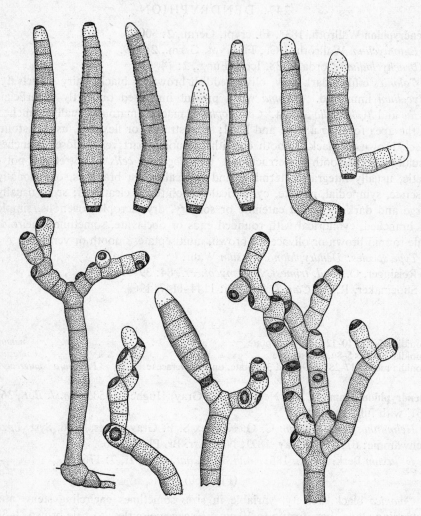

FIG. 360. *Dendryphion nanum* (× 650).

solitary, obclavate to almost cylindrical but always tapered somewhat at the ends, rounded or with a scar at the apex, truncate with a well-defined dark brown scar at the base, 5–11-septate, often slightly constricted at the septa, brown except for the end cells which are usually subhyaline, smooth or verruculose, 45–90µ long, 10–12·5µ thick in the broadest part, 4–6µ wide at the ends.

Common on dead stems of herbaceous plants, also on roots and cut tree stumps; Europe, N. America.

Dendryphion comosum Wallr., 1833, Fl. crypt. Germ., **2**: 300.

D. curtum Berk. & Br., 1851, *Ann. Mag. nat. Hist.*, II, **7**: 176.

(Fig. 361)

Colonies dark grey when young, later dark reddish brown to olive or black, velvety, with conidiophores close together, often completely encircling stems

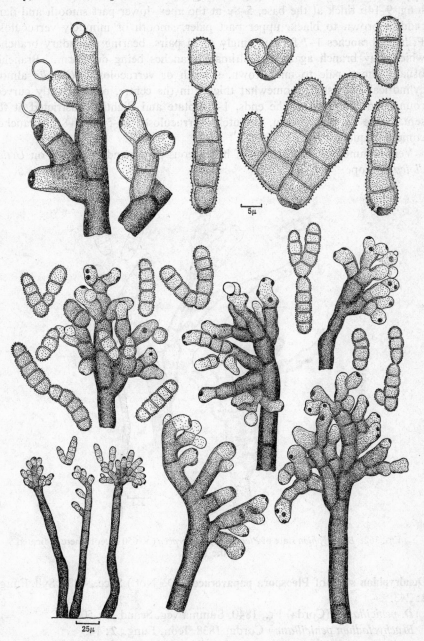

FIG. 361. *Dendryphion comosum* (× 650 except where indicated by scales).

and extending along them for 2–7 cm. *Hyphae* colourless to pale brown, 1–4μ thick swollen to 7μ at the point of origin of the conidiophores. *Stromata* often partly superficial, brown, 25–60μ wide, 6–20μ high. *Conidiophores* arising singly or in small groups, branched at the apex. Stipe almost cylindrical but usually swollen at the base, straight or slightly flexuous, 5–18-septate, 100–500μ long, 9–14μ thick at the base, 5–8μ at the apex, lower part smooth and dark reddish brown to black, upper part paler, smooth or minutely verruculose. Primary branches 1–7 formed singly or in pairs, bearing secondary branches which may branch again, the ultimate branches being doliiform. Branches usually short, pale to mid brown, smooth or verruculose. *Conidia* almost cylindrical but usually somewhat thicker in the centre, often slightly curved, rounded or flattened at the ends, 1–7-septate and slightly constricted at the septa, pale to mid brown, minutely verruculose, 10–50 × 5–8μ; branched conidia frequent.

Very common on dead stems of herbaceous plants and especially on *Urtica dioica*; Europe.

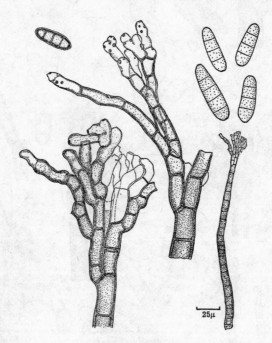

FIG. 362. *Dendryphion* state of *Pleospora papaveracea* (× 650 except where indicated by the scale).

Dendryphion state of **Pleospora papaveracea** (De Not.) Sacc., 1883, Syll. Fung., **2**: 243.

    D. penicillatum (Corda) Fr., 1840, Summa veg. Scand., **2**: 504.

    Brachycladium penicillatum Corda, 1838, Icon. Fung., **2**: 14.

    D. papaveris (Sawada) Sawada, 1959, Descr. Cat. Formosan Fung., **11**: 200.

Helminthosporium papaveris Sawada, 1917, *Trans. Formosan nat. Hist. Soc.,* **32**: 129.

(Fig. 362)

Colonies effuse, black, hairy. *Stromata* formed on stems consist of groups of a few large, dark brown cells. *Conidiophores* variable; on stems macronematous, subulate, up to 600μ long, 14–20μ thick at the base, tapering to 5–10μ, dark brown to dark blackish brown below, becoming paler upwards, smooth, with septa close together, branched at the apex; on leaves and in culture often semi-macronematous, up to 60μ long, 4–6μ thick, pale olive. *Conidia* solitary or catenate, cylindrical rounded at the ends or obclavate, pale olive, smooth; those borne on macronematous conidiophores are usually 3-septate, 17–28 × 5–9μ, but on leaves and in culture they are sometimes up to 60μ long and up to 8-septate.

On Papaveraceae; Europe, Formosa, India, N. America, S. Africa. It is the agent of a destructive disease of opium poppy affecting the plants at all stages of their growth. Seedlings sustain heavy damage from damping off, girdling of roots of older plants leads to collapse, spots develop on stems and capsules and leaves shrivel.

248. SPOROSCHISMA

Sporoschisma Berkeley & Broome, 1847, apud Berk. in *Gdnrs' Chron.,* 1847: 540 (footnote).

Colonies effuse, black. *Mycelium* immersed. *Stroma* often present but usually small. *Setae* scattered or in groups mixed with conidiophores, capitate, usually swollen at the base as well as the apex, smooth, dark brown below becoming paler upwards; apical vesicle almost hyaline. *Hyphopodia* absent. *Conidiophores* macronematous, mononematous, scattered or caespitose, unbranched, straight or flexuous, dark brown, smooth, each composed of a swollen base, a cylindrical stipe and a large, often lageniform spore sac with a long cylindrical neck. *Conidiogenous cells* monophialidic, integrated, terminal, usually determinate, rarely percurrent, lageniform. *Conidia* catenate, formed endogenously in basipetal succession, simple, cylindrical, truncate or rounded at the ends, with a few transverse septa, pale to dark brown, often somewhat darker at the septa, smooth or minutely verruculose.

Type species: Sporoschisma mirabile Berk. & Br.

Hughes, S. J., *Mycol. Pap.,* **31**: 1–23, 1949.

KEY

Conidia dark brown, smooth . . . . *mirabile*
Conidia pale brown, minutely verruculose . . . . *juvenile*

Sporoschisma mirabile Berk. & Br., 1847, apud Berk. in *Gdnrs' Chron.,* 1847: 540 (footnote).

(Fig. 363 A)

Setae up to 250μ long, 6–11μ thick at the base, tapering to 4–6μ; apical vesicle 6–9μ diam. *Conidiophores* often in groups of up to 20, mixed with capitate setae, up to 320μ long, stipes 6–11μ thick, phialides 14–20μ thick in the broadest part, neck 12–16μ wide. *Conidia* cylindrical with truncate ends, dark brown, often extra dark at the septa, smooth, almost always 3-septate, 23–45 × 10–15μ.

Common on rotten wood and bark and occasionally on dead herbaceous stems, specimens in Herb. IMI are on *Acer, Alnus, Bambusa, Betula, Carpinus,*

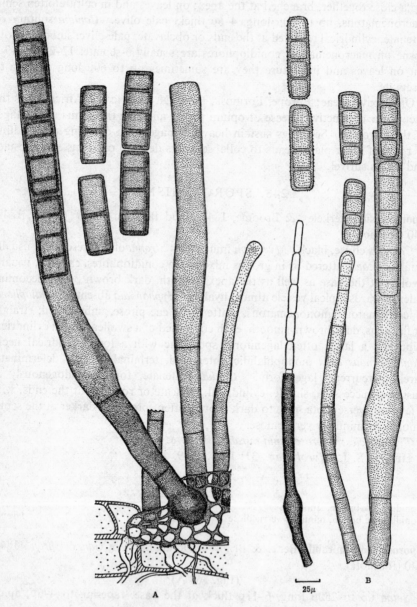

FIG. 363. A, *Sporoschisma mirabile*; B, *S. juvenile* (× 650 except where indicated by the scale).

Carya, Epilobium, Fagus, Fraxinus, Heracleum, Ilex, Quercus, Salix, Sorbus and *Ulmus*; Europe, N. America.

Sporoschisma juvenile Boudier, 1910, Icon. mycol., **3**: Pl. 589, **4**: 348 (1911).
(Fig. 363 B)

Setae up to 160μ long, 5–8μ thick at the base, tapering to 4–5·5μ; apical vesicle 6–8μ diam. *Conidiophores* usually scattered or in small groups, mixed with capitate setae, up to 240μ long, stipes 8–10μ thick, phialides 14–21μ thick in the broadest part, neck 9·5–12μ wide. *Conidia* in very long chains, often remaining hyaline and non-septate for a long time, eventually becoming 3-septate, pale brown and minutely verruculose; they measure 20–42 × 7–11μ and tend to be rounded rather than truncate at the ends.

On dead wood and bark and occasionally dead herbaceous stems, specimens in Herb. IMI are on *Alnus, Arctium, Epilobium, Fagus, Fraxinus, Hedera, Quercus, Sorbus* and *Ulmus*; Europe.

249. CHALARA

Chalara (Corda) Rabenhorst, 1844, Krypt.-Fl., **1**: 38.

Colonies effuse, grey, olive, brown or black, hairy or velvety. *Mycelium* immersed or superficial. *Stroma* none. *Setae* and *hyphopodia* absent. *Conidiophores* macronematous, mononematous, straight or slightly flexuous, unbranched, brown, smooth. *Conidiogenous cells* monophialidic, integrated, terminal, usually determinate, occasionally percurrent, cylindrical or lageniform. *Conidia* catenate, endogenous, cylindrical or oblong with truncate ends, usually colourless but sometimes brown, 0–3-septate, smooth or with the ends verruculose.

Type species: Chalara fusidioides (Corda) Rabenhorst.

Ellis, M. B., *Mycol. Pap.*, **79**: 20–22, 1961.

Henry, B. W., *Phytopathology*, **34**: 631–635, 1944.

Nag Raj, T. R.   A monograph of the genus *Chalara*, in preparation.

Chalara aurea (Corda) Hughes, 1958, *Can. J. Bot.*, **36**: 747.
(Fig. 364 A)

Colonies orange yellow; *conidiophore stipes* up to 25 × 5–6μ; *phialides* lageniform, 35–65μ long, swollen base 6–7μ wide, long cylindrical neck 3μ thick; *conidia* colourless, 0–1-septate, 12–16 × 2μ. On wood and horse chestnut fruits; Czechoslovakia, Great Britain.

Chalara cladii M. B. Ellis, 1961, *Mycol. Pap.*, **79**: 20.
(Fig. 364 B)

Colonies reddish brown to black; *conidiophore stipes* up to 860 × 3·5–8μ; *phialides* cylindrical, 40–70 × 6·5–10μ; *conidia* 1-septate, colourless to brown, end walls verruculose, 11–17 × 5–7μ. On *Cladium mariscus*; Great Britain.

Chalara cylindrica Karst., 1887, *Meddn Soc. Fauna Flora fenn.*, **14**: 108.
(Fig. 364 C)

Colonies dark brown; *conidiophore stipes* up to 65 × 4–5μ; *phialides* lageniform, 20–35μ long, swollen base 5–6μ wide, cylindrical neck 2μ thick; *conidia* colourless, 0–1-septate, 3–6 × 1μ. On dry scales and needles of *Picea excelsa*; Finland, Great Britain.

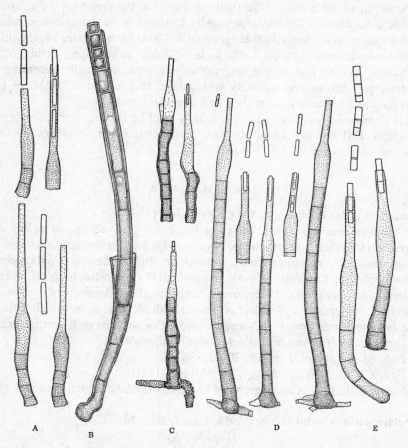

Fig. 364. *Chalara* species: A, *aurea*; B, *cladii*; C, *cylindrica*; D, *cylindrosperma*; E, *pteridina* (× 650).

Chalara cylindrosperma (Corda) Hughes, 1958, *Can. J. Bot.*, **36**: 747.
(Fig. 364 D)

Colonies blackish brown, hairy; *conidiophore stipes* up to 135 × 4–6μ, base 9–10μ; *phialides* lageniform, 22–40μ long, base 6–7μ wide, cylindrical neck 2–3μ thick; *conidia* colourless, 0-septate, 4–11 × 1–1·7μ. On dead wood, beech cupules, pine and dead stems of *Aconitum napellus*; Czechoslovakia, Great Britain.

Chalara pteridina Syd., 1912, *Annls mycol.*, **10**: 450.
(Fig. 364 E)

Colonies brown, hairy; *conidiophore stipes* up to 80 × 5–8µ; *phialides* lageniform, 50–70µ long, lower part 8–9µ wide, neck 3–4µ thick; *conidia* colourless, 1–3-septate, 9–17 × 2–3µ.   On bracken (*Pteridium aquilinum*); Austria, Germany, Great Britain.

Chalara state of **Ceratocystis fagacearum** (Bretz) Hunt, *Lloydia*, **19**: 21; syn. *C. quercina* B. W. Henry.

Colonies grey to olive, *conidiophore stipes* up to 50 × 2·5–5µ; *phialides* cylindrical, 20–40 × 2·5–5µ; *conidia* 0-septate, 4–22 × 2–4·5µ. On oak (*Quercus*) in U.S.A.; the cause of an important wilt disease.

See also *Chalara* states under *Chalaropsis* and *Thielaviopsis*.

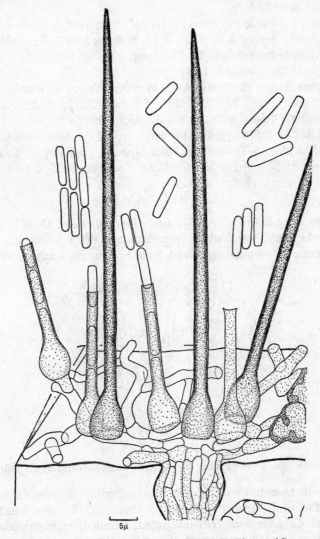

FIG. 365. *Chaetochalara bulbosa* (del. K. Pirozynski).

250. CHAETOCHALARA

Chaetochalara Sutton & Pirozynski, 1965, *Trans. Br. mycol. Soc.*, **48**: 350–354.

Colonies effuse, dark brown to black, hairy. *Mycelium* partly superficial, partly immersed. *Stroma* none but hyphae often aggregated in substomatal cavities. *Setae* present, simple, subulate, brown or dark brown. *Hyphopodia* absent. *Conidiophores* macronematous, mononematous, straight, unbranched, often swollen at base, smooth, yellowish brown. *Conidiogenous cells* monophialidic, integrated, determinate, ampulliform or cylindrical. *Conidia* catenate, endogenous, simple, cylindrical or oblong and truncate or rounded at the ends, colourless, 0-septate or occasionally 1-septate, smooth.

Type species: Chaetochalara bulbosa Sutton & Pirozynski.

Three species described:

(1) **C. bulbosa** Sutton & Pirozynski (Fig. 365). On rotting leaves of *Ilex aquifolium*, England. *Setae* 60–90 × 2·5–3·5μ, bulbous base 7–7·5μ wide; *conidiophores* (phialides) ampulliform 20–30μ long, bulbous base 7–7·5μ wide, neck cylindrical 3μ thick; *conidia* 6–12·5 × 1·5–2μ.

(2) **C. africana** Sutton & Pirozynski. On rotting leaves of *Brachystegia spiciformis*, Zambia. *Setae* 90–180 × 2·5–3·5μ, base 5·5–7μ wide; *conidiophores* 25–40μ long, base 4·5–6μ wide, neck 2–3μ thick; *conidia* 5·5–8·5 × 1·5–2μ.

(3) **C. cladii** Sutton & Pirozynski. On dead leaves of *Cladium mariscus*, England. *Setae* 80–350μ long, 5–8μ thick at base, tapering gradually to pointed apex. *Conidiophores* 30–50 × 4·5–9·5μ. *Conidia* 0–1-septate, 8–25 × 4–6μ.

251. BLOXAMIA

Bloxamia Berkeley & Broome, 1854, *Ann. Mag. nat. Hist.*, **13**: 468.

Sporodochia black, scattered or gregarious and often confluent. *Mycelium* mostly immersed. *Stroma* superficial, pale brown, prosenchymatous. *Setae*

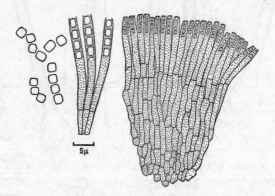

F_IG. 366. *Bloxamia truncata* (× 650 except where indicated by the scale).

and *hyphopodia* absent. *Conidiophores* macronematous, forming a close palisade over the surface of the stroma, straight or flexuous, usually unbranched, smooth, pale brown. *Conidiogenous cells* integrated, terminal, monophialidic, usually determinate, occasionally percurrent, cylindrical. *Conidia* catenate, endogenous, simple, colourless, smooth, 0-septate, oblong and truncate at the ends or almost square.

Type species: Bloxamia truncata Berk. & Br.

Pirozynski, K. A. & Morgan-Jones, G., *Trans. Br. mycol. Soc.*, **51**: 185–187, 1968.

Bloxamia truncata Berk. & Br., 1854, *Ann. Mag. nat. Hist.*, **13**: 468.

(Fig. 366)

Hyphae 2–2·5μ thick, very pale brown. *Conidiophores* up to 35μ long, 2·5–3·5μ thick. *Conidia* 2–3 × 2μ.

On wood of elm and apple; Great Britain, Italy.

252. CYSTODENDRON

Cystodendron Bubák, 1914, *Annls mycol.*, **12**: 212.

Colonies hypophyllous, effuse, pale brown, with the conidiophores aggregated in darker clumps, scarcely visible to the naked eye. *Mycelium* superficial. *Stroma* none. *Setae* and *hyphopodia* absent. *Conidiophores* macronematous or semi-macronematous, short, much branched, pale to mid brown, smooth, forming rather loose sporodochia. *Conidiogenous cells* monophialidic, discrete and also sometimes integrated and terminal, arranged penicillately, determinate, lageniform, with collarettes. *Conidia* aggregated in slimy heads, endogenous, simple, ellipsoidal, spherical or subspherical, colourless, smooth, 0-septate.

Type species: Cystodendron dryophilum (Pass.) Bubák.

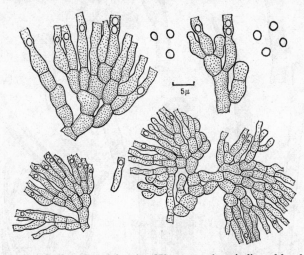

FIG. 367. *Cystodendron dryophilum* (× 650 except where indicated by the scale).

Cystodendron dryophilum (Pass.) Bubák, 1914, *Annls mycol.*, **12**: 211–212.

(Fig. 367)

Conidiophore branches 2–6μ thick. *Phialides* 10–16 × 2–4μ. *Conidia* 2–2·5μ diam.

On living leaves of *Quercus*, causing large, round or irregular, brown, sometimes zonate leaf spots; Austria.

253. CATENULARIA

Catenularia Grove in Saccardo, 1886, Syll. Fung., **4**: 303.
Psiloniella Costantin, 1888, Mucéd. simpl.: 190.

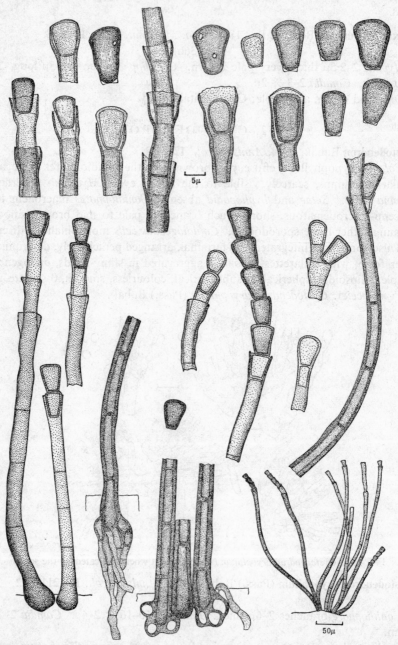

FIG. 368. *Catenularia* state of *Chaetosphaeria cupulifera* (× 650 except where indicated by scales).

Colonies effuse, dark brown, hairy. *Mycelium* immersed. *Stroma* present, but often limited to a few cells. *Setae* when present distinctly capitate. *Hyphopodia* absent. *Conidiophores* macronematous, mononematous, straight or flexuous, unbranched, brown, smooth. *Conidiogenous cells* integrated, terminal, monophialidic, percurrent, calyciform. *Conidia* catenate, endogenous, simple, cuneiform, hyaline to fairly dark brown, smooth, 0-septate.

Type species: Catenularia state of *Chaetosphaeria cupulifera* (Berk. & Br.) Sacc. = *C. cuneiformis* (Richon) Mason = *C. simplex* Grove.

Booth, C., *Mycol. Pap.*, **68**: 5–10, 1957.
Hughes, S. J., *N.Z. Jl Bot.*, **3**: 136–150, 1965.
Mason, E. W., *Mycol. Pap.*, **5**: 120–121, 1941.

Catenularia state of **Chaetosphaeria cupulifera** (Berk. & Br.) Sacc., 1883, Syll. Fung., **2**: 94.

Catenularia cuneiformis (Richon) Mason [For other synonyms see Hughes, 1965, p. 144].

(Fig. 368)

Conidiophores often in tufts, up to 300μ long, 6–8μ thick. Calyciform part of terminal conidiogenous cell 10–11μ long, 10–11μ wide at apex. *Conidia* often formed in chains of up to 10, cuneiform, 9–14μ long, 5–7μ thick at base, 9–10μ at apex, pale to fairly dark brown.

Not uncommon in Great Britain on dead wood of ash, beech, oak, etc.

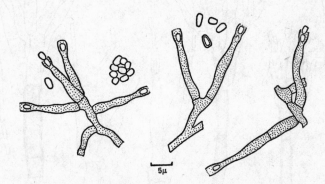

FIG. 369. Catenularia state of Chaetosphaeria myriocarpa.

Catenularia state of **Chaetosphaeria myriocarpa** (Fr.) Booth, 1957, *Mycol. Pap.*, **68**: 5.

Catenularia heimii Mangenot; excluded from the genus by Hughes, 1965.

[(Fig. 369)

Conidiophores up to 80μ long, 3–4μ thick. *Conidia* hyaline, more or less cuneiform, 2–2·5 × 1·5–2μ.

Very common in Great Britain on dead wood of many different deciduous trees.

254. PHIALOCEPHALA

Phialocephala Kendrick, 1961, *Can. J. Bot.*, **39**: 1079–1085.

Colonies effuse, dark grey, greyish olive, olivaceous brown or dark blackish brown. *Mycelium* partly superficial, partly immersed. *Stroma* none. *Setae* and *hyphopodia* absent. *Conidiophores* macronematous, mononematous, composed of a stout, erect, straight, mid to dark blackish brown, smooth stipe and a more or less complex head made up of several series of branches, the ultimate ones bearing conidiogenous cells. *Conidiogenous cells* monophialidic, discrete, arranged penicillately, determinate, lageniform or subulate, with well defined collarettes which are sometimes flared. *Conidia* often catenate, always aggregated in slimy heads, endogenous or semi-endogenous, simple, cylindrical with truncate ends, ellipsoidal or subspherical, colourless to pale olive or olivaceous brown, smooth, 0-septate.

Type species: Phialocephala dimorphospora Kendrick.

Kendrick, W. B., *Can. J. Bot.*, **39**: 1079–1085, 1961; **41**: 573–577 and 1015–1023, with key to 5 species on p. 1022; **42**: 1291–1295, 1964.

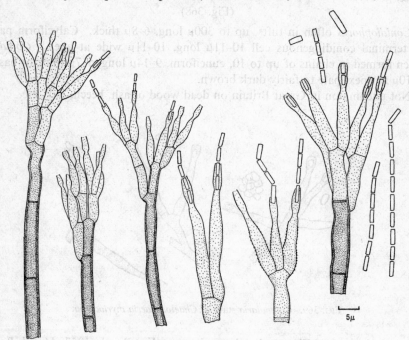

Fɪɢ. 370. *Phialocephala bactrospora* (× 650 except where indicated by the scale).

Phialocephala bactrospora Kendrick, 1961, *Can. J. Bot.*, **39**: 1083–1084.
(Fig. 370)

Conidiophore stipe up to 1 mm. long, 4–7µ thick. *Phialides* sometimes verrucose, 15–38µ long, 3–4µ thick at the base; collarettes 5–8 × 2–2·5µ. *Conidia* cylindrical, hyaline, mostly 4–7 × 1–1·5µ.

Isolated from wood of *Populus* and *Tilia*; Europe, N. America.

255. AUREOBASIDIUM

Aureobasidium Viala & Boyer, 1891, *Rev. gén. Bot.*, **3**: 369–371.

Pullularia Berkhout, 1923, De Schimmelgeschachten Monilia, Oospora en Torula. Dissert. Utrecht: 54–55. For other synonyms see W. B. Cooke, 1962.

Colonies effuse, at first white or creamy, later becoming black at least in part and usually slimy. *Mycelium* mostly immersed, often torulose, very variable in thickness, cells sometimes rounding off and separating. *Stroma* none. *Setae* and *hyphopodia* absent. *Conidiophores* micronematous, mononematous, branched, flexuous, at first hyaline, becoming mid to dark brown, smooth, thick-walled. *Conidiogenous cells* monophialidic, integrated, intercalary, determinate, cylindrical. *Conidia* aggregated in slimy masses, semi-endogenous, pleurogenous, simple, ellipsoidal or ovoid, colourless, smooth, 0-septate, each completely encased in a slimy coat; secondary conidia produced by yeast-like budding of primary conidia. Species of *Aureobasidium* have also a *Scytalidium* state.

Type species: Aureobasidium pullulans (De Bary) Arnaud = *A. vitis* Viala & Boyer.

Cooke, W. B., *Mycopath. Mycol. appl.*, **17**: 1–43, 1962.

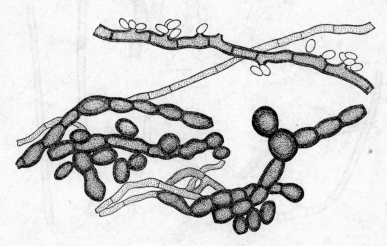

FIG. 371. *Aureobasidium pullulans* (× 650).

Aureobasidium pullulans (De Bary) Arnaud, 1918, Les Asterinées (in *Ann. Ec. Agric. Montpellier*, N.S. **16**): 39 [footnote].

(Fig. 371)

Hyphae hyaline, becoming mid to dark brown, usually 2–10μ thick but with some cells up to 15 or even 20μ. *Conidiophores* mostly 5–8μ thick, brown, with small lateral protuberances which become short, open-ended necks of phialides. *Conidia* mostly 4–6 × 2–3μ but may be up to 12 × 6μ in old colonies.

Widely distributed on dead plants and isolated from air, foodstuffs, soil, textiles, wood, etc.; cosmopolitan.

256. BAHUPAATHRA

Bahupaathra Subramanian & Lodha, 1964, *Antonie van Leeuwenhoek*, **30**: 323–325 and 329.

Colonies dense, blackish. *Mycelium* mostly superficial. *Stroma* none. *Setae* present, simple, dark, smooth. *Hyphopodia* absent. *Conidiophores* semi-macronematous, mononematous, flexuous, irregularly branched, pale brown, smooth, septate. *Conidiogenous cells* integrated, intercalary and terminal, mostly monophialidic, occasionally polyphialidic, determinate, cylindrical with pronounced projecting lateral collarettes. *Conidia* semi-endogenous, acro-pleurogenous, aggregated in slimy masses, simple, colourless or pale brown, 0-septate, smooth, spherical or subspherical.

Type species: Bahupaathra samala Subram. & Lodha.

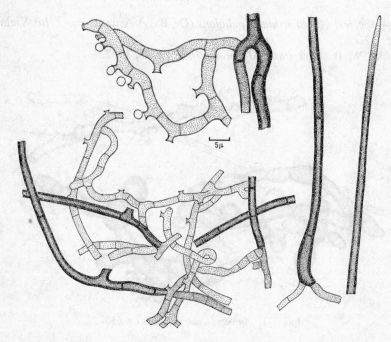

FIG. 372. *Bahupaathra samala* (× 650 except where indicated by the scale).

Bahupaathra samala Subram. & Lodha, 1964, *Antonie van Leeuwenhoek*, **30**: 323.

(Fig. 372)

Setae 3–6μ thick. *Conidiophores* up to 2 mm. long, 2–3μ thick along most of their length but often wider at the base, frequently forming loops, sometimes terminating in setae. *Conidia* 2–3μ diam.

On dung; India.

257. MAHABALELLA

Mahabalella Sutton & Patel, 1966, *Nova Hedwigia*, **11**: 203–207.

Colonies effuse, greyish brown, velvety. *Mycelium* partly superficial, partly immersed. *Stroma* none. *Setae* erect, unbranched, subulate with a bulbous base and an acute apex, mid to very dark brown, smooth. *Hyphopodia* absent. *Conidiophores* micronematous, mononematous, irregularly branched, hyaline or olivaceous, smooth. *Conidiogenous cells* monophialidic, discrete, close together forming a palisade with scattered setae, determinate, ampulliform or subspherical, with collarettes. *Conidia* aggregated in slimy masses, semi-endogenous, appendiculate with a setula at each end, cylindrical with rounded ends, hyaline, smooth, 0-septate.

Type species: Mahabalella acutisetosa Sutton & Patel.

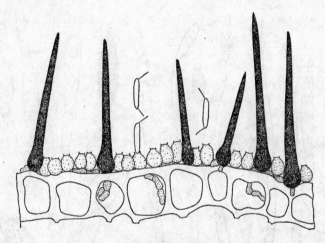

Fig. 373. *Mahabalella acutisetosa* (× 650).

Mahabalella acutisetosa Sutton & Patel, *Nova Hedwigia*, **11**: 203–207.

(Fig. 373)

Setae 30–220 × 4–6µ; bulbous base 7–9µ diam. *Phialides* 5–9 × 4–6µ. *Conidia* 13–18·5 × 2µ; setulae 4–11µ long.

On old fallen leaves of *Bambusa*; India.

258. GLIOMASTIX

Gliomastix Guéguen, 1905, *Bull. trimest. Soc. mycol. Fr.*, **21**: 240.

Basitorula Arnaud, 1953, *Bull. trimest. Soc. mycol. Fr.*, **69**: 276.

Colonies on natural substrata green, brown or black, usually with sparse, floccose superficial mycelium and abundant sporulation. In culture many species form mycelial ropes which are usually white, contrasting sharply with the dark-pigmented spore masses. *Stroma* none. *Setae* and *hyphopodia* absent. *Conidiophores* macronematous, mononematous, unbranched or occasionally

forked near the base, straight or flexuous, colourless or dark, smooth or verruculose, often rougher and darker near the apex. *Conidiogenous cells* monophialidic, integrated, terminal, determinate or sometimes percurrent, subulate;
collarette often present. *Conidia* catenate or aggregated in slimy heads, semi-
endogenous, simple, doliiform, ellipsoidal, fusiform, pyriform, spherical or

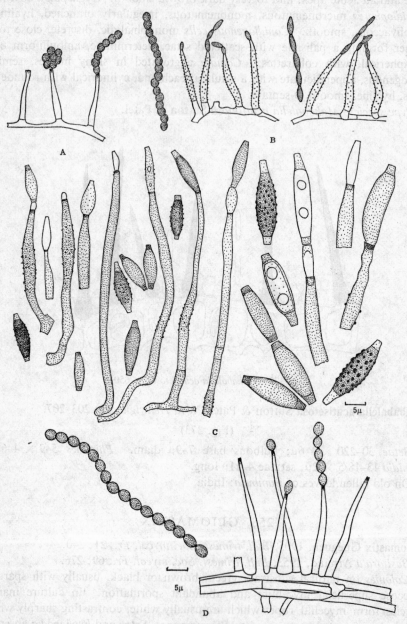

FIG. 374. *Gliomastix* species (1): A, *murorum*; B, *luzulae*; C, *elata*; D, *murorum* var. *polychroma*
(× 650 except where indicated by scales).

subspherical, pigmented, colour especially noticeable when seen in mass, smooth or rough, 0-septate.

Type species: Gliomastix murorum (Corda) Hughes = *G. chartarum* Guéguen. Dickinson, C. H., *Mycol. Pap.*, **115**: 1–24, 1968.

KEY

Conidia nearly always in chains 1
Conidia usually aggregated in slimy masses 6
1. Conidia more than 10µ long 2
   Conidia less than 10µ long 3
2. Conidia fusiform with truncate ends . . . . *elata*
   Conidia ellipsoidal or limoniform, ends not truncate . . . . . *fusigera*
3. Conidia narrowly fusiform often with dark median band . . . . *luzulae*
   Conidia not narrowly fusiform 4
4. Conidia often doliiform with truncate ends . . . . . . *musicola*
   Conidia not doliiform 5
5. Conidia ellipsoidal, small (2–4×2–3µ) with aggregations of black deposit between each
   conidium in a chain; cultures multicoloured in reverse . . *murorum* var. *polychroma*
   Conidia often subspherical, 2·5–5·5×2–4·5µ, deposit between conidia not especially noticeable; cultures not multicoloured in reverse . . . . . *murorum*
6. Phialides and conidia golden brown, smooth . . . *Wallrothiella subiculosa*
   Phialides not golden brown, frequently rough at the apex 7
7. Conidia often pyriform or narrowed to a truncate base, guttulate . . . *cerealis*
   Conidia broadly ellipsoidal, not narrowed to a truncate base . . *murorum* var. *felina*

Gliomastix murorum (Corda) Hughes, 1958, *Can. J. Bot.*, **36**: 769.
(Fig. 374 A)

Phialides frequently arising from mycelial ropes, subulate, colourless and smooth at the base, olivaceous brown and roughened with granules at the apex which sometimes has a small collarette, usually about 20–30µ long, 2–3µ thick at the base, tapering to about 1µ at the apex. *Conidia* often subspherical, dark olivaceous brown to black, 2·5–5·5 × 2–4·5µ, usually roughened with granules.

Frequently collected on plant debris and isolated from textiles, wood, soil, etc.; Europe, New Zealand, N. America.

Conidial measurements, substrata and geographical distribution of 8 other species of *Gliomastix* are given below.

G. elata Dickinson (Fig. 374 C): 17–23 × 4–7µ, on *Musa*; Sierra Leone.
G. fusigera (Berk. & Br.) Dickinson (Fig. 375 A): 12–21 × 6–9µ, on *Bambusa* and palms; Ceylon, New Caledonia.
G. luzulae (Fuckel) Mason ex Hughes (Fig. 374 B): 5–9 × 2–2·5µ, on dead plants, especially common on herbaceous stems; Europe, New Zealand, Sabah, W. Africa.
G. musicola (Speg.) Dickinson (Fig. 375 B): 5–7 × 2–3·5µ, on *Musa* and occasionally other plants, also isolated from paper and soil; Cuba, Europe, Ghana, India, N. & S. America.
G. murorum (Corda) Hughes var. **polychroma** (van Beyma) Dickinson (Fig. 374 D): 2–4 × 2–3µ, on or isolated from apple (*Malus*), banana (*Musa*), pineapple (*Ananas*) and other plants, also isolated from air, man and soil; Africa, Europe, Java, N. America, Pakistan.

G. state of **Wallrothiella subiculosa** Höhnel (Fig. 376 A): 3–8 × 2–4µ on bamboos, oil palm (*Elaeis*), *Phormium*, *Solanum*, *Theobroma*, etc. and isolated from soil; Europe, Java, Hong Kong, New Zealand, W. Africa.

G. cerealis (Karst.) Dickinson (Fig. 376 B): 3–5 × 2–3·5µ, on wood and dead herbaceous plants and isolated from sacking and soil; Europe, Venezuela.

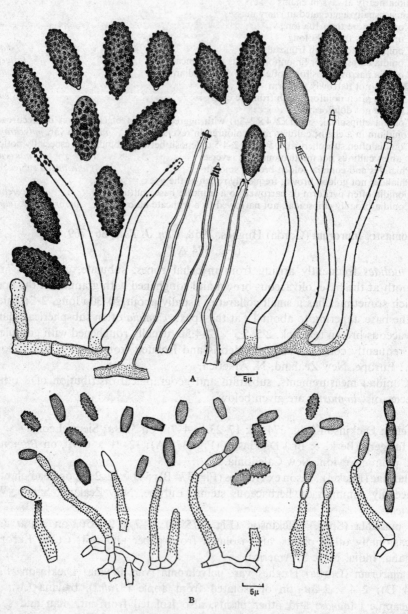

FIG. 375. *Gliomastix* species (2): A, *fusigera*; B, *musicola*.

G. murorum (Corda) Hughes var. **felina** (Marchal) Hughes (Fig. 376 C): 3–6 ×
2–4µ, common on seeds, wood and dead herbaceous stems and isolated from
air, cardboard, paint, rope and soil; Australia, Europe, Hong Kong, New
Caledonia, New Zealand, N. America.

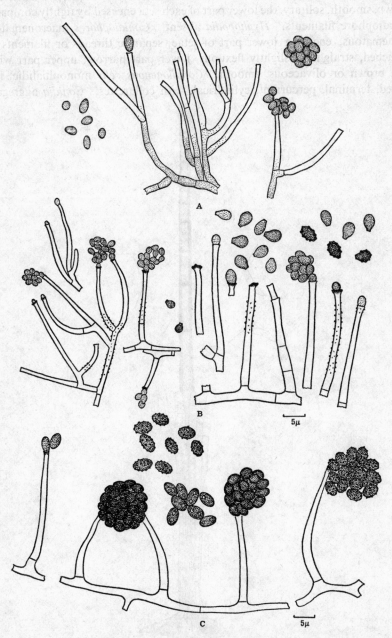

FIG. 376. *Gliomastix* species (3): A, state of *Wallrothiella subiculosa*; B, *cerealis*; C, *murorum*
var. *felina* (× 650 except where indicated by scales).

259. MENISPOROPSIS

Menisporopsis Hughes, 1952, *Mycol. Pap.*, **48**: 59–60.

Colonies hypophyllous, effuse, thin, inconspicuous. *Mycelium* partly superficial, partly immersed. *Stroma* none. *Setae* present, simple, subulate, dark brown, smooth, solitary, the lower part of each seta encased by tightly compacted conidiophore filaments. *Hyphopodia* absent. *Conidiophores* macronematous, synnematous, encasing lower part of setae; separate threads or filaments unbranched, straight or slightly flexuous, lower part narrow, upper part wider, pale brown or olivaceous, smooth. *Conidiogenous cells* monophialidic, integrated, terminal, percurrent, cylindrical, with collarettes. *Conidia* aggregated

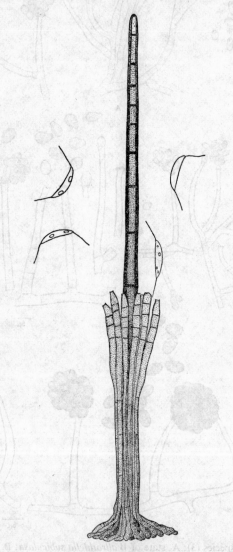

FIG. 377. *Menisporopsis theobromae* (× 650).

in slimy heads, semi-endogenous, usually appendiculate with a setula at each end, curved, cylindrical rounded at the ends or fusiform, colourless, smooth, 0–1-septate.

Type species: Menisporopsis theobromae Hughes.

Menisporopsis theobromae Hughes, 1952, *Mycol. Pap.*, **48**: 59–60.
(Fig. 377)

Setae up to 400µ long, 5–7µ thick. *Conidiophore* threads up to 120µ long, lower part 1µ thick, upper part 3µ thick. *Conidia* 0-septate, 14–18 (mostly 15–16) × 2–3µ, setulae up to 10µ long.

On dead fallen leaves of *Theobroma*; Ghana.

260. CRYPTOSTROMA

Cryptostroma Gregory & Waller, 1951, *Trans. Br. mycol. Soc.*, **34**: 593–594.

Colonies effuse, dark blackish brown, powdery. *Mycelium* immersed. *Stroma* very extensive, consisting of floor and roof separated by a cavity and connected by stromatic columns. No true *setae* but slender sticky threads lie between and hold together spore masses. *Hyphopodia* absent. *Conidiophores* arising from floor stroma, macronematous, mononematous, short, straight, unbranched, brown near base, hyaline at apex, smooth. *Conidiogenous cells* monophialidic, integrated, terminal, determinate, lageniform. *Conidia* catenate, dry, semi-endogenous, simple, ellipsoidal or obovoid, pale to mid brown, smooth.

Type species: Cryptostroma corticale (Ellis & Everh.) Gregory & Waller.

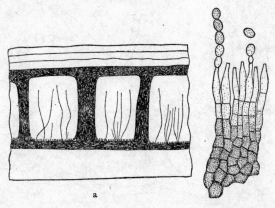

FIG. 378. *Cryptostroma corticale* (a, sketch of section; other figure × 650).

Cryptostroma corticale (Ellis & Everh.) Gregory & Waller, 1951, *Trans. Br. mycol. Soc.*, **34**: 579–597.
(Fig. 378)

Conidiophores up to 30 × 3–4µ. *Conidia* 4–6·5 × 3·5–4µ.

First found on maple logs in Canada. It is the cause of sooty bark disease of sycamore (*Acer pseudoplatanus*) in Great Britain. The outer layers of bark peel off exposing enormous dark blackish brown masses of conidia which often extend over a very large area. The fungus also causes wilt and die back.

261. PHIALOPHORA

Phialophora Medlar, 1915, *Mycologia*, **7**: 200–203.

Cadophora Lagerberg & Melin, 1928, apud Lagerberg, Lundberg & Melin in *Svenska SkogsFör. Tidskr*, 1927, Häft 2, Och 4: 263 (1928).

Colonies effuse, brown or olivaceous brown to black. *Mycelium* partly superficial, partly immersed. *Stroma* none. *Setae* and *hyphopodia* absent. *Conidiophores* semi-macronematous, mononematous, branched, straight or flexuous, pale to mid brown or olivaceous brown, smooth. *Conidiogenous cells* monophialidic, integrated and terminal or discrete, determinate, ampulliform, lageniform or subulate, with well defined collarettes. *Conidia* aggregated in slimy heads, semi-endogenous, simple, straight or curved, ellipsoidal or oblong rounded at the ends, colourless to rather pale brown or olivaceous brown, smooth, 0-septate.

Type species: Phialophora verrucosa Medlar.
Cole, G. T. & Kendrick, W. B., *Can. J. Bot.*, **47**: 779–781, 1969.
Moreau, C., *Revue Mycol.*, **28**: 260–276, 1963.
Nicot, J., *Revue Mycol.*, **32**: 28–40, 1967.
Schol-Schwarz, M. B., *Persoonia*, **6**: 59–94, 1970.
Wang, C. J. K., Fungi of pulp and paper in New York: 69–78, 1965.

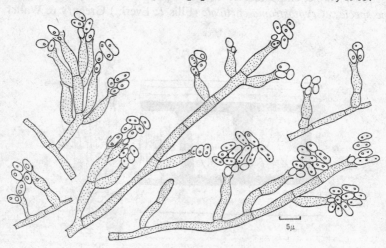

FIG. 379. *Phialophora fastigiata.*

Phialophora fastigiata (Lagerb., Lundberg & Melin) Conant, 1937, *Mycologia*, **29**: 598.

(Fig. 379)

Conidiophores very variable in length, 2–3µ thick. *Phialides* lageniform, 7–13 × 2·5–3µ. *Conidia* colourless to very pale brown, 3–7 × 1·5–2·5µ.
Isolated from air, soil, water, wood and wood pulp; Europe, N. America.

262. AGYRIELLA

Agyriella Sacc., 1884, *Atti Ist. veneto Sci.*, 6 ser., **2**: 454.

Sporodochia pulvinate, hard, gelatinous, black, shining. *Mycelium* mostly immersed. *Stroma* present, erumpent, olivaceous brown. *Setae* and *hyphopodia* absent. *Conidiophores* macronematous, colourless to rather pale brown or olivaceous brown, much branched, grouped very closely together. *Conidiogenous cells* discrete, monophialidic; phialides flask-shaped with narrow, often curved necks. *Conidia* produced in basipetal succession and forming slimy heads at the tips of phialides, allantoid, hyaline, 0-septate.

Type species: Agyriella nitida (Lib.) Sacc.

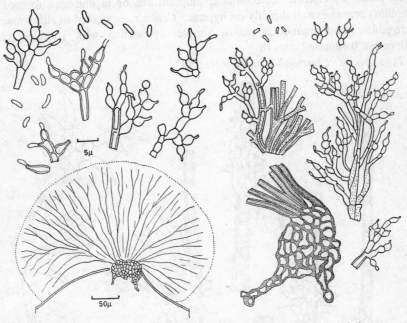

Fig. 380. *Agyriella nitida* (× 650 except where indicated by scales).

Agyriella nitida (Lib.) Sacc., 1884, *Atti Ist. veneto. Sci.*, 6 Ser., **2**: 454.
  Agyrium nitidum Lib., 1834, *Pl. crypt. Ard.*, No. 235.

(Fig. 380)

Sporodochia erumpent, pulvinate, oval or irregular, hard, gelatinous, black, shining. *Mycelium* mostly immersed; hyphae branched, septate, colourless to pale olivaceous brown, smooth, 1–4µ thick. *Stroma* pseudoparenchymatous, olivaceous brown. *Conidiophores* arising from upper part of stroma, straight or flexuous, septate, colourless to rather pale brown or olivaceous brown, up to 200µ long, 1·5–3µ thick, much branched, grouped very closely together, upper branches bearing numerous phialides. *Phialides* flask-shaped, with narrow, often curved necks, hyaline, smooth, 4–6µ long, 2–3µ thick, neck 0·3–1µ wide.

Conidia produced in basipetal succession and forming slimy heads at the tips of phialides, allantoid, hyaline, 2–4 × 1μ.

On branches of *Ribes* and *Rubus*; Austria, Belgium, France, Germany.

263. CHAETOPSINA

Chaetopsina Rambelli, 1956, *Atti Accad. Sci. Ist. Bologna Rc.*, Ser. 11, **3**: 5.

Colonies effuse, usually rather pale brown, glistening, hairy. *Mycelium* mostly superficial. *Stroma* none. *Setae* absent but conidiophores setiform. *Hyphopodia* absent. *Conidiophores* macronematous and micronematous; *macronematous conidiophores* straight, subulate, brown, smooth, septate with short lateral branches arising below the middle, apex sterile, setiform. *Conidiogenous cells* monophialidic, discrete, determinate, ampulliform or lageniform, formed on the short branches and directly on hyphae. *Conidia* aggregated in slimy masses, acrogenous and semi-endogenous, simple, cylindrical with rounded ends, colourless, 0-septate, smooth.

Type species: Chaetopsina fulva Rambelli.

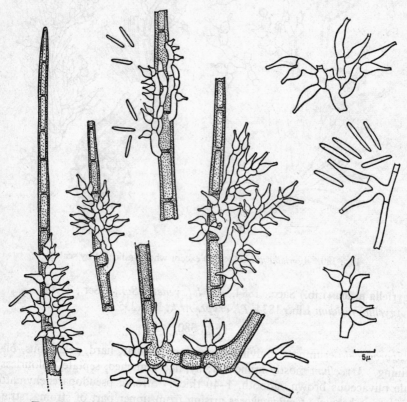

FIG. 381. *Chaetopsina fulva* (× 650 except where indicated by the scale).

Chaetopsina fulva Rambelli, 1956, *Atti Accad. Sci. Ist. Bologna Rc.*, Ser. 11, **3**: 5.

(Fig. 381)

Macronematous conidiophores up to 280 × 5–8μ, base often swollen up to 15–20μ; *conidiogenous cells* (phialides) 7–15μ long, swollen base 3–4μ wide, neck about 1μ thick; *conidia* 7–11 × 1μ.

On dead fallen leaves and in soil; Canada, Italy.

264. FUSARIELLA

Fusariella Saccardo, 1884, *Atti Ist. veneto Sci.*, 6 ser., **2**: 463.

Colonies compact or effuse, greyish green, blackish green or black. *Mycelium* superficial and immersed. *Stroma* none. *Setae* and *hyphopodia* absent. *Conidiophores* semi-macronematous, mononematous, branched irregularly or sometimes dichotomously or trichotomously, flexuous, colourless or pale brown, smooth or verruculose. *Conidiogenous cells* monophialidic, integrated and terminal, or discrete, determinate, often curved, cylindrical, subulate or lageniform, with collarettes. *Conidia* catenate, acrogenous, semi-endogenous, developing in basipetal succession and frequently hanging together in slipped chains, the tip of each conidium except the apical one being deflected laterally, simple, straight, bent or flexuous, often fusiform pointed at the apex blunt at the base but sometimes cylindrical, dumb-bell-shaped, clavate or obclavate, pale to mid olive brown, olive green or greyish green, pale greyish green, blackish green or black in mass, usually smooth, 1–3-septate.

Type species: F. atrovirens Sacc.

Hughes, S. J., *Mycol. Pap.*, **28**: 1–11, 1949.

KEY

```
Conidia all more or less the same shape  .      .       .       .      .       .       .       1
Conidia of different shapes, many short 1-septate, dumb-bell-shaped, clavate or obclavate,
  others longer, fusiform, straight or curved, 1–3-septate   .       .       .       indica
1. Conidia fusiform tapered to an acute apex, usually slightly curved, always in slipped chains;
    in mass black or blackish green      .      .      .      .      .      .      .      2
   Conidia cylindrical tapered only slightly at each end; not in slipped chains; in mass pale
    greyish green      .      .      .      .      .      .      .      .      hughesii
2. Conidiophores often verruculose and pale brown  .      .      .      .      atrovirens
   Conidiophores colourless, smooth      .      .      .      .      .      .      3
3. Conidia 25–32 (27) × 5·5–6·5μ      .      .      .      .      .      bizzozeriana
   Conidia 16–22 (19) × 3·5–4μ      .      .      .      .      .      .      concinna
   Conidia 14–20 (16) × 4·5–7  .      .      .      .      .      .      .      obstipa
```

Fusariella atrovirens Sacc., 1854, *Atti Ist. veneto Sci.*, 6 ser., **2**: 463.

(Fig. 382 A)

Colonies compact, black. *Conidiophores* up to 70 × 3μ, colourless to pale brown, often verruculose. *Phialides* curved, 15–26 × 3–4μ tapering to 2μ. *Conidia* mostly curved, fusiform, pointed at the apex, blunt at the base, olive brown, black in mass, 3-septate, sometimes slightly constricted at the septa, 20–26 (23) × 6–7μ.

On *Allium*; Europe.

Conidial measurements, substrata and distribution of 5 other species are given below; references are cited only for those published since 1949.

F. indica Roy & B. Rai, 1968, *Trans. Br. mycol. Soc.*, **51**: 333–334 (Fig. 382 B):
8–16 × 4–6μ, on *Saccharum*; India.

F. hughesii Chabelska-Frydman, 1964, *Can. J. Bot.*, **42**: 1485–1487 (Fig. 382 C):
14–25 × 2·5–3·5μ, on dead stems and leaves of *Dipsacus, Foeniculum, Lupinus,
Phalaris, Trigonella* and *Urtica*; Great Britain, Israel.

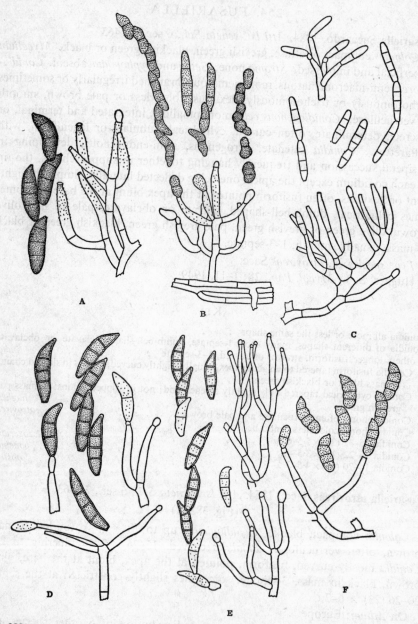

FIG. 382. *Fusariella* species: A, *atrovirens*; B, *indica*; C, hughesii; D, *bizzozeriana*; E, *concinna*;
F, *obstipa* (× 650).

F. bizzozeriana (Sacc.) Hughes (Fig. 382 D): 25–32 (27) × 5·5–6·5µ, on leaves of *Erythrina* and isolated from soil; Europe, N. America.

F. concinna (Syd.) Hughes (Fig. 382 E): 16–22 (19) × 3·5–4µ, on fallen leaves of *Agave* and *Ficus*; India.

F. obstipa (Pollack) Hughes (Fig. 382 F): 14–20 (16) × 4·5–7µ, on *Calopogonium*, *Carica*, *Colocasia*, *Heeria* and isolated from soil; France, India, New Caledonia, U.S.A., Zambia.

265. MENISPORA

Menispora Persoon, 1822, Mycol. Eur., **1**: 32.

Colonies effuse, at first greyish brown, then olivaceous or dark blackish brown, hairy or velvety. *Mycelium* partly superficial, partly immersed. *Stroma* absent or composed of a few cells. *Setae*, when present, straight, flexuous, coiled or twisted, mid to dark brown, smooth. *Hyphopodia* absent. *Conidiophores* macronematous, mononematous, crowded, unbranched or loosely branched, sometimes anastomosing, straight or flexuous, mid to dark brown, smooth, the upper part often sterile and setiform. *Conidiogenous cells* monophialidic or sometimes polyphialidic, occasionally integrated and terminal but mostly discrete and lateral, determinate or sympodial, cylindrical, in most species recurved at the tip with rather inconspicuous collarettes. *Conidia* aggregated in slimy masses, semi-endogenous, curved, cylindrical with rounded ends, often with a setula at each end, colourless, smooth, 0–3-septate.

Type species: Menispora glauca Pers.

Hughes, S. J. & Kendrick, W. B., *Can. J. Bot.*, **41**: 693–718, 1963; *N.Z. Jl Bot.*, **6**: 323–375, 1968.

KEY

Conidia 3-septate . . . . *glauca* (also *tortuosa*)
Conidia 0-septate . . . . *ciliata*

Menispora glauca Pers., 1822, Mycol. Eur., **1**: 32.
(Fig. 383 A)

Colonies at first greyish brown, later dark blackish brown. *Conidiophores* unbranched or with a few long, upwardly directed branches, up to 800µ long, 3–5µ thick, lower part fertile, upper part sterile, setose, twisted or loosely coiled. *Phialides* hyaline, cylindrical, 10–30 × 4–5µ with a narrow uncinate tip, borne directly on the conidiophore or at the ends of short branchlets. *Conidia* 3-septate, mostly 18–23 × 4–5µ; setulae up to 14µ long.

Fairly common on wood and bark of deciduous trees, including *Acer*, *Betula* and *Fagus*; Europe and N. America. **Menispora tortuosa** Corda has similar conidia but its phialides are clustered and they are not recurved at the tip.

Menispora ciliata Corda, 1837, Icon. Fung., **1**: 16.
(Fig. 383 B)

Colonies at first greyish brown, becoming olivaceous. *Conidiophores* unbranched or occasionally branched, flexuous, frequently anastomosing, up to

900μ long, 3–5μ thick, lower part fertile, upper part sterile, setose. *Phialides* hyaline or subhyaline, cylindrical, mostly 16–25 × 4–5μ, with a narrow uncinate tip, borne directly on the conidiophore or at the ends of short branchlets. *Conidia* 0-septate, 12–21 (mostly 14–18) × 3–3·5μ; setulae up to 12μ long.

Common on wood and bark of many different trees, especially common on *Fagus* and *Quercus*; Europe, N. America.

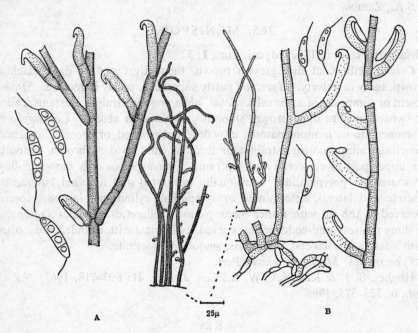

A   25μ   B

FIG. 383. A, *Menispora glauca*; B, *M. ciliata* (× 650 except where indicated by the scale).

266. GONYTRICHUM

Gonytrichum C. G. & F. Nees ex Pers.; C. G. & F. Nees, 1818, *Nova Acta Acad. Caesar. Leop. Carol.*, **9**: 244; Persoon, 1822, Mycol. eur., **1**: 19.

Colonies effuse, sometimes pulvinate, grey, olivaceous or dark blackish brown. *Mycelium* mostly immersed. *Stroma* none. Separate *setae* absent but the upper part of a conidiophore may be setiform and bear setiform sterile branches. *Hyphopodia* absent. *Conidiophores* macronematous, mononematous, branched, straight or flexuous, upper part sometimes sterile and setiform, branching systems in some species complex the branches anastomosing and difficult to tease apart; short encircling collar branches characteristic. *Conidiogenous cells* monophialidic, borne on collar branches, integrated and terminal or discrete, determinate, cylindrical or lageniform, with collarettes. *Conidia* aggregated in slimy masses, often in columns, acrogenous and semi-endogenous, simple, ellipsoidal, oblong rounded at the ends or subspherical, colourless or very pale olive, smooth, 0-septate.

Type species: Gonytrichum state of *Melanopsammella inaequalis* (Grove) Höhnel = *G. caesium* Nees ex Pers.

Hughes, S. J., *Trans. Br. mycol. Soc.*, **34**: 560–568, 1951.
Swart, H. J., *Antonie van Leeuwenhoek*, **25**: 439–444, 1959.

KEY

Conidiophore branching system complex, branches anastomosing   *Melanopsammella inaequalis*
Conidiophore branches not anastomosing, often setiform . . . . . *macrocladum*

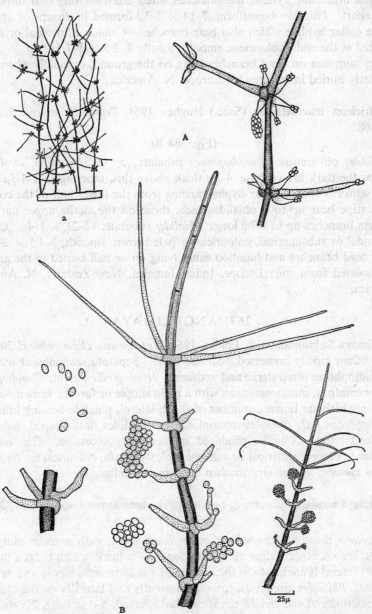

FIG. 384. A, *Gonytrichum* state of *Melanopsammella inaequalis*; B, *G. macrocladum* (a, habit sketch, other figures × 650 except where indicated by the scale).

Gonytrichum state of **Melanopsammella inaequalis** (Grove) Höhnel, 1919, *Annls mycol.*, **17**: 121.

(Fig. 384 A)

Colonies grey when sporing freely, later dark blackish brown. *Conidiophores* up to 500μ long, 2–3μ thick, dark brown, paler towards the apex, with a very complex branching system, the branches often anastomosing and difficult to tease apart. *Phialides* lageniform, 7–14 × 2–3μ formed in groups of up to 12 on the collar hyphae which also bear branches. *Conidia* ellipsoidal or oblong rounded at the ends, colourless, smooth, mostly 2–2·5 × 1–1·5μ.

Very common on dead branches lying on the ground, especially where these are partly buried in leaf mould; Europe, N. America.

Gonytrichum macrocladum (Sacc.) Hughes, 1951, *Trans. Br. mycol. Soc.*, **34**: 565–568.

(Fig. 384 B)

Colonies olivaceous. *Conidiophores* subulate, up to 350μ long, swollen to 5–8μ at the dark brown base, 4–6μ thick above this, tapering to 1–1·5μ at the paler setiform apex. Collar hyphae arising from the lower half of the conidiophore stipe bear up to 6 phialides each, those on the sterile upper part bear setiform branches up to 170μ long. *Phialides* subulate, 12–21 × 3–4μ. *Conidia* ellipsoidal or subspherical, colourless or pale brown, smooth, 3–4·5 × 2–3μ.

On dead branches and bamboo canes lying on or half buried in the ground, also isolated from soil; Europe, India, Jamaica, New Zealand, N. America, S. Africa.

267. ANGULIMAYA

Angulimaya Subramanian & Lodha, 1964, *Antonie van Leeuwenhoek*, **30**: 329.

Mycelium mostly immersed. *Stroma* none. Separate *setae* absent but ends of conidiophores often sterile and setiform. *Hyphopodia* absent. *Conidiophores* macronematous, mononematous, with a main simple or furcate, brown, septate, stipe on which are borne a number of short, lateral, phialide-bearing branches. *Conidiogenous cells* discrete, monophialidic; phialides flask-shaped, pale with dark collarettes. *Conidia* produced in basipetal succession, often hanging together in chains, spherical or subspherical, 0-septate, colourless or very pale.

Type species: Angulimaya sundara Subram. & Lodha.

Angulimaya sundara Subram. & Lodha, 1964, *Antonie van Leeuwenhoek*, **30**: 29.

(Fig. 385)

Mycelium mostly immersed. *Conidiophores* erect, with a main, simple or furcate, brown, septate stipe up to 500μ long, 4–7μ thick, which bears a number of short lateral branches 4–5μ thick; the stipe is sometimes sterile and acute at the apex. *Phialides* numerous, formed terminally and laterally on the branches and sometimes at the end of the stipe, flask-shaped, 5–14μ long, 2·5–4μ thick in the broadest part, colourless or very pale, each with a dark collarette. These phialides were incorrectly described as annellophores by the authors of the genus. *Conidia* produced in basipetal succession, emerging through the open ends of the phialides and often hanging together in quite long chains, spherical or subspherical, colourless or very pale, 2–2·5μ diam.

On cow dung; India.

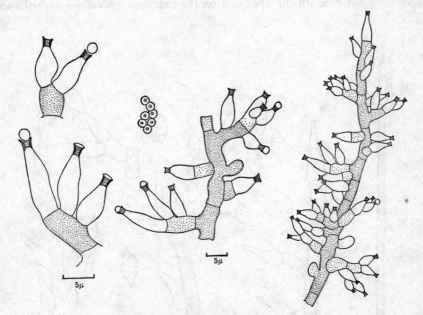

FIG. 385. *Angulimaya sundara* (× 650 except where indicated by scales).

268. ACROPHIALOPHORA

Acrophialophora J. C. Edward, 1961, *Mycologia*, **51**: 784.

Colonies effuse, felted, greyish brown. *Mycelium* superficial and immersed; hyphae narrow, colourless or pale brown. *Stroma* none. *Setae* and *hyphopodia* absent. *Conidiophores* macronematous, mononematous, simple or bearing a few branches near the apex, brown, verrucose or echinulate. *Conidiogenous cells* discrete, monophialidic. *Conidia* catenate, produced in basipetal succession at the tips of phialides, limoniform, colourless or very pale brown, echinulate or verruculose with spiral bands, 0-septate.

Type species: Acrophialophora fusispora (Saksena) M. B. Ellis = *A. nainiana* J. C. Edward.

Acrophialophora fusispora (Saksena) M. B. Ellis comb. nov.
Paecilomyces fusisporus Saksena, 1953, *J. Indian bot. Soc.*, **32**: 187.
Acrophialophora nainiana Edward, 1961, *Mycologia*, **51**: 784.
(Fig. 386)

Colonies effuse, felted, greyish brown, almost black in reverse. *Mycelium* superficial and immersed; hyphae branched, septate, at first colourless, becoming pale brown, 1·5–3·5μ thick, darker and thicker at the point of origin of the coni-diophores. *Conidiophores* arising singly terminally and laterally on the hyphae, erect or ascending, straight or slightly flexuous, almost cylindrical, but tapering very gradually towards the apex, septate, pale brown to brown, echinulate or verrucose, up to 1·5 mm. long, 2–5μ thick, simple or with a few subterminal branches, and bearing a terminal phialide and other phialides singly, in pairs or verticils just beneath the apex and on the branches. *Phialides* flask-shaped

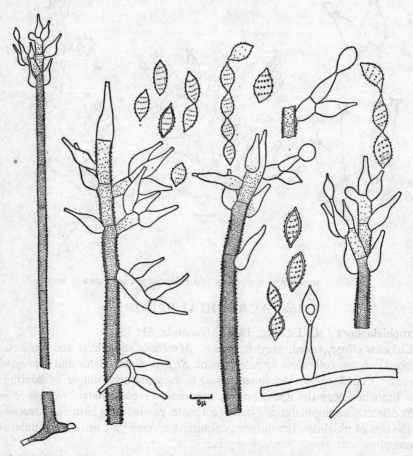

FIG. 386. *Acrophialophora fusispora* (× 650 except where indicated by the scale).

with a narrow, sometimes curved, cylindrical neck, colourless, smooth or echinulate, 9–15μ long, 3–4·5μ thick in the broadest part, neck 0·5–1μ wide. *Conidia* produced basipetally in very long (up to 300μ) chains at the tips of the phialides, limoniform, 0-septate, colourless, or very pale brown, finely echinulate with distinct spiral bands, 6–11 (9·3) × 3·5–5 (4·3)μ.

Isolated from soil, air and various plants including *Cajanus*, *Cicer*, maize and *Tephrosia*; Canada, Congo, Egypt, India, Nigeria, Pakistan, South Africa, U.S.A., It grows well and sporulates freely on p.d.a.

269. VERTICILLIUM

Verticillium Nees ex Link; Nees, 1816, Syst. Pilze Schwämme: 56–57; Link, 1824, in Linné's Sp. Pl., Ed. 4 (Willdenow's), Berlin, **6** (1): 75.

Colonies effuse, variously coloured, sometimes mid to dark blackish brown. *Mycelium* partly superficial, partly immersed; *chlamydospores* often formed in culture. *Stroma* none. *Setae* and *hyphopodia* absent. *Conidiophores* macronematous, mononematous, scattered, each composed of an erect, straight or flexuous, colourless to dark brown, smooth or verruculose stipe with branches and phialides commonly in verticils beneath the septa nearest the apex. *Conidiogenous cells* monophialidic, discrete, often arranged verticillately, determinate, ampulliform, lageniform or subulate, with well-defined collarettes. *Conidia* aggregated in slimy masses, semi-endogenous or acrogenous, simple, allantoid, ellipsoidal or cylindrical, rounded at the ends, colourless or pale brown, smooth, 0-septate.

Type species: Verticillium state of *Nectria inventa* Pethybridge = *V. tenerum* Nees ex Link.

Hughes, S. J., *Mycol. Pap.*, **45**: 1–24, 1951.

Smith, Harvey C., *N.Z. Jl agric. Res.*, **8**: 450–478, 1965.

Many species in this genus are colourless or brightly coloured; only 2 species with brown conidiophores are described here.

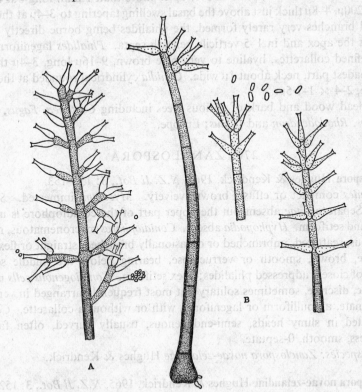

Fig. 387. A, *Verticillium cyclosporum*; B, *V. tenuissimum* (× 650).

Verticillium cyclosporum (Grove) Mason & Hughes, 1951, apud Hughes in *Mycol. Pap.*, **45**: 18–24.

Stachylidium cyclosporum Grove, 1884, *J. Bot., Lond.*, **22**: 199.

(Fig. 387 A)

Colonies pale grey to dark blackish brown, hairy. *Conidiophores* swollen at the base, dark brown below, paler near the apex, up to 500μ long, 3–7μ thick just above the basal swelling, 2–3μ at the apex. Branches formed singly, in pairs or verticils beneath the upper 4–10 septa, concolorous with the stipe; the lower branches themselves bear secondary branches and each branch bears usually 2 or 3 phialides at its end and a few also along the sides. *Phialides* subulate or lageniform, each with a distinct collarette, hyaline or very pale brown, 8–15 × 2–3μ. *Conidia* broadly ellipsoidal, hyaline, less than 2μ long.

Common on dead wood and bark of various trees including *Alnus, Betula, Castanea, Corylus, Fagus, Fraxinus, Ilex, Prunus, Quercus, Rhododendron, Sambucus* and *Ulex*; Europe.

Verticillium tenuissimum Corda, 1837, Icon. Fung., **1**: 20.

(Fig. 387 B)

V. apicale Berk. & Br., 1851, *Ann. Mag. nat. Hist.* 2, **7**: 101.

Colonies grey to dark brown, hairy. *Conidiophores* subulate, swollen at the base, dark brown below, paler above, up to more than 1 mm. long but usually about 200μ, 4–8μ thick just above the basal swelling tapering to 3–4μ at the apex. Lateral branches very rarely formed, the phialides being borne directly on the stipe at the apex and in 1–5 verticils beneath septa. *Phialides* lageniform with well-defined collarettes, hyaline to very pale brown, 9–16μ long, 3–4μ thick in the broadest part, neck about 1μ wide. *Conidia* cylindrical rounded at the ends, hyaline, 2–4 × 1–1·5μ.

On dead wood and bark of various trees including *Castanea, Fagus, Pinus, Quercus, Rhododendron* and *Ulmus*; Europe.

270. ZANCLOSPORA

Zanclospora Hughes & Kendrick, 1965, *N.Z. Jl Bot.*, **3**: 151–153.

Colonies compact or effuse, brown, velvety. *Mycelium* immersed. *Stroma* none. Separate *setae* absent but the upper part of the conidiophore is usually sterile and setiform. *Hyphopodia* absent. *Conidiophores* macronematous, mononematous, scattered, unbranched or occasionally branched, straight or flexuous, subulate, brown, smooth or verruculose, bearing below the middle several whorls of closely adpressed phialides; apex setiform. *Conidiogenous cells* monophialidic, discrete, sometimes solitary but most frequently arranged in verticils, determinate, ampulliform or lageniform, with or without a collarette. *Conidia* aggregated in slimy heads, semi-endogenous, usually curved, often falcate, colourless, smooth, 0-septate.

Type species: Zanclospora novae-zelandiae Hughes & Kendrick.

Zanclospora novae-zelandiae Hughes & Kendrick, 1965, *N.Z. Jl Bot.*, **3**: 152–165.

(Fig. 388)

Conidiophores rather thick-walled, up to 750µ long, 5–7µ thick just above the base which may be swollen. *Phialides* solitary or in whorls of 3–7, 10–18µ long, 3·5–5µ thick in the broadest part, hyaline to pale brown. *Conidia* 20–35 × 2–3·5µ.

On wood and bark of *Libocedrus*, *Nothofagus* and *Weinmannia*; New Zealand.

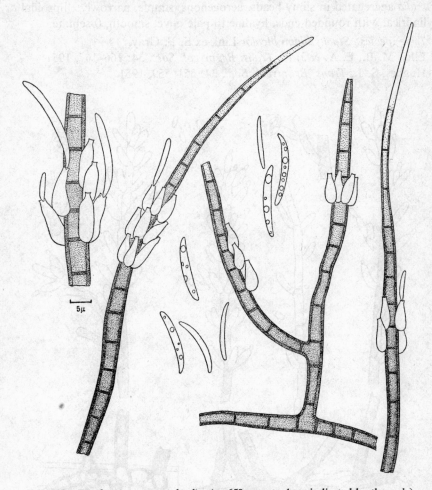

Fig. 388. *Zanclospora novae-zelandiae* (× 650 except where indicated by the scale).

271. STACHYLIDIUM

Stachylidium Link ex S. F. Gray; Link, 1809, *Magazin Ges. naturf. Freunde Berl.*, **3** (1): 15; S. F. Gray, 1821, Nat. Arr. Br. Pl.: 553.

Spondylocladium Martius ex Link; Martius, 1817, Fl. crypt. Erlang.: 354–355; Link, 1824, in Linné's Sp. Pl., Ed. 4 (Willdenow's), **6** (1): 78.

Colonies effuse, grey, olive or olivaceous brown. *Mycelium* immersed. *Stroma* often present, pseudoparenchymatous, olive to brown, immersed. *Setae* and *hyphopodia* absent. *Conidiophores* macronematous, mononematous, unbranched or branched, straight or flexuous, smooth or minutely verruculose; lower part of stipe sterile, brown or olivaceous brown, upper part and branches hyaline or olivaceous, with verticillately arranged phialides. *Conidiogenous cells* monophialidic, discrete, arranged in verticils, determinate, cylindrical rounded at the apex or narrowly ellipsoidal, with a minute opening and no collarette. *Conidia* aggregated in slimy heads, acrogenous, simple, narrowly ellipsoidal or cylindrical with rounded ends, hyaline to pale olive, smooth, 0-septate.

Type species: Stachylidium bicolor Link ex S. F. Gray.

Ellis, M. B., E. A. & J. P., *Trans. Br. mycol. Soc.*, **34**: 166–167, 1951.

Hughes, S. J., *Trans. Br. mycol. Soc.*, **34**: 551–559, 1951.

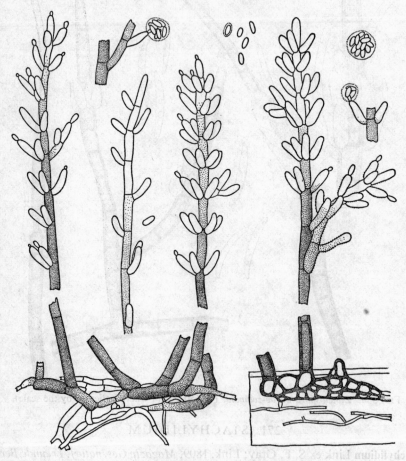

FIG. 389. *Stachylidium bicolor* (× 650).

Stachylidium bicolor Link ex S. F. Gray; Link, 1809, *Magazin Ges. naturf. Freunde Berl.*, **3** (1): 15; S. F. Gray, 1821, *Nat. Arr. Br. Pl.*: 553; Fries, 1832, *Syst. mycol.*, **3**: 391.

(Fig. 389)

Conidiophores up to 700μ long, 4–7μ thick at the base, 2·5–4μ at the apex. *Phialides* smooth or minutely verruculose, hyaline or pale olivaceous, 9–20 × 3–4μ. *Conidia* 4–8 × 2–3μ.

On dead stems, twigs, etc. of plants including *Allium*, *Ananas*, *Bambusa*, *Dioscorea*, *Gardenia*, *Heliconia*, *Heracleum*, *Hibiscus*, *Manihot*, *Musa*, *Oenanthe*, *Petasites*, *Phoenix*, *Populus*, *Pteridium*, *Sambucus*, *Solanum*, *Sporobolus*, *Theobroma*, *Urtica* and *Zea*; Europe, Ghana, Malaya, New Guinea, Rhodesia, Sabah, Sierra Leone.

272. CRYPTOPHIALE

Cryptophiale Pirozynski, 1968, *Can. J. Bot.*, **46**: 1123–1127.

Colonies effuse, rather inconspicuous, dark brown, velvety. *Mycelium* superficial, reticulate. *Stroma* none. Separate *setae* absent but the upper part of each conidiophore is sterile and setiform. *Hyphopodia* absent. *Conidiophores* macronematous, mononematous, erect, straight or flexuous, subulate, dark

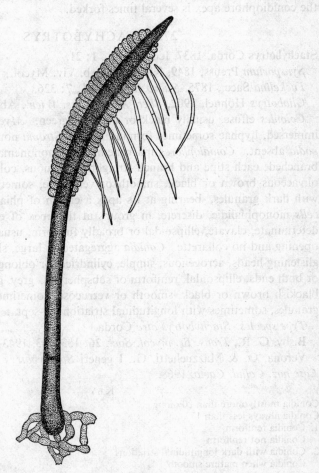

Fig. 390. *Cryptophiale kakombensis* (× 650).

brown, smooth, thick-walled, with the upper part sterile, setiform and simple or several times forked, and below this and extending to a point not quite halfway down the conidiophore a fertile region with 2 rows of phialides covered by a shield extending as a flange on each side and composed of flat, lobed, olivaceous brown cells. *Conidiogenous cells* monophialidic, discrete, in rows obscured by the shield of sterile cells, determinate, subspherical or lageniform. *Conidia* aggregated in slimy masses, semi-endogenous, simple, falcate, colourless, smooth, 1-septate.

Type species: Cryptophiale kakombensis Pirozynski.

Cryptophiale kakombensis Pirozynski, 1968, *Can. J. Bot.*, **46**: 1124–1126.
(Fig. 390)

Conidiophores up to 260µ long, 5–8µ thick, apex not forked, shield of sterile cells 90–110 × 16–20µ. *Conidia* 22–28 × 1–2µ.

On fallen leaves of *Baphia*; Tanzania. In the only other species so far described, **C. udagawae** Pirozynski & Ichinoe found on dead leaves of *Quercus* in Japan, the conidiophore apex is several times forked.

273. STACHYBOTRYS

Stachybotrys Corda, 1837, Icon. Fung., **1**: 21.

Synsporium Preuss, 1849, Klotzsch Herb. Viv. Mycol., 1285.

Fuckelina Sacc., 1875, *Nuovo G. bot. ital.*, **7**: 326.

Gliobotrys Höhnel, 1902, *Sber. Akad. Wiss. Wien.*, Abt. 1, **111**: 1048–1049.

Colonies effuse, usually black or blackish green. *Mycelium* superficial and immersed, hyphae sometimes forming ropes. *Stroma* none. *Setae* and *hyphopodia* absent. *Conidiophores* macronematous, mononematous, unbranched or branched; each stipe and branch straight or flexuous, colourless, grey, brown, olivaceous brown or black, smooth or verrucose, sometimes covered in part with dark granules, bearing at its apex a crown of phialides. *Conidiogenous cells* monophialidic, discrete, in groups at the apex of each stipe or branch, determinate, clavate, ellipsoidal or broadly fusiform, usually with a very small opening and no collarette. *Conidia* aggregated in large, slimy, often black and glistening heads, acrogenous, simple, cylindrical or oblong, rounded at one end or both ends, ellipsoidal, reniform or subspherical, grey, greenish, dark brown, blackish brown or black, smooth or verrucose, sometimes covered with dark granules, sometimes with longitudinal striations, 0-septate.

Type species: Stachybotrys atra Corda.

Bisby, G. R., *Trans. Br. mycol. Soc.*, **26**: 133–143, 1943; **28**: 11–12, 1945.

Verona, O. & Mazzuchetti, G., I generi *Stachybotrys* e *Memnoniella. Publ. Ente naz. Cellul. Carta*, 1968.

KEY

Conidia mostly more than 20µ long	*theobromae*
Conidia always less than 17µ	1
1. Conidia reniform	*nephrospora*
Conidia not reniform	2
2. Conidia with dark longitudinal striations	*cylindrospora*
Conidia when mature smooth	3
Conidia when mature verrucose	4

3. Conidia 3–5·5µ long . . . . *parvispora*
   Conidia 6–11µ long . . . . *Melanopsamma pomiformis*
4. Conidia often obliquely attenuated at the base, 8–12×4–6µ . . . *dichroa*
   Conidia not obliquely attenuated at the base 5
5. Conidia 11–15×6–8µ . . . . *kampalensis*
   Conidia 8–11×5–10µ . . . . *atra*
   Conidia 6–8×4–5µ . . . . *atra* var. *microspora*

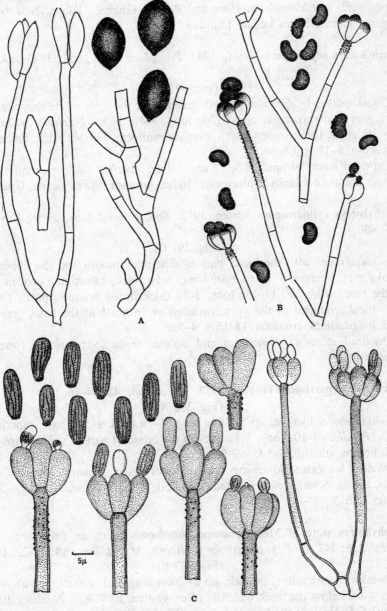

FIG. 391. *Stachybotrys* species (1): A, *theobromae*; B, *nephrospora*; C, *cylindrospora* (× 650 except where indicated by the scale).

Stachybotrys theobromae Hansf., 1943, *Proc. Linn. Soc. Lond.*, 1942–1943: 45.
(Fig. 391 A)

Conidiophores hyaline, smooth, up to 200µ long, 4–6µ thick. *Phialides* 20–27µ long, 6–8µ thick in the broadest part. *Conidia* ellipsoidal, usually with a small projecting papilla at the base, black and smooth when mature, 20–28 × 14–17µ.

Common on dead branches of *Theobroma* and occasionally found on other plants such as *Chlorophora*, *Hura* and *Persea*; Ghana, Malaya, New Guinea, Nigeria, Papua, Sierra Leone, Trinidad.

Stachybotrys nephrospora Hansf., 1943, *Proc. Linn. Soc. Lond.*, 1942–1943: 44–45.

(Fig. 391 B)

Conidiophores hyaline and smooth except near the apex where they are often dark grey and verrucose, up to 120µ long, 3–4µ thick. *Phialides* 7–12µ long, 4–6µ thick in the broadest part. *Conidia* reniform, almost black, smooth or verrucose, 8–11 × 4·5–6µ.

On dead wood of unidentified trees and on *Arachis*, *Dioscorea* and *Carica*; British Solomon Islands Protectorate, India, Jamaica, Sierra Leone, Uganda.

Stachybotrys cylindrospora Jensen, 1912, *Bull. Cornell Univ. agric. Exp. Stn*, **315**: 496.

(Fig. 391 C)

Conidiophores with the lower part hyaline and smooth and the upper part smoky grey, verrucose, up to 100µ long, 3–5µ thick, sometimes swollen to 7µ at the base. *Phialides* 11–15µ long, 5–7µ thick in the broadest part. *Conidia* cylindrical, rounded at the apex, rounded or truncate at the base, grey with dark longitudinal striations, 11–15 × 4–5µ.

On dead stems of *Heracleum* and isolated from soil; Europe, Israel, N. America.

Stachybotrys parvispora Hughes, 1952, *Mycol Pap.*, **42**: 74–76.
(Fig. 392 A)

Conidiophores hyaline, up to 160µ long, 4–7µ thick at the base, tapering to 2–3µ. *Phialides* 7–10µ long, 3–4µ thick in the broadest part. *Conidia* ellipsoidal, dark brown, smooth, 3–5·5 × 2·5–3·5µ.

On dead leaves and sometimes other parts of *Agave*, *Ananas*, *Cajanus*, *Coffea*, *Ficus*, *Hevea*, *Setaria*, etc., isolated from soil; Congo, Ghana, Malaya, Sierra Leone, U.S.A.

Stachybotrys state of **Melanopsamma pomiformis** (Pers. ex Fr.) Sacc., 1878, *Michelia*, **1**: 347. For synonymy see C. Booth, *Mycol. Pap.*, **68**: 21–22, 1957.
(Fig. 392 B)

Conidiophores hyaline, smooth, up to 250µ long, 8–11µ thick just above the base, 4–6µ below the apex which is often swollen to 7–9µ. *Phialides* 10–16µ long, 3·5–5µ thick in the broadest part. *Conidia* ellipsoidal, greenish or greyish brown, smooth, 6–11 × 4·5–7µ.

On wood of various deciduous trees including *Aesculus, Carpinus, Fagus, Fraxinus, Juglans, Populus* and *Ulmus;* Europe.

Stachybotrys dichroa Grove, 1886, *J. Bot. Lond.,* **24**: 201.
(Fig. 392 C)

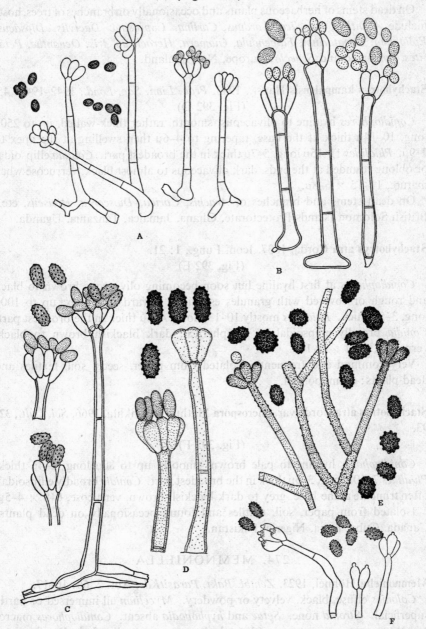

FIG. 392. *Stachybotrys* species (2): A, *parvispora*; B, state of *Melanopsamma pomiformis*; C, *dichroa*; D, *kampalensis*; E, *atra*; F, *atra* var. *microspora* (× 650).

Conidiophores hyaline, smooth, thick-walled, up to 270μ long, 8–20μ thick at the base, tapering to 3·5–7μ just below the apex then swelling again to 5–9μ. *Phialides* 9–17μ long, 3·5–6μ thick in the broadest part. *Conidia* ellipsoidal or cylindrical rounded at the ends, often obliquely attenuated at the base, olivaceous brown to almost black, verrucose when mature, 8–14 × 4–6μ.

On dead stems of herbaceous plants and occasionally on branches of trees, hosts include *Acanthus, Angelica, Carduus, Carlina, Coprosma, Dactylis, Dipsacus, Epilobium, Eupatorium, Filipendula, Gunnera, Heracleum, Iris, Oenanthe, Petasites, Sambucus* and *Senecio*; Europe, New Zealand.

Stachybotrys kampalensis Hansf., 1943, *Proc. Linn. Soc. Lond.*, 1942–1943: 45.
(Fig. 392 D)

Conidiophores hyaline to olivaceous, smooth, rather thick-walled, up to 250μ long, 10–14μ thick at the base, tapering to 4–6μ then swelling at the apex to 7–9μ. *Phialides* 11–15μ long, 5–7μ thick in the broadest part. *Conidia* ellipsoidal or oblong rounded at the ends, dark olivaceous to almost black, verrucose when mature, 11–15 × 6–8μ.

On dead stems and branches of *Arachis, Carica, Dioscorea, Hibiscus*, etc.; British Solomon Islands Protectorate, Ghana, Jamaica, Tanzania, Uganda.

Stachybotrys atra Corda, 1837, Icon. Fung., **1**: 21.
(Fig. 392 E)

Conidiophores at first hyaline but soon becoming olivaceous brown to black and rough or covered with granules especially towards the apex, up to 100μ long, 3–5μ thick. *Phialides* mostly 10–13μ long, 4–6μ thick in the broadest part. *Conidia* broadly ellipsoidal to subspherical, dark blackish brown to black, verrucose, 8–11 × 5–10μ.

Very common and frequently isolated from paper, seeds, soil, textiles and dead plants; cosmopolitan.

Stachybotrys atra Corda var. **microspora** Mathur & Sankhla, 1966, *Sci. Cult.*, **32**: 93.
(Fig. 392 F)

Conidiophores hyaline to pale brown, smooth, up to 80μ long, 2–4μ thick. *Phialides* 7–10μ long, 3–5μ thick in the broadest part. *Conidia* broadly ellipsoidal, often truncate at the base, grey to dark blackish brown, verrucose, 6–8 × 4–5μ.

Isolated from paper, soil, textiles and found occasionally on dead plants; Canada, Cuba, India, Nigeria, Pakistan.

274. MEMNONIELLA

Memnoniella Höhnel, 1923, *Zentbl. Bakt. ParasitKde*, Abt. 2, **60**: 16–17.

Colonies effuse, black, velvety or powdery. *Mycelium* all immersed or partly superficial. *Stroma* none. *Setae* and *hyphopodia* absent. *Conidiophores* macronematous, mononematous, unbranched or occasionally forked, sometimes swollen at the apex, pale to mid grey, olivaceous or brown, smooth or minutely verruculose, often covered in part with dark granules. *Conidiogenous cells* monophialidic, discrete, in groups up to 10 at the apex of the stipe, determinate, clavate, pyriform, cylindrical or ellipsoidal, usually with a small opening and no collarette. *Conidia* catenate, acrogenous, simple, spherical, subspherical sometimes slightly flattened in one plane or hemispherical, grey, dark brown or black, smooth or verrucose.

Type species: *Memnoniella echinata* (Riv.) Galloway = *M. aterrima* Höhnel.

Verona, O. & Mazzuchetti, G., I generi *Stachybotrys* e *Memnoniella*. *Publ. Ente naz. Cellul. Carta*, 1968.

KEY

Conidia verrucose, 3·5–5µ diam. . . . . *echinata*
Conidia verrucose, 6–9µ diam. . . . . *subsimplex*
Conidia smooth, often hemispherical, 4–6µ diam. . . . . . . *levispora*

Memnoniella echinata (Riv.) Galloway, 1933, *Trans. Br. mycol. Soc.*, **18**: 163–166.

(Fig. 393 A)

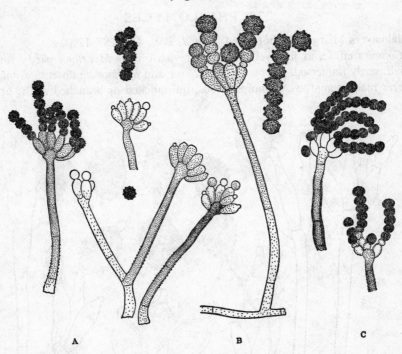

FIG. 393. *Memnoniella* species: A, *echinata*; B, *subsimplex*; C, *levispora* (× 650).

Conidiophores 50–100 × 3–4µ, grey, often covered in part with dark granules. *Phialides* mostly in groups of 4–8, 7–9 × 3–5µ. *Conidia* spherical or flattened dorsiventrally, 3·5–5µ diam.

Common on dead plants, frequently isolated from paper and textiles and occasionally from soil; cosmopolitan.

Memnoniella subsimplex (Cooke) Deighton, 1960, *Mycol. Pap.*, **78**: 5–7.
(Fig. 393 B)

Conidiophores up to 200μ but mostly 100–140μ long, 3–5·5μ thick, olivaceous or brown, minutely verruculose. *Phialides* in groups of 5–10, 8–13 × 4–6μ. *Conidia* spherical or subspherical, sometimes flattened dorsiventrally, dark brown, verrucose, with large warts widely spaced, 6–9 (mostly 6–8)μ diam.

Common on *Musa* and found occasionally on other plants; Bermuda, British Solomon Islands Protectorate, Ghana, New Zealand, Sierra Leone, Sudan Republic, U.S.A., Venezuela.

Memnoniella levispora Subram., 1954, *J. Indian bot. Soc.*, **33**: 40–42.
(Fig. 393 C)

Conidiophores up to 50 × 3–5μ, dark grey to black, often with scattered dark granules. *Phialides* 6–8 × 4–5μ. *Conidia* smooth, often hemispherical, 4–6μ diam.

On dead herbaceous stems and on twigs of *Morus* and *Sanchezia*; India, Pakistan.

275. PHIALOMYCES

Phialomyces Misra & Talbot, 1964, *Can. J. Bot.*, **42**: 1287–1290.

Colonies effuse, at first yellow, later greyish olive. *Mycelium* partly superficial, partly immersed. *Stroma* none. *Setae* and *hyphopodia* absent. *Conidiophores* macronematous, mononematous, unbranched or branched at the apex,

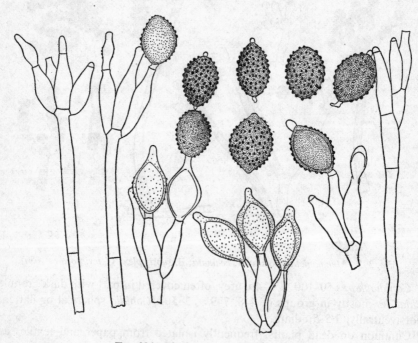

FIG. 394. *Phialomyces macrosporus* (× 650).

stipe and branch or branches each bearing terminally a small number of phialides; stipe straight or flexuous, hyaline or pale straw coloured, smooth. *Conidiogenous cells* monophialidic, discrete, determinate, lageniform. *Conidia* catenate, dry, semi-endogenous or acrogenous, simple, broadly ellipsoidal to limoniform, golden brown, verrucose, usually with a small, hyaline papilla or connective at each end, 0-septate.

Type species: Phialomyces macrosporus Misra & Talbot.

Phialomyces macrosporus Misra & Talbot, 1964, *Can. J. Bot.*, **42**: 1287–1290.
(Fig. 394)

Conidiophores up to 1 mm. long, 4–8μ thick. *Phialides* hyaline, 22–38μ long, 6–9μ thick in the broadest part, 3–4μ at the apex. *Conidia* mostly 22–27 × 16–20μ, often in long chains.

Isolated from soil; India, New Zealand.

276. ASPERGILLUS

Aspergillus Micheli ex Fries; Micheli, 1729, Nova Pl. Gen.: 212–213; Fries, 1821, Syst. mycol., **1**: XLV.

Colonies effuse, variously coloured, often green or yellowish, sometimes brown or black. *Mycelium* partly immersed, partly superficial. *Stroma* none. *Setae* and *hyphopodia* absent. *Conidiophores* macronematous, mononematous, often with a foot cell, straight or flexuous, colourless or with the upper part mid to dark brown, usually smooth, swollen at the apex into a spherical or clavate vesicle the surface of which is covered by short branches or in some species by phialides; the branches are in 1 or several series and the terminal ones in the series always bear phialides. *Conidiogenous cells* monophialidic, discrete, several arising together at the ends of terminal branches or over the surface of the vesicle, mostly determinate, rarely percurrent, ampulliform or lageniform, collarettes sometimes present. *Conidia* catenate, dry, semi-endogenous or acrogenous, spherical, variously coloured, smooth, rugose, verruculose or echinulate, sometimes with spines arranged spirally, 0-septate.

Lectotype species: Aspergillus glaucus Link ex S. F. Gray.

Raper, K. B. & Fennell, D. I., The genus Aspergillus, 1965. A comprehensive monograph.

Aspergillus niger van Tiegh., 1867, *Annls Sci. Nat.* (Bot.), Ser. 5, **8**: 240.
(Fig. 395 A)

Colonies effuse, blackish brown to black. *Mycelium* partly immersed, partly superficial; hyphae colourless to very pale yellow, 2–4μ thick. *Conidiophores* erect, straight or flexuous, often up to 3 mm. long, 15–20μ thick, colourless or with the upper part brown, swollen at the apex into a spherical vesicle which is usually 40–70μ diam. Surface of vesicle covered by closely packed, more or less clavate branches which are frequently 20–30μ long and 5–6μ thick in the broadest part. *Phialides* in groups at the apices of the branches, flask-shaped, 7–10μ long, 3–3·5μ thick, 1–1·5μ wide at the open end. *Conidia* catenate, dry, usually globose, brown, verruculose or echinulate, sometimes with the warts or spines arranged in discontinuous bands, 3–5µ (mostly 4–5µ) diam. *Conidial heads* at first globose, blackish brown to black, in age often splitting into several loose but quite well-defined columns.

On plants, plant debris, textiles, soil etc.; cosmopolitan.

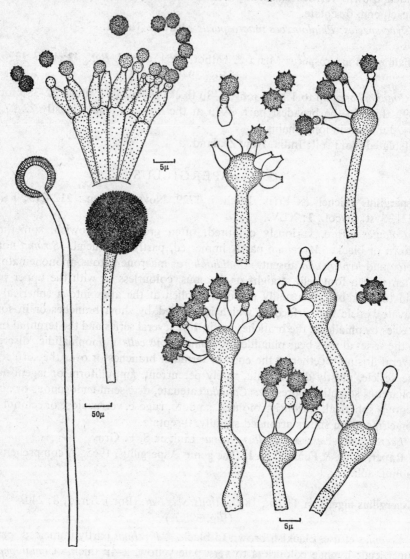

FIG. 395. A, *Aspergillus niger*; B, *A. brunneo-uniseriatus*.

277. CUSTINGOPHORA

Custingophora Stolk, Hennebert & Klopotek, 1968, apud Stolk & Hennebert in *Persoonia*, **5** (2): 195–199.

Colonies effuse, olivaceous brown to dark greyish olive. *Mycelium* partly superficial, partly immersed. *Stroma* none. *Setae* and *hyphopodia* absent. *Conidiophores* macronematous, mononematous, straight or flexuous; stipe unbranched or occasionally branched, proliferating then subapically, brown, smooth, markedly swollen at the apex over the surface of which are borne numerous phialides. *Conidiogenous cells* monophialidic, discrete, determinate, ampulliform or lageniform with distinct collarettes. *Conidia* aggregated in slimy heads, semi-endogenous, simple, oblong rounded at the ends or ovoid, hyaline or subhyaline, smooth, 0-septate.

Type species: Custingophora olivacea Stolk, Hennebert & Klopotek.

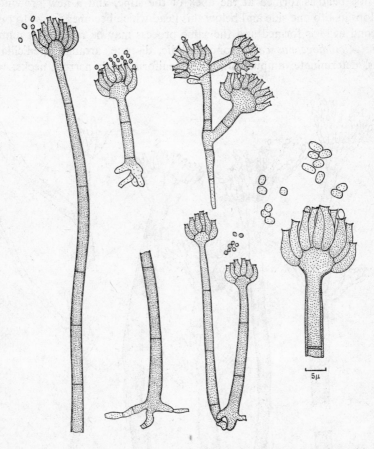

FIG. 396. *Custingophora olivacea* (× 650 except where indicated by the scale).

Custingophora olivacea Stolk, Hennebert & Klopotek, 1968, apud Stolk & Hennebert in *Persoonia*, **5** (2): 197–199.

(Fig. 396)

Conidiophores up to 250μ long, stipe 2·5–5·5μ thick, base sometimes up to 7μ; *ampullae* (vesicles) 6–14μ diam. *Phialides* 6·5–10 × 2–2·5μ. *Conidia* 2–3 × 1–1·5μ.

Isolated from compost; Germany.

278. THYSANOPHORA

Thysanophora Kendrick, 1961, *Can. J. Bot.*, **39**: 817–832.

Colonies effuse, often rather slow-growing, olivaceous, dark blackish brown or black. *Mycelium* partly superficial, partly immersed. *Stroma* none. *Sclerotia* sometimes formed. *Setae* and *hyphopodia* absent. *Conidiophores* macronematous, mononematous; stipe straight or flexuous, mid to dark brown or olivaceous brown, smooth, with a terminal head of penicillately arranged short branches and often a number of similar heads borne laterally and alternately. The first head is formed at the apex of the stipe, and a new growing point develops just to one side and below this head which becomes thrust to one side, a second head is formed and the same process may be repeated a number of times. *Conidiogenous cells* monophialidic, discrete, arranged penicillately in heads, determinate, ampulliform or lageniform, with narrow necks, without

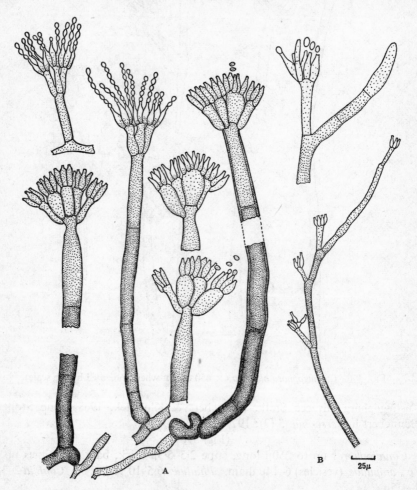

FIG. 397. A, *Thysanophora penicillioides*; B, *T. canadensis* (× 650 except where indicated by the scale).

collarettes. *Conidia* formed in dry basipetal chains, semi-endogenous or acrogenous, simple, ellipsoidal, fusiform or limoniform, colourless or pale brown, smooth or minutely verruculose or echinulate, 0-septate.

Type species: Thysanophora penicillioides (Roum.) Kendrick.

Thysanophora penicillioides (Roum.) Kendrick, 1961, *Can. J. Bot.*, **39**: 820–826.
(Fig. 397 A)

Sclerotia, when formed, at first creamy white, later pale brown. *Conidiophore* stipes dark brown below, pale at the apex, up to 950μ long, 8–16μ thick at the base, tapering to 4–10μ, branches complex or in groups of 3–8, pale olivaceous brown, 11–22 × 3–10μ. *Phialides* in groups of 3–8, 7–14 × 2·5–3·5μ. *Conidia* subspherical or ellipsoidal, smooth or minutely verruculose, 2–5 × 1·5–3μ.

Isolated from Coniferae and soil; cosmopolitan.

Thysanophora canadensis Stolk & Hennebert, 1968, *Persoonia*, **5**: 189–193.
(Fig. 397 B)

Sclerotia 250–400μ diam., dark brown when old. *Conidiophore* stipes olivaceous to brown, up to 1 mm. long, 4·5–8μ thick, branches simple, 15–30 × 3–4μ. *Phialides* in groups of 3–8, 8–17 × 2·5–4μ. *Conidia* subspherical, ellipsoidal or limoniform, subhyaline, smooth or minutely verruculose, 3–4 × 2–4μ.

On leaves of *Tsuga*; Canada.

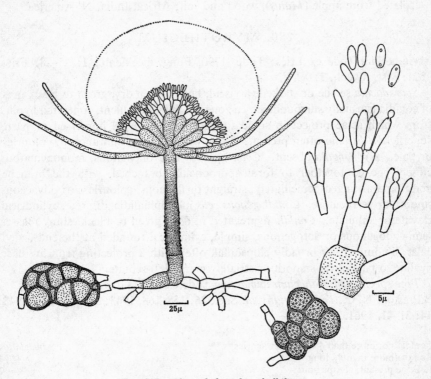

FIG. 398. *Gliocephalotrichum bulbilium.*

279. GLIOCEPHALOTRICHUM

Gliocephalotrichum J. J. Ellis & Hesseltine, 1962, *Bull. Torrey bot. Club*, **89**: 21–27.

Colonies effuse, buff to yellowish brown. *Mycelium* partly superficial, partly immersed; hyphae colourless or pale, with golden brown, multicellular, intercalary *chlamydospores*. *Stroma* none. *Setae* and *hyphopodia* absent. *Conidiophores* macronematous, mononematous, branched, with the branches restricted to the apical region forming a stipe and head; branching in the head complex, penicillate, with a whorl of typically 4 long, sterile, setiform branches below the penicillus; stipe straight or flexuous, pale to mid golden brown, smooth or verrucose. *Conidiogenous cells* monophialidic, discrete, determinate, almost cylindrical, rounded at the apex. *Conidia* aggregated in slimy heads, yellowish in mass, acrogenous, simple, cylindrical rounded at the ends or ellipsoidal, smooth, 0-septate.

Type species: Gliocephalotrichum bulbilium J. J. Ellis & Hesseltine.

Gliocephalotrichum bulbilium J. J. Ellis & Hesseltine, 1962, *Bull. Torrey bot. Club*, **89**: 21–27.

(Fig. 398)

Conidiophore stipes up to 350 × 12–20μ; setiform branches up to 220μ long, 8–13μ thick at the base, tapering to 3–4μ. *Conidia* 6–8 × 2–3μ.

Isolated from apple (*Malus*), wood and soil; Africa, India, N. America.

280. MYROTHECIUM

Myrothecium Tode ex Fries; Tode, 1790, Fung. mecklenb., **1**: 25–28; Fries, 1821, Syst. mycol., **1**: XLV.

Sporodochia sessile or stalked; a viscid, horny when dry, green to black mass of conidia is usually surrounded by a zone of white, flocculent, contorted hyphae from which setae project in some species. *Mycelium* all immersed or partly superficial. *Stroma* often present. *Setae* when present unbranched, colourless or pale. *Hyphopodia* absent. *Conidiophores* macronematous, mononematous, closely packed together to form sporodochia, branched, with the branches apical and arranged penicillately, straight or flexuous, colourless or olivaceous, smooth or verruculose. *Conidiogenous cells* monophialidic, discrete, cylindrical, clavate or subulate. *Conidia* aggregated in dark green or black, slimy masses, semi-endogenous or acrogenous, simple, cylindrical rounded at the ends, navicular, limoniform or broadly ellipsoidal, often with a projecting truncate base, hyaline to pale olive, smooth or striately marked, 0-septate.

Type species: Myrothecium inundatum Tode ex S. F. Gray.

Preston, N. C., *Trans. Br. mycol. Soc.*, **26**: 158–168, 1943; **31**: 271–276, 1948; **44**: 31–41, 1961.

KEY

Conidia not more than 4μ long, on agarics	*inundatum*
Conidia more than 4μ long	1
1. Setae present, conspicuous	2
Setae absent or very rarely seen	3

2. Setae mostly non-septate, 10–20µ thick near base . . . . . *gramineum*
   Setae septate, 2–4µ thick near base . . . . *indicum*
   Setae septate, 5–8µ thick near base . . . . *jollymannii*
3. Conidia with longitudinal striate markings 4
   Conidia without longitudinal striate markings 5
4. Conidia 7–12 × 2·5–3·5 (9·5 × 3)µ . . . . *striatisporum*
   Conidia 7–9 × 3·5–4·5 (8 × 4)µ . . . . *brachysporum*
5. Conidia slightly curved, sporodochia stalked . . . *Nectria bactridioides*
   Conidia straight, sporodochia usually sessile 6
6. Conidia cylindrical or narrowly ellipsoidal 7
   Conidia navicular, limoniform or broadly ellipsoidal 8
7. Conidia 5–7 × 1·5–2µ, usually on *Coffea* . . . . . . *advena*
   Conidia mostly 6–8 × 1·5–2·5µ . . . . *roridum*
   Conidia mostly 10–11 × 1–1·3µ . . . . *carmichaelii*
8. Conidia 6–10 × 2–4·5µ . . . . *verrucaria*
   Conidia 12–17 × 7–9µ . . . . *Nectria ralfsii*

Myrothecium inundatum Tode ex S. F. Gray, 1821, Nat. Arr. Br. Pl., **1**: 569.
(Fig. 399 A)

Sporodochia sessile, up to 1·5 mm. diam., often confluent, with black centre which becomes flat or concave and a white margin. *Setae* evanescent, hyaline, thick-walled, septate, 100–300µ long, 3–4µ thick at base, tapering to 2µ. *Phialides* 9–22 × 1–3µ. *Conidia* cylindrical, rounded at the ends, 2·5–4 × 1–1·5µ.

On dead agarics, especially *Russula adusta*; Europe.

Myrothecium gramineum Lib., 1837, Pl. crypt. Ard.: 380.
(Fig. 399 B)

Setae subulate, thick-walled, usually 0-septate, up to 400µ long, 10–20µ thick at the rounded base, tapering to a point. *Conidia* ellipsoidal or cylindrical rounded at the apex, often truncate at the base, hyaline or pale olivaceous, 7–10 × 2µ.

On grasses and occasionally other plants, also isolated from soil; Europe, Ghana, Jamaica, N. America, Sierra Leone.

Myrothecium indicum Rama Rao, 1963, *Antonie van Leeuwenhoek*, **29**: 180–182.
(Fig. 399 C)

Setae stiff, hyaline, septate, up to 250µ long, 2–4µ thick. *Conidia* cylindrical, rounded at the apex, often truncate at the base, mostly 8–11 × 1·5–2µ.

On *Saccharum* and isolated from soil; India, Nigeria, Pakistan.

Myrothecium jollymannii Preston, 1948, *Trans. Br. mycol. Soc.*, **31**: 272–275.
(Fig. 399 D)

Sporodochia sessile or short-stalked, up to 2 mm. diam., with subulate, hyaline, multiseptate, fairly thick-walled setae up to 250µ long, 5–8µ thick. *Conidia* cylindrical, rounded at the ends, pale olivaceous brown, mostly 10–12 × 2–2·5µ.

On or isolated from *Ananas, Ipomoea, Nicotiana* and *Zea*; Malawi, Rhodesia, Sarawak, Sierra Leone.

Myrothecium striatisporum Preston, 1948, *Trans. Br. mycol. Soc.*, **31**: 275–276.
(Fig. 399 E)

Sporodochia usually less than 1 mm. diam., sometimes confluent, generally without setae; brown, verrucose, sterile hyphae present. *Conidia* broadly fusiform, navicular or ellipsoidal, often slightly protuberant and truncate at the

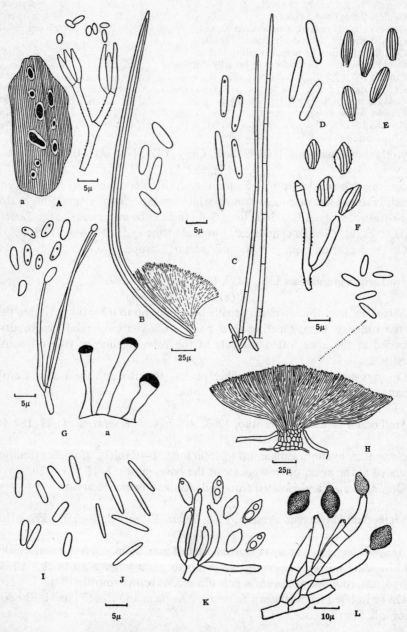

FIG. 399. *Myrothecium* species: A, *inundatum*; B, *gramineum*; C, *indicum*; D, *jollymannii*; E, *striatisporum*; F, *brachysporum*; G, state of *Nectria bactridioides*; H, *advena*; I, *roridum*; J, *carmichaelii*; K, *verrucaria*; L, state of *Nectria ralfsii* (a, habit sketches).

base, greyish or olivaceous brown, with distinct longitudinal, sometimes spirally twisted striations, 7–12 × 2·5–3·5 (9·5 × 3)μ.

Occurs on many different plants but is especially common on grasses and sedges in damp localities and is frequently isolated from soil; cosmopolitan.

Myrothecium brachysporum Nicot, 1961, *Revue gen. Bot.*, **68**: 684–685.
(Fig. 399 F)

Similar in many ways to *M. striatisporum* but the conidia are mostly shorter and fatter, 7–9 × 3·5–4·5 (8 × 4)μ, the striations widely spaced, very distinct and spirally twisted.

Isolated from soil; India, Iran.

Myrothecium state of **Nectria bactridioides** Berk. & Br., 1873, *J. Linn. Soc.*, **14**: 115.
(Fig. 399 G)

Sporodochia stalked. *Phialides* narrow, subulate, 30–45 × 0·5–1μ. *Conidia* often curved, ellipsoidal or allantoid, 4–6 × 2μ.

On dead wood, especially of *Theobroma*; Ceylon, Malaya, Uganda.

Myrothecium advena Sacc., 1908, *Annls mycol.*, **6**: 560.
(Fig. 399 H)

Similar in many ways to *M. roridum* described below but with rather smaller conidia, 5–7 × 1·5–2μ.

On *Coffea* and occasionally other plants; Ghana, India, Malaya. See Nag Raj & George in *Indian Phytopath.*, **11**: 153–158, 1958.

Myrothecium roridum Tode ex Fr., 1829, Syst. mycol., **3**: 217.
(Fig. 399 I)

Sporodochia sessile, up to 1·5 mm. diam., often confluent, at first green, later black with a white margin, without setae. *Phialides* 10–12 × 1–2μ. *Conidia* cylindrical with rounded ends, colourless to pale olive, green to black in mass, mostly 6–8 × 1·5–2·5μ.

Very common on leaves and other parts of many different kinds of plants, sometimes causing pale brown spots which may eventually drop out giving a shot-hole effect, also isolated from soil and textiles; cosmopolitan.

Myrothecium carmichaelii Grev., 1825, Scot. crypt. Fl., **3**: 140.
(Fig. 399 J)

Similar to *M. roridum* but conidia mostly 10–11 × 1–1·3μ.

On *Epilobium, Eupatorium, Iris* and *Thalictrum*; Great Britain.

Myrothecium verrucaria (Alb. & Schw.) Ditm. ex Fr., 1829, Syst. mycol., **3**: 217.
(Fig. 399 K)

Sporodochia similar to those of *M. roridum* but rough-walled hyphae are much in evidence. *Conidia* navicular, limoniform or broadly ellipsoidal, slightly protuberant and truncate at the base, distinctly coloured, 6–10 × 2–4·5μ.

Common on many different plants and frequently isolated from soil and textiles; cosmopolitan.

Myrothecium state of **Nectria ralfsii** Berk. & Br., 1854, *Ann. Mag. nat. Hist.*, 2, **13**: 467.

(Fig. 399 L)

Sporodochia black, without setae. *Phialides* 13–25 × 4–5μ. *Conidia* mostly limoniform, grey, greenish grey to black, 12–17 × 7–9μ.

On branches; Europe, N. America.

281. SACCARDAEA

Saccardaea Cavara, 1894, *Atti Ist. bot. Univ. Lab. crittogam. Pavia*, 2 Ser., **3**: 346.

Colonies effuse, grey to black; synnemata widely spaced and easily seen under a low-power binocular dissecting microscope. *Mycelium* immersed. *Stroma* usually present. *Setae* simple, hyaline or olivaceous, smooth, straight or curved, often numerous, forming a part of and projecting beyond the apex of each

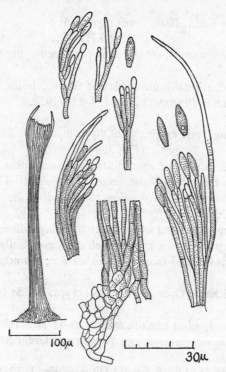

FIG. 400. Saccardaea atra.

synnema. *Hyphopodia* absent. *Conidiophores* macronematous, synnematous; *synnemata* black, cylindrical, expanded at the apex beyond which the setae project as a fringe; individual threads narrow, branched towards the tip, straight or flexuous, olivaceous, smooth. *Conidiogenous cells* monophialidic, sometimes integrated and terminal, more often discrete and arranged penicillately, determinate, cylindrical or subulate. *Conidia* aggregated in slimy heads, semi-endogenous or acrogenous, simple, fusiform or navicular, olivaceous, smooth, 0-septate.

Type species: Saccardaea echinocephala Cav.

Saccardaea atra (Desm.) Mason & M. B. Ellis, 1953, *Mycol. Pap.*, **56**: 40.
  Graphium atrum Desm., 1848, *Annls Sci. nat.*, Sér. 10, **3**: 343.
  Phaeostilbella atra (Desm.) Höhnel, 1919, *Ber. dt. bot. Ges.*, **37**: 153.
(Fig. 400)

Synnemata up to 400 × 40μ expanded to 80–200μ at the apex; individual threads 2–3μ thick; *setae* curved, projecting up to 60μ beyond the apex of the synnema, 1–2μ thick. *Phialides* up to 20 × 2μ. *Conidia* mostly 12–14 × 3–4μ.

On dead leaves of *Cynosurus*, *Arundo* and other grasses; Europe.

282. VIRGATOSPORA

Virgatospora Finley, 1967, *Mycologia*, **59**: 538–541.
  Colonies effuse, grey to blackish brown, made up of large numbers of scattered,

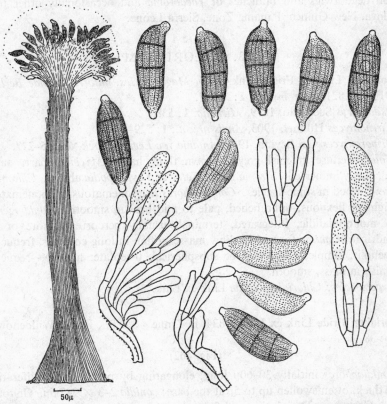

FIG. 401. *Virgatospora echinofibrosa* (× 650 except where indicated by the scale).

upright, bluish green, brown or dark blackish brown synnemata with slimy black heads. *Mycelium* immersed. *Stroma* none. *Setae* and *hyphopodia* absent. *Conidiophores* macronematous, synnematous; individual threads narrow, straight or flexuous, subhyaline to brown or blackish brown, smooth to finely echinulate, unbranched and tightly adpressed along most of their length but separating, splaying out and branching penicillately towards their apices. *Conidiogenous cells* monophialidic, sometimes integrated and terminal but mostly discrete, arranged penicillately, determinate, cylindrical to clavate. *Conidia* aggregated in large slimy heads, acrogenous, simple, straight or curved, ellipsoidal, fusiform or limoniform, papillate at each end, grey to dark brown, black in mass, with numerous very clear longitudinal ridges or striations, usually 3-septate when mature.

Type species: Virgatospora echinofibrosa Finley.

Virgatospora echinofibrosa Finley, 1967, *Mycologia*, **59**: 538–541.
(Fig. 401)

Synnemata up to 1·5 mm. high, 40–80μ thick, splaying out at apex and base up to 150μ; threads 1–3μ thick. *Phialides* 10–30 × 3–4μ. *Conidia* mostly 38–45 × 12–15μ.

On dead twigs and branches of *Theobroma* and occasionally other trees; Malaya, New Guinea, Panama Zone, Sierra Leone.

283. CHLORIDIUM

Chloridium Link ex Fries; Link, 1809, *Magazin Ges. naturf. Freunde Berl.*, **3**: 13; Fries, 1821, *Syst. mycol.*, **1**: XLVI.
Psilobotrys Saccardo, 1879, *Michelia*, **1**: 538.
Cirrhomyces Höhnel, 1903, *Annls mycol.*, **1**: 529–530.
Bisporomyces van Beyma, 1940, *Antonie van Leeuwenhoek*, **6**: 275–277.

Colonies effuse, greenish grey or brown, thinly hairy. *Mycelium* partly superficial, partly immersed. *Stroma* none. *Setae* and *hyphopodia* absent. *Chlamydospores* formed in some species. *Conidiophores* macronematous, mononematous, straight or flexuous, unbranched, pale to mid brown, smooth. *Conidiogenous cells* monophialidic, integrated, terminal, usually percurrent, more or less cylindrical. *Conidia* formed in slimy masses, often in long columns, frequently exogenous, simple, ellipsoidal or subspherical, 0-septate, hyaline, frequently greenish in mass, smooth.

Type species: Chloridium viride Link ex Link.

Chloridium viride Link ex Link, 1824, in Linné's Sp. Pl., Ed. 4 (Willdenow's), **6** (1): 38.
(Fig. 402)

Conidiophores initially 30–60μ long, elongating by percurrent proliferation, 2–3μ thick, often swollen up to 5μ at the base; *conidia* 2–3 × 2–2·5μ. On rotten wood, Germany. Description and figure based on type specimen.

Chloridium caudigerum (Höhnel) Hughes, 1958, *Can. J. Bot.*, **36**: 748.

Conidiophores 100–160 × 3–4μ; *conidia* 3–4 × 1·5–2μ. On rotten wood of beech and hornbeam, Austria. Description from Höhnel. According to F. Mangenot in his ' Recherches méthodiques sur les champignons de certains bois en décomposition ', 1952, p. 30, this species produces terminal and intercalary *chlamydospores*, 4–7 × 4–5μ, solitarily or in chains and the conidiophores may be up to 250μ long in culture.

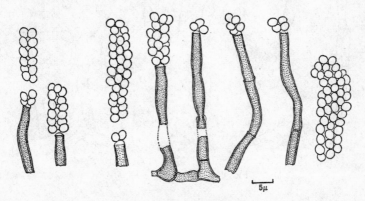

Fig. 402. *Chloridium viride.*

Chloridium chlamydosporis (van Beyma) Hughes, 1958, *Can. J. Bot.*, **36**: 748.

Conidiophores 60–180 × 2·7–3μ; *conidia* 3·7–4·7 × 2–3 (mostly 4·3–4·7 × 2·7)μ; *chlamydospores* round, brown, with a double wall, 5–6μ diam. Isolated from soil, Netherlands. Description from van Beyma. According to Mangenot the conidiophores in this species may be up to 600μ long. Barron states that this species is frequently found in soils high in organic matter.

284. EXOPHIALA

Exophiala Carmichael, 1966, *Sabouraudia*, **5**: 122–123.

Colonies effuse, slow-growing. *Mycelium* superficial and immersed. *Stroma* none. *Setae* and *hyphopodia* absent. *Conidiophores* semi-macronematous, mononematous, straight or flexuous, unbranched or irregularly branched, subhyaline to pale olivaceous brown, smooth. *Conidiogenous cells* monophialidic, integrated and terminal or discrete, determinate or percurrent, ampulliform or lageniform, without collarettes. *Conidia* aggregated in slimy masses, exogenous, simple, cylindrical rounded at the apex truncate at the base or clavate, pale to mid yellowish brown, smooth, 0–1- (occasionally 2–3-) septate.

Type species: Exophiala salmonis Carmichael.

Exophiala salmonis Carmichael, 1966, *Sabouraudia*, **5**: 122–123.

(Fig. 403)

Colonies humped, mouse-grey. *Conidiophores* very variable in length, 2–5μ thick. *Conidia* appear to arise from a column of cytoplasm which protrudes beyond the wall at the apex of the phialide as they do in *Chloridium*; they are 5–6 × 2–3·5μ.

Isolated from salmon brain; N. America.

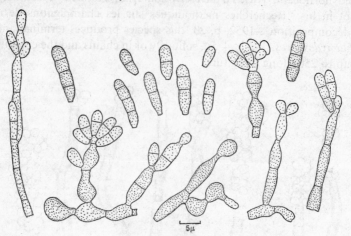

FIG. 403. *Exophiala salmonis.*

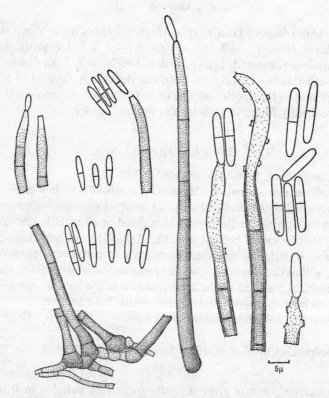

FIG. 404. *Cylindrotrichum oligospermum* (× 650 except where indicated by the scale).

285. CYLINDROTRICHUM

Cylindrotrichum Bonorden, 1851, Handb. allg. Mykol.: 88.

Colonies effuse, pale greyish brown or whitish when sporulating freely, hairy. *Mycelium* immersed and sometimes also superficial. *Stroma* none. *Setae* and *hyphopodia* absent. *Conidiophores* macronematous, mononematous, straight or flexuous, cylindrical or subulate, unbranched, mid to dark brown near the base becoming paler upwards. *Conidiogenous cells* polyphialidic, integrated, terminal, sympodial, cylindrical to lageniform, with collarettes. *Conidia* aggregated in slimy clumps, semi-endogenous, cylindrical rounded at the ends, colourless, smooth, 0–3-septate.

Lectotype species: Cylindrotrichum oligospermum (Corda) Bon.

Cylindrotrichum oligospermum (Corda) Bon., 1851, Hanb. allg. Mykol: 88. For synonymy see Hughes in *Can. J. Bot.*, **36**: 759, 1958.

(Fig. 404)

Conidiophores up to 600µ long, 5–7µ thick just above the base usually tapering to 3–4µ near the apex; sometimes swollen at the base to 9–10µ. *Conidia* 12–22 × 2·5–3µ.

Common on wood and bark of fallen branches and on dead stems and leaves of herbaceous plants; Europe.

286. CODINAEA

Codinaea Maire, 1937, *Publcions Inst. bot., Barcelona*, **3**: 15–16.

Menisporella Agnihothrudu, 1962, *Proc. Indian Acad. Sci.*, Sect. B, **56**: 97–101.

Colonies effuse, mid to dark brown or greyish brown, hairy. *Mycelium* superficial or immersed. *Stroma* sometimes present. *Setae* present or absent; when present simple, dark, smooth. *Hyphopodia* absent. *Conidiophores* macronematous, mononematous, straight or flexuous, usually unbranched, brown, smooth. *Conidiogenous cells* polyphialidic, integrated, terminal often becoming intercalary, sympodial, cylindrical, with conspicuous collarettes. *Conidia* aggregated in slimy groups, semi-endogenous, simple, cylindrical rounded at the ends, curved, often falcate, 0–3-septate, colourless, smooth; in many species there is a fine setula at each end.

Type species: Codinaea aristata Maire.

Hughes & Kendrick, *N.Z. Jl Bot.*, **6**: 331–372, 1968.

Codinaea assamica (Agnihothrudu) Hughes & Kendrick, 1968, *N.Z. Jl Bot.*, **6**: 334–335.

(Fig. 405)

Colonies effuse, dark brown, hairy, composed of setae and groups of conidiophores arising individually from flat, dark brown stromata. *Setae* thick-walled, brown, very dark near base, pale towards the apex, up to 400µ long, 4–8µ thick. *Conidiophores* brown below, pale near the apex, up to 140µ long, 3–5µ thick. *Conidia* curved, colourless, 0-septate, 14–16 × 2·5–3µ with setulae 9–14µ long.

On tea roots, Assam, India; also on *Pycnanthus angolensis*, Tanzania.

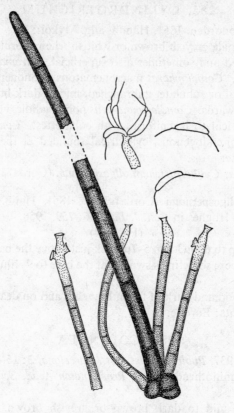

FIG. 405. *Codinaea assamica* (× 650).

287. HARPOGRAPHIUM

Harpographium Saccardo, 1880, *Michelia*, **2**: 33.

Colonies effuse, black; synnemata clearly seen under a binocular dissecting microscrope. *Mycelium* immersed. *Stroma* none seen. *Setae* and *hyphopodia* absent. *Conidiophores* macronematous, synnematous, individual threads tightly adpressed in the stipe, brown or olivaceous brown, smooth, branched towards their ends and splaying out at the top and sometimes along the sides of the synnemata. *Conidiogenous cells* polyphialidic, integrated, terminal on branches or discrete, sympodial, cylindrical, sometimes geniculate near the apex. *Conidia* aggregated in slimy heads, semi-endogenous, simple, falcate, colourless, smooth, 0-septate.

Type species: Harpographium fasciculatum Sacc.

Harpographium fasciculatum Sacc., 1880, *Michelia*, **2**: 33.

(Fig. 406)

Synnemata very variable in size, up to 700μ long and 150μ thick; individual threads 2–3μ thick. *Conidia* 11–15 × 1–2μ.

On dead wood, sometimes associated with *Peroneutypa heteracantha* (Sacc.) Berl.; Europe, N. America.

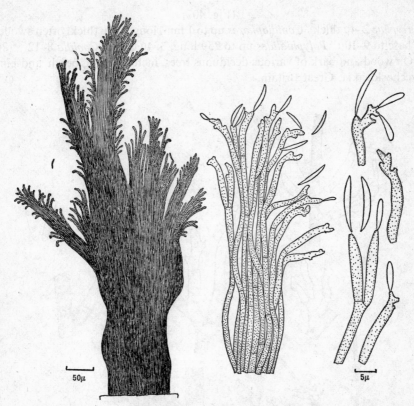

FIG. 406. *Harpographium fasciculatum* (× 650 except where indicated by the scales).

288. CHAETOPSIS

Chaetopsis Greville, 1825, *Edinb. phil. J.*, **25**: 63–4.
Chaetopsella Höhnel, 1930, *Mitt. bot. Inst. tech. Hochsch. Wien*, **7**: 44.
Colonies effuse, at first grey, later blackish brown, hairy. *Mycelium* immersed. *Stroma* none. *Setae* absent but conidiophores setiform. *Hyphopodia* absent. *Conidiophores* macronematous, mononematous, straight or flexuous, subulate, brown or dark brown, with a number of short lateral primary branches a little way above the base; upper part sterile, setiform. *Conidiogenous cells* polyphialidic, integrated, usually terminating primary and secondary branches, also sometimes discrete, sympodial, more or less cylindrical but narrower at the apex, subhyaline or pale brown. In place of polyphialides lateral branches sometimes end in sterile setae. *Conidia* aggregated in slimy bundles, acropleurogenous, semi-endogenous, simple, cylindrical with rounded ends, colourless, 0–1-septate, smooth.
Type species: Chaetopsis grisea (Ehrenb. ex Pers.) Sacc. = *C. wauchii* Grev.

Chaetopsis grisea (Ehrenb. ex Pers.) Sacc., 1880; Ehrenberg, 1818, Sylv. mycol. berol.: 12 and 23; Persoon, 1822, Mycol. eur., **1**: 15; Saccardo, 1880, *Michelia*, **2**: 26.

(Fig. 407)

Hyphae 2–4μ thick. *Conidiophores* up to 1 mm. long, 5–7μ thick, often swollen at base to 9–10μ. *Polyphialides* up to 25μ long, 3–4μ thick. *Conidia* 8–12 × 2μ.

On wood and bark of various deciduous trees, including ash, beech and elm; Czechoslovakia, Great Britain.

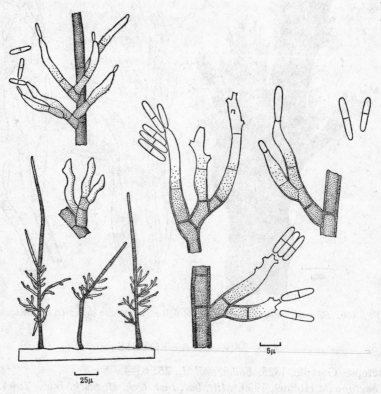

Fig. 407. *Chaetopsis grisea* (× 650 except where indicated by scales).

289. SELENOSPORELLA

Selenosporella Arnaud ex MacGarvie, 1968, *Scient. Proc. R. Dubl. Soc.*, Ser. B, **2** (16): 153–158.

Colonies effuse, inconspicuous. *Mycelium* immersed. *Stroma* none. *Setae* and *hyphopodia* absent. *Conidiophores* macronematous, mononematous, branched, with the branches in verticils; stipe straight or flexuous, subulate, with a swollen and lobed base, brown, smooth. *Conidiogenous cells* polyphialidic, integrated and terminal on stipe and branches or discrete, arranged in verticils, sympodial, cylindrical, lageniform or subulate, with protruding denticular collarettes. *Conidia* aggregated in slimy masses, semi-endogenous, simple, straight or slightly curved, acerose or almost cylindrical, rounded at the apex tapered towards the base, colourless or pale olive, smooth, 0-septate.

Type species: Selenosporella curvispora MacGarvie.

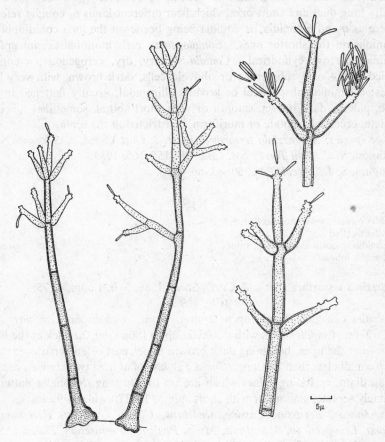

FIG. 408. *Selenosporella curvispora* (× 650 except where indicated by the scale).

Selenosporella curvispora MacGarvie, 1968, *Scient. Proc. R. Dubl. Soc.*, Ser. B, **2** (16): 153–158.

(Fig. 408)

Conidiophores up to 250μ long, 10–15μ thick at the base, 5–6μ thick immediately above the basal swelling, tapering to 2–3μ at the apex. *Phialides* 12–19μ long, 3–4·5μ thick at the base, often tapering to 1–2μ. *Conidia* 5–7 × 0·5μ.

On dead leaves of *Juncus effusus*; Ireland, France.

290. SPEGAZZINIA

Spegazzinia Saccardo, 1880, *Michelia*, **2**: 37.

Colonies discrete, orbicular, or effuse, dark blackish brown to black. *Mycelium* superficial; hyphae branched and anastomosing to form a close network.

Stroma none. *Setae* and *hyphopodia* absent. *Conidiophores* basauxic, macronematous, mononematous, arising usually singly from subspherical, ampulliform, cupulate or doliiform conidiophore mother cells, unbranched, straight or flexuous, narrow, subhyaline to brown, smooth, or verrucose; there are usually long ones and short ones which bear different kinds of conidia referred to here as *a* and *b* conidia, *a* conidia being borne on the long conidiophores, *b* conidia on the shorter ones. *Conidiogenous cells* monoblastic, integrated, terminal, narrow, cylindrical. *Conidia* solitary, dry, acrogenous; *a* conidia divided into 4 or 8 subglobose or obovoid cells, dark brown, with very long spines; *b* conidia subspherical or broadly ellipsoidal, usually flattened in one plane, pale to dark brown, smooth or with short spines, sometimes lobed or lobulate, cruciately septate or muriform, constricted at the septa.

Type species: *Spegazzinia tessarthra* (Berk. & Curt.) Sacc. = *S. ornata* Sacc. Damon, S. C., *Bull Torrey bot. Club*, **80**: 155–165, 1953.
Hughes, S. J., *Mycol. Pap.*, **50**: 62–66, 1953.

KEY

b conidia cruciately septate, 4-celled.	1
b conidia 8-celled	*deightonii*
1. *b* conidia smooth or with short spines	*tessarthra*
b conidia lobulate	*lobulata*

Spegazzinia tessarthra (Berk. & Curt.) Sacc., 1886, Syll. Fung., **4**: 758.
(Fig. 409 A)

Colonies orbicular, black, up to 2 mm. diam. *Conidiophore mother cells* 4–8 × 3–6μ. *Conidiophores* with *a* conidia up to 180μ long, 2μ thick at the base, up to 4μ at the apex, becoming dark brown, upper part often verrucose; those with *b* conidia less than 15μ long. *Conidia* always of at least two kinds: *a* conidia 12–18μ diam. excluding spines which are up to 10μ long, *b* conidia flattened, cruciately septate, smooth or with short spines, 13–17μ wide, 8–9μ thick.

On *Ananas, Borassus, Cassine, Cenchrus, Citrus, Cynodon, Heteropogon, Lantana, Lycopersicon, Mangifera, Musa, Panicum, Pennisetum, Phoenix, Saccharum, Sorghum, Theobroma, Triticum, Zea* and isolated from soil; Australia, Ghana, India, Kenya, Malaya, New Guinea, Sierra Leone, Sudan, Tanzania, Trinidad, Uganda, U.S.A., Venezuela, Zambia.

Spegazzinia deightonii (Hughes) Subram., 1956, *J. Indian bot. Soc.*, **35**: 78.
(Fig. 409 B)

Similar to *S. tessarthra* but conidia 8-celled and never cruciate; *a* conidia up to 30μ diam.; *b* conidia 23–30 × 17–22μ.

On *Andropogon, Axonopus, Borassus, Dioscorea, Euchlaena, Oryza, Pennisetum, Phragmites, Rottboellia, Saccharum, Tripsacum, Triticum, Vetiveria* and *Zea*; Cuba, Ghana, New Guinea, Nigeria, Puerto Rico, Sierra Leone.

Spegazzinia lobulata Thrower, 1954, apud McLennan, Ducker & Thrower in *Aust. J. Bot.*, **2**: 362–363.
(Fig. 409 C)

Known only in culture where colonies are effuse, dark blackish brown to black. *Conidia* of two kinds but the *a* conidia which resemble those of *S. tessarthra* are seldom formed in any quantity; *b* conidia abundant, characteristically lobulate, 16–25μ wide including lobes.

Isolated from soil; Australia.

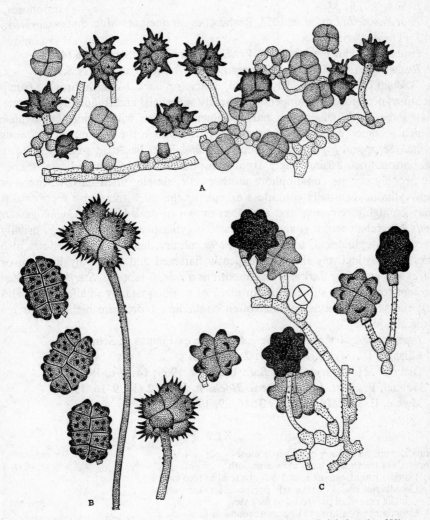

FIG. 409. *Spegazzinia* species: A, *tessarthra*; B, *deightonii*; C, *lobulata* (× 650).

291. ARTHRINIUM

Arthrinium Kunze ex Fr., 1821; Kunze in Kunze & Schmidt, 1817, Mykol. Hefte, **1**: 9; Fries, 1821, *Syst. mycol.*, **1**: XLIV.

Camptoum Link, 1824, in Linné Sp. Pl., Ed. 4 (Willdenow's), **6** (1): 44.
Goniosporium Link, 1824, *ibid.*: 45.
Sporophleum Nees ex Link, 1824, *ibid.*: 45.

Papularia Fr., 1825, Syst. Orb. veg., **1**: 195.

Gonatosporium Corda, 1839, Icon. Fung., **3**: 8.

Microtypha Speg., 1910, *An. Mus. nac. Hist. nat. B. Aires*, **20**: 432.

Tureenia Hall, 1915, *Phytopathology*, **5**: 57.

Pseudobasidium Tengwall, 1924, *Meded. phytopath. Lab. Willie Commelin Scholten*, **6**: 38.

Phaeoharziella Loubière, 1924, Recherches sur quelques Mucédinées caseicoles (Thèse) Paris: 52.

Innatospora van Beyma, 1929, *Verh. K. Akad. Wet.*, Sect. 2, **26** (4): 5.

Racemosporium Mme. & F. Moreau, 1941, *Revue Mycol.*, **6** (N.S.): 80.

Colonies compact or widely effused, black or dark blackish brown, fructifications occasionally erumpent but usually superficial and frequently pulvinate. *Mycelium* partly superficial, partly immersed, often with connecting hyphae which become very narrow where they pass through the host cuticle. *Stroma* none. *Setae* and *hyphopodia* absent. *Conidiophores* basauxic, macronematous, mononematous, arising singly from subspherical, ampulliform, barrel-shaped or broadly clavate conidiophore mother cells, simple, often narrow, more or less cylindrical, usually colourless except for the thick transverse septa which may be highly refractive and are often brown or dark brown. *Conidiogenous cells* integrated, terminal and intercalary, polyblastic, denticulate; pegs usually very short, cylindrical, truncate. *Conidia* solitary, lateral and sometimes also terminal, distinctively shaped, frequently flattened and with a hyaline rim or germ slit, brown or dark brown, smooth as a rule, 0-septate. *Sterile cells* when present terminal or subterminal in place of conidia, usually smaller, paler and not the same shape as conidia, often containing 1 or more highly refractive cubical bodies.

Type species: Arthrinium caricicola Kunze ex Ficinus & Schubert.

Ellis, M. B., *Mycol. Pap.*, **103**: 2–30, 1965.

Gjaerum, H. B., *Nytt Mag. Bot.*, **13**: 5–14, 1966; **14**: 1–6, 1967.

Höhnel, F., *Mitt. bot. Inst. tech. Hochsch. Wien*, **2** (1): 9–16, 1925.

Mason, E. W., *Mycol. Pap.*, **3**: 16–29, 1933.

KEY

Conidia truncate at base, often verruculose	*spegazzinii*
Conidia not truncate at base, always smooth	1
1. Conidia round, almost round or polygonal in face view	2
Conidia not round or regularly polygonal in face view	8
2. Conidia nearly all polygonal in face view	*puccinioides*
Conidia nearly all round or almost round in face view	3
3. Conidia lenticular	4
Conidia not lenticular	7
4. Diameter of conidia in face view usually 9–10μ	5
Diameter of conidia in face view usually 6–8μ	6
Diameter of conidia in face view usually 4–5·5μ	*euphorbiae*
5. Conidiophores 2–4μ thick	*saccharicola*
Conidiophores 1–1·5μ thick	*phaeospermum*
6. Conidiophores 1–1·5μ thick, septa numerous, brown . . .	*sacchari*
Conidiophores 0·5μ thick, septa few, hyaline . . .	*Apiospora montagnei*
7. Conidia 7–9μ diam. in face view	*sphaerospermum*
Conidia 5–6μ diam. in face view	*urticae*

Arthrinium caricicola Kunze ex Ficinus & Schubert, 1823; Kunze, 1817, in Kunze & Schmidt, Mykol. Hefte, **1**: 9; Ficinus & Schubert, 1823, *Fl. Geg. Dresd.*, *Krypt.*: 276; Fries, 1832, Syst. mycol., **3**: 376.

   A. naviculare Rostrup, 1886, *Bot. Tidsskr.*, **15**: 235.

(Fig. 410)

Colonies compact, pulvinate, round, 150–400µ diam., dark blackish brown. *Mycelium* partly superficial, partly immersed; superficial part composed of a network of branched and anastomosing septate, brown to dark brown, smooth, 2·5–6µ thick hyphae, immersed hyphae pale to dark brown 1–8µ thick. *Conidio- phore mother cells* subspherical to lageniform, 5–8 × 5–7µ. *Conidiophores* erect or ascending, simple, flexuous, cylindrical, colourless except for the thick, brown or dark brown transverse septa, up to 106µ long, 2–4·5µ thick. *Conidia* fusiform or cigar-shaped in face view, part usually dark brown, the rest paler, with a distinct hyaline rim, 30–53 (43) × 7·5–13 (9·4)µ. *Sterile cells* much smaller and paler than conidia, bicuspid or irregularly lobed.

   On various species of *Carex*; Finland, Germany, Latvia, Norway, Russia, Sweden, Switzerland.

Arthrinium phaeospermum (Corda) M. B. Ellis, 1965, *Mycol. Pap.*, **103**: 8–10 (with full synonymy). The synonym most commonly used is *Papularia sphaero- sperma* (Pers.) Höhnel.

(Fig. 411)

Colonies variable in structure. On leaves they are mostly superficial, compact, pulvinate, round or oval, 100–500µ diam., black. On most culms fructification commences beneath the epidermis which splits longitudinally to expose the shiny black spore masses, at first 2–3 mm. long and 0·5 mm. wide, later expand- ing up to 5 × 1 mm.; colonies become even more widely effused by the pro- duction of large numbers of mycelial mats on which fresh conidiophores arise. *Mycelium* partly superficial, partly immersed; superficial part composed of a network of branched and anastomosing, septate, colourless to pale brown, smooth, 2–6µ thick hyphae; immersed hyphae colourless, smooth or verru-culose, 1–4µ thick. *Conidiophore mother cells* lageniform, 5–10 × 3–5µ. *Conidio-phores* erect or ascending, simple, flexuous, cylindrical, colourless, smooth,

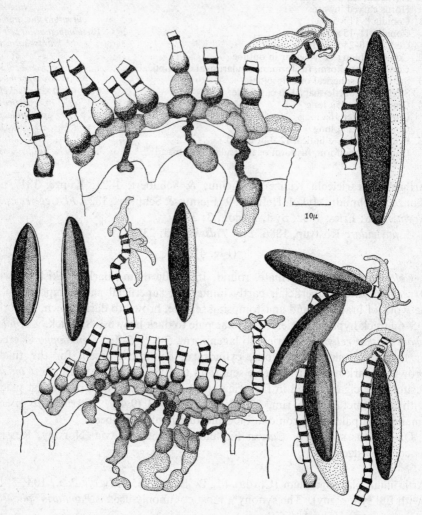

FIG. 410. *Arthrinium caricicola* (× 650 except where indicated by the scale).

with colourless highly refractive or pale brown transverse septa, up to 65µ long, 1–1·5µ thick. *Conidia* lenticular, rather dark golden brown with a hyaline band at the junction of the two sides, 8–12 (9·9)µ diam. in face view, 5–7 (5·9)µ thick. *Sterile cells* none.

Very common on bamboos, reeds and many other substrata; cosmopolitan.

Arthrinium state of **Apiospora montagnei** Sacc., 1875, *Nuovo G. bot. ital.*, **7**: 306 [see H. J. Hudson, 1963, *Trans. Br. mycol. Soc.*, **46**: 19–23]. Full synonymy given in *Mycol. Pap.*, **103**: 12. The synonym most commonly used is *Papularia arundinis* (Corda) Fr.

(Fig. 412)

Colonies when young compact and round, but soon becoming widely effused, often encircling culms and spreading along them for up to 10 cm. *Mycelium*

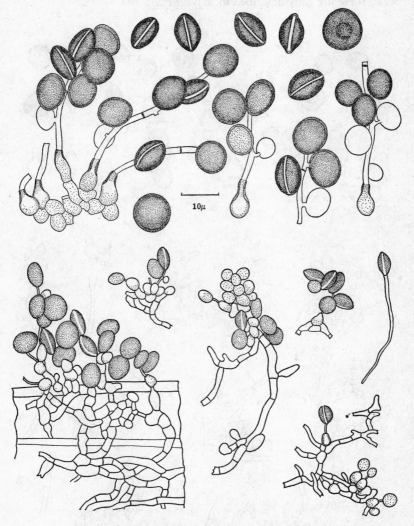

FIG. 411. *Arthrinium phaeospermum* (× 650 except where indicated by the scale).

partly superficial, partly immersed, the superficial part composed of a network of branched and anastomosing, septate, colourless to pale brown, smooth, 1–4µ thick hyphae; immersed hyphae colourless, smooth, 0·5–1µ thick. *Conidiophore mother cells* subspherical to oval or doliiform, 5–7 × 3–5µ. *Conidiophores* erect or ascending, simple, flexuous, thread-like, colourless, smooth, with a few highly refractive transverse septa, up to 50µ long, 0·5µ thick. *Conidia* lenticular, rather pale brown with a hyaline band at the junction of the two sides, 5·5–8 (6·5)μ diam. in face view, 3–4·5 (3·8)μ thick. *Sterile cells* none.

Very common on bamboos and other substrata; cosmopolitan.

Measurements of conidia and details of substrata and distribution of 16 other species and one variety of *Arthrinium* are given below; with the exception of 2 species these are fully described in *Mycol. Pap.*, **103**.

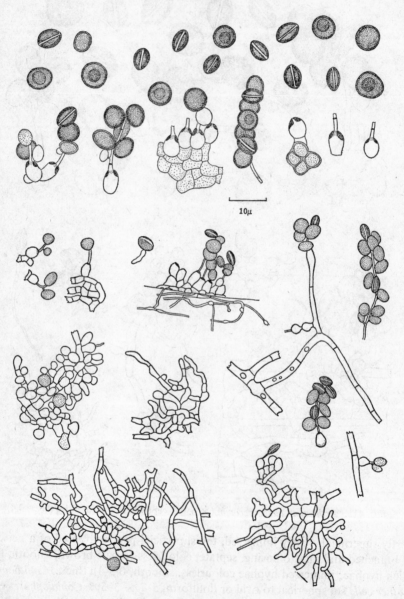

10μ

FIG. 412. *Arthrinium* state of *Apiospora montagnei* (× 650 except where indicated by the scale).

A. spegazzinii Subram. (Fig. 413 A): 5–8 (6·5) × 3–6 (4·1)μ; on sugar cane; Argentina.

A. puccinioides (DC ex Mérat) Kunze (Fig. 413 B): 9–14 (11·6)μ in diam. face view, 7–9μ thick; on *Carex*, *Deschampsia*, *Eleocharis* and *Scirpus*; Belgium, France, Germany, Great Britain, Norway, Tierra del Fuego.

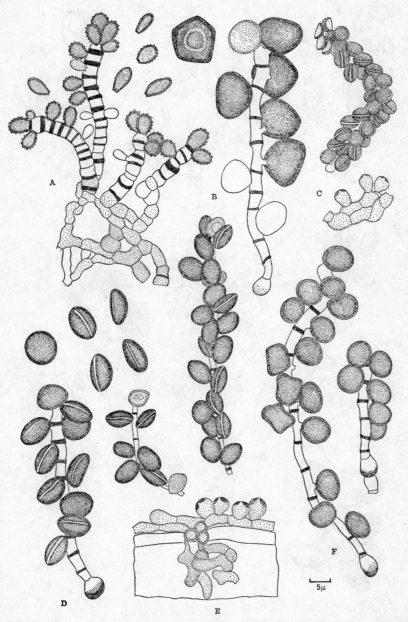

5μ

FIG. 413. *Arthrinium* species (1): A, *spegazzinii*; B, *puccinioides*; C, *euphorbiae*; D, *saccharicola*; E, *sacchari*; F, *sphaerospermum*.

A. euphorbiae M. B. Ellis (Fig. 413 C): 4–5·5 (4·7)μ diam. face view, 0·5–1μ thick; on *Euphorbia*; Zambia.

A. saccharicola Stevenson (Fig. 413 D): 7–10 (9)μ diam. face view, 4–6 (5·1)μ thick; on sugar cane; Puerto Rico, Venezuela.

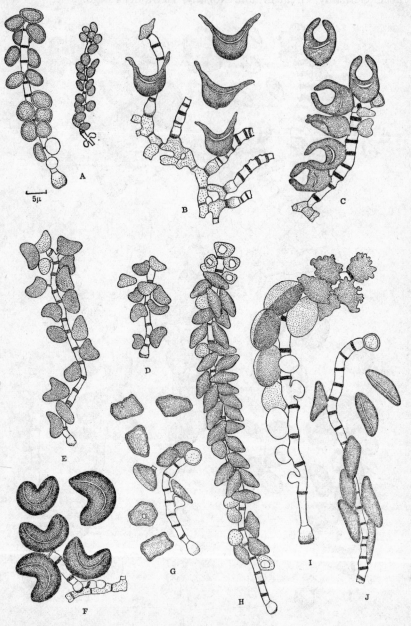

FIG. 414. *Arthrinium* species (2): A, *urticae*; B, *cuspidatum*; C, *luzulae*; D, *curvatum* var. *minus*; E, state of *Pseudoguignardia scirpi*; F, *kamtschaticum*; G, *morthieri*; H, *sporophleum*; I, *lobatum*; J, *ushuvaiense* (× 650 except where indicated by the scale).

A. sacchari (Speg.) M. B. Ellis (Fig. 413 E): 6–8 (7)μ diam. face view, 3–4 (3·8)μ thick; on sugar cane and other grasses, also on *Ananas, Borassus, Musa, Ipomoea* and *Polygonum*; Argentina, Ghana, India, Nigeria, Pakistan, Wales, Zambia.

A. sphaerospermum Fuckel (Fig. 413 F): 7–9 (7·9)μ diam.; on *Phleum pratense* and other grasses; Norway, Switzerland.

A. urticae M. B. Ellis (Fig. 414 A): 5–6 (5·6)μ diam. face view, 3–4 (3·6)μ thick; on nettle stems; England.

A. cuspidatum (Cooke & Harkn.) Tranz. (Fig. 414 B): 15–32 (26) × 7–11 (9·2)μ; on *Juncus* and *Scirpus*; Canada, England, Finland, Norway, Russia, Switzerland, U.S.A.

A. luzulae M. B. Ellis (Fig. 414 C): 18–21 (19·5) × 12–14 (13·5)μ face view, 8–11μ thick; on *Luzula*; Switzerland, U.S.A.

A. curvatum Kunze var. **minus** M. B. Ellis (Fig. 414 D): 8–11 (9) × 5–6 (5·5)μ; on *Carex, Juncus* and *Scirpus*; France, Germany, Great Britain, Norway.

A. state of **Pseudoguignardia scirpi** Gutner [**A. curvatum** Kunze] (Fig. 414 E): 11–15 (13) × 6–8 (7·2)μ; common on *Scirpus sylvaticus*, found also on *S. lacustris* and *Carex*; Austria, Belgium, England, Finland, France, Germany, Latvia, Norway, Russia, Sweden, Switzerland.

A. kamtschaticum Tranz. & Woronich. (Fig. 414 F): 18–24 × 9–11μ; on *Carex*; Kamtchatka, Norway.

A. morthieri Fuckel (Fig. 414 G): 12–16 (14·7) × 7–9 (8·3)μ in face view, 4–8μ thick; on *Carex*; Norway, Sweden, Switzerland.

A. fuckelii Gjaerum (Not figured; differs from *A. morthieri* in having bottle-shaped to cuspidate sterile cells but has the same shaped conidia); 14·5–20 (18) × 5·5–9·5 (7·8)μ in face view, 5·5–7μ thick; on *Carex*; Norway, Switzerland.

A. sporophleum Kunze (Fig. 414 H): 11–15 (12·3) × 5–7·5 (5·9)μ in face view, 5–7μ thick; on *Carex, Eriophorum, Juncus* and *Typha*; France, Germany, Great Britain, India, Norway, Portugal.

A. lobatum M. B. Ellis (Fig. 414 I): 17–20 (19·5) × 12–14 (12·4)μ; on grass, Venezuela.

A. ushuvaiense Speg. (Fig. 414 J): 17–25 (21) × 6–9 (7·2)μ in face view, 6–8μ thick; on *Luzula* and grasses; Argentina, China, Tierra del Fuego.

292. CORDELLA

Cordella Spegazzini, 1886, *An. Soc. cient. argent.*, **22**: 210.

Colonies compact or effuse, dark blackish brown or black, each made up of a close carpet of setae mixed with small groups of conidiophores. *Mycelium* partly superficial, partly immersed. *Stroma* none. *Setae* simple, subulate, brown or black. *Hyphopodia* absent. *Conidiophores* macronematous, mononematous, basauxic, arising singly from ampulliform or doliiform conidiophore mother cells, straight or flexuous, unbranched, rather narrow, cylindrical, colourless or pale brown with thick dark brown or black transverse septa, smooth or verruculose. *Conidiogenous cells* monoblastic or polyblastic, integrated, terminal and intercalary, cylindrical, denticulate; denticles short, cylindrical. *Conidia* solitary, dry, acropleurogenous, simple, lenticular, 0-septate, pale to mid brown with a hyaline band at the junction of the two sides, smooth.

Lectotype species: Cordella coniosporioides Speg.

Ellis, M. B., *Mycol. Pap.*, **103**: 30–33, 1965.

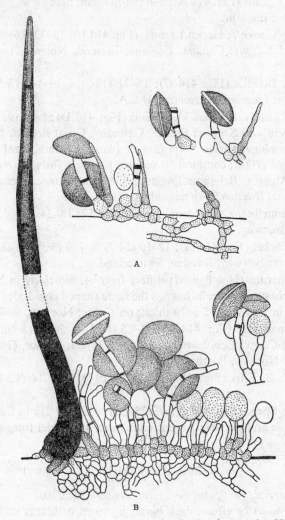

FIG. 415. A, *Cordella coniosporioides*; B, *C. johnstonii* (× 650).

Cordella coniosporioides Speg., 1886, *An. Soc. cient. argent.*, **22**: 210.

(Fig. 415 A)

Setae mid to dark brown, 15–40μ long, 3–6μ thick below, tapering to 0·5–1·5μ near the apex. *Conidiophore mother cells* 5–8 × 4·5–6μ; *conidiophores* 15–25 × 2·5–4μ. *Conidia* 17–21μ diam., 8–10μ thick.

On bamboo culms; Paraguay and Venezuela.

Cordella johnstonii M. B. Ellis, 1965, *Mycol. Pap.*, **103**: 31–33.

(Fig. 415 B)

Setae dark blackish brown to black, 200–300μ long, 8–15μ thick near base, tapering to an acute apex. *Conidiophore mother cells* 4–8 × 3–5μ; *conidiophores* 15–50 × 1·5–3μ. *Conidia* 19–22μ diam., 10–12μ thick.

On culms of *Bambusa blumeana*; Malaya.

293. PTEROCONIUM

Pteroconium Saccardo ex Grove; Sacc., 1892, *Syll. Fung.*, **10**: 570; Grove, 1914, *Hedwigia*, **55**: 146.

Sporodochia punctiform, scattered, black. *Mycelium* immersed. *Stroma* well-developed, erumpent, brown or dark brown, pseudoparenchymatous. *Setae* and *hyphopodia* absent. *Conidiophores* basauxic, macronematous, mononematous, arising singly from lageniform, ellipsoidal or doliiform conidiophore mother cells on the surface of the stroma, usually unbranched, straight or flexuous, rather narrow, cylindrical, colourless or pale brown except for the thick, mid to dark brown transverse septa, smooth or verruculose. *Conidiogenous cells* monoblastic or occasionally polyblastic, integrated, terminal and

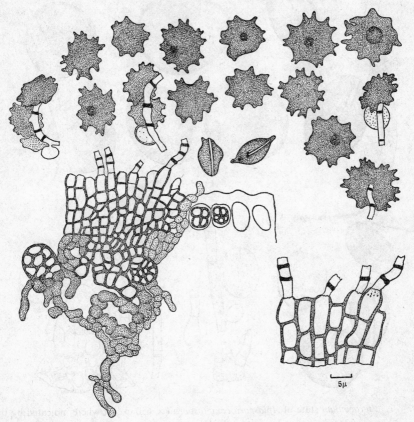

Fig. 416. *Pteroconium pterospermum* (× 650 except where indicated by the scale).

intercalary, cylindrical, denticulate; denticles usually very short, cylindrical. *Conidia* solitary, dry, acropleurogenous, simple, lenticular with a hyaline rim or germ slit, round, polygonal or irregular in face view, sometimes lobed or dentate, pale to dark brown, smooth, 0-septate.

Type species: Pteroconium pterospermum (Cooke & Massee) Grove.

KEY

Conidia on natural substrata almost all lobed or dentate . . . . *pterospermum*
Conidia on natural substrata not lobed or dentate . . . *Apiospora camptospora*

Pteroconium pterospermum (Cooke & Massee) Grove, 1914, *Hedwigia*, **55**: 146.
   Coniosporium pterospermum Cooke & Massee, 1891, apud Cooke, *Grevillea*, **19**: 90.

(Fig. 416)

Sporodochia numerous, very small, often elliptical or elongated, black. *Conidiophore mother cells* projecting only slightly above the surface of the stroma, 5–10 × 4–6µ. *Conidiophores* up to 35µ long, 2–3·5µ thick, usually with

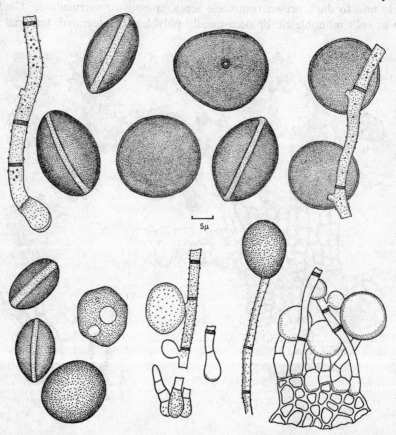

5µ

FIG. 417. *Pteroconium* state of *Apiospora camptospora* (× 650 except where indicated by the scale).

2–3 thick, dark transverse septa. *Conidia* lobed or dentate, pale to mid brown, 13–27μ diam. in face view, including lobes, 9–12μ thick.

On *Lepidospermum*; Australia.

Pteroconium state of **Apiospora camptospora** Penz. & Sacc., 1897, *Malpighia*, **11**: 398.

Papularia vinosa (Berk. & Curt.) Mason, 1933, *Mycol. Pap.*, **3**: 21, with synonyms.

(Fig. 417)

Sporodochia ellipsoidal or elongated, black. *Conidiophore mother cells* 8–12 × 4–7μ. *Conidiophores* often pale brown, verruculose, up to 140μ long, 3–4μ thick. *Conidia* lenticular, round or polygonal in face view, 20–32μ diam., 14–18μ thick, mid to dark brown with a distinct hyaline rim or germ slit.

On *Bambusa*, *Chusquea*, *Cymbopogon*, *Dendrocalamus*, *Panicum*, *Pennisetum*, *Saccharum*, *Setaria*, *Sorghum* and *Zea*; Antigua, Ceylon, Chile, Cuba, Ghana, Hong Kong, Jamaica, Kenya, Malaya, Nigeria, New Caledonia, New Guinea, Philippines, Sabah, Sierra Leone, St. Vincent, Tanzania, U.S.A., Venezuela. A cultural study of the perfect and imperfect states was made by H. J. Hudson, *Trans. Br. mycol. Soc.*, **43**: 607–611, 1960.

294. ENDOCALYX

Endocalyx Berkeley & Broome, 1876, *J. Linn. Soc.*, **15**: 84–85.

Mycelium immersed. *Stroma* superficial or immersed, erect, arising from an annulus, cylindrical below, expanding above into an open, fringed funnel packed with a black mass of conidia. *Setae* and *hyphopodia* absent. *Conidiophores* basauxic, macronematous, mononematous, slender, thread-like, flexuous, irregularly branched, colourless or pale, smooth, forming sporodochia. *Conidiogenous cells* monoblastic or polyblastic, integrated, terminal and intercalary, sympodial or determinate, cylindrical, denticulate; denticles cylindrical. *Conidia* solitary, acropleurogenous, simple, 0-septate, lenticular, elliptical or almost round in one plane, mid to dark brown or blackish brown, smooth or minutely echinulate, often with an elongated germ slit.

Type species: Endocalyx thwaitesii Berk. & Br.

Hughes, S. J., *Mycol. Pap.*, **50**: 14–17, 1953.

KEY

Fructifications stilboid, pale olivaceous, conidia echinulate   .   .   .   *thwaitesii*
Fructifications cupulate, bright greenish yellow, conidia smooth .   .   . *melanoxanthus*

Endocalyx thwaitesii Berk. & Br., 1876, *J. Linn. Soc.*, **15**: 84.
(Fig. 418 A)

Fructifications up to 1 mm. high, lower part cylindrical, 50–70μ thick expanding into a funnel containing a black mass of conidia. *Conidia* minutely echinulate, 15–22 × 13–21μ in face view, 10–11μ thick.

On *Cissus* and *Oncosperma*; Ceylon, Ghana.

Endocalyx melanoxanthus (Berk. & Br.) Petch, 1908, *Ann. Bot.*, **22**: 389.
(Fig. 418 B)

Fructifications bright greenish yellow shallow cups containing dark brown to black conidia. *Conidia* smooth, 13–16 × 10–13µ in face view, 6–9µ thick; longitudinal germ slit often well defined.

Common, especially on palms including *Borassus*, *Cocos*, *Elaeis* and *Phoenix*; Ceylon, Ghana, Jamaica, Malaya, New Guinea, Pakistan, Philippines, Sabah, Sarawak, Sierra Leone, U.S.A.

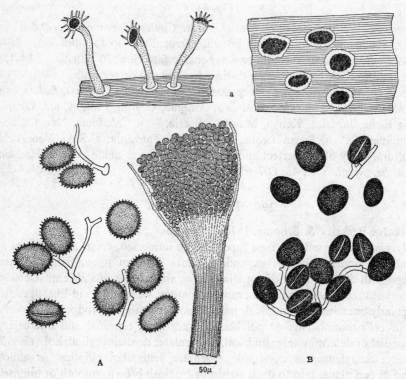

FIG. 418. A, *Endocalyx thwaitesii*; B, *E. melanoxanthus* (a, habit sketches, other figures × 650 except where indicated by the scale).

295. DICTYOARTHRINIUM

Dictyoarthrinium Hughes, 1952, *Mycol. Pap.*, **48**: 29–33.

Colonies compact or effuse, black, pulverulent, often pulvinate. *Mycelium* superficial, a close network of branched and anastomosing hyphae. *Stroma* none. *Setae* and *hyphopodia* absent. *Conidiophores* basauxic, macronematous, mononematous, arising usually singly from subspherical or cupulate conidiophore mother cells, straight or flexuous, narrow, cylindrical, colourless to pale brown except for the thick brown or dark brown transverse septa, smooth to minutely verruculose. *Conidiogenous cells* monoblastic or polyblastic, integrated, terminal and intercalary, cylindrical, denticulate; denticles usually very short, cylindrical, truncate. *Conidia* solitary, dry, acropleurogenous, simple, oblong or square rounded at the corners, spherical or subspherical, often flattened in one plane, pale to dark brown, muriform or cruciately septate, verrucose or echinulate.

Type species: Dictyoarthrinium sacchari (Stevenson) Damon = *D. quadratum* Hughes.

P. R. Rao & Dev Rao, *Mycopath. Mycol. appl.*, **23**: 23–28, 1964.

KEY

Conidia cruciately septate, 4-celled . . . . *sacchari*
Conidia muriform, 16-celled . . . . *africanum*

Dictyoarthrinium sacchari (Stevenson) Damon, 1953, *Bull. Torrey bot. Club*, **80**: 164.

Tetracoccosporium sacchari Stevenson [as '*Tetracoccosporis*'], 1917, in *J. Dep. Agric. P. Rico*, **1** (4): 225.

D. quatratum Hughes, 1952, *Mycol. Pap.*, **48**: 30–31.

(Fig. 419 A)

Colonies black, up to 1 mm. diam., sometimes coalescing. *Conidiophore mother cells* 4–6µ wide, 3–4µ long. *Conidiophores* up to 130µ long, 4–5µ thick.

FIG. 419. A, *Dictyoarthrinium sacchari*; B, *D. africanum* (× 650).

Conidia square, spherical or subspherical, flattened in one plane, cruciately septate, 4-celled, mid to dark brown, verruculose, 9–15µ diam. in face view, 7–9·5µ thick.

On dead stems and leaves of *Ananas, Bambusa, Borassus, Cassia, Cymbopogon, Dracaena, Erythrina, Phragmites, Saccharum* and *Zinnia*; Ghana, India, Malaya, Puerto Rico, Venezuela, Zambia.

Dictyoarthrinium africanum Hughes, 1952, *Mycol. Pap.*, **48**: 31–33.

(Fig. 419 B)

Colonies black, about 0·5 mm. diam. *Conidiophore mother cells* 5–6µ wide, 3–4µ long. *Conidiophores* up to 70µ long, 4–6µ thick. *Conidia* oblong or square, rounded at the corners, flattened in one plane, muriform, 16-celled, constricted at the septa, pale to mid-dark brown, verrucose, 20–28 × 18–25µ in face view, 8–10µ thick.

On *Panicum* and *Saccharum*; Ghana, Venezuela.

SUBSTRATUM INDEX

(s = state of)

PLANTS, INCLUDING FUNGI

A

Abies—Cheiromycella microscopica, Sterigmatobotrys macrocarpa. **Acacia**—Camptomeris albiziae, Helminthosporium acaciae, Melanographium cookei, Pithomyces atro-olivaceus. **Acalypha**—Botrytis s. Sclerotinia ricini. **Acanthus**—Stachybotrys dichroa. **Acer**—Bactrodesmium abruptum, B. traversianum, Brachysporium obovatum, Cacumisporium capitulatum, Centrospora acerina, Corynespora pruni, Cryptostroma corticale, Endophragmia boothii, Menispora glauca, Oedemium s. Thaxteria fusca, O. s. T. phaeostroma, Petrakia echinata, Pseudospiropes nodosus, P. simplex, Septosporium bulbotrichum, Sporidesmium altum, S. folliculatum, Sporoschisma mirabile, Stigmina negundinis, Trichocladium canadense, Virgaria nigra, Wardomyces inflatus, Xylohypha nigrescens. **Aconitum**—Chalara cylindrosperma. **Aesculus**—Chalara aurea, Haplariopsis fagicola, Stachybotrys s. Melanopsamma pomiformis. **Afzelia**—Exosporium cantareirense. **Agave**—Bahusakala olivaceonigra, Curvularia brachyspora, C. eragrostidis, Fusariella concinna, Stachybotrys parvispora. **Agropyron**—Drechslera s. Pyrenophora tritici-repentis. **Agrostis**—Drechslera s. Pyrenophora tritici-repentis, D. erythrospila, D. fugax. **Albizia**—Allescheriella crocea, Camptomeris albiziae, Pithomyces cupaniae. **Alchornea**—Hermatomyces tucumanensis. **Aleurites**—Melanographium citri. **Algae**—Teratosperma appendiculatum. **Allaeanthus**—Corynespora calicioidea. **Allium**—Acremoniella atra, Alternaria porri, Botrytis s. Sclerotinia globosa, B. s. S. porri, B. s. S. sphaerosperma, B. s. S. squamosa, B. allii, B. byssoidea, Fusariella atrovirens, Stachylidium bicolor, Stemphylium lycopersici. **Alnus**—Bactrodesmium obovatum, Brachysporium bloxami, Corynespora pruni, Helminthosporium mauritianum, Oedemium s. Thaxteria phaeostroma, Passalora alni, P. bacilligera, P. microsperma, Pseudospiropes simplex, Sporoschisma juvenile, S. mirabile, Taeniolella stilbospora, Triposporium elegans, Verticillium cyclosporum. **Alpinia**—Phaeodactylium alpiniae. **Amazonia**—Spiropes effusus. **Ammophila**—Periconia s. Didymosphaeria igniaria, Sporidesmium leptosporum, Tetraploa aristata, Thyrostromella myriana. **Ampelopsis**—Podosporium rigidum. **Anacardium**—Pithomyces acchari. **Anadelphia**—Tetraploa aristata. **Ananas**—Arthrinium sacchari, Beltrania rhombica, Curvularia eragrostidis, Dictyoarthrinium sacchari, Gliomastix murorum var. polychroma, Gyrothrix hughesii, Myrothecium jollymannii, Pithomyces sacchari, Spegazzinia tessarthra, Stachybotrys parvispora, Stachylidium bicolor, Thielaviopsis s. Ceratocystis paradoxa, Trichobotrys effusa. **Andropogon**—Curvularia deightonii, Lacellina graminicola, Lacellinopsis sacchari, Periconia atropurpurea, P. digitata, P. echinochloae, Pithomyces maydicus, P. sacchari, Spegazzinia deightonii, Tetraploa aristata, Trichobotrys effusa. **Anethum**—Cercosporidium punctum. **Angelica**—Cercosporidium angelicae, C. depressum, Stachybotrys dichroa. **Anisophyllea**—Periconiella anisophylleae, Pithomyces cupaniae. **Anthocleista**—Annellophora africana. **Anthoxanthum**—Drechslera dematioidea. **Antiaris**—Zygophiala jamaicensis. **Apium**—Alternaria radicina, Centrospora acerina, Cercospora apii, Drechslera biseptata. **Appendiculella**—Spiropes capensis, S. dorycarpus. **Aquilegia**—Haplobasidion thalictri. **Arachis**—Cercosporidium s. Mycosphaerella berkeleyi, Curvularia eragrostidis, Scopulariopsis brevicaulis, Stachybotrys kampalensis, S. nephrospora, Torula caligans. **Araucaria**—Mycoenterolobium platysporum. **Arctium**—Pleurophragmium simplex, Sporoschisma juvenile. **Areca**—Corynespora elaeidicola, Periconia atropurpurea, Pithomyces sacchari, Pseudoepicoccum cocos, Sporidesmium macrurum. **Arenga**—Blastophorella smithii, Melanographium selenioides. **Argemone**—Cercosporidium guanicense. **Aristida**—Lacellina graminicola, Pithomyces sacchari. **Aronicum**—Fuscladiella melaena. **Arrhenatherum**—Drechslera s. Pyrenophora tritici-repentis, Helicosporium s. Tubeufia helicomyces. **Artemisia**—Periconia s. Didymosphaeria igniaria. **Arthraxon**—Deightoniella bhopalensis. **Artocarpus**—Ellisiopsis gallesiae, Gyrothrix circinata, Zygosporium echinosporum, Z. gibbum, Z. masonii, Z. minus. **Arundinaria**—Cercosporidium compactum, Periconia atropurpurea, Tripospermum myrti. **Arundinella**—Pithomyces sacchari. **Arundo**—Acrodictys erecta, Periconia hispidula, Saccardaea atra.

583

Asparagus—Botrytis tulipae. **Aspilia**—Hormocephalum equadorense. **Asteridiella**—Deightoniella leonensis, Periconiella ellisii, Spiropes capensis, S. dorycarpus, S. guareicola, S. helleri, S. japonicus, S. penicillium. **Asterina**—Domingoella asterinarum, Eriocercospora balladynae, Hansfordiella asterinarum, Spiropes effusus, Tetraposporium asterinearum. **Asterinella**—Spiropes lembosiae. **Asterolibertia**—Domingoella asterinarum, Hansfordiella asterinarum. **Astragalus**—Camptomeris astragali. **Avena**—Acremoniella atra, Alternaria s. Pleospora infectoria, Drechslera s. Cochliobolus sativus, D. s. C. victoriae, D. s. Pyrenophora avenae, D. s. P. semeniperda, D. avenacea. **Averrhoa**—Beltrania africana, Hermatomyces tucumanensis, Melanographium citri, Sporidesmium brachypus, S. leptosporum, S. vagum. **Axonopus**—Gyrothrix hughesii, Spegazzinia deightonii, Tetraploa aristata.

B

Balladyna—Acrodictys balladynae, A. furcata, Eriocercospora balladynae, Spiropes balladynae. **Balladynopsis**—Acrodictys balladynae, Eriocercospora balladynae. **Bambusa**—Acremoniella atra, Acrodictys bambusicola, A. brevicornuta, A. dennisii, A. fimicola, Arthrinium s. Apiospora montagnei, A. phaeospermum, Cordella coniosporioides, C. johnstonii, Corynespora foveolata, Dictyoarthrinium sacchari, Didymobotryum rigidum, Endophragmia hyalosperma, Gliomastix s. Wallrothiella subiculosa, G. fusigera, Gonytrichum macrocladum, Haplobasidion lelebae, Mahabalella acutisetosa, Periconia lateralis, Pseudospiropes simplex, Pteroconium s. Apiospora camptospora, Pyricularia grisea, Septosporium rostratum, Spiropes caaguazuensis, Sporoschisma mirabile, Stachylidium bicolor, Stephanosporium cereale, Tetraploa aristata, Trichobotrys effusa. **Baphia**—Cryptophiale kakombensis, Periconiella angusiana. **Barbacenia**—Zygophiala jamaicensis. **Bauhinia**—Helminthosporium bauhiniae, Melanographium cookei. **Beckmannia**—Drechslera s. Pyrenophora tritici-repentis. **Berberis**—Oncopodium antoniae, Pseudospiropes nodosus. **Beta**—Acremoniella atra. **Betula**—Actinocladium rhodosporum, Bactrodesmium atrum, B. betulicola, B. obovatum, B. spilomeum, Brachysporium bloxami, B. britannicum, B. nigrum, Cacumisporium capitulatum, Corynespora cespitosa, Diplococcium spicatum, Endophragmia biseptata, E. glanduliformis, E. hyalosperma, Menispora glauca, Monodictys paradoxa, Polyscytalum fecundissimum, Pseudospiropes nodosus, P. simplex, Septonema secedens, Septotrullula bacilligera, Spadicoides bina, Sporidesmium eupatoriicola, Sporoschisma mirabile, Taeniolella exilis, Trimmatostroma betulinum, Verticillium cyclosporum, Virgaria nigra, Xylohypha nigrescens. **Bignonia**—Circinotrichum olivaceum, Piricauda paraguayensis. **Bombax**—Melanographium citri. **Borassus**—Arthrinium sacchari, Dictyoarthrinium sacchari, Endocalyx melanoxanthus, Gyrothrix hughesii, Lacellina graminicola, L. leonensis, L. macrospora, Lacellinopsis sacchari, Melanographium citri, Periconia s. Didymosphaeria igniaria, P. atropurpurea, P. digitata, Pithomyces africanus, P. sacchari, Scolecobasidium humicola, Spegazzinia deightonii, S. tessarthra, Sporidesmium macrurum, Stigmina palmivora, Zygosporium gibbum. **Borreria**—Curvularia borreriae. **Brabejum**—Periconiella velutina. **Brachiaria**—Periconia echinochloae, Pyricularia grisea. **Brachystegia**—Chaetochalara africana. **Brassica**—Alternaria brassicae, A. brassicicola, Doratomyces purpureofuscus, Endophragmia elliptica, E. hyalosperma, Graphium putredinis, Pleurophragmium simplex, Pseudobotrytis terrestris, Scopulariopsis brevicaulis. **Bridelia**—Acarocybe hansfordii, Annellophora africana, Pithomyces maydicus. **Bromus**—Drechslera s. Pyrenophora bromi, D. s. P. tritici-repentis. **Buxus**—Actinocladium rhodosporum, Brachysporium britannicum, Sporidesmium adscendens, S. altum, S. leptosporum.

C

Cactaceae—Drechslera cactivora. **Cajanus**—Acrophialophora fusispora, Mycovellosiella cajani, Periconia s. Didymosphaeria igniaria, Pithomyces sacchari, Stachybotrys parvispora. **Caladium**—Cercosporidium caladii. **Calamagrostis**—Drechslera s. Pyrenophora tritici-repentis. **Calcophyllum**—Zygophiala jamaicensis. **Calliandra**—Camptomeris calliandrae. **Calopogonium**—Fusariella obstipa, Pithomyces maydicus. **Calothyrium**—Microclava miconiae. **Calyculosphaeria**—Acrodictys obliqua. **Camellia**—Corynespora polyphragmia, Pithomyces sacchari. **Canna**—Periconiella portoricensis. **Caperonia**—Helicomina caperoniae. **Capparis**—Melanographium fasciculatum. **Capsicum**—Curvularia ovoidea. **Carapa**—Gyrothrix circinata. **Carduus**—Fusicladiella melaena, Stachybotrys dichroa. **Carex**—Arthrinium s. Pseudoguignardia scirpi, A. caricicola, A. curvatum var. minus, A. fuckelii, A. kamtschaticum, A. morthieri, A. puccinioides, A. sporophleum, Clasterosporium caricinum, C. flagellatum, Endophragmia hyalosperma, Gyrothrix podosperma, Myrothecium striatisporum, Periconia

Cyathea—Periconiella cyatheae. **Cymbopogon**—Curvularia andropogonis, C. comoriensis C. cymbopogonis, Dictyoarthrinium sacchari, Lacellinopsis sacchari, Pteroconium s. Apiospora camptospora, Tetraploa aristata. **Cynodon**—Curvularia s. Cochliobolus intermedius, Drechslera s. C. cynodontis, D. gigantea, Periconia atropurpurea, Pithomyces cynodontis, P. sacchari, Pyricularia grisea, Spegazzinia tessarthra. **Cynosurus**—Saccardaea atra. **Cyperus**—Clasterosporium cyperi, Duosporium cyperi, Tetraploa aristata. **Cytisus**—Endophragmia biseptata, E. hyalosperma, Pleiochaeta setosa.

D

Dactylis—Drechslera biseptata, Periconia s. Didymosphaeria igniaria, Stachybotrys dichroa Tetraploa aristata, T. ellisii. **Dactyloctenium**—Drechslera s. Trichometasphaeria holmii. **Dahlia**—Periconia atropurpurea. **Dalbergia**—Helminthosporium dalbergiae. **Daniellia**— Zygosporium masonii. **Daphniphyllum**—Periconiella daphniphylli, Trochophora simplex. **Datura**—Alternaria crassa. **Daucus**—Acremoniella atra, Alternaria dauci, A. radicina, Centrospora acerina, Chalaropsis thielavioides, Trichocladium asperum. **Davilla**—Spiropes davillae. **Deinbollia**—Periconiella deinbolliae. **Dendrocalamus**—Pteroconium s. Apiospora camptospora. **Dendrographium**—Annellophora dendrographii. **Deschampsia**—Arthrinium puccinioides, Curvularia protuberata, Periconia hispidula, Tetraploa aristata. **Desmanthus**— Camptomeris desmanthi, Cercosporidium desmanthi. **Desmodium**—Zygophiala jamaicensis. **Desplatzia**—Zygosporium echinosporum. **Dianthus**—Alternaria dianthi, A. dianthicola, Cladosporium s. Mycosphaerella dianthi, Zygophiala jamaicensis. **Dichanthium**—Curvularia robusta. **Dichrostachys**—Melanographium cookei, Sporidesmium brachypus. **Digitaria**— Curvularia eragrostidis, Periconia lateralis, Pyricularia grisea. **Dioscorea**—Curvularia eragrostidis, C. stapeliae, Scopulariopsis brevicaulis, Spegazzinia deightonii,Stachybotrys kampalensis, S. nephrospora, Stachylidium bicolor, Zygosporium gibbum. **Diplorrhynchus**—Virgaria nigra. **Dipsacus**—Fusariella hughesii, Stachybotrys dichroa. **Doronicum**—Fusicladiella melaena. **Dracaena**—Dictyoarthrinium sacchari, Zygosporium geminatum, Z. minus. **Drimys**— Circinotrichum olivaceum. **Duranta**—Piricauda paraguayensis. **Durio**—Curvularia affinis.

E

Echidnodes—Spiropes echidnodis. **Echinochloa**—Drechslera monoceras, Periconia echinochloae, Pyricularia grisea. **Eichhornia**—Drechslera s. Cochliobolus bicolor. **Elaeis**—Acremoniella atra, Acrodictys elaeidicola, Brachysporiella gayana, Corynespora elaeidicola, Curvularia eragrostidis, C. fallax, C. oryzae, Drechslera s. Cochliobolus carbonus, Endocalyx melanoxanthus, Exosporium stilbaceum, Gliomastix s. Wallrothiella subiculosa, Hermatomyces tucumanensis, Melanographium citri, Phaeotrichoconis crotalariae, Pithomyces elaeidicola, P. maydicus, P. sacchari, Sporidesmium adscendens, S. leptosporum, S. macrurum, S. vagum, Thielaviopsis s. Ceratocystis paradoxa, Zygosporium gibbum, Z. minus. **Eleocharis** —Arthrinium puccinioides, Coremiella cubispora. **Elettaria**—Phaeodactylium alpiniae, Xenosporium mirabile. **Eleusine**—Drechslera s. Cochliobolus nodulosus, Pyricularia grisea. **Elymus**—Drechslera s. Pyrenophora tritici-repentis. **Epilobium**—Endophragmia elliptica, Myrothecium carmichaelii, Pleurophragmium simplex, Sporoschisma juvenile, S. mirabile, Stachybotrys dichroa, Triposporium elegans. **Eragrostis**—Curvularia eragrostidis, Drechslera miyakei, Periconia atropurpurea, Pyricularia grisea. **Erianthus**—Lacellina graminicola, Lacellinopsis sacchari, Tetraploa aristata. **Eriophorum**—Arthrinium sporophleum, Periconia digitata. **Erythrina**—Dictyoarthrinium sacchari, Exosporium extensum, E. mexicanum, Fusariella bizzozeriana. **Esenbeckia**—Beltraniopsis esenbeckiae, Ellisiopsis gallesiae. **Eucalyptus**—Camposporium antennatum, Endophragmia stemphylioides, Gyrothrix macroseta, Zygosporium gibbum. **Euchlaena**—Spegazzinia deightonii, Tetraploa aristata. **Eugenia**— Exosporium coonoorense, Gyrothrix podosperma. **Euonymus**—Ellisiopsis gallesiae. **Eupatorium**—Mycovellosiella perfoliata, Myrothecium carmichaelii, Pseudospiropes nodosus, P. simplex, Sporidesmium eupatoriicola, Stachybotrys dichroa. **Euphorbia**—Arthrinium euphorbiae, Botrytis s. Sclerotinia ricini, Cercosporidium chaetomium, Deightoniella jabalpurensis, Drechslera euphorbiae.

F

Fagara—Zygosporium masonii. **Fagraea**—Stenellopsis fragraeae. **Fagus**—Arachnophora fagicola, Bactrodesmiella masonii, Bactrodesmium atrum, B. obovatum, B. pallidum, B. spilomeum, Bispora antennata, Brachydesmiella biseptata, Brachysporium bloxami, B.

britannicum, B. masonii, B. nigrum, Cacumisporium capitulatum, Catenularia s. Chaetosphaeria cupulifera, Ceratocladium microspermum, Chaetopsis grisea, Chalara cylindrosperma, Chloridium caudigerum, Corynespora pruni, Diplococcium spicatum, Endophragmia biseptata, E. catenulata, E. elliptica, E. glanduliformis, E. uniseptata, E. verruculosa, Gonatobotryum fuscum, Haplariopsis fagicola, Menispora ciliata, M. glauca, Oedemium s. Thaxteria fusca, O. s. T. phaeostroma, Pendulispora venezuelanica, Polyscytalum fecundissimum, Pseudobotrytis terrestris, Pseudospiropes nodosus, P. simplex, Septotrullula bacilligera, Spadicoides bina, S. grovei, Spondylocladiopsis cupulicola, Sporidesmium folliculatum, S. leptosporum, Sporoschisma juvenile, S. mirabile, Stachybotrys s. Melanopsamma pomiformis, Triposporium elegans, Verticillium cyclosporum, V. tenuissimum, Virgaria nigra, Virgariella atra, Xylohypha nigrescens. **Faurea**—Annellophorella faureae. **Festuca**—Acremoniella atra, Drechslera s. Pyrenophora dictyoides, Periconia hispidula. **Ficus**—Fusariella concinna, Gonatophragmium mori, Stachybotrys parvispora, Zygosporium echinosporum, Z. gibbum, Z. masonii, Z. minus. **Filipendula**—Coremiella cubispora, Endophragmia elliptica, E. hyalosperma, E. prolifera, Pleurophragmium simplex, Sporidesmium eupatoriicola, Stachybotrys dichroa, Triposporium elegans. **Fimbristylis**—Deightoniella fimbristylidis. **Finschia**—Verrucispora proteacearum. **Flagellaria**—Circinoconis paradoxa. **Foeniculum**—Cercosporidium punctum, Fusariella hughesii. **Fragaria**—Idriella lunata. **Fraxinus**—Actinocladium rhodosporum, Bactrodesmium abruptum, B. obovatum, B. pallidum, B. spilomeum, Brachydesmiella biseptata, Brachysporium bloxami, B. britannicum, B. nigrum, Catenularia s. Chaetosphaeria cupulifera, Chaetopsis grisea, Oedemium s. Thaxteria phaeostroma, Phaeoisaria clavulata, Pseudospiropes nodosus, P. simplex, Sarcinella s. Schiffnerula pulchra, Sporidesmium folliculatum, S. leptosporum, Sporoschisma juvenile, S. mirabile, Stachybotrys s. Melanopsamma pomiformis, Tripospermum myrti, Triposporium elegans, Verticillium cyclosporum, Virgaria nigra, Virgariella atra, Xylohypha nigrescens. **Funtumia**—Exosporium ampullaceum, Pithomyces cupaniae. **Furcraea**—Curvularia eragrostidis.

G

Galium—Cercosporidium galii. **Gallesia**—Ellisiopsis gallesiae. **Garcinia**—Corynespora garciniae. **Gardenia**—Acrodictys deightonii, Stachylidium bicolor. **Garrya**—Podosporiella glomerata. **Gaultheria**—Zygosporium mycophilum. **Geonoma**—Periconiella geonomae. **Gladiolus**—Botrytis s. Sclerotinia draytonii, Curvularia trifolii f. sp. gladioli, Pithomyces maydicus, P. sacchari, Stemphylium lycopersici. **Gliricidia**—Zygosporium masonii. **Gloxinia**—Thielaviopsis basicola. **Glyceria**—Helicosporium s. Tubeufia helicomyces, Periconia hispidula. **Glycine**—Corynespora cassiicola. **Glyphaea**—Acarocybella jasminicola. **Gmelina**—Allescheriella crocea, Scolecobasidium constrictum. **Gomphrena**—Phaeoramularia gomphrenae. **Gossypium**—Alternaria macrospora, Drechslera papendorfii, Scopulariopsis brevicaulis, Zygosporium masonii. **Gramineae**—Cercosporidium graminis, Cerebella andropogonis, Cladosporium spongiosum, Curvularia trifolii, Doratomyces microsporus, Drechsler s. Cochliobolus bicolor, D. s. C. cynodontis, D. s. C. sativus, D. s. C. spicifer, D. s. Pyrenophora semeniperda, D. australiensis, D. biseptata, D. halodes, D. hawaiiensis, D. papendorfii, D. rostrata, Hadrotrichum phragmitis, Myrothecium gramineum, M. striatisporum, Nigrospora panici, N. sacchari, N. sphaerica, Periconia atra, P. echinochloae, P. minutissima, Pithomyces chartarum, Saccardaea atra, Sadasivania girisa, Torula graminis, Ustilaginoidea ochracea. **Grewia**—Endophragmiopsis pirozynskii. **Gunnera**—Stachybotrys dichroa. **Gymnosporia**—Zygosporium masonii. **Gynerium**—Lacellina graminicola, Lacellinopsis sacchari, Tetraploa aristata.

H

Hakea—Verrucispora proteacearum. **Halbaniella**—Spiropes bakeri. **Hamamelis**—Gonatobotryum apiculatum. **Harungana**—Spiropes harunganae. **Hedera**—Pseudospiropes nodosus, P. simplex, Sporidesmium altum, S. cookei, S. folliculatum, Sporoschisma juvenile, Xylohypha nigrescens. **Heeria**—Fusariella obstipa, Periconia lateralis, Sporidesmium adscendens, Virgaria nigra. **Heliconia**—Hansfordia ovalispora, Stachylidium bicolor. **Helicteres**—Corynespora siwalika. **Heracleum**—Doratomyces purpureofuscus, Pleurophragmium simplex, Sporoschisma mirabile, Stachybotrys cylindrospora, S. dichroa, Stachylidium bicolor. **Heteropogon**—Periconia echinochloae, Spegazzinia tessarthra, Tetraploa aristata. **Hevea**—Allescheriella crocea, Ceratosporium productum, Corynespora cassiicola, Curvularia eragrostidis, Drechslera heveae, Periconia manihoticola, Periconiella heveae, Scolecobasidium constrictum, S. verruculosum, Stachybotrys parvispora. **Hibiscus**—Stachybotrys kampalensis, Stachylidium bicolor. **Hippocratea**—Melanographium citri, Sporidesmium adscendens. **Hippophäe**—Balanium stygium. **Homalium**—Corynespora homaliicola. **Hordeum**—Acremoniella atra, Alternaria s. Pleospora infectoria, Curvularia inaequalis, Drechslera s. Cochliobolus sativus, D. s. Pyrenophora graminea, D. s. P. teres, D. s. P. tritici-repentis. **Huernia**—Curvularia stapeliae. **Hura**—Septoidium s. Parodiopsis hurae, Stachybotrys theobromae. **Hydnocarpus**—Zygosporium gibbum. **Hymenocardia**—Periconiella angusiana, Pithomyces cupaniae, Zygophiala jamaicensis. **Hyparrhenia**—Lacellina graminicola, Trichobotrys effusa. **Hyphaene**—Pithomyces africanus, Zygosporium echinosporum, Z. gibbum. **Hypselodelphys**—Helminthosporium hypselodelphyos. **Hysterostomella**—Domingoella asterinarum.

I

Ilex—Chaetochalara bulbosa, Corynespora smithii, C. yerbae, Endophragmia biseptata, Oedemium s. Thaxteria fusca, Pseudospiropes simplex, Sporoschisma mirabile, Triposporium elegans, Verticillium cyclosporum. **Imperata**—Circinoconis paradoxa, Deightoniella africana, Pithomyces sacchari. **Indigofera**—Cercosporidium pulchellum, Sclerographium aterrimum. **Ipomoea**—Arthrinium sacchari, Curvularia eragrostidis, Drechslera s. Cochliobolus cynodontis, Myrothecium jollymannii, Trichurus spiralis. **Irenopsis**—Spiropes capensis, S. dorycarpus, S. guareicola, S. palmetto, S. pirozynskii. **Iris**—Botrytis s. Sclerotinia convoluta, Cladosporium s. Mycosphaerella macrospora, Dematophora s. Rosellinia necatrix, Drechslera iridis, Myrothecium carmichaelii, Stachybotrys dichroa. **Ischaemum**—Tetraploa aristata.

J

Jasminum—Acarocybella jasminicola, Curvularia prasadii, Pithomyces cupaniae. **Juglans**—Chalaropsis thielavioides, Stachybotrys s. Melanopsamma pomiformis, Virgaria nigra. **Juncus**—Arthrinium curvatum var. minus, A. cuspidatum, A. sporophleum, Periconia atra, P. curta, P. digitata, P. funerea, Selenosporella curvispora, Tetraploa aristata. **Juniperus**—Conoplea juniperi var. juniperi, Stigmina glomerulosa.

K

Khaya—Gyrothrix dichotoma, G. verticillata.

L

Lactuca—Alternaria sonchi, Cercosporidium scariolae. **Lagenaria**—Curvularia affinis. **Lagerstroemia**—Sporidesmium vagum. **Landolphia**—Zygosporium gibbum. **Lannea**—Beltraniella odinae, Corynespora lanneicola. **Lantana**—Periconia atropurpurea, Spegazzinia tessarthra. **Larix**—Cheiromycella microscopica, Spadicoides atra. **Laurus**—Zygosporium gibbum. **Lawsonia**—Chalaropsis punctulata. **Ledum**—Ampulliferina persimplex. **Leersia**—Drechslera s. Pyrenophora tritici-repentis. **Lembosia**—Spiropes lembosiae. **Lepidospermum**—Pteroconium pterospermum. **Leptoderris**—Corynespora leptoderridicola, Exosporium leptoderridicola, Periconiella leptoderridis. **Leucaena**—Camptomeris leucaenae, Helminthosporium mauritianum. **Libocedrus**—Zanclospora novae-zelandiae. **Lichens**—Hansfordiellopsis lichenicola, Teratosperma anacardii. **Licuala**—Sporidesmium macrurum. **Ligustrum**—Sarcinella s. Schiffnerula pulchra, Xylohypha nigrescens. **Lilium**—Botrytis elliptica. **Linotexis**—Domingoella asterinarum. **Lithocarpus**—Ellisiopsis gallesiae. **Litsea**—Mystrosporiella litseae, Zygosporium gibbum. **Livistona**—Stigmina palmivora. **Lobelia**—Schizotrichum lobeliae. **Lolium**—Drechslera s. Pyrenophora dictyoides, D. siccans. **Lomatea**—Periconiella lomateae. **Loranthus**—Melanographium citri. **Lupinus**—Chalaropsis thielavioides, Fusariella hughesii, Pleiochaeta setosa, Thielaviopsis basicola. **Luzula**—Arthrinium luzulae, A. ushuvaiense. **Lychnodiscus**—Acarocybe deightonii. **Lycopersicon**—Alternaria solani, Corynespora cassiicola, Drechslera australiensis, Fulvia fulva, Periconia s. Didymosphaeria igniaria, Spegazzinia tessarthra, Stemphylium lycopersici, S. solani, Thielaviopsis basicola, Trichurus spiralis. **Lysimachia**—Endophragmia elliptica. **Lythrum**—Coremiella cubispora.

M

Maba—Clasterosporium s. Asterina clasterosporium. **Macrolobium**—Helminthosporium bauhiniae. **Magnolia**—Corynespora pruni, Gyrothrix circinata, Spadicoides obovata. **Malus**—Bloxamia truncata, Dematophora s. Rosellinia necatrix, Drechslera s. Cochliobolus cynodontis, Gliocephalotrichum bulbilium, Gliomastix murorum v. polychroma, Spilocaea s. Venturia inaequalis. **Mangifera**—Allescheriella crocea, Gyrothrix podosperma, Spegazzinia tessarthra,

Stigmina mangiferae. **Manihot**—Corynespora cassiicola, Curvularia affinis, Helminthosporium mauritianum, Periconia manihoticola, P. shyamala, Stachylidium bicolor. **Mansonia**—Melanographium cookei. **Marsilea**—Phaeotrichoconis crotalariae. **Mascagnia**—Exosporium mexicanum. **Matthiola**—Alternaria raphani. **Mauritia**—Sporidesmium macrurum. **Meliola**—Isthmospora s. Trichothyrium asterophorum, Spiropes capensis, S. clavatus, S. deightonii, S. dorycarpus, S. fumosus, S. guareicola, S. helleri, S. japonicus, S. leonensis, S. melanoplaca, S. palmetto, S. penicillium. **Meliolineae**—Spiropes dialii, S. intricatus. **Metroxylon**—Zygosporium gibbum. **Miconia**—Exosporium insuetum, Microclava miconiae. **Microthyriaceae**—Spiropes shoreae. **Milletia**—Pithomyces cupaniae, Zygosporium echinosporum. **Miscanthidium**—Lacellina graminicola. **Morus**—Memnoniella levispora, Sirosporium mori, Sporidesmium leptosporum. **Mucuna**—Periconiella mucunae. **Musa**—Arthrinium sacchari, Cladosporium musae, Cordana musae, Curvularia fallax, Deightoniella torulosa, Drechslera s. Cochliobolus bicolor, D. musae-sapientum, Gliomastix elata, G. murorum var. polychroma, G. musicola, Gyrothrix hughesii, Haplobasidion musae, Memnoniella subsimplex, Periconia digitata, P. lateralis, Periconiella musae, Pithomyces sacchari, Pyriculariopsis parasitica, Spegazzinia tessarthra, Stachylidium bicolor, Tetraploa aristata, Thielaviopsis s. Ceratocystis paradoxa, Zygosporium gibbum, Z. masonii, Z. minus. **Mussaenda**—Annellophora mussaendae. **Myrtaceae**—Tripospermum myrti.

N

Narcissus—Botrytis s. Sclerotinia narcissicola, B. s. S. polyblastis. **Nauclea**—Sporidesmium adscendens. **Nectandra**—Hemibeltrania nectandrae. **Nephelium**—Zygosporium gibbum. **Nicotiana**—Acremoniella atra, Alternaria longipes, Drechslera papendorfii, Myrothecium jollymannii, Pithomyces sacchari, Pyricularia grisea, Scopulariopsis brevicaulis, Thielaviopsis basicola. **Nitraria**—Periconia atropurpurea. **Nothofagus**—Spiropes nothofagi, Zanclospora novae-zelandiae. **Nyctanthes**—Curvularia fallax. **Nyssa**—Sporidesmium adscendens.

O

Ochthocosmus—Exosporium phyllantheum, Sporidesmium eupatoriicola, Zygosporium gibbum. **Oenanthe**—Stachybotrys dichroa, Stachylidium bicolor. **Olax**—Melanographium citri. **Oldenlandia**—Curvularia eragrostidis. **Olea**—Acremoniella atra, Clasterosporium s. Asterodothis solaris, Curvularia brachyspora, Spilocaea oleaginea. **Olearia**—Pseudospiropes simplex. **Oncosperma**—Endocalyx thwaitesii, Pithomyces flavus. **Ophioglossum**—Curvularia crepinii. **Ophiurus**—Pithomyces sacchari. **Ornithopus**—Pleiochaeta setosa. **Oryza**—Acremoniella atra, Alternaria longissima, A. padwickii, Curvularia s. Cochliobolus intermedius, C. affinis, C. fallax, C. uncinata, Drechslera s. Cochliobolus miyabeanus, D. s. Trichometasphaeria pedicellata, D. australiensis, D. hawaiiensis, Lacellinopsis sacchari, Nakataea s. Leptosphaeria salvinii, Nigrospora s. Khuskia oryzae, Periconia digitata, P. echinochloae, P. lateralis, Phaeotrichoconis crotalariae, Pithomyces maydicus, Pyricularia oryzae, Scopulariopsis brevicaulis, S. brumptii, Spegazzinia deightonii, Ustilaginoidea virens.

P

Paeonia—Botrytis paeoniae, Cladosporium chlorocephalum. **Palmae**—Gliomastix fusigera, Kostermansinda magna, Melanographium selenioides, Pithomyces pulvinatus, Stigmina palmivora, Trichobotrys effusa. **Paludia**—Exosporium cantareirense. **Pancovia**—Zygosporium echinosporum, Z. geminatum, Z. masonii. **Panicum**—Curvularia fallax, C. leonensis, Dictyoarthrinium africanum, Drechslera frumentacei, Oncopodium panici, Periconia atropurpurea, P. digitata, P. echinochloae, P. lateralis, Pteroconium s. Apiospora camptospora, Pyricularia grisea, Spegazzinia tessarthra, Trichobotrys effusa. **Papaver**—Dendryphion s. Pleospora papaveracea. **Pappophorum**—Periconia lateralis. **Parkia**—Pseudospiropes simplex. **Parosela**—Cercosporidium caracasanum. **Passiflora**—Alternaria passiflorae, Zygosporium gibbum. **Pastinaca**—Acremoniella atra, Alternaria radicina. **Pennisetum**—Acrodictys bambusicola, Brachysporiella gayana, Curvularia penniseti, Drechslera s. Cochliobolus bicolor, D. s. C. nodulosus, Lacellinopsis spiralis, Pteroconium s. Apiospora camptospora, Pyricularia grisea, Spegazzinia deightonii, S. tessarthra. **Persea**—Circinotrichum olivaceum, Gyrothrix podosperma, Stachybotrys theobromae, Zygosporium masonii. **Petasites**—Stachybotrys dichroa, Stachylidium bicolor. **Petraea**—Sporidesmium brachypus. **Petroselinum**—Cercosporidium punctum. **Peucedanum**—Alternaria radicina, Cercosporidium depressum. **Phaeo-**

pappus—Fusicladiella phaeopappi. **Phalaris**—Drechslera s. Pyrenophora tritici-repentis, Fusariella hughesii, Helicosporium s. Tubeufia helicomyces, Periconia s. Didymosphaeria igniaria, P. hispidula. **Phaseolus**—Acremoniella atra, Phaeoisariopsis griseola, Tetraploa aristata. **Phleum**—Arthrinium sphaerospermum, Curvularia protuberata, Drechslera phlei. **Phoenix**—Annellophora phoenicis, Brachysporiella gayana, Chalaropsis s. Ceratocystis radicicola, Dwayabeeja sundara, Endocalyx melanoxanthus, Lacellina graminicola, Melanographium citri, Phragmosphathula phoenicis, Spegazzinia tessarthra, Sporidesmium macrurum, Stachylidium bicolor, Stigmina palmivora, Tetraploa aristata. **Phormium**—Gliomastix s. Wallrothiella subiculosa, Periconiella phormii, Tetraploa aristata. **Phragmites**—Arthrinium phaeospermum, Deightoniella arundinacea, Dictyoarthrinium sacchari, Hadrotrichum phragmitis, Helicosporium s. Tubeufia helicomyces, Periconia s. Didymosphaeria igniaria, P. digitata, P. hispidula, Spegazzinia deightonii, Tetraploa aristata, Trichobotrys effusa. **Phryganocydia**—Acarocybella jasminicola. **Phylica**—Brachysporium pulchrum. **Phyllanthus**—Botrytis s. Sclerotinia ricini, Exosporium phyllantheum, Helminthosporium mauritianum, Sclerographium phyllanthicola. **Picea**—Actinocladium rhodosporum, Bactrodesmium obliquum, Chalara cylindrica, Endophragmia nannfeldtii, Spadicoides atra, Sterigmatobotrys macrocarpa, Trichocladium asperum, Verticicladiella abietina. **Pimpinella**—Fusicladiella pimpinellae. **Pinus**—Acremoniella verrucosa, Brachysporium bloxami, Chalara cylindrosperma, Cheiromycella microscopica, Circinotrichum olivaceum, Curvularia tuberculata, Diplococcium spicatum, Drechslera s. Cochliobolus cynodontis, Phaeoisaria clavulata, Rhinocladiella s. Dictyotrichiella mansonii, Scytalidium lignicola, Spadicoides bina, Sporidesmium vagum, Sympodiella acicola, Verticicladium s. Desmazierella acicola, Verticillium tenuissimum. **Piper**—Cephaliophora tropica, Curvularia fallax. **Pistacia**—Clasterosporium pistaciae. **Pisum**—Acremoniella atra, A. verrucosa, Curvularia inaequalis, Thielaviopsis basicola, Trichocladium asperum. **Pithecolobium**—Camptomeris floridana, Stenella araguata. **Pittosporum**—Virgaria nigra. **Plantago**—Taeniolella plantaginis. **Platanus**—Circinotrichum maculiforme, Gyrothrix podosperma, Scytalidium lignicola, Stigmina platani. **Plectocomia**—Circinoconis paradoxa. **Poa**—Drechslera poae. **Pogonarthria**—Lacellina graminicola. **Polyalthia**—Zygosporium minus. **Polygala**—Curvularia eragrostidis. **Polygonum**—Arthrinium sacchari, Endophragmia elliptica, Pleurophragmium simplex, Sporidesmium leptosporum. **Polypodium**—Zygosporium gibbum, Z. masonii. **Polytrichum**—Drechslera biseptata. **Populus**—Brachysporium obovatum, Chalaropsis thielavioides, Endophragmia elliptica, Graphium penicillioides, Phialocephala bactrospora, Pollaccia radiosa, Pseudospiropes nodosus, P. simplex, Rhinocladiella s. Dictyotrichiella mansonii, Spadicoides atra, Sporidesmium pedunculatum, Stachybotrys s. Melanopsamma pomiformis, Stachylidium bicolor, Taeniolella stilbospora, Troposporella fumosa, Xylohypha nigrescens. **Portulaca**—Dichotomophthora portulacae. **Primula**—Thielaviopsis basicola. **Prosopis**—Gyrothrix hughesii. **Prunus**—Brachysporium bloxami, B. masonii, B. obovatum, Chalaropsis thielavioides, Cladosporium s. Venturia carpophila, Corynespora cambrensis, C. pruni, Periconia macrospinosa, Phaeoisaria clavulata, Pseudospiropes simplex, Spadicoides bina, Sporidesmium altum, S. cookei, Stigmina carpophila, Triposporium elegans, Verticillium cyclosporum. **Pseudotsuga**—Spadicoides atra. **Psidium**—Zygophiala jamaicensis. **Pteridium**—Acremoniella atra, Chalara pteridina, Drechslera biseptata, Stachylidium bicolor, Wardomyces pulvinatus, Zygosporium gibbum. **Pterocarpus**—Camposporium antennatum, Exosporium pterocarpi. **Pueraria**—Pithomyces maydicus. **Punica**—Drechslera papendorfii, Periconia atropurpurea. **Puya**—Stemphyliomma valparadisiacum. **Pycnanthus**—Codinaea assamica. **Pyrus**—Dematophora s. Rosellinia necatrix, Fusicladium s. Venturia pirina, Pseudospiropes simplex, Spilocaea s. Venturia inaequalis.

Q

Quercus—Actinocladium rhodosporum, Arthrobotryum stilboideum, Bactrodesmium abruptum, B. obovatum, B. pallidum, Beltrania querna, Bispora antennata, Brachysporium bloxami, B. britannicum, B. masonii, B. nigrum, B. obovatum, Cacumisporium capitulatum, Catenularia s. Chaetosphaeria cupulifera, Ceratophorum helicosporum, C. uncinatum, Chalara s. Ceratocystis fagacearum, Chalaropsis s. Ceratocystis variospora, Cryptophiale udagawae, Cystodendron dryophilum, Diplococcium spicatum, Endophragmia hyalosperma, Gonatobotryum fuscum, Hadronema orbiculare, Haplariopsis fagicola, Helminthosporium ahmadii, H. microsorum, Menispora ciliata, Monodictys glauca, Polyscytalum fecundissimum, Pseudospiropes simplex, Septotrullula bacilligera, Spadicoides atra, S. bina, Sporidesmium adscendens, S. leptosporum, Sporoschisma juvenile, S. mirabile, Triposporium elegans, Verticillium cyclosporum, V. tenuissimum, Virgariella atra.

R

Rapanea—Periconiella rapaneae, Pithomyces cupaniae, Trichodochium disseminatum, T. pirozynskii. **Raphanus**—Acremoniella atra, Alternaria raphani. **Rauwolfia**—Acrodictys deightonii, Exosporium ampullaceum. **Rhododendron**—Triposporium elegans, Verticillium cyclosporum, V. tenuissimum. **Rhus**—Gonatobotryum apiculatum, Podosporium rigidum. **Ribes**—Agyriella nitida. **Ricinus**—Alternaria ricini, Botrytis s. Sclerotinia ricini, Scopulariopsis brumptii. **Rosenscheldiella**—Spiropes scopiformis. **Rotala**—Phaeotrichoconis crotalariae. **Rottboellia**—Lacellinopsis sacchari, Spegazzinia deightonii. **Rubus**—Agyriella nitida, Curvulariopsis cymbisperma, Endophragmia elliptica, E. hyalosperma, Pseudospiropes nodosus, Triposporium elegans. **Ruscus**—Pseudospiropes simplex. **Russula**—Myrothecium inundatum.

S

Sabal—Melanographium citri. **Saccharum**—Arthrinium sacchari, A. saccharicola, A. spegazzinii, Curvularia brachyspora, C. eragrostidis, Deightoniella papuana, Dictyoarthrinium sacchari, D. africanum, Drechslera sacchari, Fusariella indica, Hansfordia ovalispora, Lacellina graminicola, Lacellinopsis sacchari, Myrothecium indicum, Nigrospora sacchari, Oncopodium panici, Periconia s. Didymosphaeria igniaria, P. atropurpurea, P. echinochloae, P. sacchari, Pithomyces maydicus, P. sacchari, Pseudobotrytis terrestris, Pteroconium s. Apiospora camptospora, Spegazzinia deightonii, S. tessarthra, Staphylotrichum coccosporum, Tetraploa aristata, Thielaviopsis s. Ceratocystis paradoxa, Trichurus spiralis, Zygosporium masonii. **Salix**—Endophragmia elliptica, Oedemium s. Thaxteria fusca, Polyscytalum fecundissimum, Pseudospiropes nodosus, P. simplex, Sporoschisma mirabile, Taeniolella stilbospora, Trimmatostroma salicis, Xylohypha nigrescens. **Sambucus**—Balanium stygium, Brachysporium britannicum, B. obovatum, Endophragmia hyalosperma, Gyrothrix podosperma, Phaeoisaria clavulata, Pleurophragmium simplex, Sporidesmium altum, S. cookei, S. leptosporum, Stachybotrys dichroa, Stachylidium bicolor, Verticillium cyclosporum, Xylohypha nigrescens. **Sanchezia**—Memnoniella levispora. **Santaloides**—Periconiella santaloidis. **Sasa**—Didymobotryum verruculosum. **Schiffnerula**—Acremoniula sarcinellae, Eriocercospora balladynae, Spiropes dorycarpus. **Scirpus**—Arthrinium s. Pseudoguignardia scirpi, A. curvatum var. minus, A. cuspidatum, A. puccinioides. **Scleria**—Clasterosporium scleriae. **Secale**—Alternaria s. Pleospora infectoria, Drechslera s. Cochliobolus sativus, D. s. Pyrenophora tritici-repentis, Pyricularia grisea, Stephanosporium cereale. **Sechium**—Acremoniella verrucosa. **Senecio**—Stachybotrys dichroa. **Sequoia**—Brachysporiella turbinata. **Sesamum**—Alternaria sesami, Corynespora cassiicola, Curvularia eragrostidis. **Sesleria**—Periconia hispidula. **Setaria**—Cladosporium spongiosum, Curvularia leonensis, Drechslera s. Cochliobolus setariae, Hansfordia ovalispora, Pteroconium s. Apiospora camptospora, Pyricularia grisea, Stachybotrys parvispora. **Sium**—Cercosporidium sii. **Smeathmannia**—Periconiella smeathmanniae. **Smilax**—Corynespora aterrima, Exosporium mexicanum, Hermatomyces tucumanensis, Periconiella leonensis, P. smilacis, Zygosporium gibbum. **Solanum**—Acremoniella atra, A. verrucosa, Alternaria solani, Annellophora solani, Chuppia sarcinifera, Gliomastix s. Wallrothiella subiculosa, Helminthosporium solani, Stachylidium bicolor, Stemphylium solani, Trichocladium asperum, Trichurus spiralis, Zygosporium masonii, Z. minus. **Sonchus**—Alternaria sonchi. **Sorbus**—Actinocladium rhodosporum, Corynespora cambrensis, Diplococcium spicatum, Endophragmia biseptata, Pseudospiropes nodosus, P. simplex, Septosporium bulbotrichum, Sporidesmium folliculatum, Sporoschisma juvenile, S. mirabile, Virgaria nigra. **Sorghum**—Alternaria longissima, Curvularia clavata, C. eragrostidis, C. fallax, C. leonensis, Drechslera s. Cochliobolus bicolor, D. s. C. carbonus, D. s. C. heterostrophus, D. s. Trichometasphaeria turcica, D. papendorfii, D. sorghicola, Lacellinopsis sacchari, Periconia circinata, P. digitata, P. echinochloae, Pithomyces maydicus, P. sacchari, Pteroconium s. Apiospora camptospora, Pyricularia grisea, Spegazzinia tessarthra, Tetraploa aristata. **Sorindaea**—Pithomyces cupaniae. **Spartina**—Drechslera s. Pyrenophora tritici-repentis, Periconia s. Didymosphaeria igniaria. **Spartium**—Tripospermum myrti. **Sphagnum**—Casaresia sphagnorum. **Spinacia**—Cladosporium variabile. **Sporidesmium**—Annellophora sydowii. **Sporobolus**—Curvularia eragrostidis, Drechslera ravenelii, Periconia atropurpurea, Stachylidium bicolor. **Stapelia**—Curvularia stapeliae.

T

Tabebuia—Zygosporium minus. **Taenidia**—Cercosporidium punctiforme. **Taraxacum**—Acremoniella verrucosa. **Taxus**—Cheiromycella microscopica, Endophragmia stemphylioides, E. taxi, Sporidesmium pedunculatum, Sterigmatobotrys macrocarpa. **Telopea**—Periconiella telopeae. **Tephrosia**—Acrophialophora fusispora. **Thalictrum**—Haplobasidion thalictri, Myrothecium carmichaelii. **Thea**—Beltrania rhombica, Codinaea assamica, Sporidesmium leptosporum, S. vagum. **Themeda**—Pithomyces sacchari. **Theobroma**—Annellophora borneoensis, Beltrania africana, Cephaliophora irregularis, C. tropica, Exosporium ampullaceum, Gliomastix s. Wallrothiella subiculosa, Melanographium citri, Menisporopsis theobromae, Myrothecium s. Nectria bactridioides, Spegazzinia tessarthra, Stachybotrys theobromae, Stachylidium bicolor, Thielaviopsis s. Ceratocystis paradoxa, Virgatospora echinofibrosa, Zygosporium echinosporum. **Thevetia**—Gyrothrix thevetiae, Sporidesmium brachypus. **Thysanolaena**—Annellophragmia coonoorensis. **Tilia**—Brachysporium nigrum, Cacumisporium capitulatum, Corynespora olivacea, Endophragmia hyalosperma, Exosporium tiliae, Oedemium s. Thaxteria fusca, Phialocephala bactrospora, Sporidesmium vagum, Xylohypha nigrescens. **Tinospora**—Zygophiala jamaicensis. **Toddalia**—Annellophora africana. **Trachycarpus**—Melanographium citri. **Trema**—Tomenticola trematis. **Trichilia**—Circinotrichum falcatisporum, Corynespora trichiliae. **Trichocaulon**—Curvularia stapeliae. **Trifolium**—Acremoniella atra, A. verrucosa, Botrytis s. Sclerotinia spermophila, B. anthophila, Curvularia prasadii, C. trifolii, Periconia macrospinosa, Polythrincium s. Cymadothea trifolii, Stemphylium globuliferum, S. sarciniforme. **Trigonella**—Fusariella hughesii. **Tripogon**—Curvularia clavata. **Tripsacum**—Spegazzinia deightonii. **Triticum**—Acremoniella verrucosa, Alternaria s. Pleospora infectoria, Curvularia s. Cochliobolus intermedius, C. brachyspora, C. harveyi, C. inaequalis, C. verruciformis, Drechslera s. Cochliobolus bicolor, D. s. C. sativus, D. s. Pyrenophora semeniperda, D. s. P. tritici-repentis, D. s. Trichometasphaeria pedicellata, D. avenacea, D. biseptata, D. erythrospila, Periconia circinata, P. macrospinosa, Physalidium elegans, Pithomyces sacchari, Pseudospiropes simplex, Scopulariopsis brumptii, Spegazzinia deightonii, S. tessarthra, Tetraploa aristata, Thermomyces lanuginosus. **Tropaeolum**—Acroconidiella tropaeoli. **Tsuga**—Thysanophora canadensis. **Tulipa**—Botrytis tulipae. **Typha**—Arthrinium sporophleum, Lacellina graminicola, Periconia hispidula.

U

Uapaca—Sporidesmium adscendens. **Ulex**—Actinocladium rhodosporum, Endophragmia biseptata, Pseudospiropes simplex, Triposporium elegans, Verticillium cyclosporum. **Ulmus**—Bactrodesmium obovatum, B. spilomeum, Bispora antennata, Bloxamia truncata, Chaetopsis grisea, Chalaropsis thielavioides, Dictyodesmium ulmicola, Oncopodiella trigonella, Pseudospiropes simplex, Spadicoides bina, Sporidesmium folliculatum, S. leptosporum, Sporoschisma juvenile, S. mirabile, Stachybotrys s. Melanopsamma pomiformis, Stigmina compacta, Teratosperma singulare, Verticillium tenuissimum. **Urena**—Curvularia senegalensis. **Urtica**—Arthrinium urticae, Corynesporella urticae, Dendryphion comosum, Endophragmia atra, E. elliptica, E. prolifera, Fusariella hughesii, Gyrothrix verticillata, Pleurophragmium simplex, Stachylidium bicolor. **Uvaria**—Exosporium leptoderridicola, E. mexicanum, Sporidesmium leptosporum.

V

Vaccinium—Curvularia inaequalis. **Vanda**—Pithomyces maydicus. **Vangueria**—Sporidesmium adscendens. **Vernonia**—Spilodochium vernoniae, **Vetiveria**—Lacellina graminicola, Spegazzinia deightonii. **Viburnum**—Xylohypha nigrescens. **Vicia**—Botrytis fabae, Trichurus spiralis. **Vigna**—Corynespora cassiicola, Periconia atropurpurea, Pithomyces sacchari. **Viola**—Centrospora acerina, Thielaviopsis basicola. **Vismia**—Corynespora vismiae, Exosporium cantareirense, Oedothea vismiae. **Vitex**—Pithomyces maydicus, Xylohypha nigrescens. **Vitis**—Endophragmia stemphylioides, Pseudocercospora vitis, Scolecobasidium constrictum, Scytalidium lignicola.

W

Weinmannia—Zanclospora novae-zelandiae. **Withania**—Pithomyces sacchari.

X

Xanthorrhoea—Herposira velutina. **Xanthosoma**—Curvularia fallax.

Y

Yucca—Bahusakala olivaceonigra.

Z

Zea—Acremoniella atra, A. verrucosa, Acrodictys erecta, Alternaria longissima, Ceratosporella bicornis, Curvularia s. Cochliobolus intermedius, C. eragrostidis, C. leonensis, C.

senegalensis, Drechslera s. Cochliobolus carbonus, D. s. C. heterostrophus, D. s. Trichometa-sphaeria pedicellata, D. s. T. turcica, Myrothecium jollymannii, Periconia s. Didymosphaeria igniaria, P. digitata, P. echinochloae, Pithomyces maydicus, Pteroconium s. Apiospora camptospora, Pyricularia grisea, Spegazzinia deightonii, S. tessarthra, Stachylidium bicolor, Tetraploa aristata, T. ellisii, Thermomyces lanuginosus. **Zinnia**—Dictyoarthrinium sacchari. **Ziziphus**—Mitteriella ziziphina, Sporidesmium vagum.

OTHER SUBSTRATA

Air—Acremoniella atra, Acrophialophora fusispora, Aureobasidium pullulans, Beltrania rhombica, Botryotrichum s. Chaetomium piluliferum, Cladosporium cladosporioides, C. herbarum, C. sphaerospermum, Curvularia brachyspora, C. fallax, C. lunata var. aeria, C. oryzae, C. tuberculata, C. uncinata, Dictyosporium heptasporum, Drechslera s. Cochliobolus spicifer, D. papendorfii, Epicoccum purpurascens, Gliomastix murorum var. felina, G. murorum var. polychroma, Monodictys levis, Nigrospora s. Khuskia oryzae, Periconia s. Didymosphaeria igniaria, P. echinochloae, Phaeotrichoconis crotalariae, Phialophora fastigiata, Pithomyces chartarum, Scolecobasidium constrictum, Scopulariopsis brevicaulis, S. brumptii, Stemphylium s. Pleospora herbarum, Stephanosporium cereale, Torula herbarum, Wallemia sebi. **Animals**—Curvularia verruciformis, Epicoccum purpurascens, Exophiala salmonis, Gliomastix murorum var. polychroma, Polypaecilum insolitum, Scopulariopsis brevicaulis, Thermomyces lanuginosus, Wallemia sebi. **Dung**—Acremoniella atra, Acrodictys fimicola, Angulimaya sundara, Bahupaathra samala, Cephaliophora irregularis, C. tropica, Doratomyces microsporus, D. nanus, D. purpureofuscus, D. stemonitis, Gilmaniella humicola, Tetracoccosporium paxianum, Torula terrestris. **Foodstuffs**—Alternaria alternata, Aureobasidium pullulans, Cladosporium herbarum, C. sphaerospermum, Epicoccum purpurascens, Nigrospora sphaerica, Scolecobasidium variabile, Scopulariopsis brevicaulis, Veronaea botryosa, Wallemia sebi, Wardomyces anomalus, Zygosporium echinosporum. **Herbaceous stems (dead)**—Camposporium pellucidum, Cladosporium herbarum, C. macrocarpum, C. oxysporum, Conoplea fusca, Coremiella cubispora, Cylindrotrichum oligospermum, Deightoniella infuscans, D. vinosa, Dendryphion comosum, D. nanum, Dictyosporium elegans, D. heptasporum, D. toruloides, Dicyma s. Ascotricha chartarum, Doratomyces microsporus, D. nanus, D. purpureofuscus, D. stemonitis, Endophragmia elliptica, Gliomastix cerealis, G. luzulae, G. murorum var. felina, Graphium putredinis, Gyrothrix flagella, Helminthosporium velutinum, Memnoniella echinata, Monodictys castaneae, M. levis, Periconia byssoides, P. cookei, P. minutissima, Phaeoisaria clematidis, Stachybotrys atra, S. dichroa, Stemphylium s. Pleospora herbarum, S. vesicarium, Torula herbarum, T. herbarum f. quaternella, Trichocladium opacum, Triposporium elegans, Ulocladium botrytis. **Kerosene**—Cladosporium s. Amorphotheca resinae. **Paint**—Cladosporium herbarum, C. sphaerospermum, Curvularia senegalensis, Gliomastix murorum var. felina. **Paper**—Acremoniella verrucosa, Bactrodesmium papyricola, Botryotrichum s. Chaetomium piluliferum, Dicyma s. Ascotricha chartarum, Gliomastix murorum var. felina, G. musicola, Memnoniella echinata, Pithomyces chartarum, Scopulariopsis brevicaulis, Stachybotrys atra, S. atra var. microspora, Stemphylium s. Pleospora herbarum, Stephanosporium cereale, Trichurus spiralis, Ulocladium botrytis. **Soil**—Acremoniella atra, A. verrucosa, Acrophialophora fusispora, Alternaria alternata, Aspergillus niger, Asteromyces cruciatus, Aureobasidium pullulans, Beltrania rhombica, Botryotrichum s. Chaetomium piluliferum, Cephaliophora irregularis, C. tropica, Chaetopsina fulva, Chloridium chlamydosporis, Cladosporium s. Amorphotheca resinae, C. cladosporioides, C. herbarum, C. sphaerospermum, Costantinella micheneri, C. terrestris, Curvularia affinis, C. eragrostidis, C. fallax, C. inaequalis, C. lunata var. aeria, C. prasadii, C. senegalensis, C. tuberculata, Custingophora olivacea, Doratomyces microsporus, D. phillipsii, D. purpureofuscus, Drechslera s. Cochliobolus nodulosus, D. s. C. setariae, D. s. C. spicifer, D. biseptata, D. halodes, D. hawaiiensis, D. papendorfii, D. rostrata, Echinobotryum s. Doratomyces stemonitis, Fusariella bizzozeriana, F. obstipa, Gilmaniella humicola, Gliocephalotrichum bulbilium, Gliomastix s. Wallrothiella subiculosa, G. cerealis, G. murorum, G. murorum var. felina, G. murorum var. polychroma, G. musicola, Gonatobotryum apiculatum, Gonytrichum macrocladum, Graphium putredinis, Humicola fuscoatra, H. grisea, Idriella lunata, Mammaria echinobotryoides, Memnoniella echinata, Monodictys levis, Murogenella terrophila, Myrothecium brachysporum, M. gramineum, M. indicum, M. roridum, M. striatisporum, M. verrucaria, Nigrospora s. Khuskia oryzae, N. sphaerica, Oidiodendron griseum, O. tenuissimum, Periconia s. Didymosphaeria igniaria, P. atropurpurea, P. echinochloae, P. lateralis,

P. macrospinosa, Phaeotrichoconis crotalariae, Phialomyces macrosporus, Phialophora fastigiata, Pithomyces chartarum, Pseudobotrytis terrestris, Rhinocladiella cellaris, Scolecobasidium constrictum, S. humicola, S. terreum, S. variabile, S. verruculosum, Scopulariopsis brevicaulis, S. brumptii, Scytalidium lignicola, Spegazzinia lobulata, Stachybotrys atra, S. cylindrospora, Staphylotrichum coccosporum, Stemphylium s. Pleospora herbarum, Stephanosporium cereale, Tetracoccosporium paxianum, Thermomyces lanuginosus, Thysanophora penicillioides, Torula caligans, T. terrestris, T. herbarum, Trichocladium asperum, T. opacum, T. pyriforme, Trichurus spiralis, Ulocladium botrytis, Wardomyces anomalus, W. humicola, W. inflatus, W. pulvinatus, Zygosporium masonii. **Textiles**—Alternaria alternata, Aspergillus niger, Aureobasidium pullulans, Botryotrichum s. Chaetomium piluliferum, Cladosporium cladosporioides, C. herbarum, C. sphaerospermum, Curvularia tuberculata, Drechslera halodes, D. hawaiiensis, Epicoccum purpurascens, Gliomastix cerealis, G. murorum, Memnoniella echinata, Monodictys castaneae, M. levis, Myrothecium roridum, M. verrucaria, Scopulariopsis brevicaulis, Stachybotrys atra, S. atra var.' microspora, Stephanosporium cereale, Torula herbarum, Trichurus spiralis, Ulocladium botrytis, Wallemia sebi. **Timber**—Acremoniella atra, A. verrucosa, Aureobasidium pullulans, Coniosporium s. Hysterium insidens, Curvularia lunata var. aeria, Doratomyces stemonitis, Endophragmia australiensis, Pachnocybe ferruginea, Rhinocladiella cellaris. **Wood and bark (logs, fallen branches, etc.)**—Acrodictys globulosa, A. obliqua, Acrogenospora s. Farlowiella carmichaeliana, A. sphaerocephala, Actinocladium rhodosporum, Allescheriella crocea, Alysidium resinae, Arthrobotryum stilboideum, Bactrodesmium abruptum, B. atrum, B. obovatum, B. pallidum, B. spilomeum, Balanium stygium, Berkleasmium concinnum, Bispora antennata, B. betulina, Brachysporiella gayana, B. setosa, Brachysporium bloxami, B. britannicum, B. dingleyae, B. masonii, B. nigrum, B. novae-zelandiae, B. obovatum, B. pendulisporum, Cacumisporium capitulatum, Camposporium cambrense, C. pellucidum, Catenularia s. Chaetosphaeria cupulifera, C. s. C. myriocarpa, Cephaliophora irregularis, Ceratocladium microspermum, Ceratosporium fuscescens, Chaetopsis grisea, Chalara aurea, C. cylindrosperma, Chloridium viride, Coniosporium olivaceum, Conoplea fusca, C. globosa, C. olivacea, Cordana pauciseptata, Coremiella cubispora, Corynespora smithii, Costantinella micheneri, C. terrestris, Cryptocoryneum condensatum, Cylindrotrichum oligospermum, Dactylosporium macropus, Dendryphiopsis s. Amphisphaeria incrustans, Dictyosporium elegans, D. heptasporum, D. oblongum, D. toruloides, Diplococcium spicatum, Doratomyces microsporus, D. nanus, D. purpureofuscus, D. stemonitis, Geniculosporium s. Hypoxylon serpens, Gliomastix murorum, G. murorum var. felina, Gonytrichum s. Melanopsammella inaequalis, G. macrocladum, Graphium calicioides, Hansfordia pulvinata, Haplographium s. Hyaloscypha dematiicola, Harpographium fasciculatum, Helicodendron paradoxum, Helicoma s. Lasiosphaeria pezicula, Helicoon ellipticum, Helicorhoidion botryoideum, H. pulchrum, Helicosporium s. Tubeufia cerea, Helminthosporium velutinum, Iyengarina elegans, Kumanasamuha sundara, Leptographium lundbergii, Mammaria echinobotryoides, Menispora ciliata, M. glauca, Monodictys castaneae, M. glauca, M. levis, M. putredinis, Myrothecium s. Nectria ralfsii, Nodulisporium gregarium, Oedemium s. Thaxteria fusca, O. s. T. phaeostroma, Oidiodendron tenuissimum, Paathramaya sundara, Phaeoisaria clematidis, Phialophora fastigiata, Pseudospiropes nodosus, P. simplex, Scopulariopsis brevicaulis, Spadicoides atra, S. bina, S. grovei, S. xylogena, Sporidesmium tropicale, Sporoschisma mirabile, Trichocladium opacum, Triposporium elegans, Ulocladium botrytis, Xenosporium berkeleyi.

LIST OF VERY COMMON PLURIVOROUS SPECIES

Alternaria s. Pleospora infectoria, A. alternata, A. longissima, A. tenuissima, Arthrinium s. Apiospora montagnei, A. phaeospermum, Aureobasidium pullulans, Botrytis s. Sclerotinia fuckeliana (=B. cinerea), Cladosporium cladosporioides, C. herbarum, C. macrocarpum, C. oxysporum, C. sphaerospermum, Corynespora cassiicola, Curvularia s. Cochliobolus geniculatus, C. s. C. lunatus, C. lunata var. aeria, C. pallescens, C. verruculosa, Dendryphiella vinosa, Dendryphion comosum, D. nanum, Drechslera s. Cochliobolus cynodontis, D. s. C. spicifer, D. halodes, D. hawaiiensis, D. papendorfii, D. rostrata, Epicoccum purpurascens, Gonatophragmium mori, Grallomyces portoricensis, Helminthosporium velutinum, Hiospira s. Brooksia tropicalis, Myrothecium roridum, M. striatisporum, M. verrucaria, Nigrospora s. Khuskia oryzae, N. sphaerica, Periconia byssoides, Pithomyces chartarum, Sporidesmium tropicale, Stemphylium s. Pleospora herbarum, Torula herbarum, T. herbarum f. quaternella, Ulocladium botrytis, Zygosporium oscheoides.

INDEX